VOLUME SIX HUNDRED AND THIRTY

METHODS IN ENZYMOLOGY

Nanoarmoring of Enzymes with Carbon Nanotubes and Magnetic Nanoparticles

METHODS IN ENZYMOLOGY

VOLUME SIX HUNDRED AND THIRTY

METHODS IN ENZYMOLOGY

Nanoarmoring of Enzymes with Carbon Nanotubes and Magnetic Nanoparticles

Edited by

CHALLA V. KUMAR

Department of Chemistry; Department of Molecular and Cell Biology, University of Connecticut, Storrs, CT, United States

ELSEVIER

ACADEMIC PRESS

An imprint of Elsevier

Academic Press is an imprint of Elsevier
50 Hampshire Street, 5th Floor, Cambridge, MA 02139, United States
525 B Street, Suite 1650, San Diego, CA 92101, United States
The Boulevard, Langford Lane, Kidlington, Oxford OX5 1GB, United Kingdom
125 London Wall, London, EC2Y 5AS, United Kingdom

First edition 2020

ISBN: 978-0-12-820143-5
ISSN: 0076-6879

For information on all Academic Press publications
visit our website at https://www.elsevier.com/books-and-journals

Publisher: Zoe Kruze
Acquisition Editor: Ashlie M. Jackman
Editorial Project Manager: Shellie Bryant
Production Project Manager: James Selvam
Cover Designer: Greg Harris

Typeset by SPi Global, India

Contents

Contributors

Reinu E. Abraham
Centre for Marine Bioproducts Development, College of Medicine and Public Health, Flinders University, Bedford Park, SA, Australia

Sameh S. Ali
Botany Department, Faculty of Science, Tanta University, Tanta, Egypt; Biofuels Institute, School of the Environment and Safety Engineering, Jiangsu University, Zhenjiang, China

M. Yakup Arica
Biochemical Processing and Biomaterial Research Laboratory, Gazi University, Ankara, Turkey

Gulay Bayramoglu
Biochemical Processing and Biomaterial Research Laboratory; Department of Chemistry, Faculty of Sciences, Gazi University, Ankara, Turkey

Paolo Bollella
Department of Chemistry and Biomolecular Science, Clarkson University, Potsdam, NY, United States

Nikolaos Chalmpes
Department of Materials Science & Engineering, University of Ioannina, Ioannina, Greece

Kanika Chauhan
Carbohydrate and Protein Biotechnology Laboratory, Department of Biotechnology, Punjabi University, Patiala, Punjab, India

Angela Chiew Wen Ch'ng
Institute for Research in Molecular Medicine, Universiti Sains Malaysia, Penang, Malaysia

Osama M. Darwesh
Agriculture Microbiology Department, National Research Centre, Dokki, Cairo, Egypt

Nithyakalyani Doraiswamy
Centre for Biotechnology, Anna University, Chennai, Tamil Nadu, India

Evangelia Dounousi
Department of Nephrology, Medical School, University of Ioannina, Ioannina, Greece

Tamer Elsamahy
Biofuels Institute, School of the Environment and Safety Engineering, Jiangsu University, Zhenjiang, China

Wenjiao Fan
Key Laboratory of Applied Surface and Colloid Chemistry, Ministry of Education; Key Laboratory of Analytical Chemistry for Life Science of Shaanxi Province; School of Chemistry & Chemical Engineering, Shaanxi Normal University, Xi'an, China

Dimitrios Gournis
Department of Materials Science & Engineering, University of Ioannina, Ioannina, Greece

Nicole Grobert
Department of Materials, University of Oxford, Oxford, United Kingdom

Qayyum Husain
Department of Biochemistry, Faculty of Life Sciences, Aligarh Muslim University, Aligarh, UP, India

Ankarao Kalluri
Department of Chemistry, University of Connecticut, Storrs, CT, United States

Evgeny Katz
Department of Chemistry and Biomolecular Science, Clarkson University, Potsdam, NY, United States

Sunaina Kaul
National Agri Food Biotechnology Institute, Food and Nutritional Biotechnology, Mohali, Punjab, India

Maryam Khan
Department of Biochemistry, Faculty of Life Sciences, Aligarh Muslim University, Aligarh, UP, India

Zoltán Konthur
Max Planck Institute of Colloid and Interfaces, Department of Systems Biology, Potsdam, Germany

Challa V. Kumar
Department of Chemistry; Department of Molecular and Cell Biology, University of Connecticut, Storrs, CT, United States

Lorico D.S. Lapitan Jr.
Department of Chemical Engineering and Advanced Materials Research Laboratory, Research Center for the Natural and Applied Sciences, University of Santo Tomas, Manila, Philippines

Theam Soon Lim
Institute for Research in Molecular Medicine; Analytical Biochemistry Research Centre, Universiti Sains Malaysia, Penang, Malaysia

Chenghui Liu
Key Laboratory of Applied Surface and Colloid Chemistry, Ministry of Education; Key Laboratory of Analytical Chemistry for Life Science of Shaanxi Province; School of Chemistry & Chemical Engineering, Shaanxi Normal University, Xi'an, China

Yehia A. Mahmoud
Botany Department, Faculty of Science, Tanta University, Tanta, Egypt

Mansi Malhotra
Department of Chemistry, University of Connecticut, Storrs, CT, United States

Caterina G.C. Marques Netto
Laboratório de Metaloenzimas e Biomiméticos, Departamento de Química, Universidade Federal de São Carlos (UFSCar), São Paulo, Brazil

Ashish Mathur
Amity Institute of Nanotechnology, Amity University, Noida, Uttar Pradesh, India

Ibrahim A. Matter
Agriculture Microbiology Department, National Research Centre, Dokki, Cairo, Egypt

Hamid Moghimi
Department of Microbial Biotechnology, School of Biology, College of Science, University of Tehran, Tehran, Iran

Ketan H. Mulchandani
Food Engineering and Technology Department, Institute of Chemical Technology, Mumbai, India

Abhijeet B. Muley
Food Engineering and Technology Department, Institute of Chemical Technology, Mumbai, India

Mohammad Pooya Naghshbandi
Department of Microbial Biotechnology, School of Biology, College of Science, University of Tehran, Tehran, Iran

Jagriti Narang
Department of Biotechnology, School of Chemical and Life Sciences, Jamia Hamdard, New Delhi, India

Veli C. Ozalp
Kit-Argem Research Center and Department of Bioengineering, Konya Food and Agriculture University, Konya, Turkey

Michaela Patila
Laboratory of Biotechnology, Department of Biological Applications and Technology, University of Ioannina, Ioannina, Greece

Gautam Pennathur
Centre for Biotechnology; AU-KBC Research Centre, Anna University, Chennai, Tamil Nadu, India

Megan K. Puglia
Department of Chemistry, University of Connecticut, Storrs, CT, United States

Munish Puri
Centre for Marine Bioproducts Development, College of Medicine and Public Health, Flinders University, Bedford Park, SA, Australia

Holly A. Reeve
Department of Chemistry, University of Oxford, Inorganic Chemistry Laboratory, Oxford, United Kingdom

Eduardo A. Reis
Laboratório de Metaloenzimas e Biomiméticos, Departamento de Química, Universidade Federal de São Carlos (UFSCar), São Paulo, Brazil

Wei Ren
Key Laboratory of Applied Surface and Colloid Chemistry, Ministry of Education; Key Laboratory of Analytical Chemistry for Life Science of Shaanxi Province; School of Chemistry & Chemical Engineering, Shaanxi Normal University, Xi'an, China

Jack S. Rowbotham
Department of Chemistry, University of Oxford, Inorganic Chemistry Laboratory, Oxford, United Kingdom

Rajat Sandhir
Department of Biochemistry, Panjab University, Chandigarh, India, India

Mahalakshmi Sarathi
Centre for Biotechnology, Anna University, Chennai, Tamil Nadu, India

Natarajan Saravanan
Nano and Bioelectrochemistry Research Laboratory, Department of Chemistry, School of Advanced Sciences, Vellore Institute of Technology, Vellore, India

Diego E. Sastre
Laboratório de Metaloenzimas e Biomiméticos, Departamento de Química, Universidade Federal de São Carlos (UFSCar), São Paulo, Brazil

Annamalai Senthil Kumar
Nano and Bioelectrochemistry Research Laboratory, Department of Chemistry, School of Advanced Sciences; Carbon Dioxide Research and Green Technology Centre, Vellore Institute of Technology, Vellore, India

Ram Sarup Singh
Carbohydrate and Protein Biotechnology Laboratory, Department of Biotechnology, Punjabi University, Patiala, Punjab, India

Vishal Singh
National Agri Food Biotechnology Institute, Food and Nutritional Biotechnology, Mohali, Punjab, India

Nitin Kumar Singhal
National Agri Food Biotechnology Institute, Food and Nutritional Biotechnology, Mohali, Punjab, India

Rekha S. Singhal
Food Engineering and Technology Department, Institute of Chemical Technology, Mumbai, India

Haralambos Stamatis
Laboratory of Biotechnology, Department of Biological Applications and Technology, University of Ioannina, Ioannina, Greece

Lisa A. Thompson
Department of Chemistry, University of Oxford, Inorganic Chemistry Laboratory, Oxford, United Kingdom

Manshi Tyagi
Department of Biotechnology, School of Chemical and Life Sciences, Jamia Hamdard, New Delhi, India

Kylie A. Vincent
Department of Chemistry, University of Oxford, Inorganic Chemistry Laboratory, Oxford, United Kingdom

Shikha Wadhwa
Amity Institute of Nanotechnology, Amity University, Noida, Uttar Pradesh, India

Neelam Yadav
Department of Biotechnology, Deenbandhu Chhotu Ram University of Science and Technology, Sonepat; Centre for Biotechnology, Maharishi Dayanand University, Rohtak, India

Dejian Zhou
School of Chemistry and Astbury Centre for Structural Molecular Biology, University of Leeds, Leeds, United Kingdom

Liping Zhu
Key Laboratory of Applied Surface and Colloid Chemistry, Ministry of Education; Key Laboratory of Analytical Chemistry for Life Science of Shaanxi Province; School of Chemistry & Chemical Engineering, Shaanxi Normal University, Xi'an, China

Ceren Zor
Department of Materials, University of Oxford, Oxford, United Kingdom

Preface

This volume of *Methods in Enzymology* builds on the previous volumes focused on the nanotechnology platform. Going one step further, the compilation demonstrates the central role of carbon nanotubes and magnetic particles in pushing the frontiers in this area. This volume is also the largest compilation among these volumes on nanotechnology, and the enthusiastically increased the number of authors and total number of pages unequivocally demonstrate increasing interest in this area of enzymology.

The carbon nanotubes are special due to their high surface area per gram, high electrical and thermal conductivities, essentially one-dimensional materials of significant length for entanglement in fibrous matrices, and low solubility in aqueous media (Scheme 1). In this manner, they are similar to highly porous materials in terms of their surface area, highly conducting like metals, linear molecules like double-stranded DNA/RNA but strongly hydrophobic like synthetic polymers. Thus, carbon nanotubes have multiple features that are specific to a variety of materials that enzymologists are familiar with, but all embodied in one nanomaterial. Thus, carbon nanotubes are special and provide numerous orthogonal opportunities for enzyme attachment and biocatalysis applications.

Carbon nanotubes have been around for decades and why is there a renewed interest in their use for enzymology applications? The answer lies partly in the availability of large quantities of this wonder material of the last millennium at very low cost, and hence, many research groups around the world have access for testing their utility. Second, multi-walled carbon nanotubes, in this regard, are quite inexpensive and would work in ways similar to their more expensive cousins, single-walled carbon nanotubes. Thus, the latter would be of more interest for physicists while the former would be of more interest to chemists and enzymologists. One significant advantage of multi-walled carbon nanotubes over those of their cousins, besides their low cost, is uncompromised electrical conductivity even when the top layers are damaged by chemical oxidation. More specifically, when the single-walled carbon nanotube is damaged chemically, there is a significant loss in its conductivity but when a multi-walled carbon nanotube is damaged in a similar manner, the inner tubes of the bundle still retain high conductivity. Thus, multi-walled tubes resist loss in conductivity, both thermal as well as electrical, subsequent to chemical functionalization of their surfaces.

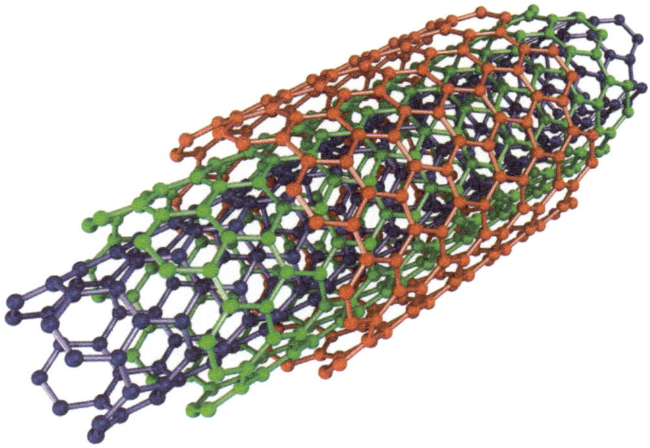

Scheme 1 Schematic structures of single-walled and multi-walled carbon nanotubes (By Eric Wieser—Own work, CC BY-SA 3.0, https://commons.wikimedia.org/w/index.php?curid=12453670).

Chemical functionalization of the surface of the carbon nanotubes is of high importance for enzymology applications, because this approach was used many authors in this volume to securely anchor enzymes. Anchored enzymes do not leach during the biocatalysis processes and the long tubes can be quickly separated from the reaction mixture simply by filtration or centrifugation in an ordinary laboratory centrifuge. Thus, carbon nanotubes are fitting the bill to anchor enzymes on nanomaterial surfaces with highly advantageous properties, and they might turn out to be the most attractive nanomaterials for this purpose.

There are several unexplored avenues of enzymology with multi-walled carbon nanotubes. For example, how does the electrical conductivity of the nanotube influence the activities of the redox enzymes? Perhaps, anchoring the active sites of the redox enzymes via these tubes might electronically couple them and potentially promote long distance electron transfer, perhaps far better than double-stranded DNA. This hypothesis is yet to be tested systematically but the vast majority of electron transfer theory predicts its success. Only systematic experiments can be the jurors to decide the final outcome.

Another aspect of multi-walled carbon nanotubes that have been exploited in this volume is their ability to undergo facile chemical functionalization, yet retain most of their fundamental material properties (Scheme 2). Majority of the chapters in this volume followed this

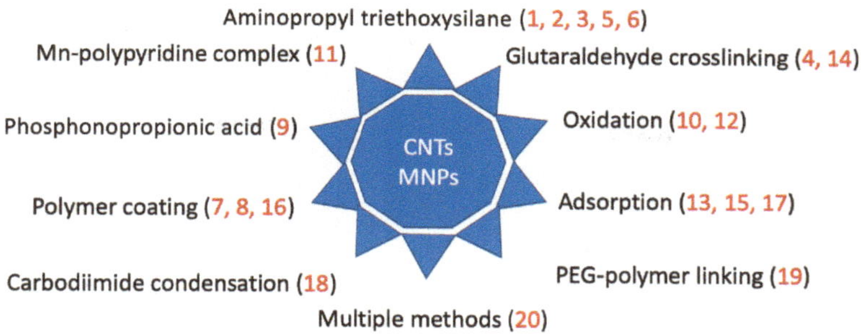

Scheme 2 Functionalization of carbon nanotube (CNT) and magnetic nanoparticle (MNP) investments in enzymology contributions (chapter numbers are in the parenthesis) of this volume.

clear-cut path and produced functional biocatalysts that are similar in activities as those of the free enzymes. Added to the ease of separation, this approach is expected to far exceed the practical applications of these carbon-based biocatalysts.

Enzymes are inherently green, sustainable, and the most efficient catalysts that are still unmatched by most chemical catalysts. However, their high selectivity and fragility make them not so user friendly. Strategies described in this volume attempt to address these issues. While examining the chapters listed in this volume, a variety of approaches have been illustrated to anchor enzymes on the carbon nanotubes and the more familiar magnetic particles. These approaches are illustrated in Scheme 2. Each chapter has a specific advantage or unique approach that led to building a different kind of biocatalysts for a particular set of enzymes or chemical transformations.

Thus, the journey of enzymology taken through nanomaterials, so far, has been gratifying, productive, illuminating and hopefully, it will also be practically useful over the alternative technologies. Hopefully, this nanotechnology series will be of significance for improvement of the larger issues of human kind, climate change, and the environment. We hope more articles along these lines will continue to address these important issues via this series.

Challa V. Kumar
Department of Chemistry,
University of Connecticut,
Storrs, CT, United States

CHAPTER ONE

Magneto-controlled enzyme reactions

Paolo Bollella*, Evgeny Katz*
Department of Chemistry and Biomolecular Science, Clarkson University, Potsdam, NY, United States
*Corresponding authors: e-mail address: pbollell@clarkson.edu; ekatz@clarkson.edu

Contents

Abstract

Various approaches to magneto-controlled biocatalytic enzyme reactions are discussed with specific example systems. Magnetic nano- and micro-size particles functionalized with enzymes or cofactors/electron transfer mediators have been used to translocate the components of the biocatalytic processes and to activate/inhibit their reactions. Magneto-induced deposition of the functionalized particles on an electrode surface resulted in activation of bioelectrocatalytic reactions. On the other hand, magneto-induced removal of the particles from the electrode surface resulted in the inhibition of the electrochemical reactions. Aggregation/disaggregation of enzyme-modified magnetic nanoparticles resulted in different mechanisms of biocatalytic cascades,

Methods in Enzymology, Volume 630
ISSN 0076-6879
https://doi.org/10.1016/bs.mie.2019.09.002

changing them reversibly between substrate diffusion and substrate channeling processes. Magnetohydrodynamic activation of bioelectrocatalytic processes allowed enhancement of a biofuel cell operation. Overall, a large variety of possible magneto-controlled enzyme reactions is briefly discussed, particularly emphasizing their applications in different bioelectronic systems.

1. Introduction: Signal-controlled artificial biomolecular systems

Signal-controlled, adaptable to environment, biological systems are highly important for life and based on sophisticated multi-level regulating mechanisms, which are difficult or even impossible to reproduce in artificial biomolecular systems (Choi, 2010; Hancock, 2010). Chemical approaches, much easier than the way used by Nature, have been applied to design artificial signal-switchable biomolecular systems of limited complexity, mostly based on stimuli-controlled biocatalytic (Shakya & Nandakumar, 2018; Wang, 2012) or biorecognition (Patolsky, Filanovsky, Katz, & Willner, 1998; Willner, Lion-Dagan, & Katz, 1996) reactions.

1.1 Molecular systems switchable by various physical signals

In order to achieve switchable/tunable activity of biomolecular systems, biomolecules were incorporated in various stimuli-responsive materials changing their properties upon variation of pH, temperature or application of magnetic field (Lee et al., 2005; Wang, Liu, & Liu, 2014). Signal-controlled variation of the environment, usually represented by polymeric matrices, resulted in changing access for substrates/cofactors to the biocatalytic/biorecognition species, thus alternating their reaction rates. A more advanced approach has been developed using modification of protein backbones with signal-switchable molecules, e.g., with photoisomerizable species (Aizawa, Namba, & Suzuki, 1977; Katz, Shipway, & Willner, 2002). Photo-induced isomerization of these species resulted in conformational changes of the proteins, thus affecting their biocatalytic activity or biorecognition affinity. The photoisomerizable molecules have been randomly attached to functional groups of the protein backbone (Lion-Dagan, Katz, & Willner, 1994) or specifically bound to enzyme active centers (Blonder, Katz, Willner, Wray, & Bückmann, 1997). The latter demonstrated much more effective light-switchable features due to conformational changes in the close vicinity to the catalytic centers. Other signal-switchable species integrated with enzyme molecules have been represented with oligonucleotides (Abrosimova et al., 2013) and complex DNA "machines" (Hu et al., 2014) allowing modulation of the enzyme activity with different signals, e.g., temperature variation. The systems mentioned above have been constructed using organic chemistry synthetic reactions, e.g., carbodiimide coupling of carboxyl-functionalized signal-switchable molecules to amino groups of the protein backbone. Another approach based on genetically modified proteins allowed construction of artificial chimeric enzymes combining a biocatalytic domain with an allosteric receptor unit able to change its conformation upon reacting with specific signal-molecules, thus resulting in the enzyme activity change induced by allosteric activation (Guo et al., 2016, 2016; Ostermeier, 2009; Stein & Alexandrov, 2015).

1.2 Molecular systems switchable by magnetic field application

Switching/tuning of enzyme-catalyzed processes with applied magnetic field represents another challenging research direction allowing biocatalytic reactions controlled at a distance. Magnetic control of enzyme reactions, particularly proceeding at electrode-interfaces, can originate from magneto-hydrodynamic effect facilitating mass-transport (Katz, Lioubashevski, & Willner, 2004; Lioubashevski, Katz, & Willner, 2007), thus stimulating bioelectrocatalytic reactions. Alternatively, magneto-switchable features can be based on translocation of enzyme-modified magnetic particles to and from electrode surfaces, thus allowing and inhibiting their redox reactions, respectively (Hirsch, Katz, & Willner, 2000; Katz, 2016). Notably, the magnetic nanoparticles translocation resulted in the change of the reacting species concentration at the electrode surface, thus affecting the interfacial reaction rates. In the recent work magnetic nanoparticles modified with enzyme-substrate pairs have been studied. Application of external magnetic field resulted in aggregation of the magnetic nanoparticles, thus, bringing the enzyme and its substrate to each other and allowing their reaction (Zacharchenko, Guz, Laradji, Katz, & Minko, 2017). It should be noted that signal-switchable enzymes have been usually studied in single-step reactions, while signal-controlled multi-step biocatalytic cascades have been studied only fragmentally (Katz, Zayats, Willner, & Lisdat, 2006; Willner, Lion-Dagan, Marx-Tibbon, & Katz, 1995). Particularly, multi-enzyme biocatalytic cascades controlled by magnetic field as a signal for modulation of biocatalytic reaction rates are very novel systems with highly important features for future applications (Filipov, Zakharchenko, Minko, & Katz, 2018).

2. Bioelectrocatalytic reactions performed with enzymes conjugated to magnetic microparticles and controlled by external magnetic field

Enzyme reactions, particularly interfacial processes in bioelectrocatalytic systems, require positioning of the enzymes, their cofactors and often electron transfer mediators close to an electrode surface. In order to control these processes by applying external magnetic field some components of the bioelectrocatalytic systems can be conjugated with nano- or micro-size magnetic particles. These may include covalent binding of redox enzymes to magnetic particles or, in the easier approach, either electron transfer mediators or cofactors can be associated with magnetic particles. The magneto-induced translocation of the functionalized magnetic particles to and from an electrode surface will result in the activation and inhibition of bioelectrocatalytic processes, respectively. This approach to the magneto-switchable bioelectrocatalytic reactions has been realized with many different magnetic particle conjugates moving them to and from electrodes (Katz, 2016).

The most representative example of such systems included the bioelectrocatalytic oxidation of glucose in the presence of soluble glucose oxidase (GOx) mediated by a ferrocene derivative bound to magnetic micro-particles (Hirsch et al., 2000; Willner & Katz, 2003), Fig. 1A. Magnetic particles (Fe_3O_4, saturated magnetization ca. $65\,emu\,g^{-1}$, ca. $1\,\mu m$ average diameter) have been used to transport the ferrocene derivative to and from the electrode surface. Since GOx cannot communicate directly with an electrode surface (Wilson & Turner, 1992), a mediating process in the presence of the ferrocene derivative is needed to activate the bioelectrocatalytic reaction (Willner & Katz, 2000). In the present example, the bioelectrocatalytic reaction was followed by cyclic voltammetry and it was activated only when the ferrocene-modified magnetic particles were associated with the electrode surface, being attracted by the magnet located below the electrode, Fig. 1B, curve a. When the particles were suspended in the solution, the ferrocene mediator was not able to transport electrons from GOx to the electrode surface, thus demonstrating the inhibited process, Fig. 1B, curve b. The reversible activation/inhibition of the bioelectrocatalytic process was observed upon stepwise translocation of

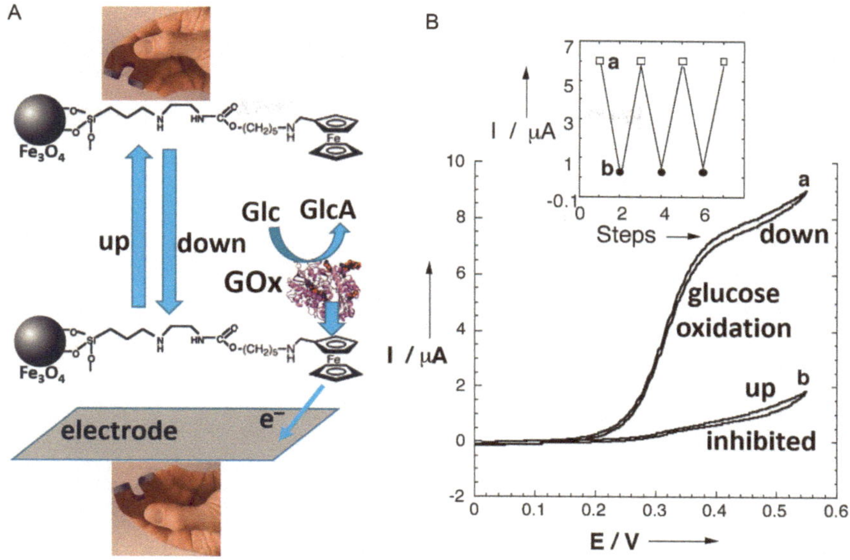

Fig. 1 (A) Reversible activation/inhibition of the bioelectrocatalytic reaction mediated by the redox species covalently bound to Fe_3O_4 magnetic micro-particles upon their vertical translocation with help of an external magnet. (B) Cyclic voltammograms recorded for the ferrocene-modified particles in the presence of GOx ($1\,mg\,mL^{-1}$) and glucose (10 mM) magnetically attracted to the electrode surface (curve a) and removed from it (curve b). Potential scan rate, $5\,mV\,s^{-1}$. Inset shows reversible variation of the bioelectrocatalytic current upon stepwise attraction of the particles to and removal from the electrode surface. *A part of this figure was reproduced from Willner, I., & Katz, E. (2003) Magnetic control of electrocatalytic and bioelectrocatalytic processes.* Angewandte Chemie, International Edition, 42, *4576–4588 with permission.*

the magnetic particles between the surface-confined and solution-suspended position by applying an external magnetic field below and above the electrode, respectively, Fig. 1B, inset.

The concept demonstrated in the magneto-switchable system described above was adapted for reversible activation of NAD^+-dependent enzymes. In the new system (Katz, Sheeney-Haj-Ichia, & Willner, 2002), the magnetic Fe_3O_4 micro-particles were silanized to yield amino groups on the surface and then stepwise functionalized, using carbodiimide coupling, with pyrroloquinoline quinone (PQQ) and then with amino-derivative of NAD^+ (N^6-(2-aminoethyl)-β-nicotinamide adenine dinucleotide (Bückmann & Wray, 1992)), Fig. 2A. The immobilized NAD^+ cofactor was needed to activate soluble NAD^+-dependent lactate dehydrogenase (LDH), while PQQ was operating as a catalyst for electrocatalytic oxidation

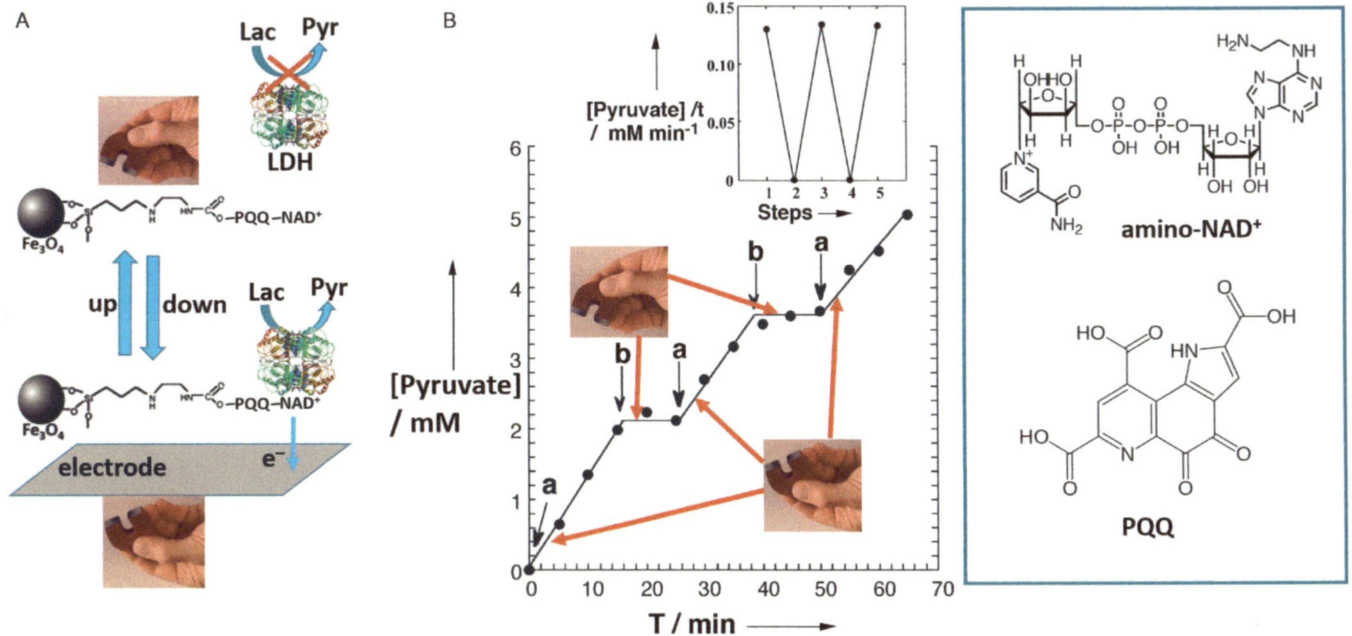

Fig. 2 (A) Magneto-switched bioelectrocatalytic oxidation of lactate in the presence of lactate dehydrogenase (LDH) and magnetic particles functionalized with PQQ-NAD^+. (B) Formation of pyruvate upon the bioelectrocatalytic oxidation of lactate (0.1 M) in the presence of LDH (2 mg/mL) and PQQ-NAD^+-functionalized magnetic particles with a potential of $E = 0.05$ V (vs. SCE) applied on the electrode. Domains started from points "a": The magnetic particles are attracted to the electrode surface. Domains started from points "b": The magnetic particles are retracted from the electrode surface. Inset: Reversible changes in the rate of the bioelectrocatalytic formation of pyruvate upon the magnetic switching the biocatalytic process ON and OFF by attraction and retraction of the magnetic particles to and from the electrode surface, respectively. The data were recorded in Tris-buffer (0.1 M, pH 7.0) and $CaCl_2$ (10 mM) under argon. *A part of this figure was reproduced from Katz, E., Sheeney-Haj-Ichia, L., & Willner, I. (2002) Magneto-switchable electrocatalytic and bioelectrocatalytic transformations.* Chemistry—A European Journal, 8, *4138–4148 with permission.*

of NADH and NAD^+ regeneration (Katz, Lötzbeyer, Schlereth, Schuhmann, & Schmidt, 1994). When the magnetic particles were suspended in the solution containing LDH and lactate (note the external magnet position above the electrode), the bioelectrocatalytic oxidation of lactate was inhibited because the PQQ-NAD^+ complex was disconnected from the electrode surface, thus being not able to regenerate NAD^+ cofactor for cyclic operation with the enzyme. When the magnet was placed below the electrode surface and the PQQ-NAD^+-functionalized particles were magnetically attracted to the surface, the bioelectrocatalytic process was initiated resulting in lactate oxidation yielding pyruvate through the enzyme reaction and the electron transport to the electrode assisted by NAD^+/NADH cofactor and PQQ electrocatalyst, Fig. 2A. While the reversible electrocatalytic current formation and inhibition were detected similarly to the systems described above, in addition to that chemical analysis was applied to follow accumulation of the pyruvate product in course of the bioelectrochemical process, Fig. 2B. The pyruvate concentration was increasing during the time-intervals when the magnetic particles were located at the electrode surface and the concentration was unchanged when the particles were suspended in the solution being disconnected from the electrode surface. The process was reversibly activated/inhibited by repositioning the external magnet below and above the electrode surface, respectively, Fig. 2B, inset. The present example of the magneto-switchable bioelectrocatalytic process is important to illustrate the multi-step process including the enzyme reaction, cofactor reduction and its electrocatalytic regeneration, all controlled by the variable magnetic field.

3. Assembling conducting nanowires from magnetic nanoparticles in the presence of external magnetic field

A different way of controlling electrode interfacial properties, allowing reversible alteration of the electrode enzyme-based bioelectrocatalytic activity, was realized by generating a "forest" of conducting nanowires in the presence of an external magnetic field (Jimenez et al., 2008). These standing nanowires were composed of Au-shell/$CoFe_2O_4$-magnetic core nanoparticles (ca. 20 nm diameter), self-assembled in the presence of an external magnetic field, to generate an array of nanoelectrodes extending the conducting electrode support and increasing the effective electrode area, Fig. 3A. This resulted in the enhancement of the

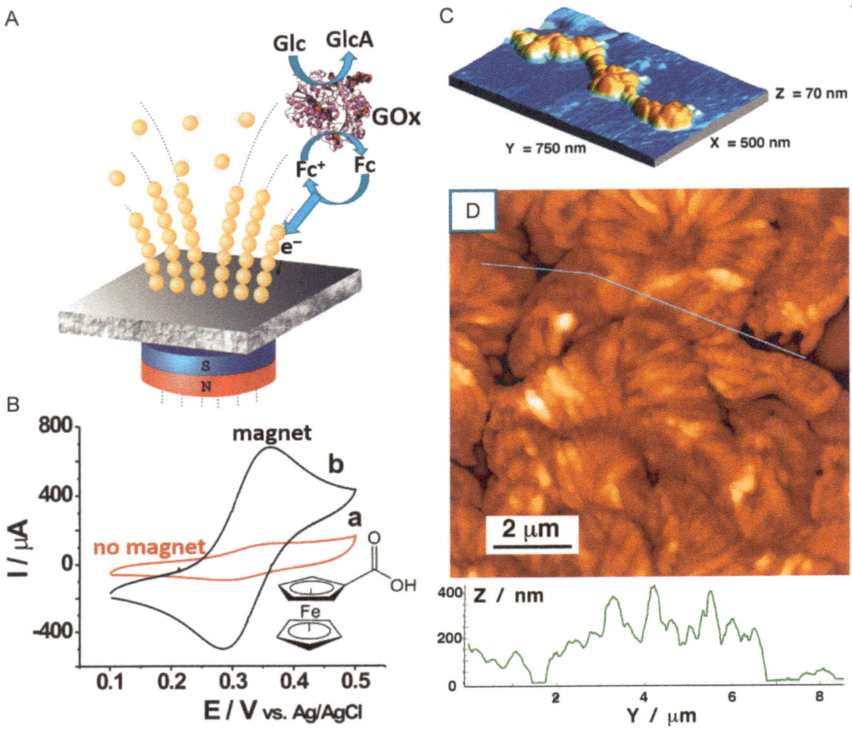

Fig. 3 (A) Formation of the conducting nanowires upon self-assembling of the Au-coated $CoFe_2O_4$ magnetic nanoparticles along the magnetic field lines and their use as a nanostructured electrode for electrochemical oxidation of ferrocene monocarboxylic acid coupled with the glucose oxidation biocatalyzed by GOx. (B) Cyclic voltammograms of ferrocene monocarboxylic acid, 0.1 mM, obtained before (a) and after (b) formation of self-assembled nanowires. Potential scan rate, $200\,mV\,s^{-1}$. (C) AFM image of an individual nanowire self-assembled from the Au-coated magnetic nanoparticles in the presence of the external magnetic field. (D) AFM image of the nanostructured array obtained on the electrode surface upon magneto-assisted self-assembling of the Au-coated magnetic nanoparticles. (Note that the AFM imaging was performed in the absence of the magnetic field when the original alignment of the nanowires along the magnetic field lines was not preserved.) The cross section of the nanostructured array is shown at the bottom. *A part of this figure was reproduced from Jimenez, J., Sheparovych, R., Pita, M., Narvaez Garcia, A., Dominguez, E. et al. (2008)* Journal of Physical Chemistry C, 112, *7337–7344 with permission.*

electrochemical response for a diffusional redox probe, ferrocene monocarboxylic acid, by about sixfold as followed by cyclic voltammetry, Fig. 3B. The magneto-assembled nanowires were collected on a surface and characterized by atomic force microscopy (AFM), Fig. 3C, D. The magneto-induced formation of nanowires was reversed when the external magnet was removed, thus switching OFF the magnetic field and resulting in disaggregation of the conducting nanowires due to thermally activated Brownian motion of the nanoparticles. The primary electrochemical reaction of the diffusional redox species was coupled with the bioelectrocatalytic oxidation of glucose in the presence of glucose oxidase (GOx), resulting in amplification of the biocatalytic cascade at the nanostructured array at the electrode surface, Fig. 3A. The magneto-switchable nanoelectrode array can operate as a general platform for electrochemical biosensors with the enhanced current output signals controlled by the structure of the self-assembled nanowires. The formation of nanowires resulting in the current enhancement can be tuned by various positioning of the magnet, thus allowing tuning the electrode sensitivity to the desired extend.

4. Magneto-controlled enzyme biocatalytic cascades: Substrate channeling vs. free diffusion

4.1 Motivation and goals

In biological systems, enzymes working cooperatively in biocatalytic pathways are organized in macromolecular complexes. The organized placement of enzymes in cells allows communication between catalytic sites without diffusion of biosynthetic intermediates into the bulk phase of the cell. The direct transport of substrates/products, called "*metabolic channeling*" is highly important for life because it provides much more effective biocatalytic processes than can be achieved in chemical systems operating through random diffusion of intermediate species (Mendes, Kell, & Welch, 1995; Winkel, 2004). Inspired by biological prototypes, many artificial biomolecular systems with organized placement of enzymes in synthetic nanoscale scaffolds have been designed (Dueber et al., 2009; Idan & Hess, 2013a). The major motivation for this research has been the enhancement of biocatalytic reactions (Idan & Hess, 2013b), including those applied in biotechnological processes (Percival Zhang, 2011). The important feature of these systems is "*substrate channeling*" resulting in effective communication between enzymes similarly to biological processes. The common approach to the organized positioning of enzymes is based on the use of macromolecular scaffolds, frequently using DNA/RNA for their assembling (Delebecque, Lindner, Silver, & Aldaye, 2011; Fu, Liu, Liu, Woodbury, & Yan, 2012; Wilner et al., 2009). In addition to static DNA scaffolds, dynamically operating DNA "machines" have been used to regulate enzyme cascades by external signals (Xin, Zhou, Yang, & Liu, 2013). Despite the fact that the direct mimicking of biological structures requires exact positioning of cooperatively working enzymes at specific locations and with specific distances, much simpler approach based on random placement of enzymes in micro-containers (Zhang et al., 2011) or on surfaces (Mathesh, Liu, Barrow, & Yang, 2017; Okhokhonin et al., 2018) has been used and demonstrated some advantages comparing with solute state of the enzymes. In addition to highly challenging experimental work, the systems demonstrating substrate channeling have been studied theoretically, modeling the biocatalytic processes and comparing them with those proceeding through bulk solution diffusion (Idan & Hess, 2013b; Zhang & Hess, 2017). The important conclusion was that the substrate channeling is not always more effective than diffusional process (Idan & Hess, 2013b) and the explanations for reaction facilitation due to substrate channeling is not always correct (Zhang, Tsitkov, & Hess, 2016). The substrate channeling results in the improved kinetics only under very specific conditions controlled by concentrations of reacting species and at specific time-intervals of biocatalytic reactions (Idan & Hess, 2013b; Welch & Easterby, 1994).

While the substrate channeling achieved by enzyme placement on nanoscale scaffolds has been demonstrated in many different systems (Delebecque et al., 2011; Dueber et al., 2009; Fu et al., 2012; Idan & Hess, 2013a; Wilner et al., 2009), switchable systems with signal-controlled features (Xin et al., 2013), particularly switched reversibly between *substrate channeling* and *substrate diffusion*, have been studied only fragmentally (Filipov et al., 2018). Since dynamic changes in molecular (e.g., DNA) scaffolds are not easy to achieve (Xin et al., 2013), the switchable systems require different approach. The system outlined in the next section (Filipov et al., 2018) aims at demonstrating a magneto-controlled biocatalytic cascade activated by two enzymes communicating through *diffusional substrates* in the absence of magnetic field when the enzyme-functionalized magnetic nanoparticles are dispersed in a solution and through *substrate channeling* in the presence of magnetic field when the nanoparticles are aggregated and the enzymes are close to each other.

4.2 Experimental realization

In the studied biocatalytic system (Filipov et al., 2018) two kinds of the magnetic nanoparticles (MNPs) were functionalized with glucose oxidase (GOx) and horseradish peroxidase (HRP), respectively, Fig. 4. In the

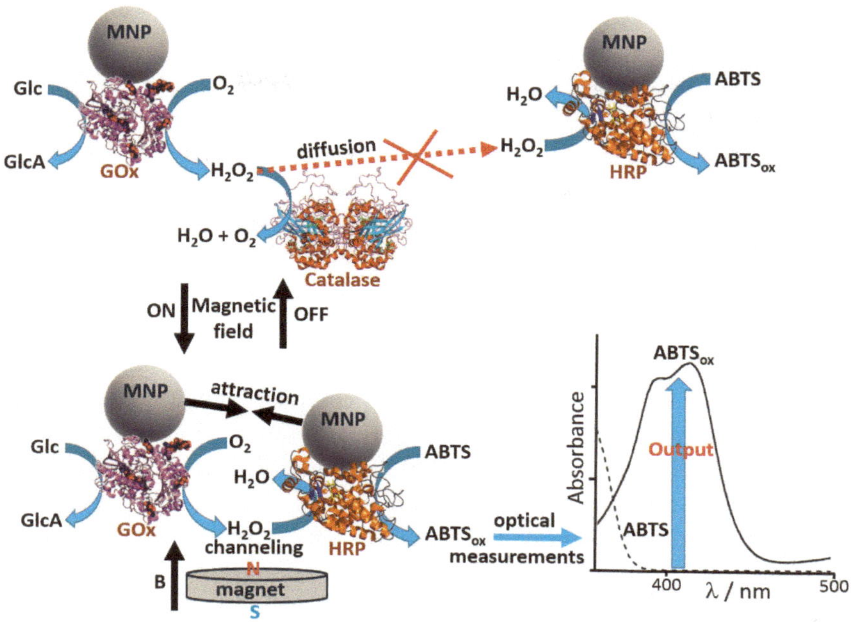

Fig. 4 Schematics of the biocatalytic cascade realized in the presence of the GOx-functionalized MNPs and HRP-functionalized MNPs operating in the dispersed and aggregated states switchable by the external magnetic field. In the aggregated state produced in the presence of the magnetic field (bottom part of the scheme), the enzymes communicate via *"substrate channeling"*—the direct transfer of the H_2O_2 intermediate between enzymes in the cascade without the release of the intermediate into the bulk solution. In the dispersed state in the absence of the magnetic field (upper part of the scheme) the H_2O_2 intermediate is released to the bulk solution where it is intercepted and catalytically decomposed by soluble catalase, thus inhibiting its reaction with HRP. Note that the reaction proceeds to the HRP-catalyzed ABTS oxidation in the absence of catalase. The spectra shown correspond to the original ABTS optical absorbance and the absorbance change upon ABTS biocatalytic oxidation (formation of $ABTS_{ox}$). The following abbreviations are used: glucose (Glc), gluconic acid (GlcA), magnetic field (B), other abbreviations are explained in the text. *A part of this figure was reproduced from Filipov, Y., Zakharchenko, A., Minko, S., & Katz, E. (2018) Magneto-controlled biocatalytic cascades with logically processed input signals—Substrate channeling versus free diffusion.* ChemPhysChem, 19, *3035–3043 with permission.*

absence of the magnetic field the MNPs were suspended in a buffer solution and H_2O_2 produced in the first reaction catalyzed by GOx was moving to the next reaction step catalyzed by HRP by diffusion through the bulk solution. Note that in this experiment catalase was not included in the solution, thus H_2O_2 was not intercepted and consumed on its way. Alternatively, in the presence of the magnetic field the MNPs were aggregated (they were collected at the walls of the cuvette, where the magnets were applied), the aggregated MNPs were visible with a naked eye. It was expected that in this state the intermediate H_2O_2 produced by GOx can be delivered to the next enzyme, HRP, by the substrate channeling due to short distances separating the enzymes bound to the MNPs. Both systems, with and without MNPs aggregation, resulted in 2,2′-azino-*bis*(3-ethylbenzothiazoline-6-sulphonic acid) (ABTS; a chromogenic reagent) oxidation followed by the optical absorbance change, Fig. 4. Surprisingly, the time-dependent absorbance changes in both cases were not much different, thus reflecting similar kinetics of the ABTS oxidation through the diffusional pathway and the substrate channeling, Fig. 5A, (it should be noted that at this point we could only assume the "substrate channeling," but it cannot be proved because of the kinetics similar to the diffusional reaction). The small difference in the kinetics of the diffusional reaction and the substrate channeling may originate from large substrate concentration and long-time of the reaction observed when the difference between two mechanisms becomes indistinguishable (Idan & Hess, 2013b).

In order to demonstrate the difference between the reactions catalyzed by the dispersed and aggregated MNPs the system composition was changed by adding catalase to the solution, Fig. 4. This enzyme, operating in the solution, rapidly decomposed the diffusional intermediate product (H_2O_2) and

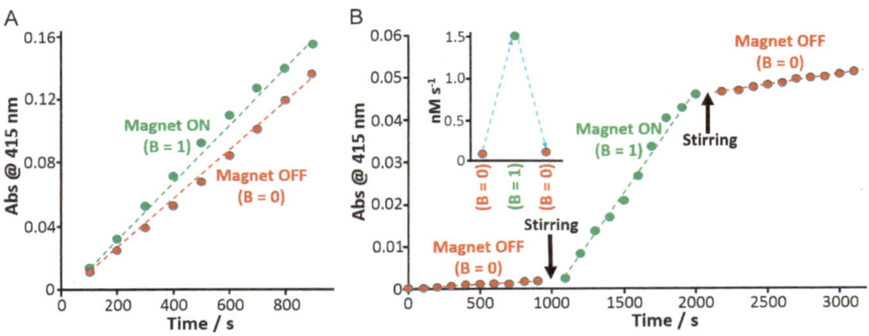

Fig. 5 (A) Absorbance increase corresponding to the biocatalytic formation of $ABTS_{ox}$ in the presence of the GOx- and HRP-functionalized MNPs and in the presence and absence of the external magnetic field. The experiments were performed in the absence of catalase. (B) The same experiments performed in the presence of catalase in the bulk solution. The inset shows reversible change of the $ABTS_{ox}$ formation reaction rate upon switching ON-OFF the magnetic field. *A part of this figure was reproduced from Filipov, Y., Zakharchenko, A., Minko, S., & Katz, E. (2018) Magneto-controlled biocatalytic cascades with logically processed input signals—Substrate channeling versus free diffusion.* ChemPhysChem, 19, *3035–3043 with permission.*

inhibited the second reaction step, which is responsible for the absorbance change, Fig. 5B (magnet OFF). The function of this enzyme in the present system is similar to the "filter" operation studied in enzyme-logic systems (Privman, Domanskyi, Mailloux, Holade, & Katz, 2014), where such reactions were applied for decreasing analog noise in binary signal processing (Melnikov, Strack, Pita, Privman, & Katz, 2009). Thus, for brevity we will keep the name "filter" in the present discussion, while the reason for its application was different. When the magnetic field was applied and the MNPs were aggregated, the second biocatalytic step in the cascade was reactivated and the corresponding absorbance increase was observed, Fig. 5B (magnet ON). This was explained by the substrate channeling, which was much less affected by the H_2O_2 decomposition catalyzed by catalase. Removing magnets and re-suspending the MNPs again resulted in the process inhibition due to H_2O_2 decomposition during the diffusional path, Fig. 5B (magnet OFF again). Therefore, the changes in the reaction mechanism were reversible upon applying and removing the magnetic field, which resulted in the aggregation and disaggregation of the MNPs, respectively, Fig. 5B, inset. The observed phenomenon is in agreement with the theoretical model (Idan & Hess, 2013b), which predicts the improved kinetics for the substrate channeling only when the substrate concentration is low (being consumed by catalase in the present experiment). It should be noted that the rate of the ABTS oxidation (the last step in the biocatalytic cascade) in the aggregated system, where we assume the substrate channeling, is still smaller than that without the "filter" (compare Fig. 5A and B with the magnet ON). This can originate from partial decomposition of the intermediate substrate H_2O_2 by catalase even in the aggregated state of the MNPs.

Overall, the magnetic nanoparticles (MNPs) functionalized with two different enzymes allowed reversible switch between the aggregated and dispersed states in the presence and absence of the magnetic field, respectively. It has been shown that the biocatalytic cascade reactions proceeding in the aggregated and dispersed state of the enzyme-functionalized MNPs may have similar or significantly different kinetics. The difference, when it was observed, was explained in the terms of the "*substrate channeling*" providing direct transfer of the intermediate substrates/products from one enzyme to another without their diffusion to the bulk solution. In order to get the difference between the reactions proceeding in the dispersed and aggregated states, the intermediate substrate/product must be intercepted/consumed in the bulk solution using the "filter" catalyzed by an additional soluble enzyme. In the absence of such filter, the biocatalytic cascades proceeding through the substrate channeling in the aggregated state and through normal bulk solution diffusion in the dispersed state are not much different. This was explained by the high concentration of the reacting species according to the general theoretical model (Idan & Hess, 2013b). The use of the filter process consuming rapidly the intermediate substrate/product allowed magneto-induced switch between the active state of the system in the aggregated state and inhibited state in the dispersed state. The novelty and significance of the present study is the use of dynamic changes induced by the magnetic field switching reversibly between *diffusional* and *substrate channeling* processes.

5. Magnetohydrodynamic effect applied to enzyme bioelectrocatalytic reactions

Application of magnetic micro-/nano-particles functionalized with enzymes to allow magneto-switchable biocatalytic and bioelectrocatalytic reactions was exemplified in the previous sections. This approach is still waiting to reach its pinnacle of future use in various biosensors and biofuel cells with switchable/tunable operation. Another approach to the magnetic control of bioelectrocatalytic enzyme reactions is based on the magnetohydrodynamic effect, which is well known for simple inorganic electrochemical processes (Fahidy, 1973; Leventis, Chen, Gao, Canalas, & Zhang, 1998; Ragsdale, Lee, Gao, & White, 1996; Ragsdale, Lee, & White, 1997), but rarely applied in bioelectrochemistry (Katz et al., 2004; Katz, Lioubashevski, & Willner, 2005; Lioubashevski et al., 2007). Recent development of the theory quantitatively describing the mechanism of the magnetohydrodynamic effect (Katz et al., 2004; Lioubashevski et al., 2007) allowed its effective application in complex bioelectrocatalytic systems and biofuel cells (Katz et al., 2005).

A depletion layer produced at an electrode interface upon electrochemical consumption of redox species generates a concentration gradient and results in the diffusion of the reactive species from the bulk solution to the electrode surface. Application of a permanent magnetic field in the direction perpendicular to the diffusion flux of the charge-transporting species results in their rotation in the magnetic field, yielding the rotation of the entire liquid phase. This phenomenon results in shrinking of the depletion layer at the electrode surface and enhances the mass-transport and generated current at the electrode. Application of a magnetic field to a biofuel cell (in the direction perpendicular to the current between the cathode and anode) results in a convective mass-transport, thus resulting in the enhancement of the current generated by the cell.

A biofuel cell composed of bioelectrocatalytic electrodes based on reconstituted enzymes was applied to demonstrate the phenomenon of the current enhancement in the presence of a constant magnetic field (Katz et al., 2005). The biofuel cell included a lactate-oxidizing anode based on reconstituted lactate dehydrogenase (LDH) (Bardea, Katz, Bückmann, & Willner, 1997), while the oxygen-reducing cathode was based on cytochrome oxidase integrated with cytochrome *c* (COx-Cyt c) (Pardo-Yissar et al., 2000), Fig. 6A. Both bioelectrocatalytic electrodes included redox enzymes interacting with the corresponding cofactors/natural mediators NAD^+ and Cyt c for LDH and COx, respectively, immobilized in an aligned monolayer configuration. An additional mediator—pyrroloquinoline quinone (PQQ) facilitated the electron transfer from the enzymatically reduced NADH to the electrode conducting support. The biofuel cell produced an open circuitry voltage, V_{oc}, of ca. 120 mV and a short circuitry current density, i_{sc}, of ca. 100 μA cm^{-2} in the absence of magnetic field. Upon application of the magnetic field, $B=0.92$ T, the voltage generated by the biofuel cell was almost unchanged, but the current output was increased by ca. threefold reflecting the mass-transport enhancement originating from the magnetohydrodynamic effect. The magneto-induced current output increase was observed at different loading resistances resulting in the shift of the polarization curve in the presence of the magnetic field, Fig. 6B, curve b, compared with the polarization curve recorded in the absence of the external magnet, Fig. 6B, curve a. The current output was dependent on the intensity of magnetic flux density, B, revealing the slope of $\log i_{sc}$ vs. $\log B$ plot, Fig. 6C, that corresponds to 0.332 as predicted by the theoretical model ($i \propto B^{1/3}$) (Katz et al., 2004). Thus, variation of the applied magnetic field allowed tuning of the current output generated by the biofuel cell. The electrical power density output, $P = i \times V$, of the biofuel cell was analyzed as a function of the external load resistance, R. The power produced by the biofuel cell at the optimum external resistance ($R = 1.2\,k\Omega$) was threefold higher upon application of the magnetic field of $B=0.92$ T. Cyclic switching ON–OFF of the magnetic field resulted in the reversible changes of the biofuel cell activity reflected in the polarization curves and power release functions. The switchable/tunable biofuel cell was controlled by external magnetic signals at a distance without direct contact with the electrodes and solutions.

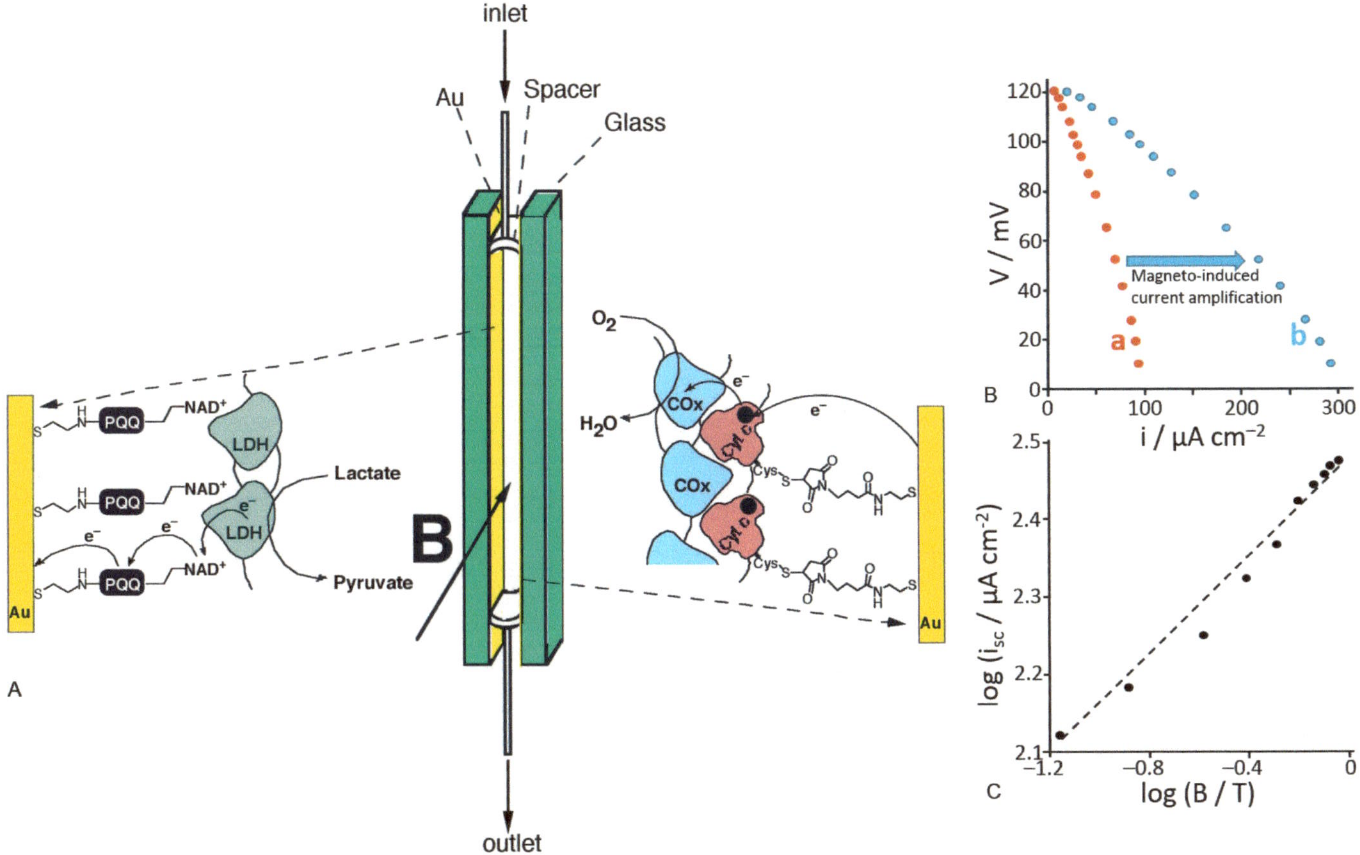

Fig. 6 (A) Configuration of the biofuel cell based on lactate-oxidizing LDH/NAD$^+$-PQQ-electrode as an anode and O_2-reducing COx/Cyt c-electrode as a cathode. (B) The polarization function measured at variable external resistances: (a) in the absence of magnetic field; (b) in the presence of magnetic field, $B = 0.92$ T. (C) The dependence of the short-circuit current density on the magnetic flux density. The biofuel cell operated upon pumping of the solution (flow rate 1 mL min^{-1}) composed of 0.1 Tris-buffer, pH 7.0, containing $CaCl_2$, 10 mM, lactate, 20 mM, and oxygen (the solution equilibrated with air). *A part of this figure was reproduced from Katz, E., Lioubashevski, O., & Willner, I. (2005) Magnetic field effects on bioelectrocatalytic reactions of surface-confined enzyme systems: Enhanced performance of biofuel cells.* Journal of the Americal Chemical Society, 127, *3979–3988 with permission.*

6. Experimental details

The present article overviews several conceptually different approaches to the magneto-controlled enzyme reactions. The experimental details for all discussed systems can be found in the corresponding references. In the present section we summarized the experimental details for the system discussed in Section 4. This selection is based on the novelty and importance of the discussed approach.

6.1 Materials and instrumentation

Enzymes: glucose oxidase from *Aspergillus niger* (GOx; EC 1.1.3.4), peroxidase from horseradish (HRP; EC 1.11.1.7), catalase from bovine liver (EC 1.11.1.6); substrates: D-(+)-glucose (Glc), 2,2′-azino-*bis*(3-ethylbenzothiazoline-6-sulfonic acid) diammonium salt (ABTS); additional materials: bovine serum albumin (BSA), polyethylene glycol (PEG), (1-ethyl-3[3-(dimethylamino)propyl]) carbodiimide (EDC), *N*-hydroxysuccinimide (NHS), sodium citrate dihydrate (citrate buffer), 2-(*N*-morpholino) ethanesulfonic acid hydrate (MES-buffer), 4-(2-hydroxyethyl)-1-piperazineethanesulfonic acid (HEPES-buffer), and other standard organic/inorganic chemicals and materials were obtained from MilliporeSigma Company (former Sigma-Aldrich) and used without further purification. All experiments were carried out in ultrapure water (18.2 MΩ cm; Barnstead NANOpure Diamond). A Shimadzu UV-2450 UV–vis spectrophotometer with 1 mL (10 mm optical path) poly (methyl methacrylate) (PMMA) cuvettes was used for optical absorbance measurements.

6.2 Synthesis of magnetite-silica-core-shell nanoparticles

(1) The superparamagnetic nanoparticles (MNPs) were synthesized using a coprecipitation method as described elsewhere (Bumb et al., 2008).

The following synthetic steps were applied:

(a) Iron chloride salts (MilliporeSigma), 4.43 g $FeCl_3$ $6H_2O$ and 1.625 g $FeCl_2$ $4H_2O$ were dissolved in 190 mL of water with a stoichiometric ratio Fe^{3+}: Fe^{2+} = 2: 1 with magnetic stirring at room temperature (20 ± 2 °C).

(b) Then, 10 mL of 25% ammonium hydroxide (MilliporeSigma) was added immediately. Formation of a black precipitate was observed.

(c) The solution was stirred for additional ten minutes, then the precipitate was separated with a magnet and rinsed three times with water using a magnetic separation.

(d) The colloidal dispersion of MNPs was stabilized with citrate ions by a rapid rinsing of the precipitate twice with a 2 M nitric acid (MilliporeSigma) aqueous solution followed by the addition of 5 mL of 0.5 M trisodium citrate (MilliporeSigma) in water while maintaining pH 2.5 using a sodium hydroxide aqueous solution.

(e) After stirring for 1.5 h, MNPs were magnetically separated, rinsed with water and diluted to obtain 100 mL (pH = 6) of the dispersion. The concentration of MNPs in the final stock dispersion was 2% wt.

(2) A modified Stöber method (Deng, Wang, Hu, Yang, & Fu, 2005) was applied to coat the MNPs with a silica layer.

The following synthetic steps were applied:

(a) 2 mL of the MNPs stock solution was diluted in a mixture of 160 mL of ethanol (MilliporeSigma) and 40 mL of water.

(b) Then, 5 mL of ammonium hydroxide (MilliporeSigma) was added to the MNPs dispersion.

(c) After 10 min of treatment in an ultrasonic bath, 1 mL of silicon tetraethoxide (MilliporeSigma) was dropwise added into the reactor. The synthesis was carried out at 0 °C and sonication for 3 h.

(d) The reaction was terminated by the addition of several droplets of 10% HCl (MilliporeSigma) followed by precipitation of silica-coated MNPs.

(e) The precipitate was collected by a magnet and rinsed three times with water.

(f) Then the precipitate was suspended in a 50 mL of water at sonication. The resulting product was a stable 2 mg/mL dispersion of the MNPs.

Fig. 7 shows transmission electron microscopy (TEM) images of the MNPs, demonstrating their small size distribution. Further structural and functional characterization of the MNPs is detailed elsewhere (Zacharchenko et al., 2017).

6.3 Modification of magnetic nanoparticles with enzymes

The following synthetic steps were applied:

(a) A suspension of the MNPs (1 mg/mL) in MES-buffer (MilliporeSigma) (0.1 M, pH 4.8) was reacted with 20 mM EDC (MilliporeSigma) and 40 mM NHS (MilliporeSigma) for 20 min at 37 °C.

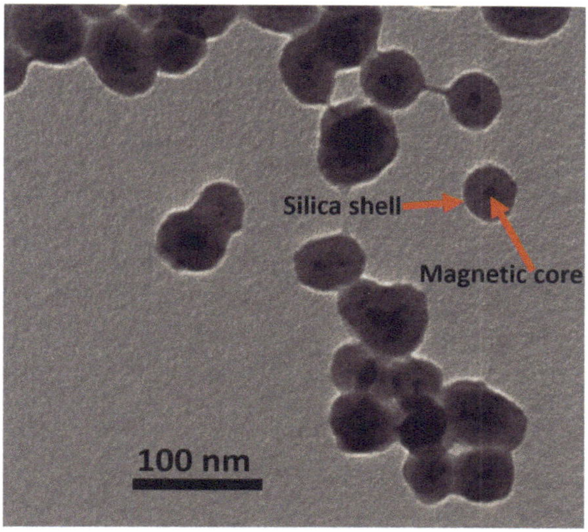

Fig. 7 Transmission electron microscopy (TEM) images of the MNPs. *The figure was reproduced from Filipov, Y., Zakharchenko, A., Minko, S., & Katz, E. (2018) Magneto-controlled biocatalytic cascades with logically processed input signals—Substrate channeling versus free diffusion.* ChemPhysChem, *19, 3035–3043 with permission.*

(b) Then it was centrifuged and the MES-buffer was replaced with HEPES-buffer (MilliporeSigma) (0.1 M, pH 7.0) containing one of the enzymes (GOx, HRP; MilliporeSigma) in the amount of 100 U with 0.01% BSA (MilliporeSigma) and 1% of PEG.

(c) The reacting suspension was placed in a shaker for 4 h at room temperature (ca. $20 \pm 2\,°C$).

(d) After that, the enzyme-functionalized MNPs were centrifuged to separate them from the non-reacted materials and rinsed with the HEPES-buffer three times.

(e) The biocatalytic activities of the enzyme-functionalized MNPs were measured according to the standard protocols recommended by Sigma-Aldrich: 0.2 U of HRP per 1 mg of MNPs (HRP enzyme assay), 0.12 U of GOx per 1 mg of MNPs (GOx enzyme assay).

6.4 Measurements

The following experimental conditions were applied:

(a) All experiments were performed at room temperature (ca. $20 \pm 2\,°C$) in Semi-Micro PMMA cuvettes (BrandTech® Scientific) with 1 mL volume of aqueous citrate buffer (MilliporeSigma) solution (100 mM, pH 5.5) containing dispersed MNPs (5 μg each kind) functionalized with the corresponding enzymes, GOx, HRP, applied as needed for realization of the biocatalytic cascade.

(b) The "filter" system was realized with catalase (MilliporeSigma) (5 U/mL).

(c) The substrates were added to activate the biocatalytic cascade: glucose (1 mM) and ABTS (MilliporeSigma) (0.25 mM). Oxygen needed for the GOx operation was always present in the system under equilibrium with air.

(d) Two magnets (NdFeB, grade N42, plating/coating: Ni-Cu-Ni, dimensions: 1/2″ × 1/4″ × 1/8″, B842 from K&J Magnetics, Inc.) were placed at both sides of the cuvette, Fig. 8, at the distance of 6 mm to create a nearly uniform (homogeneous) magnetic field in the liquid system with the intensity of 0.3 T measured in the center position between the magnets.

(e) After application or removing of the external magnetic field, the solution was stirred mechanically with a pipet to obtain a "homogeneous" dispersed system.

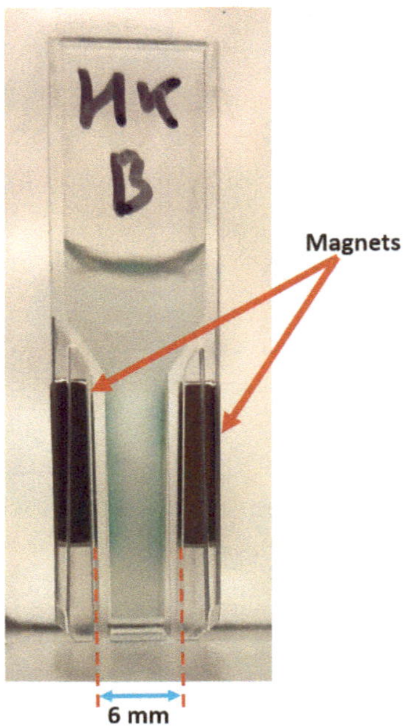

Fig. 8 The cuvette with two magnets attached at both sides. *The figure was reproduced Filipov, Y., Zakharchenko, A., Minko, S., & Katz, E. (2018) Magneto-controlled biocatalytic cascades with logically processed input signals—Substrate channeling versus free diffusion.* ChemPhysChem, 19, *3035–3043 with permission.*

(f) The optical absorbance was measured at $\lambda = 415$ nm (characteristic of the ABTS oxidized form, $ABTS_{ox}$) (Shimadzu UV-2450 UV–Vis spectrophotometer) during the reaction time. The absorbance measured was considered as the output signal generated by the biocatalytic system.

7. Conclusions

The magneto-switchable biocatalytic systems overviewed in this article show great variety of different approaches to control enzyme reactions with an external magnetic field. In addition to the fundamental scientific importance, the studied systems keep promise for numerous practical applications. The state-of-the-art in the preparation, characterization and application of biomolecule-functionalized magnetic particles and other related micro-/nano-objects allows efficient performance of various in vitro and in vivo biosensors and bioelectronic devices. Many of these biosensors are directed to biomedical applications. Many other examples outlined in the review demonstrated various aspects of magneto-controlled enzyme systems.

Acknowledgment

The work outlined in this article was performed by many students and postdoctoral researchers with participation of many collaborators. Their contribution was highly important and greatly appreciated.

References

Abrosimova, L. A., Monakhova, M. V., Migur, A. Y., Wolfgang, W., Pingoud, A., et al. (2013). Thermo-switchable activity of the restriction endonuclease SsoII achieved by site-directed enzyme modification. *IUBMB Life*, *65*, 1012–1016.

Aizawa, M., Namba, K., & Suzuki, S. (1977). Light-induced enzyme activity changes associated with the photoisomerization of bound spiropyran. *Archives of Biochemistry and Biophysics*, *182*, 305–310.

Bardea, A., Katz, E., Bückmann, A. F., & Willner, I. (1997). NAD^+-Dependent enzyme electrodes: Electrical contact of cofactor-dependent enzymes and electrodes. *Journal of the American Chemical Society*, *119*, 9114–9119.

Blonder, R., Katz, E., Willner, I., Wray, V., & Bückmann, A. F. (1997). Application of a nitrospiropyran-FAD reconstituted glucose oxidase and charged electron mediators as optobioelectronic assemblies for the amperometric transduction of recorded optical signals: Control of the "ON"-"OFF" direction of the photoswitch. *Journal of the American Chemical Society*, *119*, 11747–11757.

Bückmann, A. F., & Wray, V. (1992). A simplified procedure for the synthesis and purification of N(6)-(2-aminoethyl)-NAD and tricyclic 1.N(6)-ethanoadenine-NAD. *Biotechnology and Applied Biochemistry*, *15*, 303–310.

Bumb, A., Brechbiel, M. W., Choyke, P. L., Fugger, L., Eggeman, A., et al. (2008). Synthesis and characterization of ultra-small superparamagnetic iron oxide nanoparticles thinly coated with silica. *Nanotechnology*, *19*. art. No. 335601.

Choi, S. (Ed.), (2010). *Systems biology for signaling networks*. New York: Springer-Verlag.

Delebecque, C. J., Lindner, A. B., Silver, P. A., & Aldaye, F. A. (2011). Organization of intracellular reactions with rationally designed RNA assemblies. *Science*, *333*, 470–474.

Deng, Y. H., Wang, C. C., Hu, J. H., Yang, W. L., & Fu, S. K. (2005). Investigation of formation of silica-coated magnetite nanoparticles via sol-gel approach. *Colloids and Surfaces, A: Physicochemical and Engineering Aspects*, *262*, 87–93.

Dueber, J. E., Wu, G. C., Reza Malmirchegini, G., Moon, T. S., Petzold, C. J., et al. (2009). Synthetic protein scaffolds provide modular control over metabolic flux. *Nature Biotechnology*, *27*, 753–759.

Fahidy, T. Z. (1973). Hydrodynamic models in magnetoelectrolysis. *Electrochimica Acta*, *18*, 607–614.

Filipov, Y., Zakharchenko, A., Minko, S., & Katz, E. (2018). Magneto-controlled biocatalytic cascades with logically processed input signals—Substrate channeling versus free diffusion. *ChemPhysChem*, *19*, 3035–3043.

Fu, J., Liu, M., Liu, Y., Woodbury, N. W., & Yan, H. (2012). Interenzyme substrate diffusion for an enzyme cascade organized on spatially addressable DNA nanostructures. *Journal of the American Chemical Society*, *134*, 5516–5519.

Guo, Z., Johnston, W. A., Stein, V., Kalimuthu, P., Perez-Alcala, S., et al. (2016). Engineering PQQ-glucose dehydrogenase into an allosteric electrochemical Ca^{2+} sensor. *Chemical Communications*, *52*, 485–488.

Guo, Z., Murphy, L., Stein, V., Johnston, W. A., Alcala-Perez, S., et al. (2016). Engineered PQQ-glucose dehydrogenase as a universal biosensor platform. *Journal of the American Chemical Society*, *138*, 10108–10111.

Hancock, J. (2010). *Cell signalling* (3rd ed.). . Oxford University Press.

Hirsch, R., Katz, E., & Willner, I. (2000). Magneto-switchable bioelectrocatalysis. *Journal of the American Chemical Society*, *122*, 12053–12054.

Hu, Y., Wang, F., Lu, C.-H., Girsh, J., Golub, E., et al. (2014). Switchable enzyme/DNAzyme cascades by the reconfiguration of DNA nanostructures. *Chemistry—A European Journal*, *20*, 16203–16209.

Idan, O., & Hess, H. (2013a). Engineering enzymatic cascades on nanoscale scaffolds. *Current Opinion in Biotechnology*, *24*, 606–611.

Idan, O., & Hess, H. (2013b). Origins of activity enhancement in enzyme cascades on scaffolds. *ACS Nano*, *7*, 8658–8665.

Jimenez, J., Sheparovych, R., Pita, M., Narvaez Garcia, A., Dominguez, E., et al. (2008). *Journal of Physical Chemistry C*, *112*, 7337–7344.

Katz, E. (2016). Magneto-switchable electrodes and electrochemical systems. *Electroanalysis*, *28*, 904–919.

Katz, E., Lioubashevski, O., & Willner, I. (2004). Magnetic field effect on cytochrome c-mediated bioelectrocatalytic transformations. *Journal of the American Chemical Society*, *126*, 11088–11092.

Katz, E., Lioubashevski, O., & Willner, I. (2005). Magnetic field effects on bioelectrocatalytic reactions of surface-confined enzyme systems: Enhanced performance of biofuel cells. *Journal of the American Chemical Society*, *127*, 3979–3988.

Katz, E., Lötzbeyer, T., Schlereth, D. D., Schuhmann, W., & Schmidt, H.-L. (1994). Electrocatalytic oxidation of reduced nicotinamide coenzymes at gold and platinum electrode surfaces modified with a monolayer of pyrroloquinoline quinone. Effect of Ca^{2+} cations. *Journal of Electroanalytical Chemistry*, *373*, 189–200.

Katz, E., Sheeney-Haj-Ichia, L., & Willner, I. (2002). Magneto-switchable electrocatalytic and bioelectrocatalytic transformations. *Chemistry—A European Journal*, *8*, 4138–4148.

Katz, E., Shipway, A. N., & Willner, I. (2002). Electronic and optical transformations of photoisomerization processes at molecular- and biomolecular-functionalized surfaces. In S. Sekkat & W. Knoll (Eds.), *Photoreactive organic thin films* (pp. 219–268). San Diego, USA: Elsevier. Chapter II-7.

Katz, E., Zayats, M., Willner, I., & Lisdat, F. (2006). Controlling the directions of photocurrents by means of CdS nanoparticles and cytochrome c-mediated biocatalytic cascades. *Chemical Communications*, *13*, 1395–1397.

Lee, J., Lee, D., Oh, E., Kim, J., Kim, Y.-P., et al. (2005). Preparation of a magnetically switchable bio-electrocatalytic system employing cross-linked enzyme aggregates in magnetic mesocellular carbon foam. *Angewandte Chemie, International Edition*, *44*, 7427–7432.

Leventis, N., Chen, M. G., Gao, X. R., Canalas, M., & Zhang, P. (1998). Electrochemistry with stationary disk and ring – disk millielectrodes in magnetic fields. *The Journal of Physical Chemistry. B*, *102*, 3512–3522.

Lion-Dagan, M., Katz, E., & Willner, I. (1994). Amperometric transduction of optical signals recorded by organized monolayers of photoisomerizable biomaterials on Au-electrodes. *Journal of the American Chemical Society*, *116*, 7913–7914.

Lioubashevski, O., Katz, E., & Willner, I. (2007). Effects of magnetic field directed orthogonally to surfaces on electrochemical processes. *Journal of Physical Chemistry C*, *111*, 6024–6032.

Mathesh, M., Liu, J., Barrow, C. J., & Yang, W. (2017). Graphene-oxide-based enzyme nanoarchitectonics for substrate channeling. *Chemistry—A European Journal*, *23*, 304–311.

Melnikov, D., Strack, G., Pita, M., Privman, V., & Katz, E. (2009). Analog noise reduction in enzymatic logic gates. *The Journal of Physical Chemistry. B*, *113*, 10472–10479.

Mendes, P., Kell, D. B., & Welch, G. R. (1995). Metabolic channeling in organized enzyme systems: Experiments and models. *Advances in Molecular and Cell Biology*, *11*, 1–19.

HRP enzyme assay. https://www.sigmaaldrich.com/technical-documents/protocols/biology/enzymatic-assay-of-peroxidase.html

GOx enzyme assay. https://www.sigmaaldrich.com/technical-documents/protocols/biology/enzymatic-assay-of-glucose-oxidase.html

Okhokhonin, A. V., Domanskyi, S., Filipov, Y., Gamella, M., Kozitsina, A. N., et al. (2018). Biomolecular release from alginate-modified electrode triggered by chemical inputs processed through a biocatalytic cascade—Integration of biomolecular computing and actuation. *Electroanalysis*, *30*, 426–435.

Ostermeier, M. (2009). Designing switchable enzymes. *Current Opinion in Structural Biology*, *19*, 442–448.

Pardo-Yissar, V., Katz, E., Willner, I., Kotlyar, A. B., Sanders, C., & Lill, H. (2000). Biomaterial engineered electrodes for bioelectronics. *Faraday Discussions*, *116*, 119–134.

Patolsky, F., Filanovsky, B., Katz, E., & Willner, I. (1998). Photoswitchable antigen-antibody interactions studied by impedance spectroscopy. *The Journal of Physical Chemistry. B*, *102*, 10359–10367.

Percival Zhang, Y.-H. (2011). Substrate channeling and enzyme complexes for biotechnological applications. *Biotechnology Advances*, *29*, 715–725.

Privman, V., Domanskyi, S., Mailloux, S., Holade, Y., & Katz, E. (2014). Kinetic model for a threshold filter in an enzymatic system for bioanalytical and biocomputing applications. *The Journal of Physical Chemistry. B*, *118*, 12435–12443.

Ragsdale, S. R., Lee, J., Gao, X. P., & White, H. S. (1996). Magnetic field effects in electrochemistry. Voltammetric reduction of acetophenone at microdisk electrodes. *The Journal of Physical Chemistry*, *100*, 5913–5922.

Ragsdale, S. R., Lee, J. H., & White, H. S. (1997). Analysis of the magnetic force generated at a hemispherical microelectrode. *Analytical Chemistry*, *69*, 2070–2076.

Shakya, A. K., & Nandakumar, K. S. (2018). An update on smart biocatalysts for industrial and biomedical applications. *Journal of the Royal Society, Interface*, *15*. art. No. 20180062.

Stein, V., & Alexandrov, K. (2015). Synthetic protein switches: Design principles and applications. *Trends in Biotechnology*, *33*, 101–110.

Wang, P. (2012). Nanoscale engineering for smart biocatalysts with fine-tuned properties and functionalities. *Topics in Catalysis*, *55*, 1107–1113.

Wang, P., Liu, S., & Liu, H. (2014). Multiple stimuli-switchable bioelectrocatalysis under physiological conditions based on copolymer films with entrapped enzyme. *The Journal of Physical Chemistry. B*, *118*, 6653–6661.

Welch, G. R., & Easterby, J. S. (1994). Metabolic channeling versus free diffusion: Transition-time analysis. *Trends in Biochemical Sciences*, *19*, 193–197.

Willner, I., & Katz, E. (2000). Integration of layered redox-proteins and conductive supports for bioelectronic applications. *Angewandte Chemie, International Edition*, *39*, 1180–1218.

Willner, I., & Katz, E. (2003). Magnetic control of electrocatalytic and bioelectrocatalytic processes. *Angewandte Chemie, International Edition*, *42*, 4576–4588.

Willner, I., Lion-Dagan, M., & Katz, E. (1996). Photostimulation of dinitrospiropyran-modified glucose oxidase in the presence of DNP-antibody. A biphase switch for the amperometric transduction of recorded optical signals. *Chemical Communications*, 623–624.

Willner, I., Lion-Dagan, M., Marx-Tibbon, S., & Katz, E. (1995). Bioelectrocatalized amperometric transduction of optical signals using monolayer-modified Au-electrodes. *Journal of the American Chemical Society*, *117*, 6581–6592.

Wilner, O. I., Weizmann, Y., Gill, R., Lioubashevski, O., Freeman, R., et al. (2009). *Nature Nanotechnology*, *4*, 249–254.

Wilson, R., & Turner, A. P. F. (1992). Glucose oxidase: An ideal enzyme. *Biosensors & Bioelectronics*, 7, 165–185.

Winkel, B. S. (2004). Metabolic channeling in plants. *Annual Review of Plant Biology*, *55*, 85–107.

Xin, L., Zhou, C., Yang, Z., & Liu, D. (2013). Regulation of an enzyme cascade reaction by a DNA machine. *Small*, *9*, 3088–3091.

Zacharchenko, A., Guz, N., Laradji, A. M., Katz, E., & Minko, S. (2017). Magnetic field remotely controlled highly-selective biocatalysis. *Nature Catalysis*, *1*, 73–81.

Zhang, Y., & Hess, H. (2017). Toward rational design of high-efficiency enzyme cascades. *ACS Catalysis*, *7*, 6018–6027.

Zhang, L., Shi, J., Jiang, Z., Jiang, Y., Qiao, S., et al. (2011). Bioinspired preparation of polydopamine microcapsule for multienzyme system construction. *Green Chemistry*, *13*, 300–306.

Zhang, Y., Tsitkov, S., & Hess, H. (2016). Proximity does not contribute to activity enhancement in the glucose oxidase–horseradish peroxidase cascade. *Nature Communications*, 7. art. No. 13982.

CHAPTER TWO

Functionalization of multiwalled carbon nanotubes for enzyme immobilization

Ram Sarup Singh*, Kanika Chauhan
Carbohydrate and Protein Biotechnology Laboratory, Department of Biotechnology, Punjabi University, Patiala, Punjab, India
*Corresponding author: e-mail address: rssbt@pbi.ac.in

Contents

Abstract

A simple strategy for enzyme immobilization onto homo- and hetero-functionalized multiwalled carbon nanotubes (MWCNTs) is described. Homo-functionalization of MWCNTs can be carried out using an amino-silane compound, i.e., 3-aminopropyltriethoxysilane (APTES). It is an important silane coupling agent, which promotes interfacial behavior of nanomaterials. Whereas, hetero-functionalization of MWCNTs can be performed using APTES and cross-linking agent glutaraldehyde (GA). Key parameters for the immobilization of an enzyme on MWCNTs are selection of an efficient functionalizing agent, sonication of MWCNTs, enzyme load and enzyme coupling time. The optimal level of these factors is very important to obtain high specific activity and immobilization efficiency of the developed immobilized biocatalyst. Both homo- and hetero-functionalized MWCNTs can be used to develop an efficient immobilized

Methods in Enzymology, Volume 630
ISSN 0076-6879
https://doi.org/10.1016/bs.mie.2019.10.014

biocatalyst having good operational stability. However, hetero-functionalized MWCNTs are more potent candidate for enzyme immobilization. The present methodology provides very efficient approach for enzyme immobilization on a versatile support to achieve high enzyme selectivity and operational stability for various industrial applications.

1. Introduction

Enzymes have great potential in food, pharmaceutical and biotechnological applications. They can be used as free precursor or in immobilized form in various biocatalytic processes. The free form of an enzyme is generally not preferable at commercial level. This can be accredited to their restricted operational and storage stability, difficulty in recovery or reusability and irreversible inactivation through aggregation, high bioprocessing cost, etc. (Risso et al., 2010). These disadvantages of free precursors limit their use at industrial level. To surmount these shortcomings of free form of an enzyme, various methods have been developed for the immobilization of enzymes. Currently, immobilized biocatalysts are gaining enormous attention, due to their various advantageous features like recyclability, ease of downstream processing, high product purity and yield, use in both batch and continuous systems, reduced production cost, etc. (Neeraj, Ravi, Somdutt, Ravi, & Kumar, 2018). Continuous advancement in immobilization strategies has helped to understand the structure and functional activities of enzymes, which has further aid to increase their stability, selectivity, sustainability, etc.

The selection of a support matrix and method of immobilization are very important to develop an efficient biocatalyst. The selection of immobilization method strongly depends upon the type of enzyme and support material chosen for immobilization. The selection of support material is considered as the most crucial challenge, due to its direct impact on various significant properties of a biocatalytic system. A broad variety of materials viz., organic, inorganic, hybrid, composites, etc., can be used as support for enzyme immobilization. However, some characteristics such as formation of stable enzyme–matrix interactions, protect enzyme from adverse reaction conditions, aid enzyme to retain its catalytic activity, reduce diffusional limitation of substrate and desorption of enzyme, expose the active site of catalyst for interaction with the substrate, etc., are very pre-requisite of an insoluble support to be selected for enzyme immobilization (Zdarata, Meyer, Jesionowski, & Pinelo, 2018). Insoluble supports of various kinds such as chitosan beads (Singh, Singh, & Kennedy, 2017), duolite XAD (Singh, Saini, & Kennedy, 2010), duolite A568 (Singh, Dhaliwal, & Puri, 2008), sodium alginate (Singh & Yadav, 2013), single-walled carbon nanotubes (Davis, Coleman, Azamian, Bagshaw, & Green, 2003), silica gel (Zhang, Hegab, Lvov, Snow, & Palmer, 2016), gelatine (Jaiswal, Prakash, Talat, Hasan, & Pandey, 2012), etc., have been successfully used for the immobilization of different enzymes to develop a biocatalyst.

Now-a-days, nanostructured materials are gaining a lot of attention of researchers for enzyme immobilization. Many nanostructured materials such as TiO_2, SiO_2, ZnO, graphene sheets, magnetic nanoparticles, gold nanoparticles, etc., have been successfully used for enzyme immobilization. Among various nanostructured materials, carbon nanotubes have become most potent and captivating choice for enzyme immobilization (Singh & Chauhan, 2019; Singh, Chauhan, & Kennedy, 2019a). This can be ascribed to their advantageous features like elastic moduli, high stability, enlarged surface area, easy functionalization with chemical groups, etc. (Kuzmany et al., 2004). Functionalization of multiwalled carbon nanotubes (MWCNTs) provides an inventive aspect to develop a unique support for biomolecules. A number of chemical compounds like amino-silane compound, glutaraldehyde, carbodiimide, etc., can be used for the functionalization of MWCNTs (Neeraj et al., 2018). The functionalization of nanomaterials can be achieved using a single (homo-functionalization) and two different functionalizing agents (hetero-functionalization). Hetero-functionalization of nanomaterial is preferred, due to their higher immobilization efficiency and good operational stability. In this chapter, homo- and hetero-functionalization of MWCNTs using chemical species has been discussed for the immobilization of enzymes.

2. Multiwalled carbon nanotubes

Carbon nanotubes also referred as "nanotubes" are the cylindrical allotrope nanostructures, which were first discovered in 1993 (Iijima & Ichihashi, 1993). Since their discovery, carbon nanotubes play a central role in nanotechnology, due to their various technical superiorities over other carbon nanomaterials. Mainly three types of nanotubes i.e., single-walled, double-walled and multiwalled carbon nanotubes are manufactured. Single-walled carbon nanotubes (SWCNTs) are seamless cylinders, made of a single layer of graphene. Whereas, MWCNTs consist of multiple rolled layers of graphene. Due to their structural complexities,

MWCNTs exhibits innumerable of advantages over SWCNTs. For instance, electrical and mechanical properties of SWCNTs changes, when they are subjected to a chemical treatment. This is due to the structural defects occurred in them by the breakage of C—C double bonds. However, the intrinsic structural properties of MWCNTs do not change as only outer wall of MWCNTs get chemically modified. Double-walled carbon nanotubes (DWCNTs) are a blend of SWCNTs and MWCNTs, demonstrating the properties intermediate between the two types. DWCNTs comprise two concentric nanotubes which are precisely separated by a distance of 0.35–0.40 nm. It has been reported that if both the walls of DWCNTs shows semi-conducting behavior, they may behave as a metal (Tison, Giusca, Stolojan, Hayash, & Silva, 2008). This complication of DWCNTs has limited their use for various applications.

Among these carbon nanotubes, MWCNTs have attracted great interest of researchers belonging to different disciplines like engineering, chemistry, biology, medicine, materials, etc., due to their remarkable properties (Davis et al., 2003). Small diameter, stiffness, high strength, tenacity, high elastic modulus, semiconducting and metallic behavior, high thermal stability, better electronic and mechanical properties, ability to withstand shear stress, etc., are some good characteristics of MWCNTs, which are a boon for various applications viz., in nanosensors, logic elements, nanocomposite materials, field-emission display, gas storage, electronic devices, environmental remediation, enzyme immobilization, etc. (Kuzmany et al., 2004). Besides, they are also extensively investigated for their utilization in leading-edge electronic fabrication and pharmaceutical field for the treatment of various diseases (Prajapati, Sharma, & Banik, 2011). Among all these applications, inclination toward their use as nanocarriers for enzyme immobilization is gaining a great momentum. MWCNTs can be functionalized with a specific chemical species to add a functional group on them for the immobilization of an enzyme (Meng, Fu, & Lu, 2009). Homo-functionalization and hetero-functionalization of MWCNTs can be carried out to develop an immobilized biocatalyst. Immobilization of various enzymes such as peroxidase (Jun et al., 2019), catalase (Rafiee-Pour, Nejadhosseinian, Firouzi, & Masoum, 2019), α-glucosidase (Mohiuddin et al., 2014), inulinase (Singh et al., 2019a; Singh & Chauhan, 2019), lipase (Jamie, Alshami, Maliabari, Muataz Ali Ateih, & Al Hamouz, 2016), tannase (Ong & Annuar, 2018), etc., have been successfully performed on MWCNTs to develop an immobilized biocatalyst for various applications.

3. Materials and methods

3.1 Material

Multiwalled carbon nanotubes (Sigma-Aldrich, USA)

3.2 Equipments

1. Refrigerated centrifuge (5804R, Eppendorf, Germany)
2. Centrifuge tubes
3. Ultrasonic processor (VCX 130, Sonics and Materials Inc., USA)
4. Spectrophotometer (UV-1700 Pharmaspec, Shimadzu, Japan)
5. Tube rotator (SLM-TR-100, GeNei, India)
6. Volumetric micropipettes (1000 μL)
7. Weighing balance

3.3 Chemicals

1. 3-Aminopropyl-triethoxysilane (APTES)
2. Bovine serum albumin (BSA)
3. Deionized water
4. Ethanol
5. Enzyme
6. Folin's phenol reagent
7. Glutaraldehyde (GA)
8. Hydrogen peroxide (H_2O_2)

3.4 Homo-functionalization of MWCNTs with APTES

Homo-functionalization is the process in which insoluble support is functionalized by the addition of a single functional group. APTES is an amino-silane compound which is mostly used for the silanization of MWCNTs for various applications (Liu, Li, Li, & He, 2013). In this process, primarily hydroxylation of MWCNTs is carried out by dispersing them in a mixture of ethanol and H_2O_2 by sonication. H_2O_2 is an oxidizing agent which on sonication splits into —OH group radicals, while ethanol causes etching of walls and side ends of MWCNTs to accommodate —OH ions. After hydroxylation, silanization of nanomaterial is performed. During silanization, hydrolysis of ethoxy group ($-C_2H_5$) from APTES molecule occurs, which generates silanols (Si-O-H) on the surface of MWCNTs. These surface silanols then further condensed with APTES silanols to form

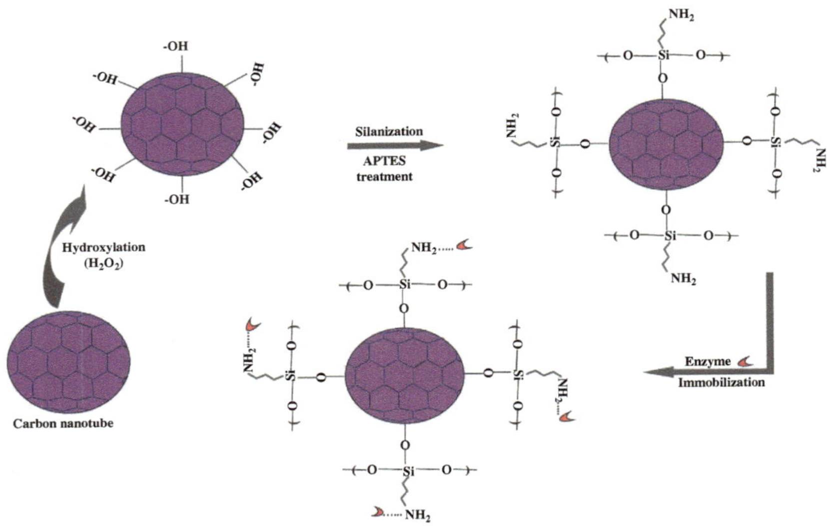

Fig. 1 Homo-functionalization of carbon nanotubes with 3-aminopropyl-triethoxysilane (APTES) for enzyme immobilization. ⟨⟩ represents the polymeric matrix formation.

siloxane networks (Si-O-Si) on MWCNTs. These networks orient themselves in such a way that amine group ($+NH_2$) always arrange away from the solid support. This causes the generation of $+NH_2$ monolayer on MWCNTs. Homo-functionalization of MWCNTs to immobilize an enzyme is illustrated in Fig. 1. Homo-functionalization of MWCNTs can be performed using the following method:

1. Disperse MWCNTs (50 mg) in ethanol (90 mL) containing 10 mL of H_2O_2 (30%) in a beaker (250 mL). Use gloves and approved respirator, while working with H_2O_2 and MWCNTs. Do not left MWCNTs open in air, disperse them immediately into liquid after weighing to prevent respiratory hazard.
2. Sonicate the mixture for 15 min using an ultrasonic processor. Always sonicate the sample by placing the beaker on an ice bath.
3. Afterward, add APTES (4%) to the sonicated mixture for silanization. Again sonicate the mixture for 3–4 h, using an ultrasonic processor. APTES on sonication generates amino ($+NH_2$) monolayer on the walls and side ends of MWCNTs.
4. After sonication, transfer the suspension to centrifuge tubes and recover the homo-functionalized MWCNTs from the suspension by centrifugation (5000 rpm, 10 min). Recover the functionalized MWCNTs after centrifugation carefully with minimal wastage.

5. After centrifugation, wash the recovered MWCNTs two to three times with buffer (which is to be used for protein of interest), to remove unbound APTES and again recover by centrifugation (5000 rpm, 10 min). Again carefully recover the functionalized MWCNTs after centrifugation.
6. Homo-functionalized MWCNTs are now ready to use for enzyme immobilization.

3.5 Hetero-functionalization of MWCNTs with APTES and GA

Hetero-functionalization is a process in which two different functional groups are added to an insoluble support. Primarily, silanization of MWCNTs with APTES is carried out to generate aminated MWCNTs as discussed in Section 3.4. Thereafter, aminated MWCNTs can be subjected to GA cross-linking. GA is a bi-functional compound, which act as spacer between MWCNTs and biocatalyst. It undergoes aldolic condensation to form α, β-unsaturated aldehyde product. This product then reacts with amino group on MWCNTs and protein to form a moiety illustrating higher stability, due to resonating effect of conjugated double C—C bonds (Richards & Knowles, 1968). That is why, enzyme immobilized onto hetero-functionalized support shows higher selectivity and operational stability during various industrial applications. A flow diagram of hetero-functionalization of MWCNTs is shown in Fig. 2. Hetero-functionalization of MWCNTs can be carried out using following method:

1. Disperse MWCNTs (50 mg) in ethanol (90 mL) containing 10 mL of H_2O_2 (30%) in a 250 mL of beaker. Do not left MWCNTs open in air to prevent respiratory problems. Besides, always wear gloves and respirator, while working with H_2O_2 and MWCNTs.
2. Sonicate the mixture for 15 min using an ultrasonic processor. Use an ice bath, while sonication to prevent heating of the sample.
3. Thereafter, add APTES (4%) to the sonicated mixture for silanization and again sonicate the mixture for 3–4 h, using an ultrasonic processor.
4. After sonication, transfer the sonicated suspension to centrifuge tubes and recover MWCNTs from the suspension by centrifugation (5000 rpm, 10 min). Cautiously recover the functionalized MWCNTs after centrifugation with nominal wastage.
5. After centrifugation, properly wash the recovered MWCNTs two to three times with buffer (which is to be used for protein of interest), to remove unbound APTES.

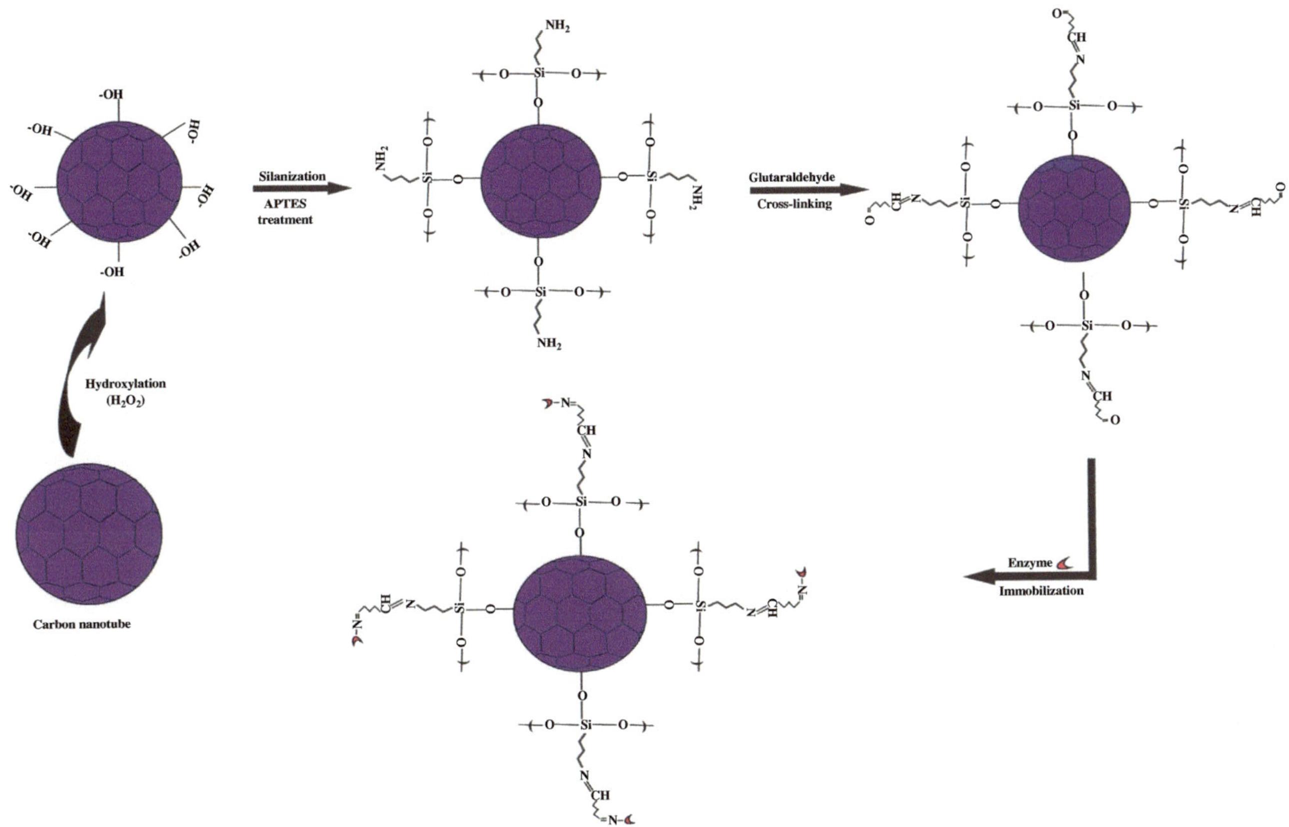

Fig. 2 Hetero-functionalization of carbon nanotubes with 3-aminopropyl-triethoxysilane (APTES) and glutaraldehyde (GA) for enzyme immobilization. —( )— represents the polymeric matrix formation.

6. After every washing, recover MWCNTs from the suspension by centrifugation (5000 rpm, 10 min). Again cautiously recover the centrifuged functionalized MWCNTs.
7. Transfer the recovered MWCNTs into glass vials/tubes with caps and re-suspend in aqueous solution of GA (2%) for 1–2 h, under gentle rotation using a tube rotator. Aldehyde group (−CHO) of GA bonds with amino terminated walls and side ends of MWCNTs to generate hetero-functionalized MWCNTs.
8. Transfer the derivatized MWCNTs to centrifuge tubes and recover the hetero-functionalized MWCNTs by centrifugation (5000 rpm, 10 min).
9. Properly wash the recovered hetero-functionalized MWCNTs two to three times with buffer (which is to be used for protein of interest), to remove unbound GA and again recover by centrifugation (5000 rpm, 10 min).
10. Hetero-functionalized MWCNTs can be now used for the immobilization of an enzyme.

3.6 Enzyme immobilization onto homo- and hetero-functionalized MWCNTs

Enzyme immobilization onto homo- or hetero-functionalized MWCNTs can be performed using the following steps:

1. Re-suspend homo- or hetero-functionalized MWCNTs in 2–3 mL of buffer (which is to be used for protein of interest) in glass vials/tubes with caps. Carefully re-suspend the hetero-functionalized MWCNTs in the buffer without any wastage.
2. Add appropriate amount of enzyme of interest using a volumetric pipette into the above mentioned suspension in glass vials/tubes with caps. Always use fresh tips for pipetting the enzyme using a volumetric pipette to prevent its mixing with other samples.
3. Incubate enzyme-MWCNTs mixture for 2–3 h, under gentle rotation using a tube rotator for enzyme coupling. Always use gentle rotation during coupling of enzyme to the MWCNTs.
4. After coupling of enzyme, recover the immobilized biocatalyst by centrifugation (5000 rpm, 10 min, 4 °C) and wash it two to three times with buffer (which is to be used for protein of interest), to remove uncoupled enzyme. Cautiously recover the developed immobilized biocatalyst after centrifugation with nominal wastage.

5. Thereafter, determine enzyme activity and protein content in supernatant to examine immobilization efficiency of immobilized enzyme in terms of activity yield and immobilization yield.

3.7 Enzyme assay

A standard enzyme assay protocol can be followed to determine the activity of enzyme of interest.

3.8 Protein estimation

Protein content can be estimated according to the method given by Lowry, Rosebrough, Farr, and Randall (1951) using the following steps:

1. Prepare the following solutions:
   Solution A: Prepare by adding 2.0% (w/v) sodium carbonate in 0.1 N NaOH solution.
   Solution B: Prepare by adding 0.5% copper sulfate in 1.0% sodium potassium tartarate solution.
   Solution C: Prepare it freshly by mixing 49 mL of solution A with 1 mL of solution B.
   Solution D: Prepare it freshly by mixing Folin-Ciocalteau's phenol reagent and deionized water in 1:1 ratio.
2. In a test tube, properly mix 5 mL of Solution C with 1 mL of properly diluted protein sample.
3. Allow them to stand at room temperature for 10 min.
4. Afterward, add 0.5 mL of Solution D using a volumetric pipette (1000 μL) and again incubate the mixture for 25 min at room temperature.
5. After incubation, take absorbance of resultant solution at 660 nm, using a spectrophotometer.
6. Estimate the protein content in resultant solution using the standard curve plotted using bovine serum albumin as standard.

3.9 Activity yield and immobilization yield

Activity yield of immobilized biocatalyst can be calculated as percentage of total enzyme activity loaded onto MWCNTs using the following equation:

$$\text{Activity yield }(\%) = \text{Enzyme activity in supernatant/Enzyme activity loaded onto MWCNTs} \times 100$$

Immobilization yield of immobilized biocatalyst can be calculated as percentage of total protein content loaded onto MWCNTs using the following equation:

$$\text{Immobilization yield } (\%) = \text{Amount of protein in supernatant/Protein loaded onto MWCNTs} \times 100$$

4. Results and discussion

Surface modification of MWCNTs with chemical species is considered an important step for enzyme immobilization. However in addition to functionalization of MWCNTs with chemical species, many other factors like sonication, enzyme load, enzyme coupling time, etc., also plays a pivotal role for enzyme immobilization. An optimal level of these factors is very important to obtain high immobilization efficiency. For instance, MWCNTs homo-functionalized with APTES, under optimal conditions (APTES concentration 4%; sonication time 4 h; enzyme coupling time 1.5 h and enzyme load 15 IU), supported 60.7% and 74.4%, inulinase activity yield and immobilization yield, respectively (Singh et al., 2019a). The developed immobilized biocatalyst showed operational stability for 10 batch cycles, during fructose production in a batch system. Use of these factors above/below optimal level poses inhibitory effect on enzyme immobilization efficiency. Sonication of MWCNTs above optimal level causes disorientation of native structure of nanomaterials, which further influences enzyme immobilization onto them. Whereas, sonication below optimal level does not elicit proper functionalization of MWCNTs for enzyme immobilization. An optimal concentration of chemical compounds like APTES/GA is also very important for the functionalization of MWCNTs. The use of APTES above its optimal concentration results in the formation of self-condensed by-products. These by-products aggregate with each other to form thick layer on the surface of MWCNTs, which prevents interaction between biocatalyst and insoluble support (Jung, Moon, & Lee, 2012). GA is an important bi-functional compound, which acts as spacer between biocatalyst and MWCNTs. At optimal concentration, it forms an α, β-unsaturated product, which further interacts with amino group of protein to form a moiety exhibiting more stability, due to resonating effect of conjugated double C—C bonds (Barbosa et al., 2014). Enzyme load and enzyme coupling time are other significant factors, which also impose a strong influence on enzyme immobilization. Higher enzyme load causes aggregation of enzyme, while lesser enzyme load would not be able to fill all the etched channels in MWCNTs, thus resulting in low immobilization efficiency. On the contrary, enzyme coupling time for an extended time period can cause loss in enzyme activity (Singh et al., 2017), while enzyme coupling for lesser time period does not result in good interactions between biocatalyst and MWCNTs, thus consequently resulting in low immobilization yield (Singh, Chauhan, & Kennedy, 2019b).

Similarly, MWCNTs hetero-functionalized with APTES and GA showed high inulinase activity yield (73.9%) and immobilization yield (84.9%), under optimal conditions viz., APTES concentration 4%; sonication time 4 h; GA concentration 2%; GA activation time 75 min; enzyme load 27.5 IU and enzyme coupling time 2.75 h (Singh & Chauhan, 2019). This again manifests the crucial role of optimal level of different factors during functionalization of MWCNTs and enzyme immobilization. However, inulinase immobilized onto hetero-functionalized MWCNTs showed an immense improvement in its operational stability (43 batch cycles) during fructose production in a batch system. The increase in immobilization efficiency and operational stability of inulinase after immobilization onto hetero-functionalized MWCNTs, demonstrates the good candidature of hetero-functionalized MWCNTs for enzyme immobilization.

5. Concluding remarks

Mechanical and structural superiorities of MWCNTs over other carbon nanomaterials made them most potent candidate for enzyme immobilization. Both homo- and hetero-functionalization of MWCNTs is an efficient strategy for enzyme immobilization. The immobilized biocatalyst developed by hetero-functionalization of MWCNTs exhibits better selectivity and reactivity. Hetero-functionalization of MWCNTs increases the enzyme activity yield and the developed immobilized biocatalyst also shows higher operational stability, which is significant for various industrial applications. This chapter illustrates a simple strategy for the immobilization of an enzyme on MWCNTs.

References

Barbosa, O., Ortiz, C., Berenguer-Murcia, A., Torres, R., Rodrigues, R. C., & Fernandes-Lafuente, R. (2014). Glutaraldehyde in bio-catalysts design: A useful crosslinker and a versatile tool in enzyme immobilization. *Research Society of Chemistry Advances*, *4*, 1583–1600.

Davis, J. J., Coleman, K. S., Azamian, B. R., Bagshaw, C. B., & Green, M. L. H. (2003). Chemical and biochemical sensing with modified single walled carbon nanotubes. *Chemistry—A European Journal, 9*, 3732–3739.

Iijima, S., & Ichihashi, T. (1993). Single-shell carbon nanotubes of 1-nm diameter. *Nature, 363*, 603–605.

Jaiswal, N., Prakash, O., Talat, M., Hasan, S. H., & Pandey, R. K. (2012). α-Amylase immobilization on gelatine: Optimization of process variables. *Journal, Genetic Engineering & Biotechnology, 10*, 161–167.

Jamie, A., Alshami, A. S., Maliabari, Z. O., Muataz Ali Ateih, M. A., & Al Hamouz, O. C. S. (2016). Immobilization and enhanced catalytic activity of lipase on modified MWCNT for oily wastewater treatment. *Environmental Progress & Sustainable Energy, 35*, 1441–1449.

Jun, L. Y., Mubarak, N. M., Yon, L. S., Bing, C. H., Khalid, M., Jagdish, P., et al. (2019). Immobilization of peroxidise on functionalized MWCNTs-buckypaper/polyvinyl alcohol nanocomposite membrane. *Scientific Reports, 9*, 2215.

Jung, H.-S., Moon, D.-S., & Lee, J.-K. (2012). Quantitative analysis and efficient surface modification of silica nanoparticles. *Journal of Nanomaterials, 1*, 593471.

Kuzmany, H., Kukovecz, A., Simon, F., Holzweber, M., Kramberger, C., & Pichler, T. (2004). Functionalization of carbon nanotubes. *Synthetic Metals, 141*, 113–122.

Liu, Y., Li, Y., Li, X.-M., & He, T. (2013). Kinetics of (3-aminopropyl)triethoxylsilane (APTES) silanization of superparamagnetic iron oxide nanoparticles. *Langmuir, 29*, 15275–15282.

Lowry, O. H., Rosebrough, N. J., Farr, A., & Randall, R. J. (1951). Protein estimation with Folin-phenol reagent. *Journal of Biological Chemistry, 193*, 265–275.

Meng, L., Fu, C., & Lu, Q. (2009). Advanced technology for functionalization of carbon nanotubes. *Progress in Natural Science, 19*, 801–810.

Mohiuddin, M., Arbain, D., Shafiqul Islam, A. K. M., Rahman, M., Ahmad, M. S., & Ahmad, M. N. (2014). Covalent immobilization of α-glucosidase enzyme onto amine functionalized multi-walled carbon nanotubes. *Current Nanoscience, 10*, 730–735.

Neeraj, G., Ravi, S., Somdutt, R., Ravi, S. K., & Kumar, V. V. (2018). Immobilized inulinase: A new horizon of paramount importance driving the production of sweetener and prebiotics. *Critical Reviews in Biotechnology, 38*, 1–14.

Ong, C.-B., & Annuar, M. S. M. (2018). Immobilization of cross-linked tannase enzyme on multiwalled carbon nanotubes and its catalytic behaviour. *Preparative Biochemistry and Biotechnology, 48*, 181–187.

Prajapati, V., Sharma, P. K., & Banik, A. (2011). Carbon nanotubes and its applications. *International Journal of Pharmaceutical Sciences and Research, 2*, 1099–1107.

Rafiee-Pour, H. L., Nejadhosseinian, M., Firouzi, M., & Masoum, S. (2019). Catalase immobilization onto magnetic multi-walled carbon nanotubes: Optimization of crucial parameters using response surface methodology. *New Journal of Chemistry, 43*, 593–600.

Richards, F. M., & Knowles, J. R. (1968). Glutaraldehyde as protein cross-linkage reagent. *Journal of Molecular Biology, 37*, 231–233.

Risso, F. V. A., Mazutti, M. A., Costa, F., Treichel, H., Costa, F., Maugeri, F., et al. (2010). Comparative studies of the stability of free and immobilized inulinase from *Kluyveromyces marxianus* NRRL Y-7571 in aqueous-organic solutions. *Food Technology and Biotechnology, 48*, 143–150.

Singh, R. S., & Chauhan, K. (2019). Immobilization of inulinase on aminated multiwalled carbon nanotubes by glutaraldehyde cross-linking for the production of fructose. *Catalysis Letters, 149*, 2718–2727.

Singh, R. S., Chauhan, K., & Kennedy, J. F. (2019a). Fructose production from inulin using fungal inulinase immobilized on 3-aminopropyl-triethoxysilane functionalized multiwalled carbon nanotubes. *International Journal of Biological Macromolecules, 125*, 41–52.

Singh, R. S., Chauhan, K., & Kennedy, J. F. (2019b). Immobilization of fungal inulinase on hetero-functionalized carbon nanofibers for the production of fructose from inulin. *LWT-Food Science and Technology*, *116*, 108569. Epub ahead of print.

Singh, R. S., Dhaliwal, R., & Puri, M. (2008). Development of a stable continuous flow immobilized enzyme reactor for the hydrolysis of inulin. *Journal of Industrial Microbiology and Biotechnology*, *35*, 777–782.

Singh, R. S., Saini, G. K., & Kennedy, J. F. (2010). Covalent immobilization and thermodynamic characterization of pullulanase for the hydrolysis of pullulan in batch system. *Carbohydrate Polymers*, *81*, 252–259.

Singh, R. S., Singh, R. P., & Kennedy, J. F. (2017). Immobilization of yeast inulinase on chitosan beads for the hydrolysis of inulin in a batch system. *International Journal of Biological Macromolecules*, *95*, 87–93.

Singh, R. S., & Yadav, M. (2013). Enhanced production of recombinant aspartase of *Aeromonas media* NFB-5 in a stirred tank reactor. *Bioresource Technology*, *145*, 217–223.

Tison, Y., Giusca, C. E., Stolojan, V., Hayash, Y., & Silva, S. R. P. (2008). The inner shell influence on the electronic structure of double-walled carbon nanotubes. *Advanced Materials*, *20*, 189–194.

Zdarata, J., Meyer, A. S., Jesionowski, T., & Pinelo, M. (2018). A general overview of support materials for enzyme immobilization: Characteristics, properties, practical utility. *Catalysts*, *8*, 92.

Zhang, D., Hegab, H. E., Lvov, Y., Snow, L. D., & Palmer, J. (2016). Immobilization of cellulose on a silica gel substrate modified using a 3-APTES self-assembled monolayer. *Springerplus*, *5*, 48.

CHAPTER THREE

Immobilization of enzymes on iron oxide magnetic nanoparticles: Synthesis, characterization, kinetics and thermodynamics

Abhijeet B. Muley, Ketan H. Mulchandani, Rekha S. Singhal*
Food Engineering and Technology Department, Institute of Chemical Technology, Mumbai, India
*Corresponding author: e-mail addresses: rsinghal7@rediffmail.com; rs.singhal@ictmumbai.edu.in

Contents

Methods in Enzymology, Volume 630
ISSN 0076-6879
https://doi.org/10.1016/bs.mie.2019.10.016

Abstract

Immobilization of enzymes on matrices like iron oxide magnetic nanoparticles provides a new technique against the stagnant conventional approaches for stabilization of enzymes. Along with ease in synthesis and modification, the advantages of using these iron oxide nano-metallic matrices as enzyme support matrices are three dimensional *viz.* structural and thermal stability, storage stability, and reusability. This book chapter details the protocol(s) for synthesis of iron oxide magnetic nanoparticles via chemical co-precipitation method, its surface modification and functionalization with 3-aminopropyl-triethoxysilane (APTES), followed by immobilization of enzymes using glutaraldehyde as a cross linker. A detailed protocol for instrumental characterization is also described. Additionally, the chapter also elucidates the basic characterization of enzyme(s), effect of operating conditions (pH, temperature) on the enzyme activity, thermal inactivation kinetics, thermodynamic parameters, storage stability, and reusability of the free and immobilized enzymes. Further, the results of all the indicated protocols and stability studies are thoroughly discussed with the findings on immobilization of cellulase in a tri-enzyme mixture on iron oxide magnetic nanoparticles. The optimum parameters for immobilization were 30 mM glutaraldehyde with 3 h incubation and 3:1 (w/w) ratio of nanoparticles:protein content. Immobilization altered the kinetic constants (K_m and V_{max}) marginally but enhanced the thermal stability as evident from inactivation kinetic constants (k_d, $t_{1/2}$ and D-value) and thermodynamic parameters (E_d, $\Delta H°$, $\Delta G°$ and $\Delta S°$) within 55–75 °C. The immobilized cellulase retained 71.68 ± 3.48% activity during 21-day storage, and 81.15 ± 5.27% activity till fifth reusability cycle.

1. Introduction

Enzymes are biologically active molecules consisting of folded polypeptide(s) made up of the chains of amino acids which fundamentally perform a range of biological functions (Silva, Martins, Jing, Fu, & Cavaco-Paulo, 2018). These are produced by all living organisms and are stated as 'functional catalytic proteins' as they accelerate the rates of reactions by reducing the activation energy without themselves being eternally consumed or altered by the reaction (Soozanipour & Taheri-Kafrani, 2018). They also enhance the rates of reactions without altering the equilibrium among the reactants and products (Mohamad, Marzuki, Buang, Huyop, & Wahab, 2015). These homogeneous biocatalysts may not only reduce the steps involved during biotransformation process but also minimize the requirements of hazardous solvents besides making the process economical and eco-friendly (Zdarta, Meyer, Jesionowski, & Pinelo, 2018). These facets have enabled focused application of enzymes in applied fields ranging from biosensors, health care and pharmaceuticals to foods, chemical and allied industrial segments (Muley, Chaudhari, Bankar, & Singhal, 2019; Munde et al., 2017). Fig. 1 illustrates the enzyme applications in different segments.

1.1 Advantages and drawbacks of enzymes

The enzyme assisted catalytic processes offer several advantages over conventional non-enzymatic processes as these are highly selective, specific, generates less by products, less toxic, and requires mild operating conditions (Chaudhari & Singhal, 2017; Sheldon, 2019). Enzymes display regio-, chemo- and enantio-selectivity making them important biocatalysts for industrial process (Stepankova et al., 2013). Although enzymes have tremendous biocatalytical potential, their implementation in industrial processes is still restricted due to limitations under operating conditions at industrial scale (difficulties in reusability, sensitivity to varying temperature and pH, inadequate stability under processing conditions, and low storage stability) as well as unfavorable process economics (limited availability, economical production and downstream processing cost) (Fortes, Daniel-da-Silva, Xavier, & Tavares, 2017; Khoshnevisan, Poorakbar, Baharifar, & Barkhi, 2019). Majority of the enzymes have very lower catalytic performance when used in organic solvent system *vis-à-vis* aqueous solvent system (Stepankova et al., 2013). The major issue with use of enzymes is due to challenges in

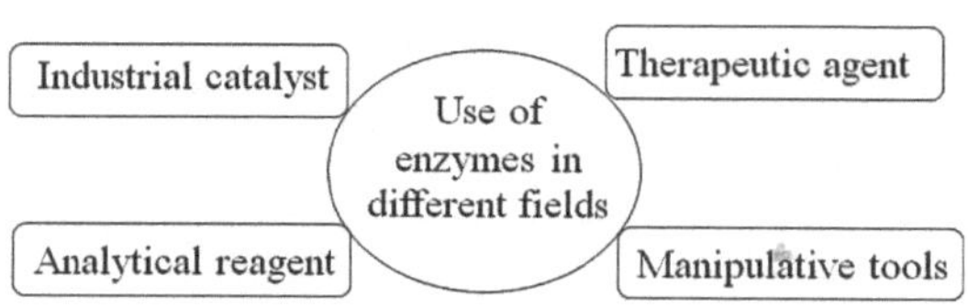

Fig. 1 Uses of enzymes in distinct fields.

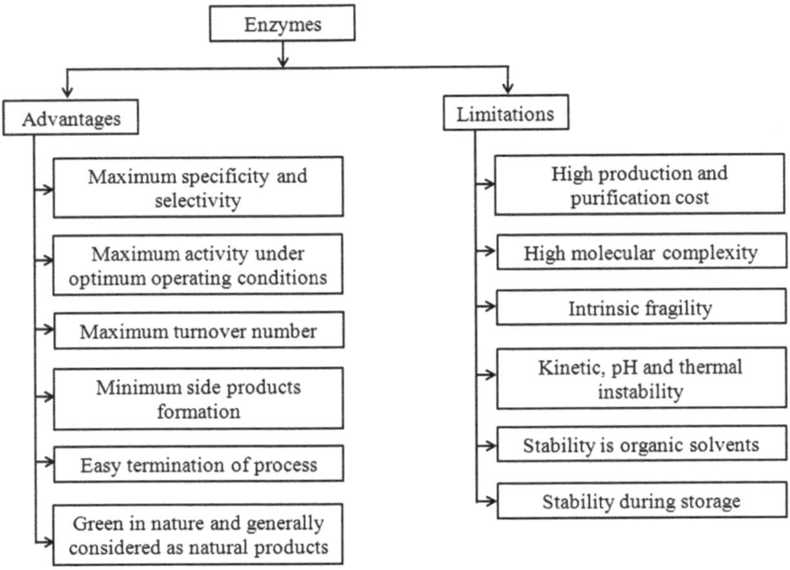

Fig. 2 Major advantages and limitations of use of enzymes in industrial biotransformations.

their cost-effective separation and purification at production scale and reusability (Bedade, Muley, & Singhal, 2019). The major advantages and limitations of application of enzymes in several biotransformation processes are illustrated in Fig. 2. The requirement and use of enzymes in different biotransformation reactions has led to extensive investigations on enhancement of the enzyme properties and their stabilization (Barbosa et al., 2013).

1.2 Stabilization of enzymes

Enzymes can lose activity through covalent and non-covalent structural alterations which alter their conformational stability. The stability of the enzymes can be enhanced by either chemical modification, conjugating them with other biopolymers, cross linking them with each other, addition of additives like ligands, sugars, polyols, and/or immobilizing on solid support matrix (Iyer & Ananthanarayan, 2008). Chemical modifications like PEGylation, glycosylation, and propanol rinsed enzyme preparations leads to surface modification of enzymes and further alter its properties (Stepankova et al., 2013). Conjugation with other biomolecules likes sugars, polysaccharides, biological peptides, and proteins can be either covalent or non-covalent (Muley, Chaudhari, & Singhal, 2017). These interactions have led to increased pH and thermal stability of several enzymes due to the rigidity and hydration provided by these biopolymers to the enzymes (Ladole, Muley, Patil, Talib, & Parate, 2014). Other factors that may stabilize the enzymes are macromolecular confinement and molecular crowding (Muley, Chaudhari, Mulchandani, & Singhal, 2018). The most widely practiced method for stabilization of enzymes is their immobilization on a solid support matrix which not only stabilizes the structure of enzyme but also maintains its catalytic performance (Silva et al., 2018).

1.3 Immobilization of enzymes

Enzymes are very expensive and like traditional chemical catalysts, they are dissolved in a suitable solvent system like water for the specified reaction (Silva et al., 2018). The recovery of these enzymes from the solvent system is a challenging task as they are soluble in solvent system (Ladole, Mevada, & Pandit, 2017). The enzymes can be separated by ammonium sulfate precipitation or thermal denaturation (only if the enzyme and the product are heat stable) or ultrafiltration. These methods generally increase the cost of downstream processing and may lead to inactivation of enzyme (Robinson, 2015). This issue can be tackled with immobilization on the solid support matrix that can be recovered via sedimentation, centrifugation or filtration (Muley, Thorat, Singhal, & Babu, 2018). Furthermore, immobilization converts the enzymes from the homogenous form to the heterogeneous form, and also enhances characteristic features such as selectivity, stability, and reduction of inhibition (Khoshnevisan et al., 2019). In addition to the ease in recovery and reusability for subsequent catalytic cycles, immobilization of enzymes also enhances the operational and storage stability (Ladole, Nair, Bhutada, Amritkar, & Pandit, 2018). Various approaches used for the immobilization of enzymes are illustrated in Fig. 3, and these include adsorption, entrapment, microencapsulation, covalent binding, cross linked enzyme crystals and/or aggregates, and retention on ultrafiltration membranes (Chaudhari, Kar, & Singhal, 2015).

A broad array of materials of different origins such as inorganic, organic, hybrid and/or composites can be used as support matrix for immobilization of enzymes. The materials used must not only maintain the structural conformation of enzymes, but also create a stable interaction with the enzyme (Zdarta et al., 2018). The selection of suitable support matrix is associated with the type of enzyme and the process in which these immobilized system is proposed to be used.

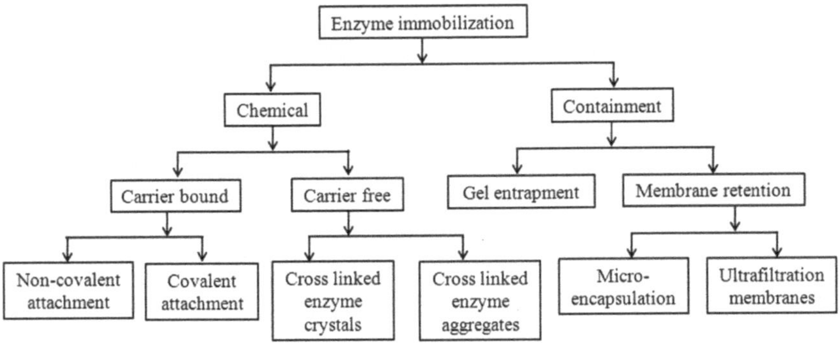

Fig. 3 Different techniques used for immobilization of enzymes.

2. Magnetic nanoparticles

Among the numerous support matrices used for immobilization of enzymes, iron oxide magnetic nanoparticles have received tremendous interest due to simple techniques of their preparation, desired physical properties of unique size and shape, low toxicity, good capacity for transfer of charge, good enzyme loading capacity, capability of surface modification, less mass transfer resistance, and strong magnetic properties which facilitates their recovery from the reaction mixture (Atacan, Çakıroğlu, & Özacar, 2016; Muley, Thorat, et al., 2018). The surface of the nanoparticles need to be modified to protect against oxidation on exposure to air, improve stability under acidic conditions (Roto, 2018), and prevent agglomeration due to magnetic dipole moments between the nanoparticles, small size, high chemical reactivity, and high specific surface area (Atacan et al., 2016; Khoshnevisan et al., 2019). Table 1 illustrates the techniques used for surface modification of nanoparticles. These include silica functionalization, amino functionalization, composite functionalization, carbohydrate functionalization, and polymer functionalization (Zhu et al., 2018).

2.1 Preparation of iron oxide magnetic nanoparticles

Protocol

1. Dissolve the anhydrous salts of ferric chloride (16 g) and ferrous chloride (8 g) in deionized water (100 mL), and then heat the mixture at $90 \pm 2\,°C$ for 10 min.
2. Add conc. HCl (2.5 mL) to the reaction mixture with continuous stirring for 8–10 min.

Table 1 Methods for surface modification iron oxide magnetic nanoparticles.

Functionalization method	Agents used	Reference
Silica functionalization	3-Mercaptopropyltrimethoxysilane	Koc, Karakus, Rajar, and Alveroglu (2017)
	Tetraethoxythoxysilane, vinyltriethoxysilane, octadecyltrimethoxysilane	Wu, Wu, Yu, Jiang, and Kim (2015)
Amino functionalization	3-Aminopropyltriethoxysilane, 3-aminopropyltrimethoxysilane, *p*-dimethylaminobenzaldehyde, aminothioamidoanthraquinone	Cashin, Eldridge, Yu, and Zhao (2018), and Zhan et al. (2018)
Composite functionalization	Metal oxides	Wu et al. (2015)
	Graphine oxide	de Souza Lima et al. (2019)
	Poly(methyl methacrylate)	Han et al. (2018)
Polymer functionalization	Polyethylene glycol	Anbarasu, Anandan, Chinnasamy, Gopinath, and Balamurugan (2015)
	Polyvinyl alcohol	Ardiyanti, Suharyadi, Kato, and Iwata (2016)
	Polydopamine	Ma et al. (2013)
	Polyethylenimine	Calatayud et al. (2013)
	Polyvinylpyrrolidone	Rose et al. (2013)
	Poly lactic acid polyetherimide	Yang et al. (2016)
	Polyamidoamine	Liu, Guo, Odusote, Qu, and Priestley (2013)
Carbohydrate functionalization	Chitosan	Díaz-Hernández et al. (2018)
	Dextran	Hauser, Mathias, Anderson, and Hilt (2015)
	Alginate	Mohammadi, Daemi, and Barikani (2014)

3. Add NaOH pellets (12 g) to the reaction mixture till a black colored precipitate is formed and the pH of solution reaches 12.
4. Heat the reaction mixture continuously at 90 ± 2 °C for 30 min and then cool it to 28 ± 2 °C.
5. Collect the precipitate settled at bottom by decanting the supernatant.
6. Wash the precipitate with 50% ethanol until the wash solution shows a pH 7 ± 0.2.
7. Dry the resultant precipitate at 80–85 °C for 90–120 min.
8. Scrub the synthesized iron oxide magnetic nanoparticles, and crush them with a mortar and pestle.
9. Store these nanoparticles at 28 ± 2 °C till further use.
10. The schematic representation of preparation of iron oxide magnetic nanoparticles is shown in Fig. 4.

2.2 Surface modification of iron oxide magnetic nanoparticles

Protocol

1. Disperse iron oxide magnetic nanoparticles (1.24 g) in 50% ethanol (45 mL) with continuous stirring at 600 rpm.
2. Add 3-aminopropyltriethoxysilane solution (5 mL) drop wise to it.

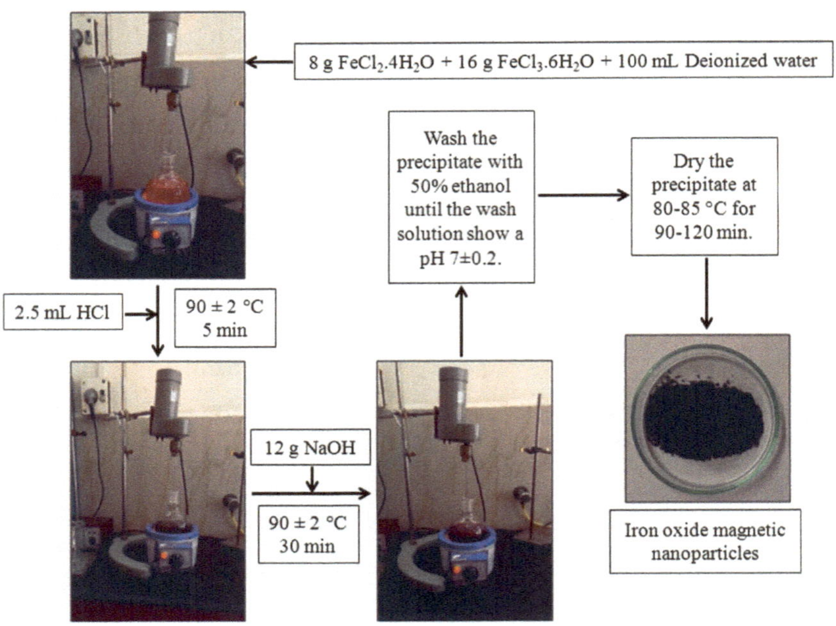

Fig. 4 Schematic representation of preparation of iron oxide magnetic nanoparticles.

3. Stir the solution so obtained vigorously at 1000 rpm at 50 ± 2 °C for 3 h.
4. Centrifuge the solution at 7500–8000 × *g* at 28 ± 2 °C for 10 min and collect the pellet.
5. Wash the pellet with deionized water 5–6 times.
6. Dry the obtained particles at 50 ± 2 °C for 6 h.
7. Store these surface modified nanoparticles at 28 ± 2 °C till further use.
8. The schematic representation for surface modification of iron oxide magnetic nanoparticles with 3-aminopropyltriethoxysilane is shown in Fig. 5.

2.3 Immobilization of enzymes on iron oxide magnetic nanoparticles

Immobilization of enzymes involves precipitation of enzymes followed by their binding/loading on the iron oxide magnetic nanoparticles with a bi-functional cross linking agent like glutaraldehyde, N-hydroxy succinimide-esters and/or imidoester.

2.3.1 Precipitation of enzymes

The enzymes present in the crude enzyme mixture or filtered fermentation broth are recovered in active state after precipitation with the chilled miscible organic solvents (methanol, ethanol, isopropanol, *n*-propanol, *n*-butanol), salts (ammonium sulfate), non-ionic hydrophilic polymers (polyethylene glycol and dextran), and/or flocculation by polyelectrolytes (alginate, tannic acid, carboxymethyl cellulose, polyacrylic acid) (Sheldon & van Pelt, 2013).

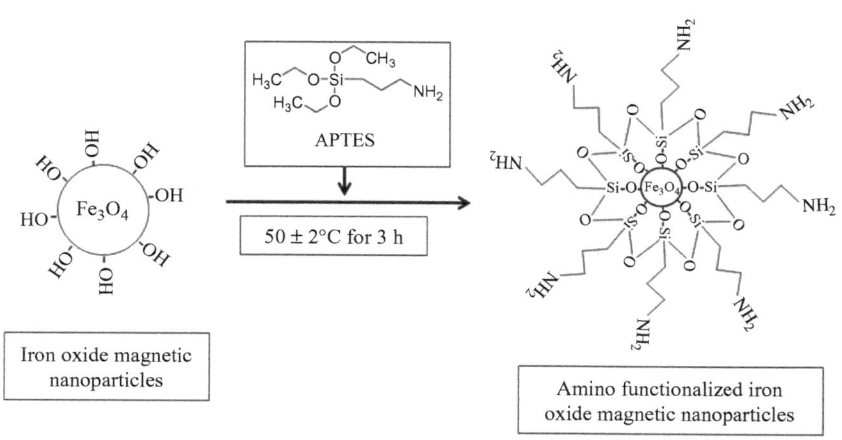

Fig. 5 Schematic representation of surface modification of iron oxide magnetic nanoparticles with 3-aminopropyltriethoxysilane (APTES).

2.3.1.1 Ammonium sulfate precipitation

Protocol

1. To the enzyme solution (10 mL), add specific amount of ammonium sulfate to achieve saturation from 10% to 90% with continuous stirring for 1 h at 4 ± 2 °C.
2. Further let this mixture stand for 1 h.
3. Centrifuge this mixture at 10,500–11,000 × *g* for 12–15 min at 4 ± 2 °C to separate the precipitate from the solution.
4. The amount of ammonium sulfate to be added for achieving specific saturation is calculated as

$$g = \frac{G_{sat} \times (S_2 - S_1)}{[1 - (P \times S_2)]}$$

where G_{sat} = grams of ammonium sulfate contained in 1 L of the saturated solution. For example, at 0 °C, G_{sat} = 515.35, S_1 and S_2 = fractions of complete saturation; for example, a 20% saturation is expressed as 0.2 and P is given by

$$P = \frac{\text{sp.vol} \times G_{sat}}{1000}$$

The percent saturation concentration of ammonium sulfate which show maximum activity recovery and fold purity is taken for the further experiments.

$$\text{Recovery of sample (\%)} = \frac{\text{Enzyme activity of precipitate}}{\text{Initial enzyme activity}} \times 100$$

$$\text{Fold purity} = \frac{\text{Specific activity of enzyme}}{\text{Initial specific activity of free enzyme}}$$

5. Determine the protein content and activity of enzyme in free form and precipitated form.
6. Calculate the activity recovery (%) as

$$\text{Activity recovery (\%)} = \frac{\text{Activity of recovered enzymes}}{\text{Activity of enzymes in free form}} \times 100$$

2.3.1.2 Solvent precipitation

Protocol

1. To the enzyme solution (10 mL), add equal volume of ice cold organic solvents drop wise followed by continuous gentle shaking.

2. Incubate this mixture for 15 min at 4 ± 2 °C.
3. Centrifuge this mixture at 10,500–11,000 × *g* for 8–10 min.
4. Determine the protein content as well as activity of enzyme in free form and precipitated form.
5. Calculate the activity recovery (%) as mentioned above.

2.3.2 Loading/binding of enzyme(s) on iron oxide magnetic nanoparticles

Protocol

1. Suspend 500 mg nanoparticles in the enzyme precipitate and add a suitable buffer system (1 mL) with continuous shaking at 180 rpm and ambient temperature.
2. Add glutaraldehyde in the reaction mixture and continue mixing at ambient temperature for a specific duration of time.
3. Recover the immobilized enzymes after incubation by the applying an external magnetic field.
4. Wash it 3–4 times with suitable buffer system, and simultaneously estimate the protein content after each wash.
5. Calculate the amount of enzyme immobilized on nanoparticles by the difference between the initial protein content in the immobilization medium and the protein recovered in the supernatant as

$$\text{Immobilization (\%)} = \frac{P_i \times V_i - P_f \times V_f}{P_f \times V_f} \times 100$$

where P_i and P_f are initial and final protein content (mg/mL), whereas V_i and V_f are initial and final volume of reaction mixture (mL).

6. The activity recovery of enzyme can be calculated and expressed in terms of activity recovery (%) as follows

$$\text{Specific activity recovery (\%)} = \frac{\text{Activity of immobilized enzyme}}{\text{Activity of free enzyme}} \times 100$$

7. The maximum immobilization and best possible recovery of enzyme activity is attained by optimizing various factors such as (i) the ratio of protein content to magnetic nanoparticles, (ii) glutaraldehyde concentration, and (iii) cross linking time, respectively, by either one factor at a time approach or statistical approach like response surface methodology.
8. A schematic representation of immobilization of enzymes on iron oxide magnetic nanoparticles is shown in Fig. 6.

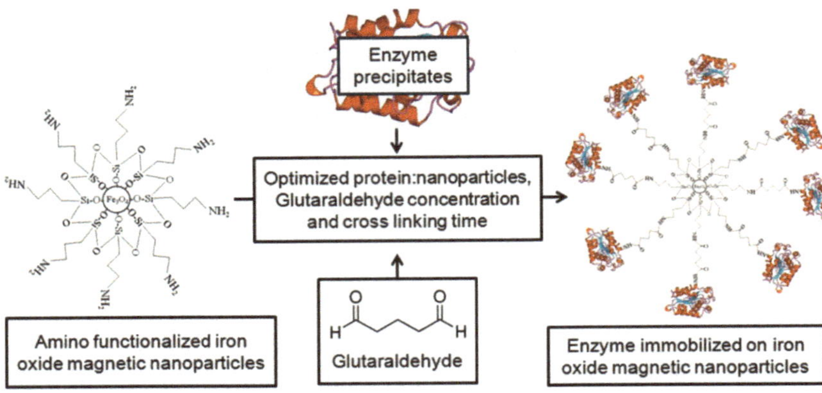

Fig. 6 Schematic representation for immobilization of enzymes on iron oxide magnetic nanoparticles.

3. Instrumental characterization of nanoparticles before and after immobilization of enzymes

Instrumental characterization plays a critical role in determining the characteristics of the materials such as the size and shape, its crystalline properties, presence of different functional groups, thermal and magnetic characteristics, and surface properties, after their synthesis, modifications, and immobilization of enzymes, respectively. Table 2 illustrates various analytical techniques used for characterization along with the possible information that can be derived from them. The characterization of nanoparticles is based in terms of shape, size, structure, chemistry, crystallography pattern and many more by one or more instrumental techniques (Borase, Muley, Patil, & Singhal, 2019; Mourdikoudis, Pallares, & Thanh, 2018).

The details of few other analytical techniques along with the protocols to determine the specific properties associated with them are discussed below.

3.1 Brunauer-Emmett-Teller (BET) surface area analysis

Protocol

1. Dry the samples by keeping them in hot air oven at 60–65 °C for 8–10 h.
2. Individually place 100 mg of iron oxide nanoparticles (plain, surface modified and enzyme immobilized) in the U-tube sample holder.
3. Attach the sorbitometer with the instrument and U-tube and ensure that it is free from any leakage by applying vacuum grease.
4. Degas the samples by heating them between 100 and 110 °C for 120 min.

Table 2 Analytical techniques for characterization of plain and enzyme immobilized nanoparticles and the information derived therefrom.

Analytical technique	Possible information derived
Dynamic light scattering	Determination of hydrodynamic size and formation of agglomerates of particles
Zeta potential	Determination of surface charge
UV-vis spectrophotometer	Determination of maximum wavelength (λ_{max}), optical properties, concentration, agglomeration state and hints on nanoparticles shape
Brunauer-Emmett-Teller (BET)	Determination of surface area
Fourier transform infrared spectroscopy	Determination of functional groups
Scanning electron microscopy	Surface morphology and cross-sections of coatings deposit
Energy-dispersive X-ray spectroscopy	Determination of elemental analysis
Transmission electron microscopy	Determination of crystallite size, grain size, atomic arrangement in the material, aggregation state and formation of new phases, and study growth kinetics
Vibrating sample magnetometry	Determination of magnetic moments as a function of magnetic field and temperature

5. Attach the Deward container to the assembly.
6. Fill the Deward container with 2–3 L of liquid nitrogen.
7. Determine the amount of gas adsorbed on the surface of nanoparticles.
8. Calculate the surface area of the nanoparticles from the amount of gas adsorbed.

3.2 Confocal laser scanning microscopy

Protocol

1. Add 50 mg of enzyme immobilized on iron oxide magnetic nanoparticles in 100 mM phosphate buffer solution (pH 7.0, 1 mL).
2. Add 0.5 mM carbonate buffer (pH 9.0, 1.5 mL) with vigorous shaking to the mixture.
3. Add freshly prepared fluorescein isothiocyanate (FITC) solution (0.3 mL) to the mixture with continuous shaking for 2 h at 25 ± 2 °C.

4. Preparation of FITC solution: Dissolve FITC (1 mg) in 0.5 mM carbonate buffer (pH 9.0, 1 mL).
5. Recover the enzyme immobilized on iron oxide magnetic nanoparticles by application of external magnetic field.
6. Wash the recovered sample with 0.5 mM carbonate buffer (2 mL) solution to remove the unbound FITC.
7. Observe the samples under confocal laser scanning microscope at an excitation and emission wavelengths of 488 and 535 nm, respectively.

3.3 XRD analysis

Protocol

1. Individually place 100–200 mg of plain, surface modified and enzyme immobilized iron oxide magnetic nanoparticles on the disk of XRD.
2. Collect the 2θ data over 10°–90°, at a regular step size (e.g., 1.2–1.5°/min).
3. Calculate the average crystallite mean size of plain, surface modified and enzyme immobilized iron oxide magnetic nanoparticles using the Debye Scherrer's formula as

$$\text{Crystallite mean size (nm)} = \frac{K \times \lambda}{\beta \times \cos\theta}$$

where λ is the wavelength of the X-ray wave, K is the shape factor, β is the full width at half maximum of the strongest diffraction peak, and θ is the angle of diffraction of the peak, respectively.

3.4 Thermo-gravimetric analysis

Protocol

1. Individually place 10 mg of plain, surface modified and enzyme immobilized iron oxide magnetic nanoparticles on the ceramic pan.
2. Determine the weight loss after heating the samples from 30 to 700 °C at a heating rate of 10 °C/min in nitrogen atmosphere (nitrogen flow rate 50 mL/min).
3. Determine the temperature at which 0.75%, 1.50%, 2.25%, 3.0% and 4.5% weight loss has occurred.
4. Determine the char yield of plain, surface modified and enzyme immobilized iron oxide magnetic nanoparticles after the end of heating.
5. Calculate the limiting oxygen index with the Van Krevelen and Hoftyzer equation as explained by Hatami (2018)

$$\text{Limited oxygen index} = 17.5 + 04 \times \text{char yield}$$

4. Determination of operating parameters

Among the various processing parameters (pH, temperature, ionic strength, solvent type and time) that have an influence on the rate of enzyme catalyzed reactions through the reversible/irreversible changes in the structure of enzyme, the effects of pH and temperature on enzyme are generally studied. The optimization of pH and temperature of enzymes is eventually carried out during the designing of a biotransformation process (Talekar et al., 2017).

4.1 Optimum pH

Protocol

1. Prepare substrate solutions by dissolving them in 50 mM buffers with different pH values.
2. Add 10 mg of enzyme immobilized on iron oxide nanoparticles in the substrate solutions with varying pH, incubate them at ambient temperature for a defined time period.
3. Determine the activity of immobilized enzyme at each pH.
4. Perform a similar set of experiment for free enzymes.
5. Plot a profile of enzyme activity against pH of substrate.
6. Determine the optimum pH.

4.2 Optimum temperature

Protocol

1. Add 10 mg of enzyme immobilized on iron oxide nanoparticles in the substrate solutions at optimum pH.
2. Incubate this mixture at varying temperature domain for a defined time period.
3. Determine the activity of immobilized enzyme at each temperature.
4. Perform a similar set of experiment for free enzymes.
5. Plot a profile of enzyme activity against temperature.
6. Determine the optimum temperature.

5. Enzyme constants

5.1 Kinetic constants (K_m and V_{max})

Protocol

1. Add 10 mg of enzyme immobilized on iron oxide nanoparticles in the substrate [S] solutions with varying concentration at optimum pH.

2. Incubate the reaction mixture at ambient temperature for a defined time period, and then determine the activity [V].
3. Perform a similar set of experiment for free enzymes samples.
4. Determine the K_m and V_{max} of enzyme by using Michaelis-Menten as

$$V = \frac{V_{max} \times [S]}{K_m + [S]}$$

5. Plot a double reciprocal Lineweaver-Burk plot of 1/[V] vs 1/[S] to deduce V_{max} and K_m from the slope and intercept, respectively.

$$\frac{1}{V} = \frac{K_m}{V_{max}} \times \frac{1}{[S]} + \frac{1}{V_{max}}$$

where S is the substrate concentration, V_{max} is the maximum rate of reaction, and K_m is the amount of substrate required for achieving half the maximum rate.

It may be noted that the substrate concentration used in the study is very important and also indicates the order of an enzyme catalyzed reaction. For instance, at $[S] \lll K_m$, $[S] \sim K_m$, and $[S] \ggg K_m$, the order of the enzyme catalyzed reaction could be first, fractional and zero, respectively (Punekar, 2018).

5.2 Mass transfer limitations

Protocol

1. Add 10 mg of enzyme immobilized on iron oxide nanoparticles in the substrate solutions with substrate concentration in the range below the K_m values of free enzymes at optimum pH.
2. Incubate the reaction mixture at ambient temperature for a defined time period, and then determine the activity.
3. The extent of mass transfer for free and enzyme immobilized on iron oxide nanoparticles for enzyme kinetics is expressed as effectiveness factor (%) and calculated as

$$\text{Effectiveness factor (\%)} = \frac{V_{\text{immobilized enzymes}}}{V_{\text{free enzymes}}} \times 100$$

where $V_{\text{immobilized enzymes}}$ and $V_{\text{free enzymes}}$ are the reaction rates catalyzed by enzymes in immobilized and free form, respectively.

6. Thermal inactivation kinetics

6.1 Inactivation constants

Protocol

1. Take 5 mg of enzyme immobilized on iron oxide nanoparticles in a set of test tubes and add 1 mL of 50 mM buffer with optimum pH to them.
2. Incubate these tubes at varying temperature regimes.
3. Remove individual tubes at 15 min intervals for 90 min and cool it quickly.
4. Carry out respective enzyme assay and calculate the enzyme activity.
5. Perform a similar set of experiment for free enzymes.
6. Plot a semi-log graph of percent residual activity against time from initial enzyme activity (A_0) and enzyme activity (A_t) of samples collected at each time duration and determine the deactivation rate constant (k_d, min^{-1}) as the slope and expressed by first order reaction rate as

$$\frac{dA}{dt} = -k_d \times A$$

The equation for deactivation rate constant is

$$ln\left(\frac{A_t}{A_0}\right) = -k_d \times t$$

$$ln\,(A_t) = -k_d t + ln\,(A_0)$$

7. The time required for the enzyme activity to reduce to half to its original activity is $t_{1/2}$ value and is calculated as

$$t_{1/2} = \frac{ln\,(2)}{k_d}$$

8. The decimal reduction time required to pre-incubate the enzyme at a given temperature to retain 10% residual activity at each temperature is D-value (min) and is calculated as

$$D - \text{value} = \frac{ln\,(10)}{k_d}$$

9. The increase in temperature required to reduce the D-value by one logarithmic cycle is the z-value and this is evaluated from the slope of the graph log (D-value) against temperature (°C) as

$$\text{Slope} = \frac{-1}{z}$$

6.2 Thermodynamic constants

Protocol

1. Plot a graph of $\ln(K_d)$ against 1/T (i.e., reciprocal of absolute temperature in K) and determine the activation energy E_d (kJ/mol) for denaturation of enzymes (in free and immobilized form) from Arrhenius equation by using the slope as

$$\text{Slope} = \frac{-E_d}{R}$$

2. The changes in ΔH°, ΔG° and ΔS° for the thermal denaturation of enzymes in free and immobilized form as caused by heat treatment are evaluated by the Eyring's transition state theory as

$$\Delta H^\circ = E_d - R \times T$$

$$\Delta G^\circ = -R \times T\ ln\left(\frac{k_d}{K_B} \times \frac{h}{T}\right)$$

$$\Delta S^\circ = \frac{\Delta H^\circ - \Delta G^\circ}{T}$$

where E_d is the activation energy for denaturation, T is the corresponding absolute temperature (K), R is gas constant (8.314 J/mol/K), h is Planks constant (11.04×10^{-34} J min), K_B is the Boltzman constant (1.38×10^{-23} J/K) and k_d is deactivation rate constant (min^{-1}).

7. Storage stability

Protocol

1. Store the free and immobilized enzymes at 4 ± 1 °C.
2. Determine the activity of free and immobilized enzymes after a specified time interval (e.g., 2–6 days) for 24–40 consecutive days.
3. Calculate the relative activity of free and immobilized enzymes by assuming the activity at the beginning of storage as 100%.

8. Reusability of immobilized enzymes

Protocol

1. Add 10 mg of enzyme immobilized on iron oxide magnetic nanoparticles in the substrate solutions dissolved in 50 mM buffer of optimum pH, at ambient temperature for a defined time period, under batch operation mode, and then determine the activity.
2. After the first cycle, recover the immobilized enzyme by the application of external magnetic field and wash with a suitable buffer.
3. Re-suspend the recovered immobilized enzyme in fresh substrate system.
4. Calculate the residual activity of free and immobilized enzymes after each cycle by assuming the activity of first cycle as 100%.

9. Results and discussion

9.1 Precipitation and recovery of enzymes

Precipitation is also the first step for the binding/immobilization of enzymes on iron oxide magnetic nanoparticles, as it facilitates the transformation of free form of enzymes to the aggregated form (Nadar, Muley, Ladole, & Joshi, 2016). For the precipitation of enzymes present in the fermentation broth, remove the cells in fermentation broth by centrifugation and/or filtration and collect the supernatant and/or filtrate to recover the enzymes by ammonium sulfate precipitation or solvent precipitation. While most enzymes can be precipitated by ammonium sulfate, a few enzymes may get denatured by solvent precipitation and lose their activity. Hence the choice of the method is specific to the enzyme.

9.1.1 Ammonium sulfate precipitation

Precipitation with ammonium sulfate is the most routinely used method and also called as "salting out." The addition of ammonium sulfate compresses the solvation layer around enzymes and increases the interactions among them. As the ammonium sulfate concentration in the solution increases, the ionogenic functional groups on the surface of enzyme interact with the salts and not water, and thus expose the hydrophobic functional groups on the enzyme surface resulting in the aggregation thereof. Precipitation of cellulase with 50–60% saturation of ammonium sulfate yielded maximum activity of $63.10 \pm 3.20\%$ (Fig. 7).

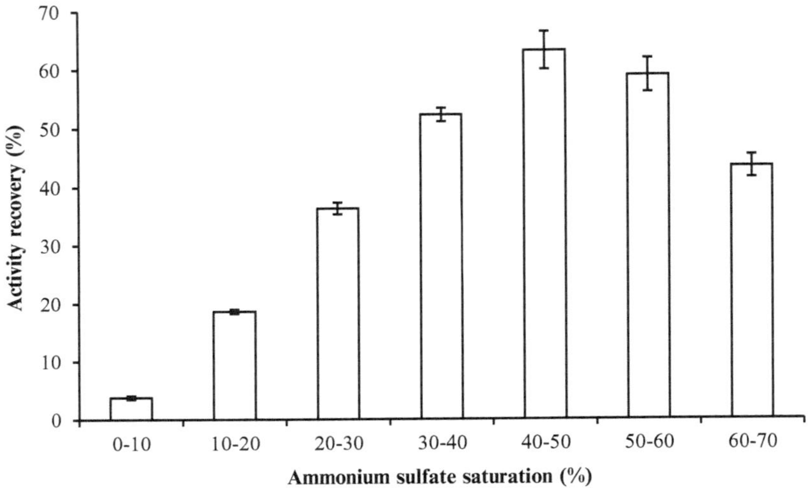

Fig. 7 Effect of ammonium sulfate saturation on the activity recovery of cellulase.

9.1.2 Solvent precipitation

Methanol, ethanol, isopropanol, *n*-propanol, *t*-butanol, *n*-butanol, acetonitrile and acetone (under ice cold conditions) are some of the miscible organic solvents generally used for precipitation of soluble enzymes. These organic solvents reduce the solvation layer around the enzymes as water displaces from the surface of enzymes. Due to small hydration layers, the enzymes get aggregated by the attractive dipole and electrostatic forces. Screening of different chilled organic solvents for precipitation of cellulase showed isopropanol to recover maximum activity of cellulase (83.62 ± 2.96%) from the fermentation broth (Fig. 8).

9.2 Loading/binding of enzyme(s) on iron oxide magnetic nanoparticles

The degree of loading/binding of enzyme(s) on the iron oxide magnetic nanoparticles is not only associated with the physico-chemical characteristics of the support matrix but also with the immobilization conditions such as the ratio of protein content to the nanoparticles, concentration of cross linker, and duration of cross linking used during immobilization, respectively (Muley et al., 2019). Conditions such as pH and temperature during cross linking have been studied in a few cases but have not seemed to show a significant effect (Chaudhari & Singhal, 2017).

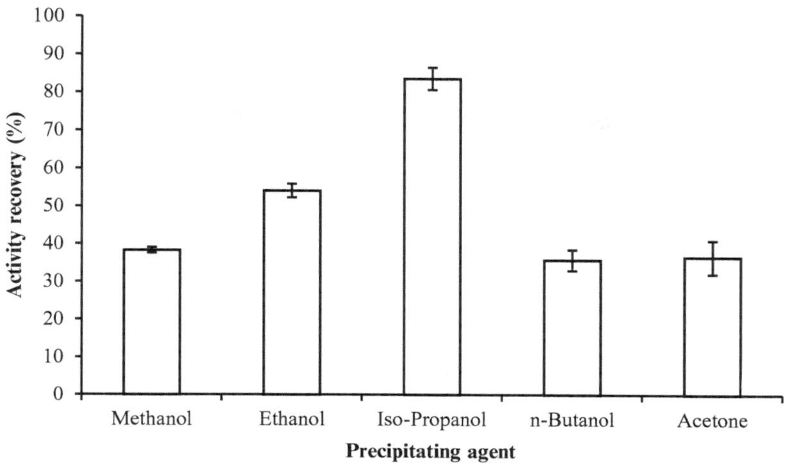

Fig. 8 Effect of precipitating agent on the activity recovery of cellulase.

The enzyme/s can be immobilized on iron oxide magnetic nanoparticles using glutaraldehyde as a cross linking agent. Glutaraldehyde is a bi-functional cross linker which reacts with amino groups of enzyme and the functionalized nanoparticles which firmly anchors the enzyme/s in their active form on the nanoparticles and further insolubilizes it. The optimum conditions of immobilization of cellulase on iron oxide magnetic nanoparticles were 30 mM glutaraldehyde concentration (Fig. 9A), 3 h cross linking time (Fig. 9B), and 3:1 (w/w) ratio of nanoparticles to protein content (Fig. 9C) resulting in an activity recovery of $87.76 \pm 1.93\%$, $78.99 \pm 1.28\%$, $80.25 \pm 1.03\%$, respectively.

9.3 Instrumental characterization

Instrumental characterization helps in determining the physico-functional properties of the synthesized nanoparticles such as size and shape, crystallinity pattern, functional groups attached, thermal and magnetic features, and surface characteristics (Sapsford et al., 2013). The same are applicable to nanoparticles that are modified and used for immobilization of enzymes.

9.3.1 Particle size analysis

The size distribution patterns of the synthesized normal and enzyme immobilized iron oxide magnetic nanoparticles were analyzed by dynamic light scattering measurements (Fig. 10A and B). An average size of 68.8 nm was noted for plain iron oxide magnetic nanoparticles which increased to

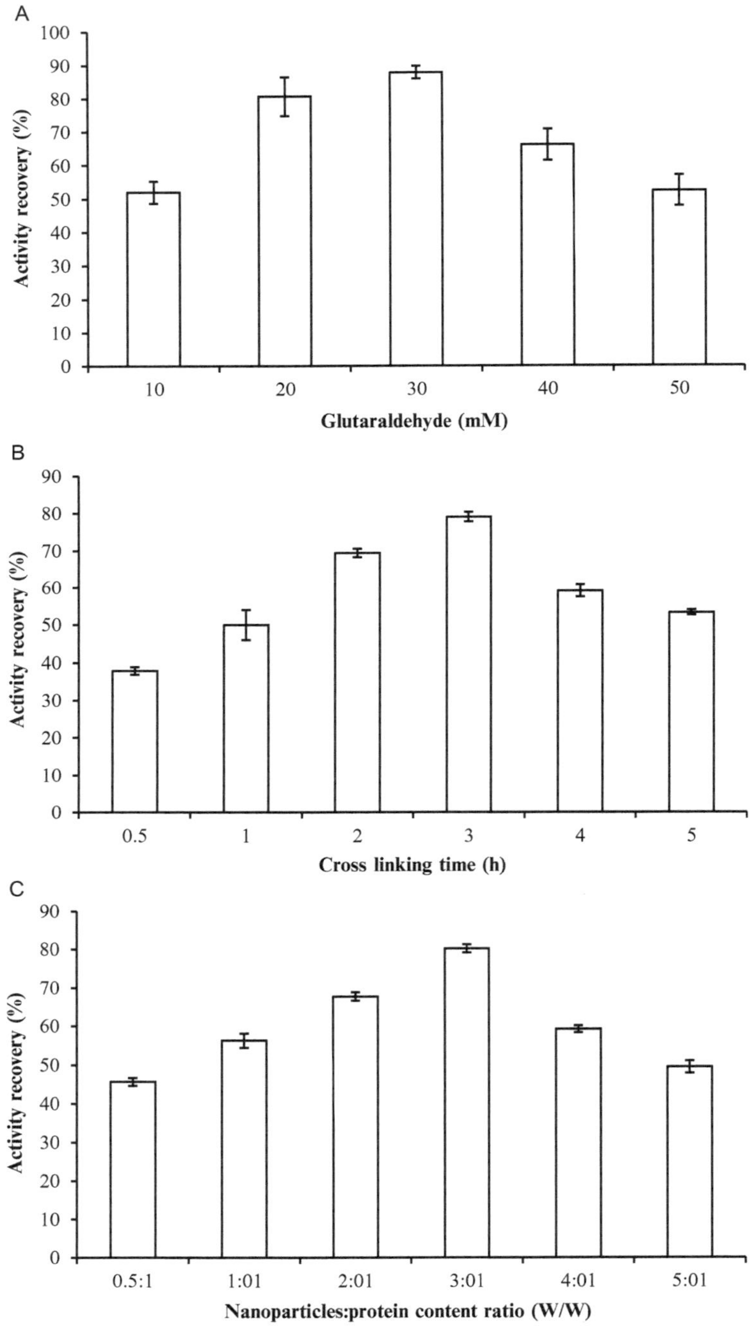

Fig. 9 Effect of (A) glutaraldehyde concentration, (B) cross linking time, and (C) nanoparticles to protein ratio on the activity recovery of cellulase.

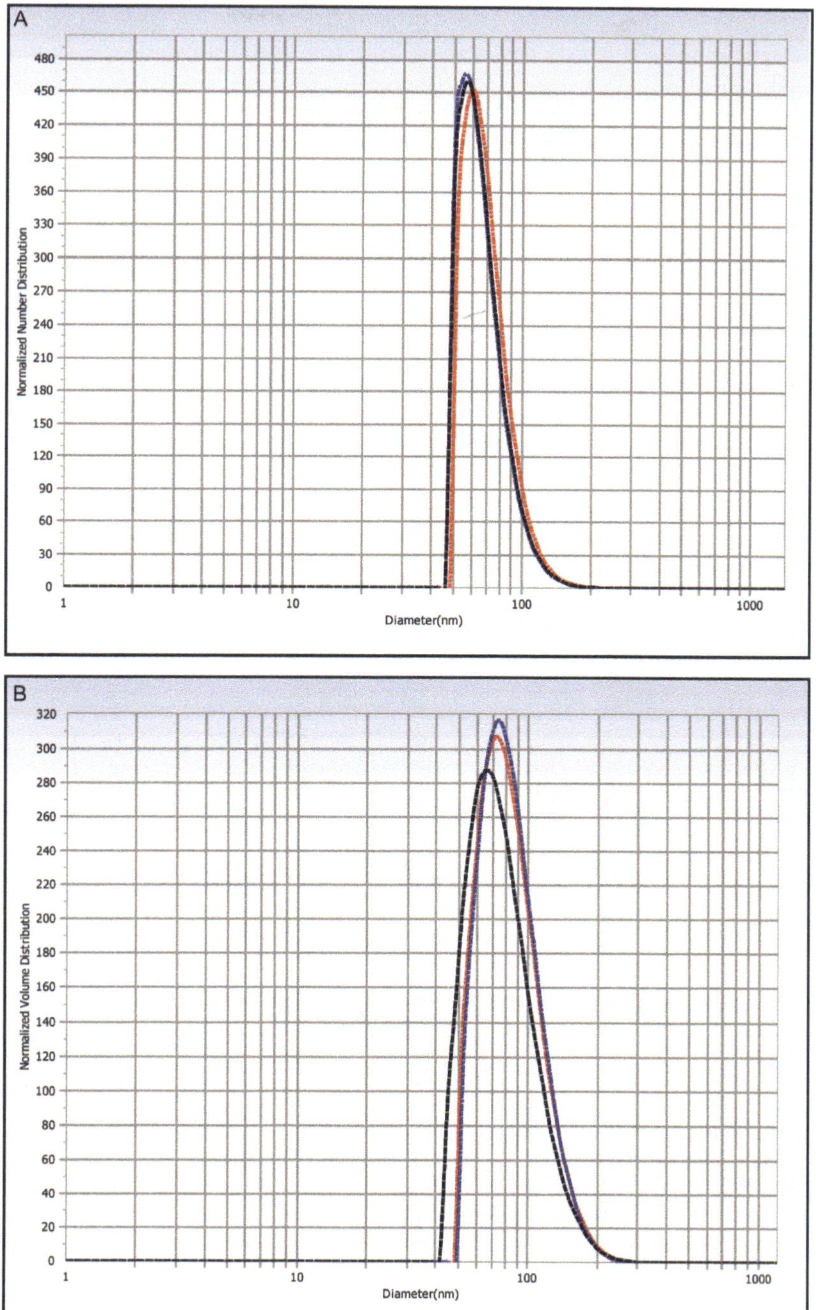

Fig. 10 Particle size distribution of (A) iron oxide magnetic nanoparticles, and (B) enzyme loaded iron oxide magnetic nanoparticles.

and 106.4 nm after immobilization of enzymes on their surface. Smaller the particle size, larger is the surface area for immobilization of celullase on nanoparticles and lower is the restriction for diffusion of substrate and product.

9.3.2 Brunauer-Emmett-Teller (BET) analysis

The BET surface area of the iron oxide magnetic nanoparticles was 13.86 m^2/g (Unpublished lab data). This value is in close agreement with the surface area of 15.63 m^2/g iron oxide magnetic nanoparticles as reported by Feng, Cao, Ma, Zhu, and Hu (2012). The surface area so obtained would be beneficial for the modification with APTES and further immobilization of enzymes thereon.

9.3.3 Confocal laser scanning microscopy

The surface modification using 3-aminopropyltriethoxysilane and immobilization of enzyme on iron oxide magnetic nanoparticles can be confirmed by labeling it with the FITC dye and then observing under confocal laser microscopy (Kadam, Jang, & Lee, 2017). The FITC dye binds with the functional $-NH_2$ groups of 3-aminopropyltriethoxysilane coated iron oxide nanoparticles and enzyme immobilized iron oxide nanoparticles, which were further visualized under confocal laser scanning microscope at excitation and emission wavelengths of 488 and 535 nm, respectively. The iron oxide magnetic nanoparticles does not show any fluorescence (Fig. 11A). However, the amino surface modified iron oxide magnetic nanoparticles as well as enzyme immobilized on iron oxide magnetic nanoparticles reflected a clear green fluorescence emission which clearly confirmed successful coating of APTES on the iron oxide magnetic nanoparticles (Fig. 11B) and also attachment of enzymes thereon (Fig. 11C).

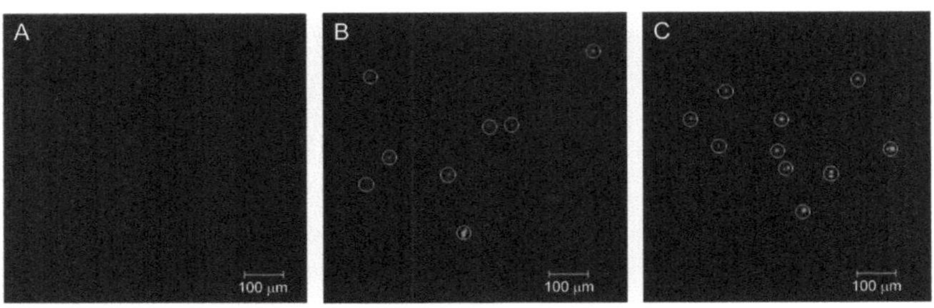

Fig. 11 Confocal laser scanning microscope images (A) iron oxide magnetic nanoparticles, (B) amino surface modified iron oxide magnetic nanoparticles, and (C) enzyme immobilized on iron oxide magnetic nanoparticles.

9.3.4 XRD analysis

The crystalline nature, crystalline mean size, crystal structure, composition and purity of the plain, surface modified and enzyme immobilized iron oxide magnetic nanoparticles are determined by X-ray diffractometer. The XRD peaks of iron oxide magnetic nanoparticles and enzyme immobilized iron oxide magnetic nanoparticles are presented in Fig. 12. The distinct diffraction peaks of Fe_3O_4 at $2\theta = 30.2°$, 35.5°, 43.2°, 57.5° and 62.7° can be indexed to the (2 2 0), (3 1 1), (4 0 0), (4 2 2) and (5 1 1) and similar peak pattern is also observed after immobilization. This indicates pure phase of the Fe_3O_4 structure and its retention after immobilization of enzyme/s on them. The average crystallite mean size of iron oxide magnetic nanoparticles and enzyme immobilized iron oxide magnetic nanoparticles can be estimated from FWHM (full width at half maximum) data of strongest diffraction peak (3 1 1) using the Scherrer formula. Muley, Chaudhari, et al. (2018) and Muley, Thorat, et al. (2018) reported an average crystallite mean size of approximately 16.95 nm for plain iron oxide magnetic nanoparticles and 12.75 nm after immobilization of three enzymes on iron oxide magnetic nanoparticles. The reduction in crystallite size after immobilization of enzyme mixture on iron oxide magnetic nanoparticles can be ascribed to inhibition of crystal growth by APTES

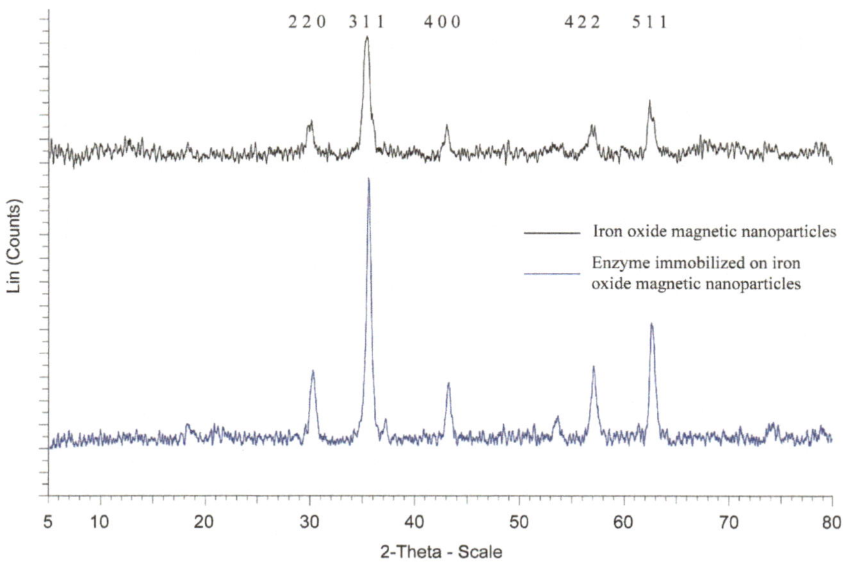

Fig. 12 XRD pattern iron oxide magnetic nanoparticles and enzyme immobilized on iron oxide magnetic nanoparticles.

coating and immobilization of enzymes. The consistent diffraction pattern of the peaks revealed iron oxide magnetic nanoparticles to be in the same phase with similar crystallinity even after immobilization of the enzymes.

9.3.5 Thermo-gravimetric analysis

The percent weight loss in plain, surface modified, and enzyme immobilized iron oxide magnetic nanoparticles, as well as the mass of surface modifying agents, and immobilized enzyme are determined with thermo-gravimetric analysis. Thermo-gravimetric analysis helps to confirm the amino modification of surface of nanoparticles by 3-aminopropyltriethoxysilane and immobilization of enzymes on iron oxide magnetic nanoparticles. The thermogram of iron oxide magnetic nanoparticles and enzyme immobilized iron oxide magnetic nanoparticles showed similar weight loss profile in a temperature domain of 70–180 °C (Fig. 13), reveling the loss of water molecules that is physically adsorbed. With a further increase in temperature to 600 °C, the total weight loss of iron oxide magnetic nanoparticles was around ~2.90% while that of enzyme immobilized iron oxide magnetic nanoparticles was ~7.42%. The char yield (%) of free and enzyme immobilized magnetic nanoparticles at 600 °C was 97.09% and 92.58%, respectively.

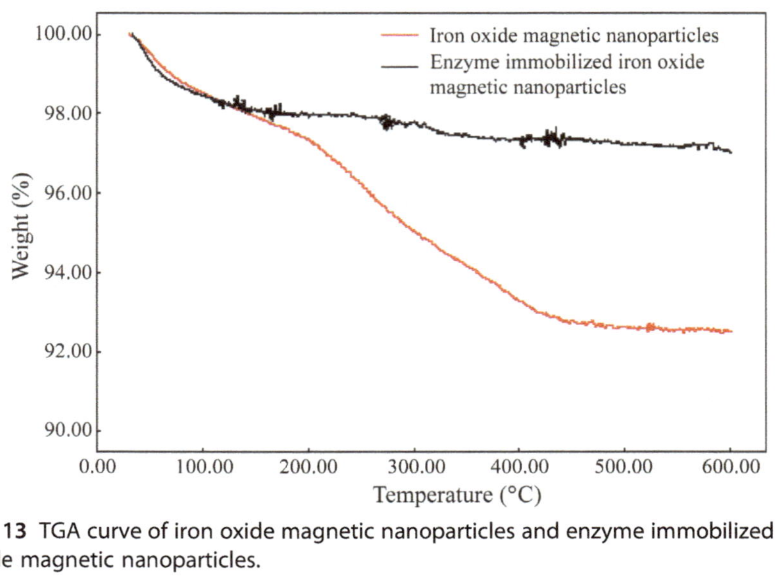

Fig. 13 TGA curve of iron oxide magnetic nanoparticles and enzyme immobilized iron oxide magnetic nanoparticles.

9.3.6 Vibrating sample magnetometry analysis

The magnetic moment and supramagnetic properties of iron oxide magnetic nanoparticles and enzyme immobilized iron oxide magnetic nanoparticles after the application of magnetic field was determined by vibrating sample magnetometry (Fig. 14). The saturation magnetization of plain and enzyme immobilized iron oxide nanoparticles have been reported to be 60.92 and 51.13 emu/g (Muley et al., 2019), 38.31 and 34.40 emu/g (Muley, Chaudhari, et al. 2018; Muley, Thorat, et al. 2018), and 53.24 and 45.48 emu/g (Bedade et al., 2019), respectively. A marginal decline in the saturation index of iron oxide magnetic nanoparticles after immobilization was seen which may be ascribed to an increase in the thickness of shell layer on the iron oxide magnetic nanoparticles surface by protein deposition. This may also reduce the external magnetic field force on nanoparticles, and consequently hamper the magnetic dominions alignment in the iron oxide magnetic nanoparticles.

9.3.7 SEM analysis

The surface morphology of iron oxide magnetic nanoparticles before and after enzyme immobilization is generally studied by the SEM analysis. The iron oxide magnetic nanoparticles pre- and post- immobilization of

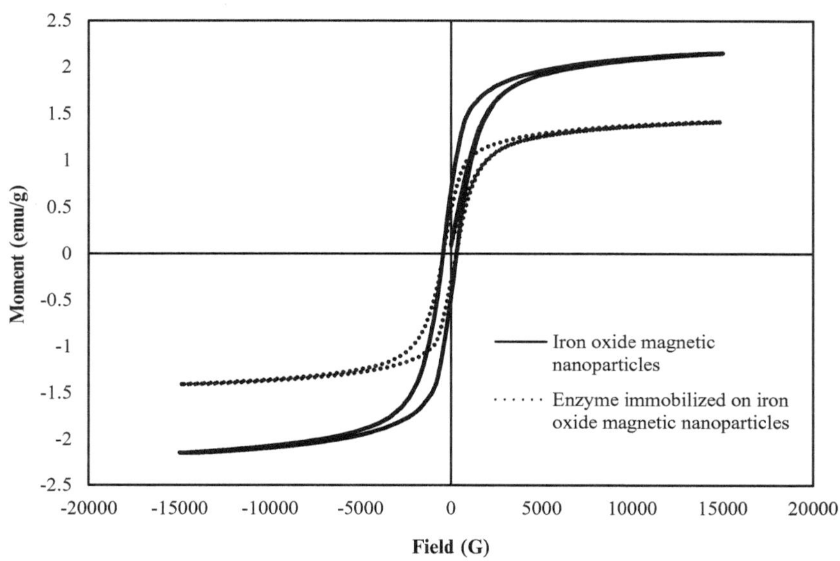

Fig. 14 VSM analysis of iron oxide magnetic nanoparticles and cellulase immobilized iron oxide magnetic nanoparticles.

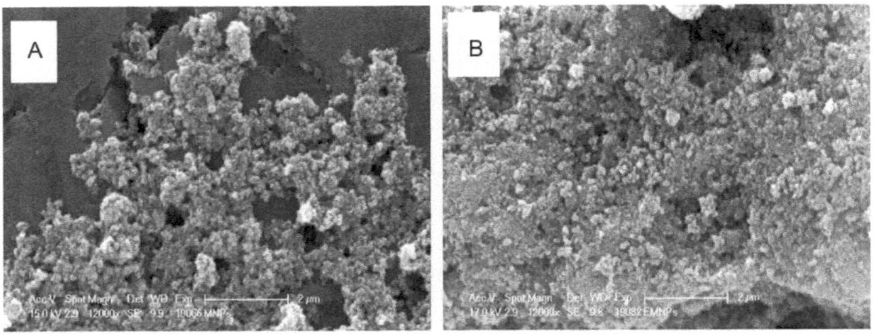

Fig. 15 SEM image of (A) iron oxide magnetic nanoparticles, and (B) enzyme immobilized oxide magnetic nanoparticles at 12,000 × magnification.

enzyme/s were monodispersed, possessed spherical geometry with a smooth surface, and in the aggregated form due to magnetic dipole moments within the particles (Fig. 15A and B). The size was found to be <70 nm. Muley, Chaudhari, et al. (2018), Muley, Thorat, et al. (2018) and Ladole et al. (2017) noticed that the shape of iron oxide magnetic nanoparticles remained nearly spherical even after coating of APTES and immobilization of enzyme on their surface.

9.4 Operating parameters

The enzyme assisted reactions are generally carried under mild operating conditions. Among the various operating conditions, pH and temperature optima of enzymes is eventually studied during the design of a biotransformation process. The optimization of pH and temperature is a mandate not only to achieve maximum activity but also to achieve superior yield of the products.

9.4.1 Optimum pH

Most of the enzymes have a distinctive optimum pH where the velocity of the biotransformation reaction is at its maximum, beyond which the reaction velocity declines. A change in pH can alter the ionization of functional groups at the active site of enzyme as well as the substrate, which alters the rate of substrate binding with the active site of enzyme (Robinson, 2015). Immobilization of cellulase on iron oxide magnetic nanoparticles did not alter the stability in the pH regime studied over that of free enzyme, nor did it change the optimum pH around 6.0–6.5 (Fig. 16).

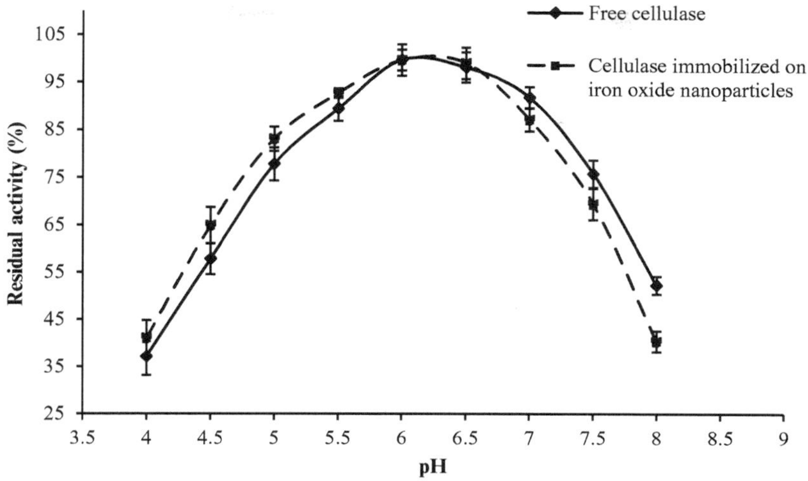

Fig. 16 Effect of pH of activity of free cellulase and cellulase immobilized on iron oxide magnetic nanoparticles (incubation time 30 min).

9.4.2 Optimum temperature

The determination of 'optimum temperature' is dependent on the exposure time. The influence of temperature on enzymes can be explained as two simultaneous forces acting constructively but in opposite ways. As the temperature increases, there is an increase in molecular movement and the rate of reaction, but with this there is also consistent denaturation causing inactivation of the enzyme (Muley, Thorat, et al., 2018). There was a change in optimum temperature of cellulase from 55 to 60 °C after immobilization as shown in Fig. 17. Similarly, a shift in optimum temperature from 30–60 to 30–65 °C for acrylamidase (Bedade et al., 2019), and from 50 to 60 °C for cellulase (Ladole et al., 2017) have been noticed after immobilization on the iron oxide magnetic nanoparticles.

9.5 Enzyme constants

9.5.1 Kinetic constants (K_m and V_{max})

The kinetic constants indicate the affinity of an enzyme with its substrate from the Michaelis-Menten constant (K_m) and speed of reactions catalyzed by enzymes from maximal velocity (V_{max}). The K_m of enzyme helps to predict the relation between the rate of formation of product from the availability of substrates relative to substrate concentration under the operating conditions (Chaudhari et al., 2015). The K_m of enzyme is an inverse measure of its affinity to the substrate. The V_{max} of enzyme is the maximum rate of

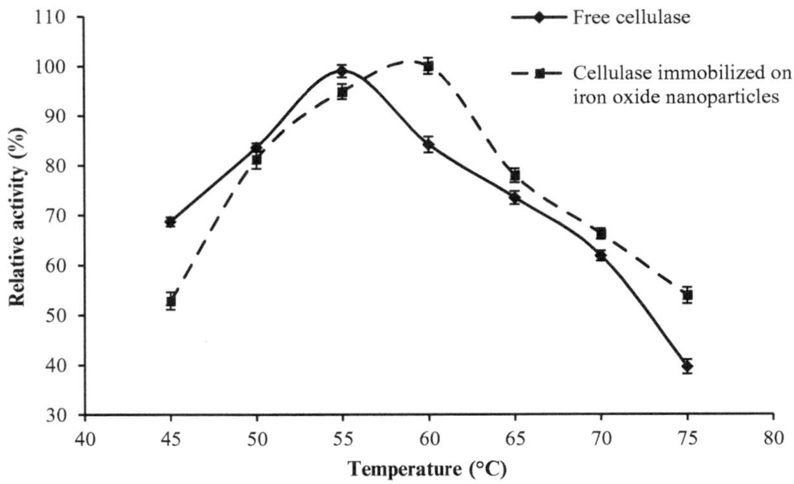

Fig. 17 Effect of temperature of activity of free cellulase and cellulase immobilized on iron oxide magnetic nanoparticles (incubation time 30 min).

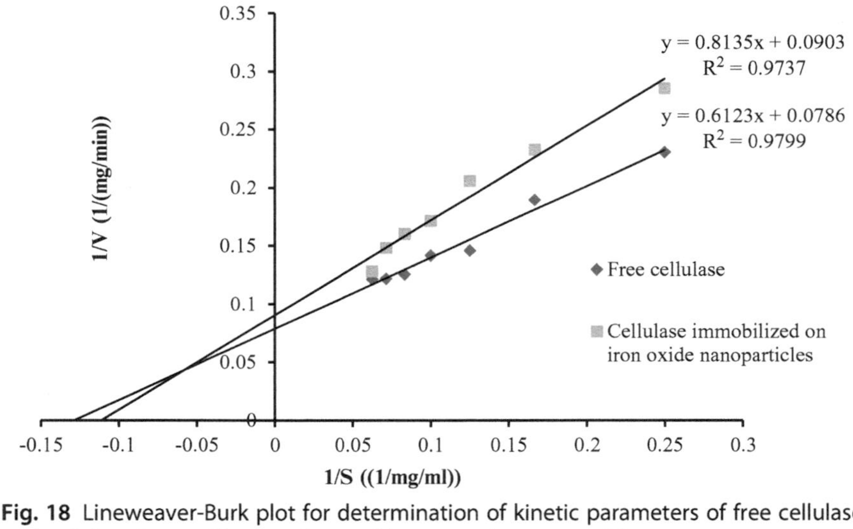

Fig. 18 Lineweaver-Burk plot for determination of kinetic parameters of free cellulase and cellulase immobilized on iron oxide magnetic nanoparticles.

reaction and is a characteristic feature of a specific enzyme at a specific concentration of substrate. A Lineweaver-Burk plot was used to calculate the kinetic constants of cellulase in free and immobilized form (Fig. 18). The K_m and V_{max} values of free cellulase were 7.85 mg/mL and 12.82 mg/(mL/min). Immobilization of cellulase in the tri-enzyme mixture on iron oxide nanoparticles increased the K_m value by 13.14% and reduced the V_{max} by 13.33% which indicated a marginal reduction in the affinity of enzyme toward the substrate and slightly reduced catalytic efficiency after immobilization.

It may be noted that the substrate concentration used in the study is very important and also indicates the order of an enzyme catalyzed reaction. For instance, at $[S] \lll K_m$, $[S] \sim K_m$, and $[S] \ggg K_m$, the order of the enzyme catalyzed reaction could be first, fractional and zero, respectively (Punekar, 2018).

9.5.2 *Mass transfer limitations*

In the Michaelis-Menten kinetics, the mass transfer of large molecular weight substrates in the enzyme is a crucial aspect for restriction of the activity after immobilization. The degree of mass transfer of enzymes after immobilization for enzyme kinetics is stated as effectiveness factor, which is the retention of relative activity of enzyme after immobilization on nanoparticles over its free form. The effectiveness factor of free and immobilized cellulase was calculated by varying the substrate concentration in the range of 2–6 mg/mL (Table 3). There were no noticeable variations in the effectiveness factor even after changing the concentration of substrate from a lower level (2 mg/mL) to higher level (6 mg/mL). This indicates that the limitation for mass transfer of large substrates was not significant (Muley, Thorat, et al., 2018).

9.6 Thermal inactivation kinetics

9.6.1 *Inactivation constants*

There is an irreversible reduction in catalytic potential of enzymes upon exposure to higher temperature which is due to loss of structural distortion and breakdown of forces, linkages and bonds required in maintaining their structural integrity. The thermal inactivation kinetics data (k_d, $t_{1/2}$, D-value and z-value) helps to correlate the structure and function of enzyme in a specified temperature domain, and guides in understanding the ability of enzyme in resisting its unfolding process. Thermal inactivation kinetics of free and immobilized cellulase revealed a steady decline in the activity with time (Fig. 19A and B). The inactivation rate constant (k_d) of cellulase immobilized on iron oxide nanoparticles was 25%, 14.29%, 20%, 25% and 19.23% lower than the free cellulase in the tri-enzyme mixture at 55, 60, 65, 70 and 75 °C, respectively, indicating enhanced thermal stability over the free cellulase. The data for the same is presented in Table 4.

Table 3 Effectivity factor of free and immobilized cellulase.

Substrate (mg/mL)	Mass transfer constant	Free cellulase	Cellulase immobilized on iron oxide nanoparticles
2	V_{max} (μmol/min)	1.96 ± 0.03	1.01 ± 0.03
	η (%)	100	51.35 ± 0.88
3	V_{max} (μmol/min)	2.14 ± 0.07	1.12 ± 0.04
	η (%)	100	52.10 ± 0.07
4	V_{max} (μmol/min)	2.89 ± 0.11	1.26 ± 0.04
	η (%)	100	43.30 ± 0.27
5	V_{max} (μmol/min)	3.11 ± 0.08	1.419 ± 0.05
	η (%)	100	45.66 ± 2.75
6	V_{max} (μmol/min)	3.36 ± 0.07	1.59 ± 0.08
	η (%)	100	47.29 ± 3.23

The determination of $t_{1/2}$ and D-value are more dependable and precise, when the stability of enzymes are to be estimated in the varying temperature domain. The $t_{1/2}$ and D-value of cellulase increased marginally at each studied temperature after immobilizing on iron oxide magnetic nanoparticles and thus imparting elevated thermal stability to the immobilized cellulase. The $t_{1/2}$ values of immobilized cellulase were 1.33, 1.17, 1.25, 1.33 and 1.24-folds higher, while the D-values improved by 33.33%, 16.67%, 24.99%, 33.33% and 23.81% at 55, 60, 65, 70 and 75 °C, respectively, over the free cellulase (Table 4). The enhancement in the thermal stability against increasing temperature might be due to protection and retention of the structural conformation of immobilized cellulase by a rigid covalent linkage among amino groups on iron oxide nanoparticles and enzymes coupled with glutaraldehyde.

9.6.2 Thermodynamic constants

The thermodynamic constants (ΔH°, ΔG° and ΔS°) represent fundamental basics of the functionality and stability of enzymes during the operating conditions, and are enumerated from the thermal inactivation kinetic constants (k_d and E_d). The stability of enzymes is a combined effect of equilibrium

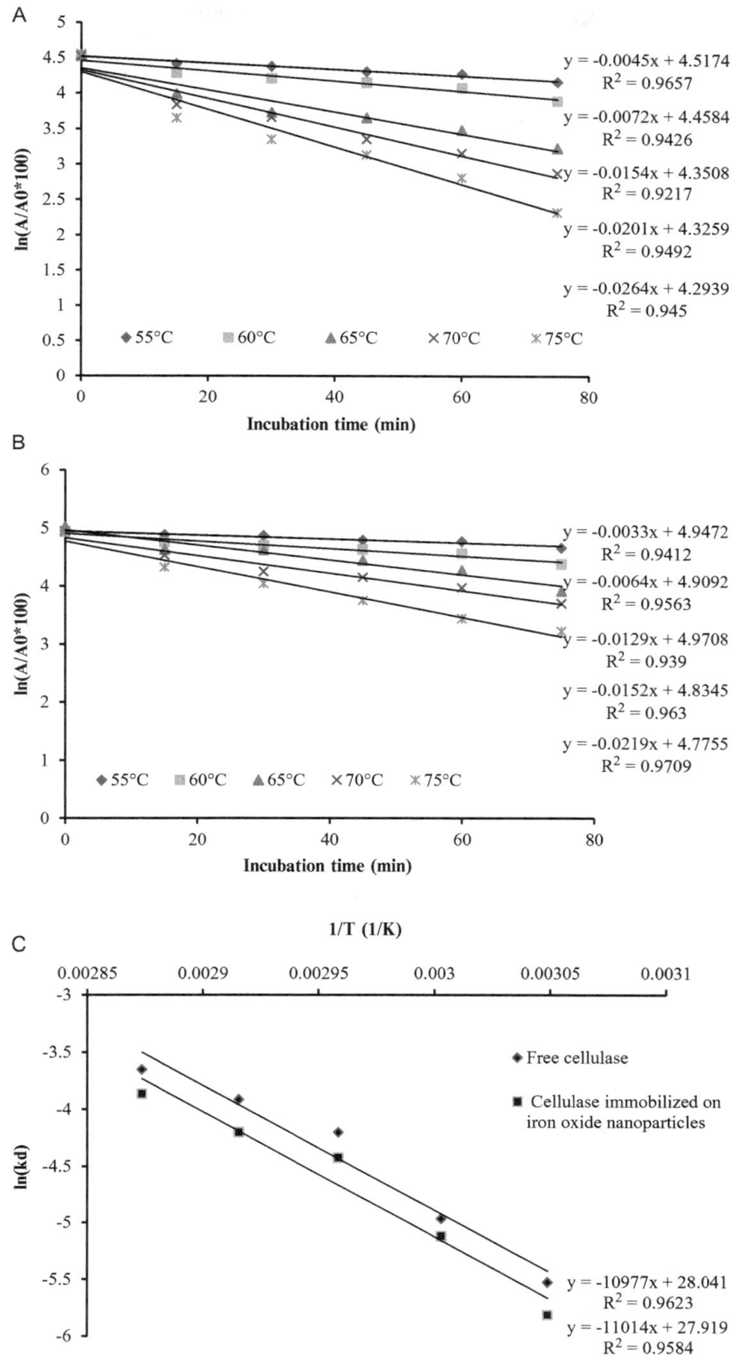

Fig. 19 Thermal deactivation kinetics of (A) free cellulase, (B) cellulase immobilized on iron oxide nanoparticles, and (C) Arrhenius plot of free cellulase and cellulase immobilized on iron oxide nanoparticles.

Table 4 Thermal inactivation kinetics of free and immobilized cellulase.

Temperature (°C)	Thermal inactivation kinetics parameter	Free cellulase	Cellulase immobilized on iron oxide nanoparticles
55	K_d (min^{-1})	0.004 ± 0.0005	0.003 ± 0.0005
	$t_{1/2}$ (min)	173.25 ± 13.61	231.00 ± 13.61
	D-value (min)	575.65 ± 26.11	767.53 ± 26.11
60	K_d (min^{-1})	0.007 ± 0.0005	0.006 ± 0.0005
	$t_{1/2}$ (min)	99.00 ± 2.91	115.50 ± 5.44
	D-value (min)	328.94 ± 5.60	383.76 ± 10.44
65	K_d (min^{-1})	0.015 ± 0.0012	0.012 ± 0.0005
	$t_{1/2}$ (min)	46.20 ± 1.10	57.75 ± 1.75
	D-value (min)	153.51 ± 1.49	191.88 ± 2.38
70	K_d (min^{-1})	0.020 ± 0.0005	0.015 ± 0.0006
	$t_{1/2}$ (min)	34.65 ± 0.61	46.20 ± 1.10
	D-value (min)	115.13 ± 0.82	153.51 ± 3.65
75	K_d (min^{-1})	0.026 ± 0.0005	0.021 ± 0.0005
	$t_{1/2}$ (min)	26.65 ± 0.39	33.00 ± 0.61
	D-value (min)	88.56 ± 0.52	109.65 ± 0.82
E_d (kJ/mol)		91.26 ± 2.28	107.19 ± 3.03

among the stabilizing and destabilizing forces which are not only linked with different bonds/linkages but also the folding/unfolding condition of the enzyme molecules. The activation energy (E_d, kJ/mol) is the minimal amount of energy needed for triggering the denaturation process for inactivating the enzymes. Enthalpy ($\Delta H°$, kJ/mol) is the amount of energy needed for the denaturation of enzymes at elevated temperature. Gibbs free energy ($\Delta G°$, kJ/mol) is the amount of utilizable energy that arises during the deactivation of enzyme and it also specifies the spontaneity of thermal inactivation process caused by conformational distortion and disruption of several linkages. Entropy ($\Delta S°$, J/mol/K) is the energy per degree required for the transition of enzymes native to denatured form or local disordering between the transition and ground state. The thermodynamic constants of enzymes in free and immobilized form are calculated by the conjugation of inactivation kinetic constants data and the Eyring's transition state theory.

The E_d values of immobilized cellulase increased from 91.26 ± 2.28 to 107.19 ± 3.03 kJ/mol for free cellulase in a temperature domain of 55–75 °C (Fig. 19C; Table 4). This clearly indicated that some excess amount of energy would be required to disrupt the natural conformation of cellulase immobilized on iron oxide nanoparticles over the free cellulase. The ΔH° values reduced gradually with an increase in temperature which again indicates lower energy requirement for denaturation of free cellulase over that of immobilized form at elevated temperature. The higher and positive values of E_d and ΔH° after immobilization of cellulase indicated higher thermal stability during the inactivation process. Further, ΔH° is also a measure of disruption of different linkages linked with the stabilization of transition state *viz.* from catalytically active native form to randomly coiled inactive denatured form. The ΔG° value of cellulase immobilized on magnetic nanoparticles was marginally higher than the free cellulase specifying the hindrances generated toward the thermal unfolding process after heat treatment. The negative ΔS° value of free and immobilized cellulase are due to the aggregation of enzyme molecules after cross linking with glutaraldehyde and immobilization on iron oxide magnetic nanoparticles. The negative ΔS° values can be associated with a decline in disorder or randomness of the enzyme system. The ΔH°, ΔG°, and ΔS° values of immobilized cellulase were 0.34–0.35%, 0.45–0.86%, and 73–76% higher, as compared to the free cellulase at 55, 60, 65, 70 and 75 °C, respectively (Table 5).

9.7 Storage stability

The process economics of numerous biotransformation reactions in industries is dictated by the stability of enzymes during storage under ambient conditions (Muley et al., 2019). Enumeration of storage stability also helps in determining the shelf life of enzymes so that they can be utilized within a given time span, as there may be a gradual decline in the activity of enzymes with the progress of storage (Chaudhari et al., 2015). The relative activity of free cellulase and cellulase immobilized on iron oxide nanoparticles after 36 days of incubation was found to reduce to 40.74 ± 1.73% and 71.68 ± 3.48%, respectively (Fig. 20). This retention in activity of cellulase immobilized on iron oxide magnetic nanoparticles during prolonged storage can be attributed to maintenance of active sites on the enzyme and resistance to disruption of enzyme active sites which was generated by cross linking with glutaraldehyde and subsequent immobilization on iron oxide magnetic nanoparticles.

Table 5 Thermodynamic parameters of free and immobilized cellulase.

Temperature (°C)	Thermal inactivation kinetics parameter	Free cellulase	Cellulase immobilized on iron oxide nanoparticles
55	ΔH° (kJ/mol)	88.54 ± 2.23	88.84 ± 1.44
	ΔG° (kJ/mol)	94.25 ± 1.29	95.04 ± 1.35
	ΔS° (J/mol/K)	−17.43 ± 2.45	−69.98 ± 3.60
60	ΔH° (kJ/mol)	88.49 ± 1.29	88.80 ± 1.66
	ΔG° (kJ/mol)	94.18 ± 0.45	94.61 ± 0.75
	ΔS° (J/mol/K)	−17.09 ± 3.65	−65.64 ± 4.24
65	ΔH° (kJ/mol)	88.45 ± 1.82	88.76 ± 0.0005
	ΔG° (kJ/mol)	93.50 ± 0.48	94.13 ± 0.60
	ΔS° (J/mol/K)	−14.93 ± 2.64	−60.65 ± 3.15
70	ΔH° (kJ/mol)	88.41 ± 1.96	88.72 ± 1.21
	ΔG° (kJ/mol)	94.10 ± 0.32	94.92 ± 0.44
	ΔS° (J/mol/K)	−16.59 ± 1.63	−70.18 ± 2.89
75	ΔH° (kJ/mol)	88.37 ± 2.13	88.68 ± 1.73
	ΔG° (kJ/mol)	94.76 ± 0.75	95.38 ± 0.85
	ΔS° (J/mol/K)	−18.36 ± 2.47	−75.79 ± 3.75

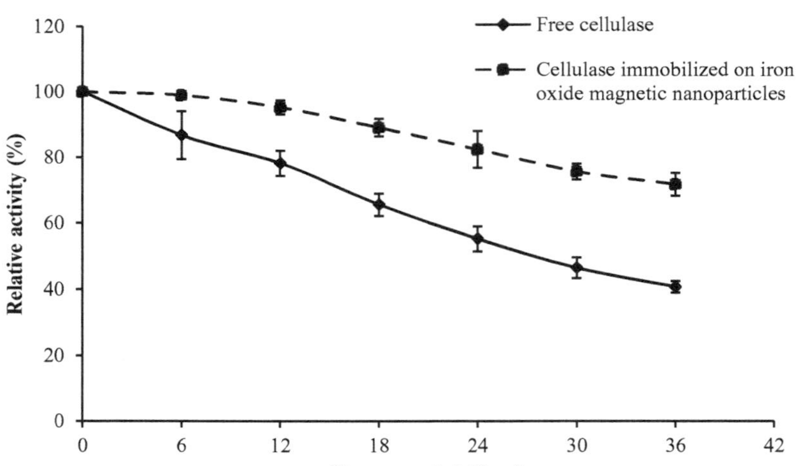

Fig. 20 Stability of cellulase in free and immobilized form during storage at 4 ± 1 °C.

9.8 Reusability of immobilized enzymes

The operational costs and economic feasibility of the process associated with the biotransformation process are decided by the quantity and reusability of enzymes (Ladole et al., 2014; Talekar et al., 2017). Therefore, an investigation of reusability of enzymes is generally carried out as there may be reduction in catalytic efficiency and performance during continuous use in subsequent cycles. After the fifth reusability batch, the cellulase immobilized on iron oxide nanoparticles retained $81.15 \pm 5.27\%$ of its original activity, which reduced steadily in successive cycles (Fig. 21). The continuous reduction in activity of cellulase after successive cycles could be due to mechanical damage over continuous repeated use.

10. Summary and conclusion

The use of iron oxide magnetic nanoparticles as an enzyme immobilization matrix has emerged in recent times due to its simple and energy effective preparation method, favorable process economics, and many desirable physical and chemical properties. To maximize the potential of these interesting biocatalysts, the process parameters of their manufacture as well as immobilization need to be optimized. In this chapter, the detailed protocols for synthesis of iron oxide magnetic nanoparticles, their surface modification and functionalization, followed by immobilization of enzymes are

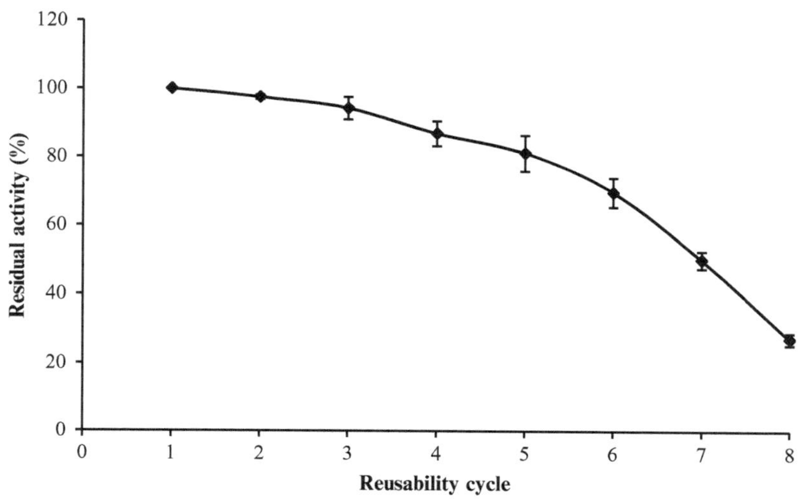

Fig. 21 Stability of free and immobilized cellulase during reusability (50 mM phosphate buffer, pH 5.5).

provided. The chapter further provides an insight of operating conditions, stability studies, enzyme and thermal inactivation kinetics, thermodynamic parameters, storage stability and reusability studies. Additionally, the chapter also deals with the basic instrumental analysis and characterization which enables confirmation of synthesis and modification of the nanoparticles as well as the successful binding of enzymes thereon for appropriate applications. Further, the findings of all the mentioned protocols and stability studies are thoroughly elaborated by using the findings of immobilization of cellulase on iron oxide magnetic nanoparticles.

Acknowledgment

The authors gratefully thank and dedicate this chapter to Hon. Dr. A.P.J. Abdul Kalam, whose ideas and thoughts guided and inspired us to move into the research field.

References

Anbarasu, M., Anandan, M., Chinnasamy, E., Gopinath, V., & Balamurugan, K. (2015). Synthesis and characterization of polyethylene glycol (PEG) coated Fe_3O_4 nanoparticles by chemical co-precipitation method for biomedical applications. *Spectrochimica Acta Part A: Molecular and Biomolecular Spectroscopy*, *135*, 536–539. https://doi.org/10.1016/j.saa.2014.07.059.

Ardiyanti, H., Suharyadi, E., Kato, T., & Iwata, S. (2016). Crystal structures and magnetic properties of magnetite (Fe_3O_4)/polyvinyl alcohol (PVA) ribbon. *AIP Conference Proceedings*, *1725*(1), 020007. https://doi.org/10.1063/1.4945461.

Atacan, K., Çakıroğlu, B., & Özacar, M. (2016). Improvement of the stability and activity of immobilized trypsin on modified Fe_3O_4 magnetic nanoparticles for hydrolysis of bovine serum albumin and its application in the bovine milk. *Food Chemistry*, *212*, 460–468. https://doi.org/10.1016/j.foodchem.2016.06.011.

Barbosa, O., Torres, R., Ortiz, C., Berenguer-Murcia, A., Rodrigues, R. C., & Fernandez-Lafuente, R. (2013). Heterofunctional supports in enzyme immobilization: From traditional immobilization protocols to opportunities in tuning enzyme properties. *Biomacromolecules*, *14*(8), 2433–2462. https://doi.org/10.1021/bm400762h.

Bedade, D. K., Muley, A. B., & Singhal, R. S. (2019). Magnetic cross-linked enzyme aggregates of acrylamidase from *Cupriavidus oxalaticus* ICTDB921 for biodegradation of acrylamide from industrial waste water. *Bioresource Technology*, *272*, 137–145. https://doi.org/10.1016/j.biortech.2018.10.015.

Borase, H. P., Muley, A. B., Patil, S. V., & Singhal, R. S. (2019). Nano-eco toxicity study of gold nanoparticles on aquatic organism *Moina macrocopa*: As new versatile ecotoxicity testing model. *Environmental Toxicology and Pharmacology*, *68*, 4–12. https://doi.org/10.1016/j.etap.2019s.02.013.

Calatayud, M. P., Riggio, C., Raffa, V., Sanz, B., Torres, T. E., Ibarra, M. R., et al. (2013). Neuronal cells loaded with PEI-coated Fe_3O_4 nanoparticles for magnetically guided nerve regeneration. *Journal of Materials Chemistry B*, *1*(29), 3607–3616. https://doi.org/10.1039/c3tb20336k.

Cashin, V. B., Eldridge, D. S., Yu, A., & Zhao, D. (2018). Surface functionalization and manipulation of mesoporous silica adsorbents for improved removal of pollutants: A review. *Environmental Science: Water Research & Technology*, *4*(2), 110–128. https://doi.org/10.1039/c7ew00322f.

Chaudhari, S. A., Kar, J. R., & Singhal, R. S. (2015). Immobilization of proteins in alginate: Functional properties and applications. *Current Organic Chemistry*, *19*(17), 1732–1754.
Chaudhari, S. A., & Singhal, R. S. (2017). A strategic approach for direct recovery and stabilization of *Fusarium* sp. ICT SAC1 cutinase from solid state fermented broth by carrier free cross-linked enzyme aggregates. *International Journal of Biological Macromolecules*, *98*, 610–621. https://doi.org/10.1016/j.ijbiomac.2017.02.033.
de Souza Lima, J., Costa, F. N., Bastistella, M. A., de Araújo, P. H. H., & de Oliveira, D. (2019). Functionalized kaolin as support for endoglucanase immobilization. *Bioprocess and Biosystems Engineering*, *42*(7), 1–9. https://doi.org/10.1007/s00449-019-02113-w.
Díaz-Hernández, A., Gracida, J., García-Almendárez, B. E., Regalado, C., Núñez, R., & Amaro-Reyes, A. (2018). Characterization of magnetic nanoparticles coated with chitosan: A potential approach for enzyme immobilization. *Journal of Nanomaterials*, 1–11. https://doi.org/10.1155/2018/9468574.
Feng, L., Cao, M., Ma, X., Zhu, Y., & Hu, C. (2012). Superparamagnetic high-surface-area Fe_3O_4 nanoparticles as adsorbents for arsenic removal. *Journal of Hazardous Materials*, *217–218*, 439–446. https://doi.org/10.1016/j.jhazmat.2012.03.073.
Fortes, C. C., Daniel-da-Silva, A. L., Xavier, A. M., & Tavares, A. P. (2017). Optimization of enzyme immobilization on functionalized magnetic nanoparticles for laccase biocatalytic reactions. *Chemical Engineering and Processing: Process Intensification*, *117*, 1–8. https://doi.org/10.1016/j.cep.2017.03.009.
Han, J., Luo, P., Wang, Y., Wang, L., Li, C., Zhang, W., et al. (2018). The development of nanobiocatalysis via the immobilization of cellulase on composite magnetic nanomaterial for enhanced loading capacity and catalytic activity. *International Journal of Biological Macromolecules*, *119*, 692–700. https://doi.org/10.1016/j.ijbiomac.2018.07.176.
Hatami, M. (2018). Production of polyimide ceria nanocomposites by development of molecular hook technology in nano-sonochemistry. *Ultrasonics Sonochemistry*, *44*, 261–271. https://doi.org/10.1016/j.ultsonch.2018.02.032.
Hauser, A. K., Mathias, R., Anderson, K. W., & Hilt, J. Z. (2015). The effects of synthesis method on the physical and chemical properties of dextran coated iron oxide nanoparticles. *Materials Chemistry and Physics*, *160*, 177–186. https://doi.org/10.1016/j.matchemphys.2015.04.022.
Iyer, P. V., & Ananthanarayan, L. (2008). Enzyme stability and stabilization-aqueous and non-aqueous environment. *Process Biochemistry*, *43*(10), 1019–1032. https://doi.org/10.1016/j.procbio.2008.06.004.
Kadam, A. A., Jang, J., & Lee, D. S. (2017). Supermagnetically tuned halloysite nanotubes functionalized with aminosilane for covalent laccase immobilization. *ACS Applied Materials & Interfaces*, *9*(18), 15492–15501. https://doi.org/10.1021/acsami.7b02531.
Khoshnevisan, K., Poorakbar, E., Baharifar, H., & Barkhi, M. (2019). Recent advances of cellulase immobilization onto magnetic nanoparticles: An update review. *Magnetochemistry*, *5*(2), 36. https://doi.org/10.3390/magnetochemistry5020036.
Koc, K., Karakus, B., Rajar, K., & Alveroglu, E. (2017). Synthesis and characterization of ZnS@ Fe_3O_4 fluorescent-magnetic bifunctional nanospheres. *Superlattices and Microstructures*, *110*, 198–204. https://doi.org/10.1016/j.spmi.2017.08.044.
Ladole, M. R., Mevada, J. S., & Pandit, A. B. (2017). Ultrasonic hyperactivation of cellulase immobilized on magnetic nanoparticles. *Bioresource Technology*, *239*, 117–126. https://doi.org/10.1016/j.biortech.2017.04.096.
Ladole, M. R., Muley, A. B., Patil, I. D., Talib, M. I., & Parate, V. R. (2014). Immobilization of tropizyme-P on amino-functionalized magnetic nanoparticles for fruit juice clarification. *Journal of Biochemical Technology*, *5*(4), 838–845.
Ladole, M. R., Nair, R. R., Bhutada, Y. D., Amritkar, V. D., & Pandit, A. B. (2018). Synergistic effect of ultrasonication and co-immobilized enzymes on tomato peels for lycopene extraction. *Ultrasonics Sonochemistry*, *48*, 453–462. https://doi.org/10.1016/j.ultsonch.2018.06.013.

Liu, R., Guo, Y., Odusote, G., Qu, F., & Priestley, R. D. (2013). Core–shell Fe_3O_4 polydopamine nanoparticles serve multipurpose as drug carrier, catalyst support and carbon adsorbent. *ACS Applied Materials & Interfaces*, *5*(18), 9167–9171. https://doi.org/10.1021/am402585y.

Ma, Y. R., Zhang, X. L., Zeng, T., Cao, D., Zhou, Z., Li, W. H., et al. (2013). Polydopamine-coated magnetic nanoparticles for enrichment and direct detection of small molecule pollutants coupled with MALDI-TOF-MS. *ACS Applied Materials & Interfaces*, *5*(3), 1024–1030. https://doi.org/10.1021/am3027025.

Mohamad, N. R., Marzuki, N. H. C., Buang, N. A., Huyop, F., & Wahab, R. A. (2015). An overview of technologies for immobilization of enzymes and surface analysis techniques for immobilized enzymes. *Biotechnology & Biotechnological Equipment*, *29*(2), 205–220. https://doi.org/10.1080/13102818.2015.1008192.

Mohammadi, A., Daemi, H., & Barikani, M. (2014). Fast removal of malachite green dye using novel superparamagnetic sodium alginate-coated Fe_3O_4 nanoparticles. *International Journal of Biological Macromolecules*, *69*, 447–455. https://doi.org/10.1016/j.ijbiomac.2014.05.042.

Mourdikoudis, S., Pallares, R. M., & Thanh, N. T. (2018). Characterization techniques for nanoparticles: Comparison and complementarity upon studying nanoparticle properties. *Nanoscale*, *10*(27), 12871–12934. https://doi.org/10.1039/c8nr02278j.

Muley, A. B., Chaudhari, S. A., Bankar, S. B., & Singhal, R. S. (2019). Stabilization of cutinase by covalent attachment on magnetic nanoparticles and improvement of its catalytic activity by ultrasonication. *Ultrasonics Sonochemistry*, *55*, 174–185. https://doi.org/10.1016/j.ultsonch.2019.02.019.

Muley, A. B., Chaudhari, S. A., Mulchandani, K. H., & Singhal, R. S. (2018). Extraction and characterization of chitosan from prawn shell waste and its conjugation with cutinase for enhanced thermo-stability. *International Journal of Biological Macromolecules*, *111*, 1047–1058. https://doi.org/10.1016/j.procbio.2008.06.004.

Muley, A. B., Chaudhari, S. A., & Singhal, R. S. (2017). Non-covalent conjugation of cutinase from *Fusarium* sp. ICT SAC1 with pectin for enhanced stability: Process minutiae, kinetics, thermodynamics and structural study. *International Journal of Biological Macromolecules*, *102*, 729–740. https://doi.org/10.1016/j.ijbiomac.2017.04.072.

Muley, A. B., Thorat, A. S., Singhal, R. S., & Babu, K. H. (2018). A tri-enzyme co-immobilized magnetic complex: Process details, kinetics, thermodynamics and applications. *International Journal of Biological Macromolecules*, *118*, 1781–1795. https://doi.org/10.1016/j.ijbiomac.2018.07.022.

Munde, P. J., Muley, A. B., Ladole, M. R., Pawar, A. V., Talib, M. I., & Parate, V. R. (2017). Optimization of pectinase-assisted and tri-solvent-mediated extraction and recovery of lycopene from waste tomato peels. *3 Biotech*, 7(3), 206. https://doi.org/10.1007/s13205-017-0825-3.

Nadar, S. S., Muley, A. B., Ladole, M. R., & Joshi, P. U. (2016). Macromolecular cross-linked enzyme aggregates (M-CLEAs) of α-amylase. *International Journal of Biological Macromolecules*, *84*, 69–78. https://doi.org/10.1016/j.ijbiomac.2015.11.082.

Punekar, N. S. (2018). Henri–Michaelis–Menten equation. In *ENZYMES: Catalysis, kinetics and mechanisms* (pp. 155–176). Singapore: Springer. https://doi.org/10.1007/978-981-13-0785-0_15.

Robinson, P. K. (2015). Enzymes: Principles and biotechnological applications. *Essays in Biochemistry*, *59*, 1–41. https://doi.org/10.1042/BSE0590001.

Rose, P. A., Praseetha, P. K., Bhagat, M., Alexander, P., Abdeen, S., & Chavali, M. (2013). Drug embedded PVP coated magnetic nanoparticles for targeted killing of breast cancer cells. *Technology in Cancer Research & Treatment*, *12*(5), 463–472. https://doi.org/10.7785/tcrt.2012.500333.

Roto, R. (2018). Surface modification of Fe_3O_4 as magnetic adsorbents for recovery of precious metals. In M. A. Chowdhury (Ed.), *Advanced surface engineering research* (pp. 127–145). UK: IntechOpen Publishers. https://doi.org/10.5772/intechopen.79586.

Sapsford, K. E., Algar, W. R., Berti, L., Gemmill, K. B., Casey, B. J., Oh, E., et al. (2013). Functionalizing nanoparticles with biological molecules: Developing chemistries that facilitate nanotechnology. *Chemical Reviews*, *113*, 1904–2074. https://doi.org/10.1021/cr300143v.

Sheldon, R. A. (2019). CLEAs, combi-CLEAs and 'smart' magnetic CLEAs: Biocatalysis in a bio-based economy. *Catalysts*, *9*(3), 261. https://doi.org/10.3390/catal9030261.

Sheldon, R. A., & van Pelt, S. (2013). Enzyme immobilisation in biocatalysis: Why, what and how. *Chemical Society Reviews*, *42*(15), 6223–6235. https://doi.org/10.1039/c3cs60075k.

Silva, C., Martins, M., Jing, S., Fu, J., & Cavaco-Paulo, A. (2018). Practical insights on enzyme stabilization. *Critical Reviews in Biotechnology*, *38*(3), 335–350. https://doi.org/10.1080/07388551.2017.1355294.

Soozanipour, A., & Taheri-Kafrani, A. (2018). Enzyme immobilization on functionalized graphene oxide nanosheets: Efficient and robust biocatalysts. *Methods in Enzymology*, *609*, 371–403. https://doi.org/10.1016/bs.mie.2018.06.010.

Stepankova, V., Bidmanova, S., Koudelakova, T., Prokop, Z., Chaloupkova, R., & Damborsky, J. (2013). Strategies for stabilization of enzymes in organic solvents. *ACS Catalysis*, *3*(12), 2823–2836. https://doi.org/10.1021/cs400684x.

Talekar, S., Joshi, A., Kamble, S., Jadhav, S., Nadar, S., & Ladole, M. (2017). A tri-enzyme magnetic nanobiocatalyst with one pot starch hydrolytic activity. *Chemical Engineering Journal*, *325*, 80–90. https://doi.org/10.1016/j.cej.2017.05.054.

Wu, W., Wu, Z., Yu, T., Jiang, C., & Kim, W.-S. (2015). Recent progress on magnetic iron oxide nanoparticles: Synthesis, surface functional strategies and biomedical applications. *Science and Technology of Advanced Materials*, *16*, 023501. https://doi.org/10.1088/1468-6996/16/2/023501.

Yang, G., Zhang, B., Wang, J., Wang, M., Xie, S., & Li, X. (2016). Synthesis and characterization of poly (lactic acid)-modified superparamagnetic iron oxide nanoparticles. *Journal of Sol-Gel Science and Technology*, 77(2), 335–341. https://doi.org/10.1007/s10971-015-3858-2.

Zdarta, J., Meyer, A., Jesionowski, T., & Pinelo, M. (2018). A general overview of support materials for enzyme immobilization: Characteristics, properties, practical utility. *Catalysts*, *8*(2), 92. https://doi.org/10.3390/catal8020092.

Zhan, Y., Zhang, J., Wan, X., Long, Z., He, S., & He, Y. (2018). Epoxy composites coating with Fe_3O_4 decorated graphene oxide: Modified bio-inspired surface chemistry, synergistic effect and improved anti-corrosion performance. *Applied Surface Science*, *436*, 756–767. https://doi.org/10.1016/j.apsusc.2017.12.095.

Zhu, N., Ji, H., Yu, P., Niu, J., Farooq, M., Akram, M., et al. (2018). Surface modification of magnetic iron oxide nanoparticles. *Nanomaterials*, *8*(10), 810. https://doi.org/10.3390/nano8100810.

Further reading

Lima, J. S., Araújo, P. H. H., Sayer, C., Souza, A. A. U., Viegas, A. C., & de Oliveira, D. (2017). Cellulase immobilization on magnetic nanoparticles encapsulated in polymer nanospheres. *Bioprocess and Biosystems Engineering*, *40*, 511–518. https://doi.org/10.1007/s00449-016-1716-4.

CHAPTER FOUR

Strategies to rationalize enzyme immobilization procedures

Diego E. Sastre, Eduardo A. Reis, Caterina G.C. Marques Netto*
Laboratório de Metaloenzimas e Biomiméticos, Departamento de Química, Universidade Federal de São Carlos (UFSCar), São Paulo, Brazil
*Corresponding author: e-mail address: caterina@ufscar.br

Contents

Abstract

Enzyme immobilization is a widespread empiric technology to achieve more stable, active and reusable enzymes. The empiricism can be reduced by the application of rational design procedures employing bioinformatic tools, engineered-proteins and detailed analyses of existent data. In this chapter, we describe relevant approaches to rationalize the design of enzyme immobilization protocols, with special attention to the modulation of immobilization pH to regulate the operational stability of glutaraldehyde cross-linked enzymes and the coating of iron-containing supports to preserve the integrity of iron-sensitive enzymes. Other strategies, such as the use of factorial planning, optimization of specific enzyme orientation through protein engineering and the use of mathematical algorithms and *in silico* prediction tools are also described to reduce the classical empiricism. Finally, a public repository creation is proposed as a new promising tool to develop an improvement on future rational design procedures of enzyme immobilization.

Methods in Enzymology, Volume 630
ISSN 0076-6879
https://doi.org/10.1016/bs.mie.2019.09.003

1. Introduction

The immobilization of enzymes on solid supports is widely used in numerous applications, including biosensors, food packaging materials, and biofuel production (Homaei, Sariri, Vianello, & Stevanato, 2013; Liu, Guo, Sun, & Liu, 2013; Mohamad, Marzuki, Buang, Huyop, & Wahab, 2015). This technology is used to achieve stable, reusable and more active enzymes by the simple act of fixing an enzyme on a support surface. Stabilization is often ascribed as a result from mutual spatial fixation against autolysis, proteolysis and aggregation due to the increase in conformational rigidity (Janecˇek, 1993). The restricted degrees of freedom of an immobilized protein can also lead to a higher protection against inactivators, contributing to the obtainment of controllable industrial biocatalysts resistant to denaturation (Minteer, 2016). Therefore, this is a successful strategy to achieve high reaction yields at low costs, enabling large scale application of enzymes for chemical synthesis (Corici et al., 2016).

The availability of several immobilization methods associated to the inherent amount of variables contribute to a low predictability of the outcome of an enzyme immobilization procedure (Garcia-Galan, Berenguer-Murcia, Fernandez-Lafuente, & Rodrigues, 2011). Hence, despite of the increasing amount of reported studies on enzyme immobilization over 50 years (Fig. 1), this technique is still known as an empiric technology, in which the observed phenomena are rationalized retrospectively, based on optimization methodologies rather than predicting the results (Rosevear, 2008). The generation of this compilation of facts does not contribute to the creation of true scientific knowledge (Feynman, 1969), in which the qualification of enzyme immobilization as a science should include the ability of scientists to answer some critical questions such as: What is the influence of the support on an immobilization? What is the best immobilization method for a determined enzyme? Can we predict if the enzyme will be stabilized or destabilized by a certain immobilization method? Or how the enzyme structure is affected upon immobilization? Certainly, some of these issues were readily answered, such as the effects of support hydrophilicity (Grimaldi, Radhakrishna, Kumar, & Belfort, 2015), immobilization method (Jesionowski, Zdarta, & Krajewska, 2014), pore size (Bayne, Ulijn, & Halling, 2013), rigidification (Rodrigues, Ortiz, Berenguer-Murcia, Torres, & Fernandez-Lafuente, 2013), and enzyme crowding (Zaak et al., 2017). However, it is evident that

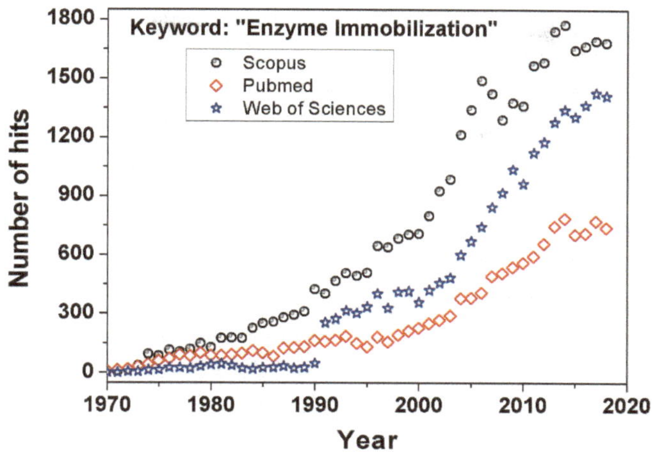

Fig. 1 Enzyme immobilization reports through the last 50 years. The initial exponential growth observed at the beginning is suffering a deceleration process and the last 5 years reach a *plateau*, most probably due to the lack of organization of the data collected on this field.

it is still necessary to improve the knowledge and control of the interactions between support and enzyme (Santos et al., 2015), since it is a common sense that the enzyme structure should not be destabilized (Zdarta, Meyer, Jesionowski, & Pinelo, 2018), but almost always is modified upon immobilization (Rodrigues et al., 2013). For instance, hydrophobic surfaces can induce the protein to loose α-helical structures and gain β-sheet ones (Moskovitz & Srebnik, 2014), whereas magnetic fields can change the tertiary structure of an enzyme (Fraga, Valerio, de Oliveira, Di Luccio, & de Oliveira, 2019) and limited availability of waters can hinder thermal denaturation (Pechkova, Sivozhelezov, & Nicolini, 2007). Hence, neglecting any conformational change suffered on the tertiary structure of the enzyme is the same as ignoring a major factor that influences enzyme activity and stability. Moreover, in order to reduce costs of immobilization protocols, there is a need to replace trial-error methodologies by the application of rational methodologies based on specific properties and special characteristics of enzymes, supports and the immobilization conditions.

The rational design procedure should include a large quantity of variables of an immobilization process related to the method, the support and the enzyme (Torres-Salas et al., 2011). For instance, the method should ascribe for immobilization time, enzyme loading, pH, ionic strength, temperature, buffer, cross-linker type and concentration, whereas, the support can impact on the immobilization by their type, size, porosity, shape, surface hydrophobicity/hydrophilicity, charge, aggregation and concentration. In addition, the enzyme is also responsible for several other variables such as its size, oligomeric state, concentration, pI, surface charges, inhibitors, active site composition, hydrophobic sites, thermal stability, pH stability and ionic strength stability. On that aspect, only few efforts have been done for the design of efficient enzyme immobilized systems, such as "freezing" an enzyme structure with activators to achieve higher activities (Rodrigues et al., 2013), interlocking enzymes on BSA-Paper support to enhance enzyme stability retaining a high activity (Riccardi, McCormick, Kasi, & Kumar, 2018) and generation of specific enzyme mutant forms to orientate the immobilization (Ryan & O'Fagain, 2007), but the most employed protocols are still based on random immobilization (Rodrigues et al., 2013). These studies indicate that the rational design of enzyme immobilization is still a real challenge.

Consequently, the lack of a systematic rationale on the changes of microenvironment, conformation and primary structure during an immobilization method should be eradicated by the use of analytical methods to detect advantages and flaws of different immobilization protocols.

In this chapter we describe approaches to a rational planning of improved procedures for biocatalysts immobilization consisting on the analysis of preexisting data, use of factorial planning, employment of bioinformatics tools, engineering a specific enzyme orientation on a support and the development of an online enzyme immobilization database.

2. Methods

Immobilization methods are optimized for a specific biocatalytic reaction, exhibiting a profound influence in the catalytic behavior of an enzyme. However, some characteristics of immobilized systems seem to be constant, indicating a higher stability or instability under certain methodologies. In this section, we describe methods attempting to decrease the empiricism of immobilization procedures.

2.1 Modulation of the immobilization pH to regulate the operational stability of glutaraldehyde cross-linked enzymes

Analysis of the immobilized enzyme contributes to the development of active and stable enzyme preparations for a given biotechnological application. In order to increase the rationalization of immobilization protocols it is important to comprehend the data obtained from the previous reports to avoid the repetition of unfavorable protocols.

The available methodologies to investigate conformational changes of an enzyme are based on conventional spectroscopic techniques such as UV–vis, FTIR, FT-Raman, circular dichroism, fluorescence, solid state nuclear magnetic resonance, neutron reflection and surface plasmon resonance (SPR) (Secundo, 2013). However, when dealing with magnetic nanoparticles as supports, the available methodologies are reduced to FTIR, FT-Raman and perhaps fluorescence and SPR, becoming a challenge the characterization of these systems. For instance, the analysis of FTIR spectra can aid in diagnosing the covalent bonds formed between an enzyme and a support, which can indicate if this system will or not be stable (Mei, Miller, Gao, & Gross, 2003).

There are several protocols to achieve covalent bonds between enzymes and supports, but glutaraldehyde cross-linking, despite of its disadvantages, such as polymerization and lack of reproducibility owing to several conjugate forms, is by far the most used method in enzyme immobilization (Zucca & Sanjust, 2014), forming distinct cross-linking bonds, such as imine, Michael-type and pyridine (Fig. 2A). Each of these bonds can affect the stability of the immobilized enzyme, with distinct roles in the reuse stability (or operational stability) (Barbosa et al., 2014). We have recently proposed that the glutaraldehyde cross-linking mode influences the operational stability of an immobilized enzyme (Modenez, Sastre, Moraes, & Marques Netto, et al., 2018) (Fig. 2B), in which basic pHs during the immobilization protocol induced the formation of Michael-type addition bonds, leading to a residual aldehyde portion in the immobilized system. This free aldehyde would react with the enzyme structure during the recycle experiments, leading to conformational changes of the enzyme and consequently to a reduced operational stability. In our case, despite of the use of a buffer solution at pH 7 for the immobilization protocol, we still have observed the formation of the aldehyde band (Fig. 2C), which indicated that other reagents were acting to increase the microenvironment pH around the nanoparticle. This hypothesis arouse due to the presence of surfactant impurities on the nanoparticle surface, which could lead to a chameleon effect (Priebe et al., 2012; Tondo et al., 2007), increasing the concentration of OH^- ions near the surface of the nanoparticle and allowing the Michael-type addition to occur. This is in agreement with findings of other groups in which non-ionic surfactants exhibited better results of catalysis of an immobilized enzyme than ionic ones (Mahmood, Ahmad, Chen, & Huizhou, 2013; Shome, Roy, &

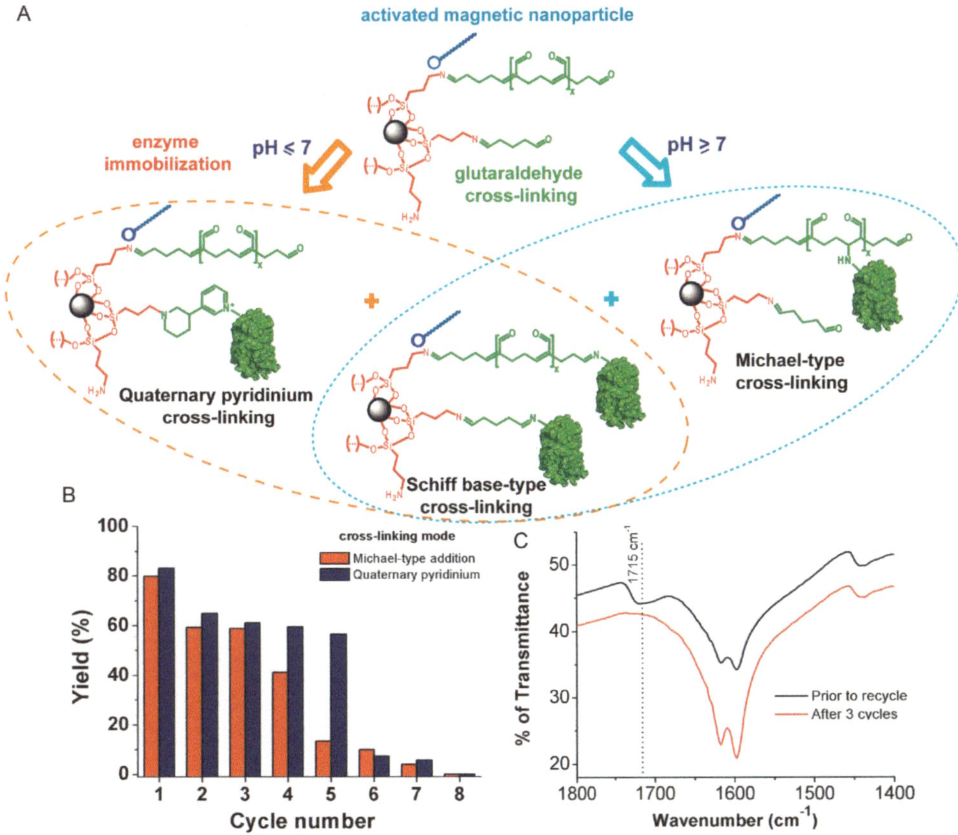

Fig. 2 Effect of pH on cross-linking modes and operational stability of immobilized enzymes using glutaraldehyde. (A) Possible cross-linking modes between an enzyme and a support when using glutaraldehyde as a cross-linking agent. (B) Recyclability of immobilized CAL-B on different supports, in which the lower operational stability is observed for the system that presents a band in 1715 cm^{-1} in the FTIR spectra. (C) FTIR spectra of immobilized CAL-B on magnetic nanoparticles via glutaraldehyde cross-linking, before the catalytic reaction and after three reaction cycles, indicating the loss of the band at 1715 cm^{-1} after a possible reaction between the aldehyde and the immobilized enzyme. *Adapted from Modenez, I. A., Sastre, D. E., Moraes, F. C., & Marques Netto, C. G. C. (2018). Influence of glutaraldehyde cross-linking modes on the recyclability of immobilized lipase B from* Candida antarctica *for trans-esterification of soy bean oil.* Molecules, *23(9), 2230–2246. doi: 10.3390/molecules23092230.*

Das, 2007) and that reverse micelle stabilized enzymes after immobilization also exhibited bands at the aldehyde region (Thudi et al., 2012). Several reports also observed lower operational stability independently of the enzyme when the aldehyde band was present after the immobilization. For example, the maintenance of aldehyde moieties in the immobilized system of Horseradish peroxidase on carbon nanotubes had a direct impact on its operational stability, resulting in loss of 50% of its initial activity during five reuse cycles (Zhai et al., 2013). A similar trend was observed for xylanase immobilized on polymethyl methacrylate nanofibers, in which the aldehyde band is still present in the FTIR and the activity of the immobilized enzyme is continuously decreasing after the first reuse, exhibiting a 20% lost of its activity on the fifth cycle (Kumar et al., 2013). Other enzymes, such as ficin, also immobilized via glutaraldehyde cross-linking, presented a residual aldehyde band after the immobilization protocol, and in this case there was an abrupt decrease in the observed activity after the second reuse cycle, going from 100% to 40% of activity. On the fifth cycle, the enzyme presents almost no activity (10%) (Rojas-Mercado, Moreno-Cortez, Lucio-Porto, & Pavon, 2018). This trend was also observed for papain immobilized on PVA membranes, exhibiting only 10% of residual activity on the fifth reuse cycle (Moreno-Cortez et al., 2015), indicating that a Michael-type addition cross-linking mode could indeed produce an enzyme immobilized system with lower operational stability.

2.1.1 Protocol

When dealing with aldehyde cross-linking agents, it is important to control the pH of immobilization and possible impurities that might alter the local pH of the support, in order to avoid the formation of Michael-type bonds, which generate unstable immobilized systems with reduced operational stability.

2.1.1.1 Equipment and reagents

General equipment

1. Safety: Goggles, laboratory coat, latex or nitrile gloves, ventilation hood.
2. Waste management: General aqueous waste receptacle.
3. Ice.
4. Orbital shaker.
5. Microtube (3 mL).
6. Volumetric micropipettes (20, 100, 200, and 1000 μL).
7. A FTIR spectrophotometer.

8. Analytical balance.
9. pH meter.
10. Vacuum pump.
11. Desiccator.
12. Ultrasound bath.
13. Micro centrifuge and/or permanent magnet.

2.1.1.2 Enzyme immobilization via glutaraldehyde method

1. Lipase B from *Candida antarctica* (powder, beige, ~9 U/mg, from Sigma-Aldrich®) (as a general rule it is possible to employ any enzyme).
2. Glutaraldehyde solution 25% (Sigma-Aldrich®).
3. Aminated support (such as APTS-modified magnetic nanoparticles).
4. Phosphate buffer (100 mM, pH 7 or lower). *Tip*: do not use pH lower than 4, since it can denaturate the enzyme.

2.1.1.3 Characterization via FTIR

1. KBr (FTIR grade, ≥99%)
2. Dried immobilized enzyme
3. Agar pestle and mortar
4. Pressure press for FTIR
5. Hydraulic press

2.1.1.4 Method of immobilization via glutaraldehyde and characterization via FTIR

(a) Make a suspension of the desired support (~10 gL^{-1}). For that, weigh 10 mg of the support into a microtube and suspend in 900 μL of 100 mM phosphate buffer (pH ≤ 7). *Tip*: if the particles are too agglomerated, use a small spatula to grind them and then place the microtube in an ultrasound bath. It is important to avoid the particles precipitation in the bottom of the microtube.

(b) Meanwhile the particles are in the ultrasound bath, weigh 1 mg of the Lipase (or the desired enzyme) and dissolve it in 100 mM phosphate buffer (pH 7) to reach 1 mL, obtaining a solution of 1 gL^{-1} concentration. Keep this solution in ice bath.

(c) To the support suspension, add 10 μL of glutaraldehyde (25%) followed by the addition of 100 μL of the lipase (or desired enzyme) solution (1 gL^{-1}).

(d) Incubate for 10 min at 0 °C (on ice bath) and 180 rpm of agitation in an orbital shaker. *Tip*: For a better agitation of the microtube in the orbital shaker, place it in the horizontal position on the ice bath.

(e) Remove the supernatant with a pipette and wash the immobilized system with 500 μL of the buffer solution ($pH \leq 7$) three times. If using magnetic nanoparticles, an external permanent magnet is required to confine them. If support is not magnetic, centrifugation is needed.

(f) To dry the immobilized enzyme, place the microtube with the lid open (covered with a needle-perforated parafilm) in a desiccator attached to a vacuum pump. Carefully open the desiccator to the vacuum. Once in full vacuum, let it dry for at least 2 h at room temperature. *Tip*: To put the desiccator under vacuum without disturbing the suspension of immobilized enzyme, do it in four steps, waiting 10 min before increasing the vacuum opening.

(g) Make a KBr pellet grinding 100 mg of KBr and 1 mg of the immobilized enzyme. Pour the solids into a pellet press and apply pressure using a hydraulic press. Pull the pressure level until tight. Keep the sample under pressure for 10 min. Depressurize the system and then remove the KBr pellet from the pellet press. *Tip*: If possible, use vacuum during the pressurization step, to remove residual solvents.

(h) Insert the KBr pellet in a solid FTIR sampler and analyze it in a FTIR spectrophotometer, using 50 scans at $2\,cm^{-1}$ of resolution. *Tip*: To enhance the resolution of the spectrum in this range purge the FTIR chamber with N_2 gas.

(i) Analyze the FTIR spectrum, looking for a band or a shoulder in the region of $1715\,cm^{-1}$. If this band is present, decrease the pH of the buffer solution of immobilization and re-start the protocol following steps (a)–(i) again.

2.2 Selection of stabilizing supports for iron-sensitive enzymes

Catalysis occurs in the active site of an enzyme, and depending on the reactivity of the aminoacid residues, the immobilization might affect negatively on the enzyme's catalytic activity (Aissaoui, Landoulsi, Bergaoui, Boujday, & Lambert, 2013). Therefore, it is important to understand which support is better for a determined active site composition. For example, we have observed that enzymes bearing a cysteine residue in their active site, such as oxidoreductases, isomerases, transferases, some hydrolases and phosphatases, would be less stable in presence of Fe^{3+} (Marques Netto,

Andrade, & Toma, 2018), which is typically present in magnetic nanoparticles. Here we will use the yeast alcohol dehydrogenase I from *Saccharomyces cerevisiae* (YADH) as an example, owing to the presence of two cysteines (C43, C153) and one histidine (H66) coordinating a zinc atom in its active site. The oxidation of the cysteine residues (−270 to −125 mV) by Fe^{3+} (771 mV) would induce the metal release from the active site, as shown in Fig. 3, inhibiting its redox activity. Therefore, the use of magnetite nanoparticles as supports requires a non-oxidative coating, in which silica shells were shown to protect the YADH active site from Fe^{3+} oxidation, achieving a 2.4-fold higher operational stability (in 10 successive cycles) than the uncoated magnetite nanoparticles (Marques Netto, Andrade, Freitas, & Toma, 2015). Other authors also observed that silica coating was successful to stabilize this enzyme, retaining at least 80% of activity for six consecutive cycles (Jiang et al., 2016). Other materials for the immobilization of YADH, such as TiO_2 (Ghannadi, Abdizadeh, Miroliaei, & Saboury, 2019) and magnetic graphene oxide nanocomposites exhibited better storage stability and reusability than Fe_3O_4 nanoparticles. In the case of the graphene nanocomposites, the YADH immobilized enzyme retained 80% of its initial activity, whereas YADH immobilized on Fe_3O_4

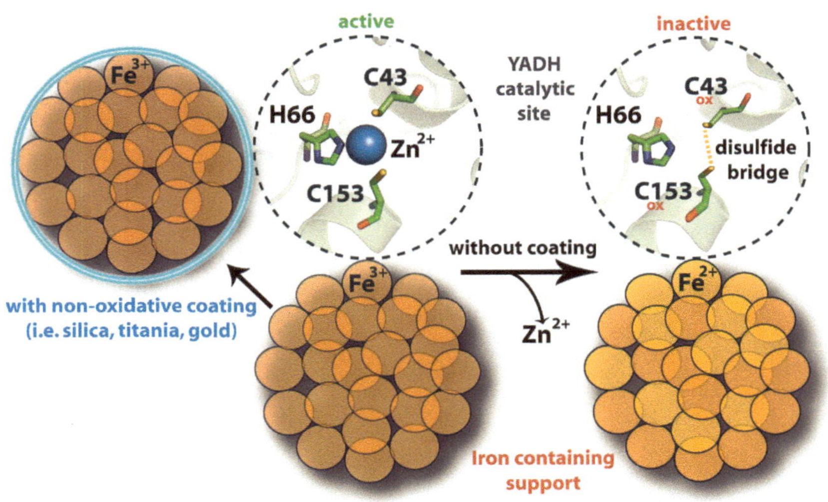

Fig. 3 Representation of the possible oxidation of the cysteine residues present in YADH active site by the Fe^{3+} of the magnetic nanoparticles. The cysteine oxidation is expected to form a disulfide bridge with consequent release of the Zn^{2+} from the active site, inactivating this enzyme. The non-oxidative coating with silica, titania or gold prevents this reaction to occur, maintaining the immobilized YADH active.

nanoparticles presented only 20% of residual activity after 10 cycles (Liu, Yu, & Chen, 2015). Maghemite nanoparticles used as supports for ADH were also shown to not stabilize this enzyme, since a constant decrease of conversion was observed for the immobilized system under continuous operation (Šalić, Pindrić, Podrepšek, Leitgeb, & Zelić, 2013). On the other hand, gold core-shell magnetic nanoparticles were shown to stabilize immobilized YADH used as ethanol biosensor, retaining 90% of the initial current response for ethanol, after repeated use for over 2 weeks (Samphao et al., 2015).

The non-oxidative coating might also be favorable for enzymes that are somehow inhibited by iron, such as D-xylose isomerases (Häusler, Weber, & Stütz, 2006), since iron lixiviation occur at acid pHs during the immobilization, inhibiting its activity. For instance, when we immobilized D-xylose isomerase from *Coraliomargarita akajimensis* on uncoated magnetite nanoparticles at pHs lower than 7, we have seen its full catalytic activity inhibition, suggesting that both non-oxidative coating or immobilization at basic pHs would be essential to avoid the inhibition by iron.

2.2.1 Protocol

In summary, the analysis of the active site of the proposed enzyme to be immobilized is essential for a proper selection of support, in order to achieve high operational activities of these immobilized systems.

(a) Analyze the composition of the active site of the desired enzyme to be immobilized using the molecular viewer UCSF Chimera (https://www.rbvi.ucsf.edu/chimera/) (Huang, Meng, Morris, Pettersen, & Ferrin, 2014) or PyMOL (DeLano Scientific LLC) (http://www.pymol.org/), looking for cysteine residues into the active site or near it. Verify if the enzyme is sensitive to iron, to avoid its inhibition by iron-containing supports.

(b) If any cysteine residue is present in the enzyme active site or if the enzyme is iron-sensitive, choose non-oxidizing supports or immobilize your enzyme using supports containing non-oxidizing shells, such as silica, gold and titania.

For instance, for a silica shell, make a suspension of magnetic nanoparticles in water (2.8 mg/mL) and adjust the pH to 10 with NH_4OH. Add a 1% v/v ethanolic solution of tetraethyl orthosilicate (TEOS) under stirring. After 1 h of reaction, magnetically confine these particles and wash them with dry methanol.

(c) Immobilize the enzyme by the chosen method.

2.3 Factorial planning applied to enzyme immobilization procedures

The optimization of the parameters in the immobilization process is fundamental to obtain the best performance of a biocatalyst (Sheldon, 2007). The conventional method for optimization evaluate the independent response of each variable by one-factor-at-a-time and in spite of directing to the optimized condition, can often be misleading because this approach does not consider the possible interactions between the variables (Siffer, Ponche, Fioux, Schultz, & Roucoules, 2005). Therefore, the factorial planning methodology provides more reliable process information, reducing the empiricism present in trial-error techniques. This methodology allows the development of an efficient and rational immobilization protocol, since it considers and understands the relationship between all variables. Factorial planning has been applied in diverse research fields (Mandenius & Brundin, 2008), with the main intention of minimizing the number of experiments while studying the interaction effect of two or more variables, reducing the time spent in optimization process (Lavine & Workman, 2013). In this sense, the use of simple chemometric methods becomes relevant in the optimization of enzymatic immobilization (Brereton, 2003).

Usually, the use of the factorial design in enzymatic immobilization is performed using full or fractional factorial design (Uliana, Riccardi, Tognolli, & Yamanaka, 2008). The full factorial design is preferentially selected when there is a limited number of variables, otherwise, the number of experiments become impracticable. In addition, in this design it is possible to study every possible combination between factors, usually applying it to two or three variables. For instance, the applicability of the full factorial for the immobilization of the enzyme cyclodextrin glycosyl transferase was verified by applying this tool for qualitative factors (support and method of immobilization) and a quantitative factor (temperature) (Sobral et al., 2003). Curiously, they were able to select the best method and the most efficient support, even when dealing with qualitative factors, which cannot be fully exploited.

On the other hand, four parameters from an immobilization procedure: effect of pH, spacer, glutaraldehyde and enzyme concentration in the immobilization of the laccase on nylon fabric was also studied (Silva, Silva, Zille, Guebitz, & Cavaco-Paulo, 2007). In contrast to Sobral et al., the factors were analyzed for three different responses: half-life time, protein retention and immobilization yield, determining the most significant factor for each response. A simpler approach was used by evaluating only two quantitative factors: cross-linking and protein concentration (González Siso et al., 1997). In this study, the optimum parameters were investigated for the immobilization of α-amylase and invertase. In both cases the enzymatic activity increased when the cross-linking agent concentration decreased and porosity of chitosan microbeads increased, facilitating diffusion of substrate and products.

The use of the fractional factorial design is an alternative to the full factorial design to evaluate a higher amount of variables. In this type of design, information about the best conditions for immobilization is extracted after analysis of the principal combination between the variables (Bruns, Scarminio, & Neto, 2006). Typically, to visualize the best combination of factors and find the values that will produce the desired response, it is necessary to use a multivariate statistic technique (Teófilo & Ferreira, 2006). In enzyme immobilization, most researchers use the response surface methodology (RSM), that is a collection of mathematical and statistical techniques based on the fit of a polynomial equation to the experimental data, which must describe the behavior of a data set with the objective of making statistical previsions (Bezerra, Santelli, Oliveira, Villar, & Escaleira, 2008). Recently, a three-level factorial design was used, in which the central point consisted in the mean value of the levels of all variables (Jun et al., 2019). The variables chosen to optimize the immobilization process of the Jicama peroxidase enzyme on bulky paper/polyvinyl alcohol were: pH, initial loading of the enzyme, and immobilization time. Using RSM, they have observed that the increase of pH had a notable impact on the enzyme immobilization efficiency, but a further increase would lead to a decrease in the immobilization efficiency. Additionally, they revealed that the low initial enzyme loading (0.1 U/mL) was sufficient to cover all pore sites on the surface of the BP/PVA membrane. In conclusion, RSM was demonstrated to be a very useful tool to optimize the operating parameters of immobilization efficiency of this particular enzyme. These examples indicate the importance of factorial planning in enzyme immobilization protocols to decrease costs and labor time.

2.3.1 Protocol

There are several possible designs to investigate the factors that can influence an immobilization protocol, but factorial planning starts basically from the same principle, in which, the factors to be evaluated must be initially chosen.

(a) After selecting the most important factors, it is recommended to select the type of factorial design (full or fractional), accordingly with the number of variables.
The next step consists of choosing the high and low level for each factor and codify them as −1 and +1 (or − and +), for low and high level, respectively, to use a standard design. The design matrices that are symmetric around 0 are almost always easier to computationally handle (Bruns et al., 2006). Also, it is possible to estimate a reasonable model and verify if there is a lack of adjustment (Teófilo & Ferreira, 2006). One example of a matrix is shown in Table 1 for a full factorial design considering three enzyme immobilization factors (temperature, pH and time of immobilization). Note that every column contains exactly four high (−1) and four low (+1) levels and, excluding the first column, there are equal numbers of experiments for the others columns at each combination of low (−1) and high (+1) levels.

(b) Perform the planned experiments and monitor the selected response suitable for the chosen system of factors.

(c) Calculate the coefficients of variable interactions using statistical softwares and define the most significant factor, considering that the minimal level of significance should be at least 10% ($P < 0.1$) (Rodrigues & Iemma, 2014). Several softwares are currently available such as MATLAB®, Octave, PTC MathCAD and Microsoft Excel for the analyses of chemometric data (Lavine & Workman, 2010).

Table 1 Three factor, two level full factorial design (2^3).

Experiment number	Factor 1: Temperature(°C)	Factor 2: pH	Factor 3: Time of immobilization(min)
1	−1	−1	−1
2	−1	−1	+1
3	−1	+1	−1
4	−1	+1	+1
5	+1	−1	−1
6	+1	−1	+1
7	+1	+1	−1
8	+1	+1	+1

(d) Analyze the variance (ANOVA) to verify the model quality and, calculate the optimal conditions using the adjusted model (Pereira & Pereira-Filho, 2018)

2.4 Engineering enzymes to a rational design of an oriented immobilization method

Protein engineering is used to design rational experimental approaches of enzyme immobilization systems and to improve its efficiency (Antikainen & Martin, 2005). The modulation of biochemical and physicochemical properties of engineered and/or immobilized enzymes solved problems of stability, selectivity, and substrate or solvent tolerance of enzymes for biotechnological processes (Datta, Christena, & Rajaram, 2013; Rueda et al., 2016). Targeted random mutagenesis, site-directed mutagenesis and in vitro recombination are the most employed methods to successfully redesign existing enzymes on the level of their primary structure (Arnold, 2001; Nannemann, Birmingham, Scism, & Bachmann, 2011; Poluri & Gulati, 2017). Although there is no universal method to modulate a particular property of an enzyme, the combination of protein engineering techniques and properly designed immobilization of enzymes seems to be a very powerful strategy to improve the efficiency and stability of enzymes for industrial processes (Singh, Tiwari, Singh, & Lee, 2013). The drawback of site-specific protein engineering is the limitation of available sites on the protein surface and the possibility of affecting enzyme structure and function. Hence, the use of structural information or homology modeling can improve the design of mutants with enhanced catalytic activity to tune the interaction with the support surface, by means of increasing immobilization yields (Arnold, 2001).

By genetic engineering it is possible to introduce linkers, tags and/or to perform specific conservative or radical aminoacid replacements to enable the protein to specifically attach on a nanomaterial surface by affinity bonds (Guzik, Hupert-Kocurek, & Wojcieszynska, 2014; Ryan & O'Fagain, 2007). The use of tag sequences could improve immobilization rates and protein orientation on the support, allowing a homogeneously distributed and well-defined position on the nanomaterial (Putzbach & Ronkainen, 2013). The site-specific enzyme immobilization is in contrast to multipoint immobilization protocols, since it can dictate the distance between the active site of an enzyme and the surface of the nanoparticle. However, even though the activity could be higher on genetically engineered-enzymes due to the specific orientation of the immobilization, this strategy has not been so well explored since it can affect protein stability and activity before and after immobilization (Hitaishi et al., 2018; Singh et al., 2013).

Polyhistidine tags, for instance, are most often employed at the N-ter or C-terminus of the protein, allowing it to chelate metals like Ni^{2+} or Co^{2+} and favor the electrostatic interactions toward hydrophilic surfaces (Arnold, 1991). However, other parts of the enzyme could also be targeted for enzyme engineering in order to enable the correlation between the position of immobilization to the enzyme activity and stability. Peptide linkers are much less explored in enzyme immobilization than chemical linkers which are frequently reported to improve enzyme flexibility and activity. Recently, flexible peptide-linker-assisted noncovalent immobilization of the bacteriolytic enzyme lysostaphin (Lst) was reported to generate anti-*Staphylococcus aureus* surfaces via silica-binding peptides (SiBPs) onto silica surface and His-tag fused at the C-terminus of Lst to immobilize it on nickel nitrilotriacetic acid (Ni-NTA) agarose beads (Wu, Fraser, Zha, & Dordick, 2018). Interestingly, the immobilization of lysostaphin without a peptide linker was less active and stable. Molecular modeling revealed that the presence of the peptide linkers enhanced the molecular flexibility of the proximal Lst binding domain, which correlated with an increased antimicrobial activity (Wu et al., 2018).

On the other hand, cellobiose dehydrogenase from *Myriococcus thermophilum*, which lacks cysteine residues on its surface, has been shown to be a good candidate for site-directed mutagenesis. Hence, cysteine residues were introduced at specific surface locations in order to orientate the immobilization of this enzyme via thiol-ene click chemistry between the cysteine moieties and vinyl groups grafted on a support (Al-Lolage, Meneghello, Ma, Ludwig, & Bartlett, 2017). This approach enabled a specific orientation of the active site upon immobilization, in which they reported an increase in enzyme electrocatalytic activity for site-directed covalent immobilization compared to physical adsorption, indicating a higher enzyme loading on the electrode by using the specific orientation design method. Cysteine residues were also essential to orientate the immobilization of formate dehydrogenase (FDH) from *Lodderomyces elongisporus* NRRL YB-4239 on polydopamine-coated iron oxide nanoparticles (Gao, Ni, Zhao, Ren, & Wei, 2014). The FDH enzyme was immobilized through site-directed mutagenesis of three specific cysteine residues in close proximity to the active site (C242A, C275V, C363V), and the introduction of a cysteine at the C-terminus (K389C) (mutant C-c). Mutant C-c exhibited a threefold higher activity than the FDH wild-type and it was immobilized with 70% yield, whereas the single, double and triple mutants presented lower activities than mutant C-c and lower immobilization yields (50–65%) (Gao et al., 2014). Moreover, Liu and coworkers achieved the specific immobilization of 6-phospho-β-galactosidase (β-Gal) through a unique cysteinyl residue, on a self-assembled monolayer (SAM) containing maleimide end groups and oligo(ethylene glycol) spacer segments (Liu, Guo, et al., 2013; Liu, Ogorzalek, et al., 2013). Interestingly, they performed a biophysical characterization of the interfacial orientation of immobilized enzymes by using a combination of sum frequency generation (SFG) vibrational spectroscopy and attenuated total reflectance (ATR) Fourier transform infrared spectroscopy (FTIR)-spectroscopy. As expected, the oriented immobilized β-Gal was not significantly changed in structure and activity, since it was similar to the free-enzyme in solution. In contrast, the immobilization of β-Gal by hydrophobic bonds onto SAMs appeared to partially denature the enzyme, exhibiting a significantly reduced activity, indicating that the rational design enzyme orientation constituted an improvement to the enzyme immobilization procedure. Remarkably, The ATR-FTIR and SFG vibrational spectroscopies were also successfully employed to study (evaluate) the molecular orientation of some immobilized peptides and proteins on different solid supports (Ding & Chen, 2012; Wang, Paszti, Clarke, Chen, & Chen, 2007; Yang, Boughton, Homan, & Chen, 2013).

A high specificity is desirable to rationally design a defined orientation with high uniformity of the enzyme immobilization. In that aspect, the high affinity and specificity of biotin toward (strept)avidin can be a powerful way to achieve specific orientation of protein immobilization (Dundas, Demonte, & Park, 2013). Particularly, the enzymatic biotinylation process is a fast and specific procedure to attach biotin to one specific lysine residue within a certain protein sequence (Fairhead & Howarth, 2015). The biotinylation performed by an ATP-dependent biotin ligase from *Escherichia coli* (BirA) in vivo or in vitro, is highly specific in covalently attaching biotin to a 15 amino acid peptide termed AviTag or acceptor peptide, to generate a homogeneously biotinylated protein (Fairhead & Howarth, 2015). AviTag serves as a recognition site for BirA and can conveniently be added genetically at the N or C-terminus or in exposed internal loop of a protein of interest. BirA-biotinylated enzymes have been applied in a wide range of areas of biochemistry and biotechnology, including for the immobilization of enzymes in quantum dots and magnetic nanoparticles (Jain & Cheng, 2017). A summary of these approaches is illustrated in Fig. 4.

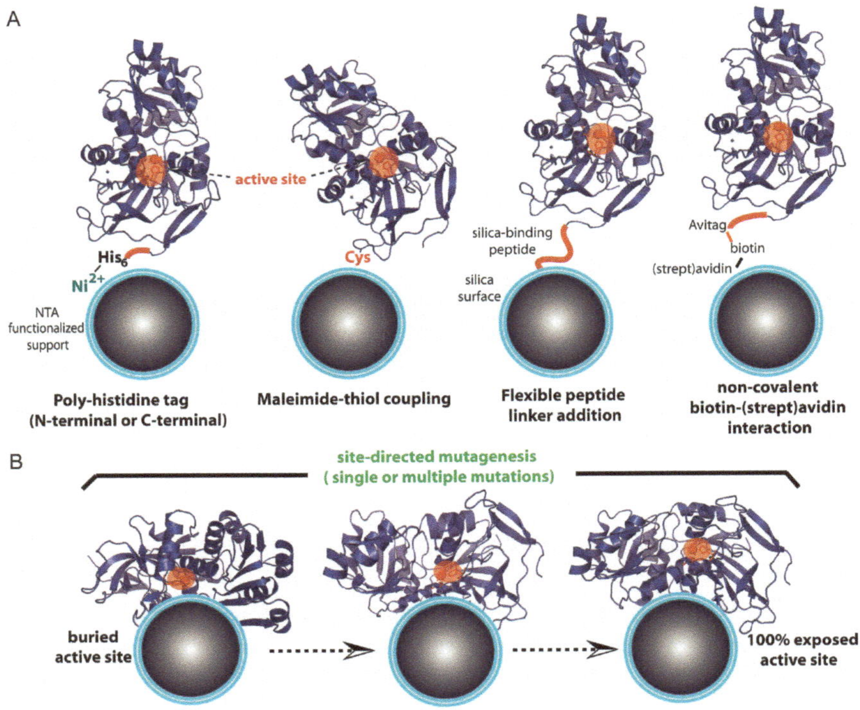

Fig. 4 Strategies to obtain specific orientations of immobilized enzymes. (A) Engineered-enzymes could be immobilized through diverse and specific peptide tags or linkers, or though noncovalent biotin-streptavidin interaction or selected cysteine residues to sulfhydryl-reactive cross-linkers like maleimides. The active site is colored in red. (B) Application of single or multiple site-directed mutagenesis can be useful to improve the active site exposition and to obtain more active immobilized enzymes.

2.4.1 Guidelines

In summary, a rationally designed specific orientation of immobilized enzymes attempt to generate enzymes with improved properties and activities based on in-depth knowledge of structural biology and enzymatic catalysis characteristics.

This strategy requires the following steps:

(a) Bioinformatics analysis and molecular modeling of enzymes to be immobilized. PDB file from PDB database or automatic modeling of protein of interest using SwissModel (Arnold, Bordoli, Kopp, & Schwede, 2006) or Protein Homology/analogY Recognition Engine V 2.0 (Phyre2) online servers (Kelley, Mezulis, Yates, Wass, & Sternberg, 2015).

(b) Molecular cloning of the gene encoding the protein of interest to be immobilized. A synthetic gene codon-optimized for expression in *E. coli* encoding the enzyme of interest can be obtained commercially or amplified by PCR employing specific oligonucleotide primers using genomic DNA (prokaryotic genes, or eukaryotic genes with no introns) or full-length cDNA (eukaryotes) as template. Subcloning into the expression vector containing an N or C-terminal His-tag, or another selected specific tag, peptide or linker (like silica-binding peptide, Avitag, GST, etc.)

(c) Site-directed mutagenesis of genes encoded enzymes to be immobilized. It is also possible to perform deletions and insertions of protein fragments (loops, domains) to regulate enzyme activity and/or design the orientation on the support to be immobilized. The aminoacid replacement can be made to introduce a specific residue in a particular region of the protein of interest by using a modified Quick Change™ Site-Directed Mutagenesis System developed by Stratagene (La Jolla, CA) protocol (Liu & Naismith, 2008).

(d) Expression vectors containing the gene of interest can be transformed into *E. coli* BL21 cells. Heterologous overexpression and purification of the chosen enzyme can be performed by standard methods (Structural Genomics Consortium et al., 2008).

(e) Enzyme immobilization protocol via linkers, His-tag, maleimide-cysteine coupling, biotin-(strept)avidin, etc., onto a solid support.

(f) Evaluation of the orientation by ATR-FTIR and SFG vibrational spectroscopies and determination of the activity and stability of immobilized engineered-enzymes through specific catalytic reaction protocols.

2.5 Enzyme immobilization *in silico* prediction

With the growing attention paid to the use of *in silico* technologies in various fields of sciences, bioinformatics models can be designed to validate the probability of success and the efficiency of the immobilization process before starting the immobilization. Hence, generation of bioinformatics tools trying to predict an immobilized system performance selecting the best conditions for the synthesis are useful for the optimization of a rational design immobilization process and complement experimental screening (del Monte-Martínez et al., 2013; Torres-Salas et al., 2011). For this purpose, was created the computer software named *RDID1.0* (©2010 Enzyme

Technology Group, Center for Protein Studies, Faculty of Biology, University of Havana, Cuba). This software is able to calculate though mathematical algorithms the maximum protein quantities for a specific support, and also predict the most probable configurations of the protein-support system after the immobilization (Del Monte-Martinez, Cutino-Avila, & Gonzalez-Bacerio, 2018).

The variety of functionally relevant conformations present in proteins native ensemble (Narayan & Naganathan, 2017) is also taken in consideration by the RDID 1.0 software. The distinct conformations with subtle differences among each other are not observed by crystallography and cryo-electron microscopy structures, since these techniques are snapshots of trapped conformations along the reaction coordinate pathway and do not give details of the dynamic system (Pietrantonio, Pandey, Gould, Hasabnis, & Prosser, 2019). Basically, these enzyme conformations are distinguished by side chain configurations and the positioning of hydrogen-bonded networks that can impact the enzyme activity in immobilization protocols (Zhang et al., 2015). This perspective is profoundly important to understand enzyme function in solution and in the immobilized form, since the conformational state of the immobilized protein is probably a mixture of stabilized and destabilized complexes coexisting (Denisov, 1992) and the overall observed activity of an immobilized enzyme is governed by several distinct enzymatic conformations. Hence, this software is able to predict the cluster of probable states of enzyme conformation after an immobilization process, which could had been a useful tool to predict the multiple immobilized ensemble of glutamate-dehydrogenase enzyme experimentally observed by our group (Marques Netto et al., 2016).

Moreover, the variation of pH in an immobilization protocol can induce preferential conformations of the enzymes, increasing its stability. Therefore, the selection of the optimum immobilization protocol has to notice that the most probable configuration must be catalytically competent at the pH of immobilization. In that respect, the software (RDID 1.0) was able to predict that 80% to 90% of the population of the lipase enzyme from *C. rugosa* was immobilized in a competent catalytic conformation on Eupergit C and Glyolxyl-Sepharose CL 4B supports, respectively. However, a very low activity was observed for the immobilized biocatalysts, due to the low stability of this enzyme at $pH > 7$ (Del Monte-Martinez et al., 2018). Hence, it is expected that if a lower pH is used, all population of the immobilized enzymes would be catalytically competent and stable. This novel methodology was successfully validated comparing the predicted values of the immobilization parameters with the ones determined experimentally for at least 15 proteins in different laboratories (del Monte-Martinez, Cutiño-Avila, Gonzalez-Bacerio, Planes, & Brito, 2014). It should be notice, that these predictions are based on the geometry of the coordinate .pdb file of each protein, obtained in particular conditions, which are different from the aqueous conditions and can sometimes not correspond with the reality.

2.5.1 Guidelines

The integration of the structural information on the protein and the support can be used to predict the interaction between both and calculate the optimal conditions of immobilization. The ***RDID1.0*** **software** is available under request to adelmonte@fbio.uh.cu and a detailed description of the software parameters is described elsewhere (Del Monte-Martinez et al., 2018).

This computer program can be used to obtain information such as:

(a) estimative of the protein diameter,
(b) maximum load of proteins per unit of support,
(c) most probable configurations of the protein-support system,
(d) catalytic competition and residual activity of the immobilized enzyme and
(e) optimal pH of immobilization.

3. Proposition of an enzyme immobilization database

In the last 12 years were created around six databases for nanomaterials (Tropsha, Mills, & Hickey, 2017). Most of these databases are curated by bioinformatics and shares the main nanomaterials available; however, the current nanomaterials databases are small and lack standards in procedures and characterizations (Tropsha et al., 2017). Besides, these databases are restrict to specific nanomaterials and do not have any information regarding their association with biocatalysts.

On the other hand, the main collection of enzyme functional data is freely available online to the scientific community (www.brenda-enzymes.org) or as an in-house database for commercial users by requests to the distributor geneXplain. The BRENDA enzyme database system founded in 1987 (Schomburg et al., 2017) is currently curated and hosted at the Institute of Biochemistry and Bioinformatics at the Technical University of Braunschweig in Germany. In this database, the information on enzyme function are extracted directly from the primary literature and critically evaluated and checked manually by qualified scientists holding a degree in Biology or Chemistry. It is clear that the data of enzymes studies are inherently very difficult to collect, interpret and standardize as they are highly distributed among journals from different fields and are subject to particular experimental conditions. The enzymatic data, including about of 6500 enzymes and only few immobilized systems reports, is being developed into a metabolic network information system with links to enzyme expression and regulation information. Based on this situation, we notice the lack of a specific scientific database for the field of Enzyme Immobilization. Therefore, we propose the development of such database as a strategy to solve the problems of empiricism, reproducibility and also to identify duplicative materials in the enzyme immobilization experimental research.

An excellent model for experimental data sharing is the Protein Data Bank (PDB), which supports basic and applied research in structural and functional proteomics and it has been created and nurtured by the structural biology community (Rose et al., 2017). This robust database managed by the Research Collaboratory for Structural Bioinformatics (RCSB) has been functioning for decades as a primary depository of protein structure determination by X-ray crystallography or NMR macromolecular structural data. Other good examples of peer-reviewed public repository are PubChem supporting research in chemical biology and drug discovery (Kim et al., 2019), NCBI Genome Database, supporting genome sequences from all three domains of life, and the microarray databases Gene Expression Omnibus (GEO) from NCBI (Clough & Barrett, 2016) or ArrayExpress from EBI, (Parkinson et al., 2007) both contain microarray gene expression data. These databases adhere to academic or industry standards and provide streamlined data-deposition capabilities and assist the development of instructional data models to guide focused experimental research. These examples show the feasibility of implementing informatics-driven strategies for prospective data collection in the enzyme immobilization field.

An organized collection of data could be accessed for scientific inquiry and long-term stewardship and should represent an important tool for classification and comparison purposes leading to an improvement in data quality, helping researchers to organize and share information. Accuracy and validation after data-deposition process will be important to ensure data quality reports. It is expected that the development of such scientific database will possibly help to choose the better immobilization protocol, support, cross-linker, etc., for a specific enzyme.

The Enzyme Immobilization database should aim at the load of details of each immobilization protocol reported in the literature, describing all of the important variables of each enzyme immobilization protocol. Hence, scientific journals should require an accession number of the data-deposition of complete enzyme immobilization procedures and results into the freely accessible Enzyme Immobilization database as a prerequisite for accepting a peer-reviewed manuscript for publication. As an example, we believe that the information needed to the Enzyme Immobilization online Database to build the effective data-submission should include:

1. Method of enzyme immobilization (adsorption, entrapment or covalent attachment)
   1.1 pH, cross-linker, reaction time, temperature
2. Support (type, size, composition, porosity, morphology, etc.)
3. Enzyme (type, size, characteristics)
4. Enzyme loading (in μg/mg)
5. Conformation of the Enzyme (native, active form, rigid, compact, crystal, etc.)
6. Enzyme activity (relative to the free enzyme in the same conditions)
7. Leaching or desorption of enzymes (desorption occur or no desorption)
8. Applications of the immobilized enzyme
9. Operational stability (number of reuses in specific conditions)

4. Concluding remarks

The immobilization of enzymes is still an empiric technology, generating loads of data that are hard to be compared and used for further experiments. In addition, changes in the enzyme structure upon immobilization are neglected, which could indicate whether a specific methodology would be favorable or not. We have described some strategies that could be employed to guide researches in immobilization of enzymes to achieve more stable and active biocatalysts. We highlight that glutaraldehyde cross-linking methods can generate different bonds between enzyme and support in a pH-dependent manner, influencing the enzymatic operational stability. For this reason, we indicate that FTIR is a powerful tool to analyze the immobilized enzyme stability. Besides, we described how to control the stability of iron-sensitive immobilized enzymes through the coating of oxidizing supports. The use of factorial planning was described as a tool to reduce experimental time and to increase the rationalization of the optimization of enzyme immobilization protocols. In addition, we discussed the importance of engineering enzymes to obtain a specific and homogeneous orientation on supports to expose the active site and achieve an increase of enzyme activity. Moreover, we rely on the experiment-driven optimization of immobilization protocols could also be complemented by *in silico* predictions, combining the properties of enzymes, supports and immobilization methods. Finally, we think that the creation of a database of Enzyme Immobilization systems to share and store the most important findings of this field will be important to reduce the empiricism of enzyme immobilization procedures, increasing the data quality and reducing the duplicative materials in the literature.

Acknowledgments

This work was supported by the Universidade Federal de São Carlos (UFSCar), São Paulo, Brazil. E.A.R. acknowledges a master fellowship from CAPES (88882.332744/2019-01). We also acknowledge the support from CAPES (project 001).

The authors draw attention to the current Brazilian Science funding crisis which endangers the continuity of science in the country (please see: Andrade, 2019).

Conflict of interest

The authors declare no competing financial interests.

References

Aissaoui, N., Landoulsi, J., Bergaoui, L., Boujday, S., & Lambert, J. F. (2013). Catalytic activity and thermostability of enzymes immobilized on silanized surface: Influence of the crosslinking agent. *Enzyme and Microbial Technology*, *52*(6–7), 336–343. https://doi.org/10.1016/j.enzmictec.2013.02.018.

Al-Lolage, F. A., Meneghello, M., Ma, S., Ludwig, R., & Bartlett, P. N. (2017). A flexible method for the stable, covalent immobilization of enzymes at electrode surfaces. *ChemElectroChem*, *4*(6), 1528–1534. https://doi.org/10.1002/celc.201700135.

Andrade, R. O. (2019). Brazil's budget cuts threaten more than 80,000 science scholarships. *Nature*, *572*, 575–576. https://doi.org/10.1038/d41586-019-02484-w.

Antikainen, N. M., & Martin, S. F. (2005). Altering protein specificity: Techniques and applications. *Bioorganic & Medicinal Chemistry*, *13*(8), 2701–2716. https://doi.org/10.1016/j.bmc.2005.01.059.

Arnold, F. H. (1991). Metal-affinity separations: A new dimension in protein processing. *Biotechnology (N Y)*, *9*(2), 151–156. https://doi.org/10.1038/nbt0291-151.

Arnold, F. H. (2001). Combinatorial and computational challenges for biocatalyst design. *Nature*, *409*(6817), 253–257. https://doi.org/10.1038/35051731.

Arnold, K., Bordoli, L., Kopp, J., & Schwede, T. (2006). The SWISS-MODEL workspace: A web-based environment for protein structure homology modelling. *Bioinformatics*, *22*(2), 195–201. https://doi.org/10.1093/bioinformatics/bti770.

Barbosa, O., Ortiz, C., Berenguer-Murcia, A., Torres, R., Rodrigues, R. C., & Fernandez-Lafuente, R. (2014). Glutaraldehyde in bio-catalysts design: A useful crosslinker and a versatile tool in enzyme immobilization. *RSC Advances*, *4*, 1583–1600. https://doi.org/10.1039/C3RA45991H.

Bayne, L., Ulijn, R. V., & Halling, P. J. (2013). Effect of pore size on the performance of immobilised enzymes. *Chemical Society Reviews*, *42*(23), 9000–9010. https://doi.org/10.1039/c3cs60270b.

Bezerra, M. A., Santelli, R. E., Oliveira, E. P., Villar, L. S., & Escaleira, L. A. (2008). Response surface methodology (RSM) as a tool for optimization in analytical chemistry. *Talanta*, *76*(5), 965–977. https://doi.org/10.1016/j.talanta.2008.05.019.

Brereton, R. G. (2003). *Chemometrics: Data analysis for the laboratory and chemical plant*. West Sussex, England: John Wiley & Sons, Ltd.

Bruns, R., Scarminio, I., & Neto, B. B. (2006). Statistical design—Chemometrics. *Vol. 25* (1st ed.). Amsterdam: Elsevier.

Clough, E., & Barrett, T. (2016). The gene expression omnibus database. *Methods in Molecular Biology*, *1418*, 93–110. https://doi.org/10.1007/978-1-4939-3578-9_5.

Corici, L., Ferrario, V., Pellis, A., Ebert, C., Lotteria, S., Cantone, S., et al. (2016). Large scale applications of immobilized enzymes call for sustainable and inexpensive solutions: Rice husks as renewable alternatives to fossil-based organic resins. *RSC Advances*, *6*(68), 63256–63270. https://doi.org/10.1039/c6ra12065b.

Datta, S., Christena, L. R., & Rajaram, Y. R. (2013). Enzyme immobilization: An overview on techniques and support materials. *3 Biotech*, *3*(1), 1–9. https://doi.org/10.1007/s13205-012-0071-7.

del Monte-Martínez, A., Cutiño-Avila, B., Gómez, D., Pereda, I., Díaz, J., & Rojas, J. (2013). Computational mathematic model for the immobilization of cells and proteins on charged solid surfaces by electrostatic interactions. In *Paper presented at the V Latin American congress on biomedical engineering CLAIB 2011 May 16–21, Habana, Cuba*.

Del Monte-Martinez, A., Cutino-Avila, B. V., & Gonzalez-Bacerio, J. (2018). Rational design strategy as a novel immobilization methodology applied to lipases and phospholipases. *Methods in Molecular Biology*, *1835*, 243–283. https://doi.org/10.1007/978-1-4939-8672-9_14.

del Monte-Martinez, A., Cutiño-Avila, B., Gonzalez-Bacerio, J., Planes, M. A. C., & Brito, J. D. (2014). Diseño racional de la inmovilización de proteínas: Aplicaciones en cromatrografía de afinidad y bioconversión enzimática. *Revista Anales de la Academia de Ciencias de Cuba*, *4*(1), 1–14.

Denisov, I. G. (1992). Thermal stability of proteins in intermolecular complexes. *Biophysical Chemistry*, *44*(1), 71–75. https://doi.org/10.1016/0301-4622(92)85036-4.

Ding, B., & Chen, Z. L. (2012). Molecular interactions between cell penetrating peptide Pep-1 and model cell membranes. *The Journal of Physical Chemistry. B*, *116*(8), 2545–2552. https://doi.org/10.1021/jp209604m.

Dundas, C. M., Demonte, D., & Park, S. (2013). Streptavidin-biotin technology: Improvements and innovations in chemical and biological applications. *Applied Microbiology and Biotechnology*, *97*(21), 9343–9353. https://doi.org/10.1007/s00253-013-5232-z.

Fairhead, M., & Howarth, M. (2015). Site-specific biotinylation of purified proteins using BirA. *Methods in Molecular Biology*, *1266*, 171–184. https://doi.org/10.1007/978-1-4939-2272-7_12.

Feynman, R. P. (1969). What is science. *The Physics Teacher*, 7(6), 313–320. https://doi.org/10.1119/1.2351388.

Fraga, F. C., Valerio, A., de Oliveira, V. A., Di Luccio, M., & de Oliveira, D. (2019). Effect of magnetic field on the Eversa(R) Transform 2.0 enzyme: Enzymatic activity and structural conformation. *International Journal of Biological Macromolecules*, *122*, 653–658. https://doi.org/10.1016/j.ijbiomac.2018.10.171.

Gao, X., Ni, K., Zhao, C., Ren, Y., & Wei, D. (2014). Enhancement of the activity of enzyme immobilized on polydopamine-coated iron oxide nanoparticles by rational orientation of formate dehydrogenase. *Journal of Biotechnology*, *188*, 36–41. https://doi.org/10.1016/j.jbiotec.2014.07.443.

Garcia-Galan, C., Berenguer-Murcia, Á., Fernandez-Lafuente, R., & Rodrigues, R. C. (2011). Potential of different enzyme immobilization strategies to improve enzyme performance. *Advanced Synthesis & Catalysis*, *353*(16), 2885–2904. https://doi.org/10.1002/adsc.201100534.

Ghannadi, S., Abdizadeh, H., Miroliaei, M., & Saboury, A. A. (2019). Immobilization of alcohol dehydrogenase on titania nanoparticles to enhance enzyme stability and remove substrate inhibition in the reaction of formaldehyde to methanol. *Industrial and Engineering Chemistry Research*, *58*(23), 9844–9854. https://doi.org/10.1021/acs.iecr.9b01370.

González Siso, M. I., Lang, E., Carrenõ-Gómez, B., Becerra, M., Otero Espinar, F., & Blanco Méndez, J. (1997). Enzyme encapsulation on chitosan microbeads. *Process Biochemistry*, *32*(3), 211–216. https://doi.org/10.1016/s0032-9592(96)00064-7.

Grimaldi, J., Radhakrishna, M., Kumar, S. K., & Belfort, G. (2015). Stability of proteins on hydrophilic surfaces. *Langmuir*, *31*(3), 1005–1010. https://doi.org/10.1021/la503865b.

Guzik, U., Hupert-Kocurek, K., & Wojcieszynska, D. (2014). Immobilization as a strategy for improving enzyme properties-application to oxidoreductases. *Molecules*, *19*(7), 8995–9018. https://doi.org/10.3390/molecules19078995.

Häusler, H., Weber, H., & Stütz, A. E. (2006). D-xylose (D-glucose) isomerase (Ec 5.3.1.5): Observations and comments concerning structural requirements of substrates as well as mechanistic features. *Journal of Carbohydrate Chemistry*, *20*(3–4), 239–256. https://doi.org/10.1081/car-100104860.

Hitaishi, V., Clement, R., Bourassin, N., Baaden, M., de Poulpiquet, A., Sacquin-Mora, S., et al. (2018). Controlling redoz enzyme orientation at planar electrodes. *Catalysts*, *8*(5), 192. https://doi.org/10.3390/catal8050192.

Homaei, A. A., Sariri, R., Vianello, F., & Stevanato, R. (2013). Enzyme immobilization: An update. *Journal of Chemical Biology*, *6*(4), 185–205. https://doi.org/10.1007/s12154-013-0102-9.

Huang, C. C., Meng, E. C., Morris, J. H., Pettersen, E. F., & Ferrin, T. E. (2014). Enhancing UCSF Chimera through web services. *Nucleic Acids Research*, *42*(W1), W478–W484. https://doi.org/10.1093/nar/gku377.

Jain, A., & Cheng, K. (2017). The principles and applications of avidin-based nanoparticles in drug delivery and diagnosis. *Journal of Controlled Release*, *245*, 27–40. https://doi.org/10.1016/j.jconrel.2016.11.016.

Janecˇek, S. T. (1993). Strategies for obtaining stable enzymes. *Process Biochemistry*, *28*(7), 435–445. https://doi.org/10.1016/0032-9592(93)85026-c.

Jesionowski, T., Zdarta, J., & Krajewska, B. (2014). Enzyme immobilization by adsorption: A review. *Adsorption*, *20*(5–6), 801–821. https://doi.org/10.1007/s10450-014-9623-y.

Jiang, X. P., Lu, T. T., Liu, C. H., Ling, X. M., Zhuang, M. Y., Zhang, J. X., et al. (2016). Immobilization of dehydrogenase onto epoxy-functionalized nanoparticles for synthesis of (R)-mandelic acid. *International Journal of Biological Macromolecules*, *88*, 9–17. https://doi.org/10.1016/j.ijbiomac.2016.03.031.

Jun, L. Y., Mubarak, N. M., Yon, L. S., Bing, C. H., Khalid, M., Jagadish, P., et al. (2019). Immobilization of peroxidase on functionalized MWCNTs-buckypaper/polyvinyl alcohol nanocomposite membrane. *Scientific Reports*, *9*(1), 2215. https://doi.org/10.1038/s41598-019-39621-4.

Kelley, L. A., Mezulis, S., Yates, C. M., Wass, M. N., & Sternberg, M. J. (2015). The Phyre2 web portal for protein modeling, prediction and analysis. *Nature Protocols*, *10*(6), 845–858. https://doi.org/10.1038/nprot.2015.053.

Kim, S., Chen, J., Cheng, T., Gindulyte, A., He, J., He, S., et al. (2019). PubChem 2019 update: Improved access to chemical data. *Nucleic Acids Research*, *47*(D1), D1102–D1109. https://doi.org/10.1093/nar/gky1033.

Kumar, P., Gupta, A., Dhakate, S. R., Mathur, R. B., Nagar, S., & Gupta, V. K. (2013). Covalent immobilization of xylanase produced from *Bacillus pumilus* SV-85S on

electrospun polymethyl methacrylate nanofiber membrane. *Biotechnology and Applied Biochemistry*, *60*(2), 162–169. https://doi.org/10.1002/bab.1072.
Lavine, B., & Workman, J. (2010). Chemometrics. *Analytical Chemistry*, *82*(12), 4699–4711. https://doi.org/10.1021/ac101202z.
Lavine, B. K., & Workman, J., Jr. (2013). Chemometrics. *Analytical Chemistry*, *85*(2), 705–714. https://doi.org/10.1021/ac303193j.
Liu, Y., Guo, C., Sun, X. T., & Liu, C. Z. (2013). Improved performance of Yarrowia lipolytica lipase-catalyzed kinetic resolution of (R,S)-2-octanol by an integrated strategy of interfacial activation, bioimprinting and immobilization. *Bioresource Technology*, *142*, 415–419. https://doi.org/10.1016/j.biortech.2013.05.045.
Liu, H., & Naismith, J. H. (2008). An efficient one-step site-directed deletion, insertion, single and multiple-site plasmid mutagenesis protocol. *BMC Biotechnology*, *8*, 91. https://doi.org/10.1186/1472-6750-8-91.
Liu, Y., Ogorzalek, T. L., Yang, P., Schroeder, M. M., Marsh, E. N., & Chen, Z. (2013). Molecular orientation of enzymes attached to surfaces through defined chemical linkages at the solid-liquid interface. *Journal of the American Chemical Society*, *135*(34), 12660–12669. https://doi.org/10.1021/ja403672s.
Liu, L., Yu, J., & Chen, X. (2015). Enhanced stability and reusability of alcohol dehydrogenase covalently immobilized on magnetic graphene oxide nanocomposites. *Journal of Nanoscience and Nanotechnology*, *15*(2), 1213–1220. https://doi.org/10.1166/jnn.2015.9024.
Mahmood, I., Ahmad, I., Chen, G., & Huizhou, L. (2013). A surfactant-coated lipase immobilized in magnetic nanoparticles for multicycle ethyl isovalerate enzymatic production. *Biochemical Engineering Journal*, *73*, 72–79. https://doi.org/10.1016/j.bej.2013.01.017.
Mandenius, C. F., & Brundin, A. (2008). Bioprocess optimization using design-of-experiments methodology. *Biotechnology Progress*, *24*(6), 1191–1203. https://doi.org/10.1002/btpr.67.
Marques Netto, C. G. C., Andrade, L. H., Freitas, R. S., & Toma, H. E. (2015). Association of yeast alcohol dehydrogenase with superparamagnetic nanoparticles: Improving the enzyme stability and performance. *Journal of Nanoscience and Nanotechnology*, *15*(12), 9482–9487. https://doi.org/10.1166/jnn.2015.10317.
Marques Netto, C. G. C., Andrade, L. H., & Toma, H. E. (2018). Carbon dioxide/methanol conversion cycle based on cascade enzymatic reactions supported on superparamagnetic nanoparticles. *Anais da Academia Brasileira de Ciências*, *90*(Suppl. 1), 593–606. https://doi.org/10.1590/0001-3765201720170330.
Marques Netto, C. G. C., da Silva, D. G., Toma, S. H., Andrade, L. H., Nakamura, M., Araki, K., et al. (2016). Bovine glutamate dehydrogenase immobilization on magnetic nanoparticles: Conformational changes and catalysis. *RSC Advances*, *6*(16), 12977–12992. https://doi.org/10.1039/c5ra24637g.
Mei, Y., Miller, L., Gao, W., & Gross, R. A. (2003). Imaging the distribution and secondary structure of immobilized enzymes using infrared microspectroscopy. *Biomacromolecules*, *4*(1), 70–74. https://doi.org/10.1021/bm025611t.
Minteer, S. D. (2016). Cell-free biotechnologies. In C. A. Eckert & C. T. Trinh (Eds.), *Biotechnology for biofuel production and optimization* (pp. 433–448). Amsterdam: Elsevier.
Modenez, I. A., Sastre, D. E., Moraes, F. C., & Marques Netto, C. G. C. (2018). Influence of glutaraldehyde cross-linking modes on the recyclability of immobilized lipase B from *Candida antarctica* for transesterification of soy bean oil. *Molecules*, *23*(9), 2230–2246. https://doi.org/10.3390/molecules23092230.
Mohamad, N. R., Marzuki, N. H., Buang, N. A., Huyop, F., & Wahab, R. A. (2015). An overview of technologies for immobilization of enzymes and surface analysis techniques for immobilized enzymes. *Biotechnology and Biotechnological Equipment*, *29*(2), 205–220. https://doi.org/10.1080/13102818.2015.1008192.

Moreno-Cortez, I. E., Romero-Garcia, J., Gonzalez-Gonzalez, V., Garcia-Gutierrez, D. I., Garza-Navarro, M. A., & Cruz-Silva, R. (2015). Encapsulation and immobilization of papain in electrospun nanofibrous membranes of PVA cross-linked with glutaraldehyde vapor. *Materials Science & Engineering. C, Materials for Biological Applications*, *52*, 306–314. https://doi.org/10.1016/j.msec.2015.03.049.

Moskovitz, Y., & Srebnik, S. (2014). Conformational changes of globular proteins upon adsorption on a hydrophobic surface. *Physical Chemistry Chemical Physics*, *16*(23), 11698–11707. https://doi.org/10.1039/c4cp00354c.

Nannemann, D. P., Birmingham, W. R., Scism, R. A., & Bachmann, B. O. (2011). Assessing directed evolution methods for the generation of biosynthetic enzymes with potential in drug biosynthesis. *Future Medicinal Chemistry*, *3*(7), 809–819. https://doi.org/10.4155/fmc.11.48.

Narayan, A., & Naganathan, A. N. (2017). Tuning the continuum of structural states in the native ensemble of a regulatory protein. *Journal of Physical Chemistry Letters*, *8*(7), 1683–1687. https://doi.org/10.1021/acs.jpclett.7b00475.

Parkinson, H., Kapushesky, M., Shojatalab, M., Abeygunawardena, N., Coulson, R., Farne, A., et al. (2007). ArrayExpress-a public database of microarray experiments and gene expression profiles. *Nucleic Acids Research*, *35*, 747–750. https://doi.org/10.1093/nar/gkl995. Database issue.

Pechkova, E., Sivozhelezov, V., & Nicolini, C. (2007). Protein thermal stability: The role of protein structure and aqueous environment. *Archives of Biochemistry and Biophysics*, *466*(1), 40–48. https://doi.org/10.1016/j.abb.2007.07.016.

Pereira, F. M. V., & Pereira-Filho, E. R. (2018). Aplicação de programa computacional livre em planejamento de experimentos: Um tutorial. *Quimica Nova*, *41*(9), 1061–1071.

Pietrantonio, C. D., Pandey, A., Gould, J., Hasabnis, A., & Prosser, R. S. (2019). Understanding protein function through an ensemble description: Characterization of functional states by 19F NMR. In A. J. Wand (Ed.), *Methods in enzymology biological NMR, part B*. Amsterdam: Elsevier.

Poluri, K. M., & Gulati, K. (2017). *Protein engineering techniques: Gateways to synthetic protein universe*. Switzerland: Springer.

Priebe, J. P., Souza, F. D., Silva, M., Tondo, D. W., Priebe, J. M., Micke, G. A., et al. (2012). The chameleon-like nature of zwitterionic micelles: Effect of cation binding. *Langmuir*, *28*(3), 1758–1764. https://doi.org/10.1021/la2043735.

Putzbach, W., & Ronkainen, N. J. (2013). Immobilization techniques in the fabrication of nanomaterial-based electrochemical biosensors: A review. *Sensors (Basel)*, *13*(4), 4811–4840. https://doi.org/10.3390/s130404811.

Riccardi, C., McCormick, S., Kasi, R., & Kumar, C. (2018). A modular approach for interlocking enzymes in Whatman paper. *Angewandte Chemie, International Edition*, *57*(32), 10158–10162.

Rodrigues, M. I., & Iemma, A. F. (2014). *Experimental design and process optimization*. Boca Raton: CRC Press.

Rodrigues, R. C., Ortiz, C., Berenguer-Murcia, A., Torres, R., & Fernandez-Lafuente, R. (2013). Modifying enzyme activity and selectivity by immobilization. *Chemical Society Reviews*, *42*(15), 6290–6307. https://doi.org/10.1039/c2cs35231a.

Rojas-Mercado, A. S., Moreno-Cortez, I. E., Lucio-Porto, R., & Pavon, L. L. (2018). Encapsulation and immobilization of ficin extract in electrospun polymeric nanofibers. *International Journal of Biological Macromolecules*, *118*(Pt. B), 2287–2295. https://doi.org/10.1016/j.ijbiomac.2018.07.113.

Rose, P. W., Prlic, A., Altunkaya, A., Bi, C., Bradley, A. R., Christie, C. H., et al. (2017). The RCSB protein data bank: Integrative view of protein, gene and 3D structural information. *Nucleic Acids Research*, *45*(D1), D271–D281. https://doi.org/10.1093/nar/gkw1000.

Rosevear, A. (2008). Immobilised biocatalysts—A critical review. *Journal of Chemical Technology and Biotechnology Biotechnology, 34*(3), 127–150. https://doi.org/10.1002/jctb.280340302.
Rueda, N., Dos Santos, J. C., Ortiz, C., Torres, R., Barbosa, O., Rodrigues, R. C., et al. (2016). Chemical modification in the design of immobilized enzyme biocatalysts: Drawbacks and opportunities. *Chemical Record, 16*(3), 1436–1455. https://doi.org/10.1002/tcr.201600007.
Ryan, B. J., & O'Fagain, C. (2007). Arginine-to-lysine substitutions influence recombinant horseradish peroxidase stability and immobilisation effectiveness. *BMC Biotechnology, 7*, 86. https://doi.org/10.1186/1472-6750-7-86.
Šalić, A., Pindrić, K., Podrepšek, G. H., Leitgeb, M., & Zelić, B. (2013). NADH oxidation in a microreactor catalysed by ADH immobilised on γ-Fe2O3 nanoparticles. *Green Processing and Synthesis, 2*(6), 569–578. https://doi.org/10.1515/gps-2013-0084.
Samphao, A., Kunpatee, K., Prayoonpokarach, S., Wittayakun, J., Švorc, Ľ., Stankovic, D. M., et al. (2015). An ethanol biosensor based on simple immobilization of alcohol dehydrogenase on Fe3O4@Au nanoparticles. *Electroanalysis, 27*(12), 2829–2837. https://doi.org/10.1002/elan.201500315.
Santos, J. C. S. d., Barbosa, O., Ortiz, C., Berenguer-Murcia, A., Rodrigues, R. C., & Fernandez-Lafuente, R. (2015). Importance of the support properties for immobilization or purification of enzymes. *ChemCatChem*, 7(16), 2413–2432. https://doi.org/10.1002/cctc.201500310.
Schomburg, I., Jeske, L., Ulbrich, M., Placzek, S., Chang, A., & Schomburg, D. (2017). The BRENDA enzyme information system—From a database to an expert system. *Journal of Biotechnology, 261*, 194–206. https://doi.org/10.1016/j.jbiotec.2017.04.020.
Secundo, F. (2013). Conformational changes of enzymes upon immobilisation. *Chemical Society Reviews, 42*(15), 6250–6261. https://doi.org/10.1039/c3cs35495d.
Sheldon, R. A. (2007). Enzyme immobilization: The quest for optimum performance. *Advanced Synthesis & Catalysis, 349*(8–9), 1289–1307. https://doi.org/10.1002/adsc.200700082.
Shome, A., Roy, S., & Das, P. K. (2007). Nonionic surfactants: A key to enhance the enzyme activity at cationic reverse micellar interface. *Langmuir, 23*(8), 4130–4136. https://doi.org/10.1021/la062804j.
Siffer, F., Ponche, A., Fioux, P., Schultz, J., & Roucoules, V. (2005). A chemometric investigation of the effect of the process parameters during maleic anhydride pulsed plasma polymerization. *Analytica Chimica Acta, 539*(1–2), 289–299. https://doi.org/10.1016/j.aca.2005.02.072.
Silva, C., Silva, C. J., Zille, A., Guebitz, G. M., & Cavaco-Paulo, A. (2007). Laccase immobilization on enzymatically functionalized polyamide 6,6 fibres. *Enzyme and Microbial Technology, 41*(6–7), 867–875. https://doi.org/10.1016/j.enzmictec.2007.07.010.
Singh, R. K., Tiwari, M. K., Singh, R., & Lee, J. K. (2013). From protein engineering to immobilization: Promising strategies for the upgrade of industrial enzymes. *International Journal of Molecular Sciences, 14*(1), 1232–1277. https://doi.org/10.3390/ijms14011232.
Sobral, K. A., Rodrigues, R. O., Oliveira, R. D., Olivo, J. E., de Moraes, F. F., & Zanin, G. M. (2003). Evaluation of supports and methods for immobilization of enzyme cyclodextringlycosyltransferase. *Applied Biochemistry and Biotechnology, 105–108*(1–3), 809–819. https://doi.org/10.1385/ABAB:108:1-3:809.
Structural Genomics Consortium, China Structural Genomics Consortium, Northeast Structural Genomics Consortium, Graslund, S., Nordlund, P., Weigelt, J., et al. (2008). Protein production and purification. *Nature Methods, 5*(2), 135–146. https://doi.org/10.1038/nmeth.f.202.
Teófilo, R. F., & Ferreira, M. M. C. (2006). Quimiometria II: Planilhas eletrônicas para cálculos de planejamentos experimentais, um tutorial. *Quimica Nova, 29*(2), 338–350.

Thudi, L., Jasti, L. S., Swarnalatha, Y., Fadnavis, N. W., Mulani, K., Deokar, S., et al. (2012). Enzyme immobilization on epoxy supports in reverse micellar media: Prevention of enzyme denaturation. *Journal of Molecular Catalysis B: Enzymatic*, *74*(1–2), 54–62. https://doi.org/10.1016/j.molcatb.2011.08.014.

Tondo, D. W., Priebe, J. M., Souza, B. S., Priebe, J. P., Bunton, C. A., & Nome, F. (2007). The chameleon-like nature of zwitterionic micelles. Control of anion and cation binding in sulfobetaine micelles. Effects on acid equilibria and rates. *The Journal of Physical Chemistry B*, *111*(41), 11867–11869. https://doi.org/10.1021/jp075208m.

Torres-Salas, P., del Monte-Martinez, A., Cutiño-Avila, B., Rodriguez-Colinas, B., Alcalde, M., Ballesteros, A. O., et al. (2011). Immobilized biocatalysts: Novel approaches and tools for binding enzymes to supports. *Advanced Materials*, *23*(44), 5275–5282. https://doi.org/10.1002/adma.201101821.

Tropsha, A., Mills, K. C., & Hickey, A. J. (2017). Reproducibility, sharing and progress in nanomaterial databases. *Nature Nanotechnology*, *12*(12), 1111–1114. https://doi.org/10.1038/nnano.2017.233.

Uliana, C. V., Riccardi, C. S., Tognolli, J. O., & Yamanaka, H. (2008). Optimization of an amperometric biosensor for the detection of hepatitis C virus using fractional factorial designs. *Journal of the Brazilian Chemical Society*, *19*(4), 782–787. https://doi.org/10.1590/s0103-50532008000400024.

Wang, J., Paszti, Z., Clarke, M. L., Chen, X., & Chen, Z. J. (2007). Deduction of structural information of interfacial proteins by combined vibrational spectroscopic methods. *Journal of Physical Chemistry B*, *111*(21), 6088–6095. https://doi.org/10.1021/jp070383o.

Wu, X., Fraser, K., Zha, J., & Dordick, J. S. (2018). Flexible peptide linkers enhance the antimicrobial activity of surface-immobilized bacteriolytic enzymes. *ACS Applied Materials & Interfaces*, *10*(43), 36746–36756. https://doi.org/10.1021/acsami.8b14411.

Yang, P., Boughton, A., Homan, K., & Chen, Z. J. (2013). Membrane orientation of $G\alpha_i\beta_1\gamma_2$ and $G\beta_1\gamma_2$ determined via combined vibrational spectroscopic studies. *Journal of the American Chemical Society*, *135*(13), 5044–5051. https://doi.org/10.1021/ja3116026.

Zaak, H., Siar, E.-H., Kornecki, J. F., Fernandez-Lopez, L., Pedrero, S. G., Virgen-Ortíz,-J. J., et al. (2017). Effect of immobilization rate and enzyme crowding on enzyme stability under different conditions. The case of lipase from *Thermomyces lanuginosus* immobilized on octyl agarose beads. *Process Biochemistry*, *56*, 117–123. https://doi.org/10.1016/j.procbio.2017.02.024.

Zdarta, J., Meyer, A., Jesionowski, T., & Pinelo, M. (2018). A general overview of support materials for enzyme immobilization: Characteristics, properties, practical utility. *Catalysts*, *8*(2), 92. https://doi.org/10.3390/catal8020092.

Zhai, R., Zhang, B., Wan, Y., Li, C., Wang, J., & Liu, J. (2013). Chitosan–halloysite hybrid-nanotubes: Horseradish peroxidase immobilization and applications in phenol removal. *Chemical Engineering Journal*, *214*, 304–309. https://doi.org/10.1016/j.cej.2012.10.073.

Zhang, L., Xiao, X., Yuan, Y., Guo, Y., Li, M., & Pu, X. (2015). Probing immobilization mechanism of alpha-chymotrypsin onto carbon nanotube in organic media by molecular dynamics simulation. *Scientific Reports*, *5*, 9297. https://doi.org/10.1038/srep09297.

Zucca, P., & Sanjust, E. (2014). Inorganic materials as supports for covalent enzyme immobilization: Methods and mechanisms. *Molecules*, *19*(9), 14139–14194. https://doi.org/10.3390/molecules190914139.

CHAPTER FIVE

Fibrous polymer functionalized magnetic biocatalysts for improved performance

Veli C. Ozalp[a,*], Gulay Bayramoglu[b,c], M. Yakup Arica[b]
[a]Kit-Argem Research Center and Department of Bioengineering, Konya Food and Agriculture University, Konya, Turkey
[b]Biochemical Processing and Biomaterial Research Laboratory, Gazi University, Ankara, Turkey
[c]Department of Chemistry, Faculty of Sciences, Gazi University, Ankara, Turkey
*Corresponding author: e-mail address: cengizozalp@gmail.com

Contents

Abstract

Although enzymes are known for their excellent catalytic performance, industrial, medical or biotechnological applications should overcome some drawbacks like long-term stability under specific conditions of the application. Immobilized enzymes have offered advantages over soluble counterparts in many industrial and laboratory scale applications by increasing operational stability and reusability. When the immobilization matrix has magnetic properties, an additional advantage is obtained as simpler processing.

Methods in Enzymology, Volume 630
ISSN 0076-6879
https://doi.org/10.1016/bs.mie.2019.09.004

Iron-based superparamagnetic nano-sized particles has large surface area for biocompatible applications are especially in focus. Adding nanofibrous polymers to magnetic nanoparticles has been an excellent way to increase efficiency of biocatalyst immobilization by further increasing loading capacity. This chapter explains various magnetic enzyme-nanoparticles based preparations with potential for future industrial applications like invertase, lipase and as model studies and focus on the nanofibrous polymer brush grafting as a way to increase catalytic efficiency of magnetic nanoparticles.

1. Introduction

Commercial use of biocatalysis are industrial enzyme applications such as chemical production, pharmaceutical processes, biofuels, food-related applications and other consumer productions (Prier & Kosjek, 2019). Industrial biocatalysts applications have extensively employed immobilization of enzymes for improving functional stability and reusability in industrial applications for a long time. Many inert matrices were utilized for immobilization of enzymes and this proved to increase protein stability under changing conditions of pH or temperature. Another advantage is that the enzymes are maintained attached to an insoluble material, which can easily be separated from products as an efficient purification step. Thus, convenient, economically advantageous and stable biocatalysts can be obtained through immobilization. Many techniques have successfully been exploited for immobilization from simple adsorption or entrapment in a matrix to affinity capture and covalent bonding (Monsan & Combes, 1988).

1.1 Magnetic nanoparticles as biocatalyst

Magnetic nanoparticles (MNP) represents unique opportunities in enzyme immobilization applications. Miniaturation to microsystems are popular due to obvious advantages, including faster reaction times, lower reagent consumption and better automation. However, there are also disadvantages in purification steps of production. Therefore, magnetic separation in industrial applications provides a rapid and simple tool for better performing biocatalysts. MNP provides external manipulation of the location of catalyst by permanent magnets, allowing simple bioreactor processes.

Specifically polymer brush functionalized MNPs are considered as favorable immobilization matrix with combined advantages of improved loading efficiency, easy manipulation with magnets and reusability (Hong, Liu, Li, & Chen, 2019). Polymer brushes provide multiple attachment sites on the nanoparticles, extending higher functional surface area immensely. For example, poly(2-vinyl-4,4-dimethylazlactone) (PVDMA) was used as brush polymer for Fe_3O_4 nanoparticles and asparaginase immobilization. Asparagase is an enzyme used in drug transformation studies in pharmaceutical applications (Mu, Qiao, Qi, Dong, & Ma, 2014). Immobilization capacity on polymer brushes systems increased with the length of polymer chain. The efficiency of enzyme activity was also increased with higher affinity to its target and the activity was shown to retain more than 72.6% of its activity after 10 weeks of storage, indicating to great improvement compared to free enzyme applications with asparaginase. In another example, glycidyl methacrylate (GMA) was used as polymer brush on silica MNPs and subsequent covalent immobilization of invertase (Bayramoglu, Doz, Ozalp, & Arica, 2017). Enzyme immobilization on Fe_3O_4@SiO_2@p(GMA) was at the level of 36.1 mg/g and Km was 39.4 mM, similar to free form immobilization amount (34.3 mM). This study demonstrated that immobilization procedure did not have any negative effect on the enzymatic activity.

1.2 Enzyme-nanoparticle conjugation techniques

Enzyme immobilization should control orientation of the enzyme for a favorable location of active site to have free access to substrates and reaction intermediates (Zhang, Ge, & Liu, 2015). In multi-enzyme application, orientation challenge might be more complicated; however, immobilization of single enzymes on nanoparticles should consider steric hinderance for substrates and thus active site should face away from the surface of the nanoparticles. The most important consideration is to use mild immobilization reaction conditions to maintain delicate internal biocatalysis specific activity, most of the time, maintaining protein conformational structure intact. Protection of the active site is also important and protective groups during immobilization can be used. Thus, selection of immobilization method has important consequences on the application (Singh, Mishra, Sharma, & Singh, 2019). There are four main categories of immobilization: (i) adsorption, (ii) entrapment, (iii) direct covalent bonding and (iv) crosslinker-mediated attachment.

Adsorption is the simplest immobilization technique, but the outcome is mostly random. The orientation of the immobilized enzyme is not easy to control. Physical adsorption usually changes protein microenvironment by multipoint interactions with immobilization matrix (Johnson, Wang, & Arnold, 1996). Adsorption is achieved by mixing nanoparticles with the enzyme of interest in the right conditions like enzyme concentration, pH, temperature and ionic strength. The adsorption mechanism is based on hydrophobic, charge attraction, van der Waals and hydrogen bonding, and thus it is reversible (Rao, Anderson, & Bachas, 1998). After washing, the adsorbed enzyme is ready for use, however, unwanted aggregation or loose attachment is common challenges (Spahn & Minteer, 2008). In entrapment, the mobility of the enzyme is restricted by confining in a molecular cage, like porous materials. The limiting factor in this method is that the substrate and product should be easily able to move through the pores of the matrix, thus only small molecules can be involved.

Covalent immobilization involves direct bonding between support matrix and the enzyme from chemical groups, which does not have a role in catalytic process. Amino acid side chains located on surface of the molecule will be more accessible to aqueous content when they are polar amino acids like lysine, arginine, aspartic acid or histidine residues or functional groups like imidazole, phenolic hydroxyl groups. Some modifications of peptides are also used for better control on orientation of immobilization (Fu, Reinhold, & Woodbury, 2011). Covalent attachment can be through a spacer arm to give more flexibility of orientation like cyanogen bromide containing carbohydrates or glutaraldehyde, which sometimes provide thermal stability (Hartmann & Kostrov, 2013). Mostly, enzyme immobilization is achieved through a linker molecule, then the properties of this linker is important in the final immobilized enzyme activity. All the parameters about the linker such as length, size, conformation, charge level or hydrophobicity play a direct role in the stability of the enzyme. Finally, cross-linking refers to attachment of single or multi-enzymes to each other.

1.3 Magnetic biocatalyst

Nanoparticles constitute a superior immobilization support matrix for enzymes mostly due to large surface area for efficient loading (Chen, Zeng, Xu, Lai, & Tang, 2017). Nanoparticles provide dispersion in solution with Brownian movements, which improves enzymatic efficiency compared to unbound enzyme. Additionally, magnetic nanoparticles provide easy separation method by an external magnet. For example, immobilization of alpha-amylase on magnetic poly (2-hydroxyethyl methacrylate-N-methacryloyl-(l)-phenylalanine) particles increased substrate affinity with 85% specific activity after 10 times reuse (Uygun, Öztürk, Akgöl, & Denizli, 2012). Cellulase immobilization on superparamagnetic nanoparticles resulted in less specific activity than free enzyme, but higher activity at elevated temperature of 80 °C (Khoshnevisan et al., 2011). Trypsin immobilization on silica based magnetic nanoparticles by amine bonding increased protein identification events compared to free trypsin in mass spectroscopy applications (Lee et al., 2011). In a similar study, magnetic silica nanoparticles were used as immobilization carrier for glucose oxidase and ferrocene to develop an efficient biosensor of glucose (Qiu, Peng, & Liang, 2007). Among many other examples, chitosan based fibrous magnetic beads were used for lysozyme capture grafted with poly (cation-exchange) groups and invertase immobilization on fibrous polymer silica magnetic nanoparticles for invertase immobilization (Bayramoglu et al., 2017; Bayramoglu, Tekinay, Ozalp, & Arica, 2015).

2. Experimental protocols

2.1 Chemicals and equipment

Chitosan, glycidyl methacrylate (GMA), bromoacetyl bromide, glutaraldehyde, triethylamine, polyvinyl alcohol (PVA; MW: 50,000), bipyridine, CuBr, and tetrahydrofuran were obtained from Sigma-Aldrich (St. Louis, USA) and used as received. All other chemicals were of analytical grade and were purchased from Merck AG (Darmstadt, Germany).

The following instruments were employed in the major characterization experiments: UV–vis spectrophotometer (AAS; AA 6800, Shimadzu, Japan), Fourier Transfer Infrared Spectroscopy (FTIR) (Perkin–Elmer Inc., Norwalk, CT, USA), Brunauer–Emmett–Teller (BET) analysis (QuantachromeNova 2200 E, USA), X-ray Diffraction (XRD) (MiniFlex 600, Rigaku), Vibrating Sample Magnetometer (VSM, model 155, Digital MeasurementSystem, Inc., Westwood, MA, USA), Scanning Electron Microscopy (SEM, ZEIZZ, Evo 50), Atomic Force Microscopy (AFM, PSIA, XE 100 E model).

2.2 Synthesis of polymer brushed magnetic nanoparticles

Fig. 1 explains an overall synthesis strategy of magnetic nanoparticles grafted with polymer brush. Fe_3O_4 nanoparticles were first silanized with (3-aminopropyl) triethoxysilane (APTES), functionalized with Br^- ends and finally grafted with glycidyl methacrylate (GMA) brush via atom transfer radical polymerization (ATRP) initialization.

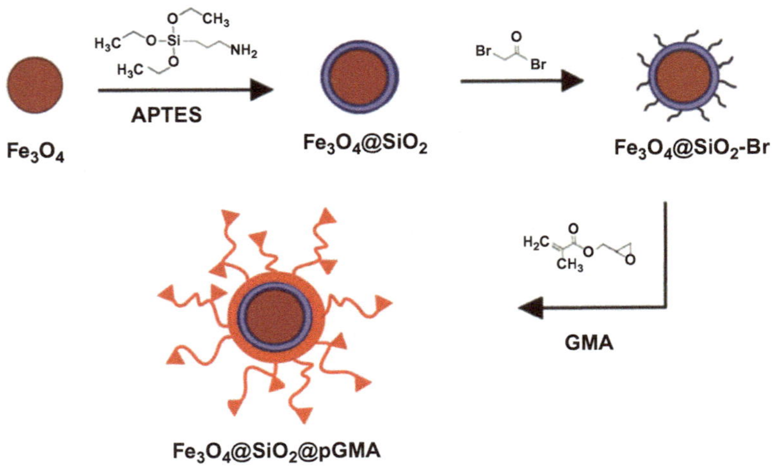

Fig. 1 Strategy of polymer brush grafting of magnetic nanoparticles (Bayramoglu et al., 2017).

2.2.1 Synthesis of magnetic nanoparticles

The magnetic nanoparticles can be synthesized by co-precipitation of 2:1 molar ratio of Fe_3 and Fe_2 ions.

1. Mix 0.03 mol $FeCl_3{\cdot}6H_2O$ and 0.01 mol $FeSO_4{\cdot}7H_2O$ in 100 mL distilled water under nitrogen gas.
2. Put in a reactor and add aqueous ammonia (50 mL, 25% w/v).
3. Stir and reflux under nitrogen gas at 40 °C for 2.0 h, and then repeat the same procedure at 80 °C for 1.0 h.

2.2.2 Surface silanization of magnetic nanoparticles

Fe_3O_4 nanoparticles surface was functionalized with silica for subsequent polymer brush grafting. There are alternative surface functionalization methods using other materials, depending on the requirements of the application.

1. Add 0.45 mL APTES in 15.55 mL toluene containing 1.0 g Fe_3O_4 nanoparticles as prepared in Section 2.2.1.
2. Reflux the mixture at 70 °C for 14.0 h with continuous stirring.
3. Separate the resultant $Fe_3O_4@SiO_2$ nanoparticles by an external magnet.
4. Wash with methanol and distilled water.

2.2.3 Br^- end functionalization

The free amino groups on $Fe_3O_4@SiO_2$ nanoparticles can be reacted with bromoacetyl bromide to initiate SI-ATRP reactions

1. Mix Fe_3O_4@SiO_2 nanoparticles (1.0 g) and 2.0 mL tetrahydrofuran, and triethylamine in a flask.
2. Stir under nitrogen atmosphere at 10.0 °C.
3. Sonicate Fe_3O_4@SiO_2 nanoparticles for 10 min to disperse them well in the solution.
4. Cool down the mixture to 0 °C.
5. Add dropwise bromoacetyl bromide (2.0 mL) within 30 min.
6. Rotate the flask on a shaker at 25 °C for 20 h.
7. Collect the product with an external magnet.
8. Wash with acetone and distilled water to obtain Br-end functionalized nanoparticles.

2.2.4 Grafting GMA brushes by ATRP

The Br-end functionalized nanoparticles from Section 2.2.3 were used for the grafting of p(GMA) brushes.

1. Mix Fe_3O_4@SiO_2-Br nanoparticles (1.0 g) in a flask, with monomer GMA (11.3–67.8 mmol), CuBr (0.1 g), bipyridine (0.5 g), and dioxane (5.0 mL).
2. Purge with N_2 gas and seal the flask.
3. Leave the mixture for 12 h at 65 °C for grafting reaction.
4. Filter the product and wash with THF (100 mL), methanol (50 mL) and ether (50 mL).
5. Remove copper residues from the Fe_3O_4@SiO_2@p(GMA) by stirring in acetic acid and THF mixture (2:1) at 150 rpm for 2.0 h.

2.3 Immobilization of enzyme

Fig. 2 demonstrates functionalization of polymer brushes on Fe_3O_4@SiO_2@p(GMA) with amino groups and covalent bonding with the enzyme. Hexamethylenediamine (HMDA) made epoxy ring opening reaction on Fe_3O_4@SiO_2@p(GMA) surface. The free amino groups on the Fe_3O_4@SiO_2@p(GMA) were activated by glutaraldehyde treatment for covalent coupling of the enzyme. The initial protein concentration and loading capacity are known to be related and it was determined by varying invertase concentration between 0.5 and 4.0 mg/mL.

2.3.1 Amination and activation of GMA brushes

1. Mix Fe_3O_4@SiO_2@p(GMA) nanoparticles (2.0 g) with HMDA (1.0%, 20.0 mL) at pH 11.0 in a flask.
2. Incubate at 65 °C for 6 h.

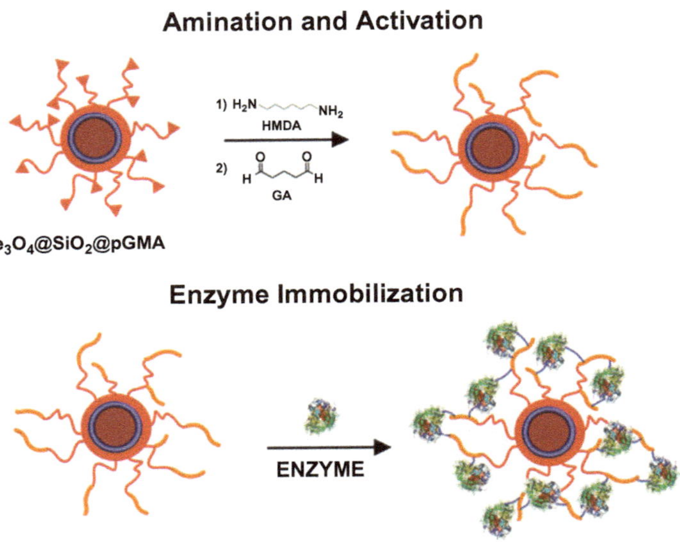

Fig. 2 Activation of magnetic nanoparticles and enzyme immobilization (Bayramoglu et al., 2017).

3. Wash the product with distilled water and methanol.
4. Dry at 45 °C for 24 h.
5. Mix Fe_3O_4@SiO_2@p(GMA) (about 2.0 g) with 2.0% glutaraldehyde solution in Tris–HCl buffer (50 mol/L, pH 8.0) in a flask.
6. Incubate the mixture at 65 °C for 3.0 h.
7. Wash Fe_3O_4@SiO_2@-pGMA-GA particles with 0.1 mol/L acetic acid solution, and 0.1 mol/L Tris–HCl buffer at pH 8.0.

2.3.2 Covalent immobilization of the enzyme

1. Incubate Fe_3O_4@SiO_2@p(GMA)-GA particles with different chain length (about 0.5 g).
2. Incubated in phosphate buffer (pH 8.0, 50 mmol/L) for 2.0 h.
3. Transfer the mixture into enzyme solution (20 mL, 2.0 mg enzyme/mL) in the same buffer.
4. Incubate the mixture at 22 °C for 6.0 h by stirring.
5. Add ethanolamine (0.5 mL ethanolamine in 25 mL buffer) for blocking unreacted aldehyde groups.
6. Apply external magnetic force to collect the enzyme-nanoparticles conjugates.
7. Wash with acetate buffer, and keep at 4 °C.

2.4 Characterization methods for magnetic nanoparticles

The amount of enzyme immobilization can simply be determined by measuring protein amount in solution before and after immobilization procedure. Bradford's method has been commonly employed by using an UV–vis spectrophotometer (Bradford, 1976). The amount of immobilized enzyme (AE, %) on the nanoparticles can be calculated by the following equation:

$$AE\ (\%) = [(C_0 - C_t)/C_0] \times 100$$

where C_0 and C_t are enzyme concentrations in the samples determined before and after the immobilization procedure, respectively. The immobilization efficiency (IE, %) was determined according to following equation:

$$IE\ (\%) = (A_r/A_i) \times 100$$

where A_r is the retained activity of the immobilized invertase, and A_i is the initial activity of free enzyme which is added in the immobilization medium.

Fourier transfer infrared spectroscopy (FTIR) analysis is based on the vibration of functional chemical groups upon absorbing infrared light. The functional groups have a constant absorption and vibration pattern regardless of the rest of the material. This phenomenon provides a direct and fast method to identify chemical groups on materials like nanoparticles. Thus, FTIR analysis can be used for determination of the success of epoxy group functionalization on the Fe_3O_4@SiO_2@pGMA nanoparticles (Bayramoğlu & Arıca, 2008). Poly(GMA) is known to have unique stretching vibration band for hydrogen-bounded alcohol at ~3480 cm^{-1}, methylene vibration at ~2960 cm^{-1}, esther at 1736 cm^{-}1, and carbonyl configuration at 1152 cm^{-1}. GMA has bands assigned to the –CH_2 scissoring band of both GMA at 1459 cm^{-1}, anti-symmetric and symmetric stretching band of carbonyl groups of GMA at 1395 and 1267 cm^{-1}. The epoxide group has a characteristic band at 971 cm^{-1} (due to epoxy ring vibrations) and Fe_3O_4 has a characteristic band at 617 cm^{-1}. Therefore, an FTIR scanning between 600 and 3500 cm^{-1} can determine the presence of GMA, magnetic core and epoxy groups, proving a tool for easy monitoring of the experimental procedures.

Studying weight gain can be a way to monitor the average degree of the GMA unit on the polymer chains on the nanoparticles. The weight increase data can be interpreted for understanding how many repeating unit of the monomer dependent ATRP reaction was added by using following equation:

$$PD = \left(\left(W_g/MW_{GMA}\right)/\left(1 + W_g\right) \times 103\right)/Br_i$$

where PD is average degree of polymerization, W_g is fractional weight gain, MW_{GMA} is molar mass of GMA and Br_i represents initial bromide content of magnetic nanoparticles.

2.5 Enzyme immobilization

Invertase enzyme were fixed on Fe_3O_4@SiO_2@p(GMA)-GA nanoparticles at a specific p(GMA) grafting density (3, Table 1). Fe_3O_4@SiO_2@p(GMA) nanoparticles can be used for control experiments.

1. Mix and incubate Fe_3O_4@SiO_2@p(GMA)-GA-3 particles (about 0.5 g) with invertase solution (2.0 mg enzyme/mL) in 20 mL phosphate buffer (pH 8.0, 50 mmol/L) for 2.0 h.
2. Repeat the immobilization with a range of enzyme concentrations, for example, between 0.5 and 4.0 mg/mL.
3. Incubate at 22 °C for 6 h by stirring.
4. Transfer the reaction mixture into ethanolamine solution for blocking free aldehyde groups on nanoparticles.
5. Use an external magnetic force to collect enzyme immobilized nanoparticles, and wash with acetate buffer, and keep at 4 °C.

The enzymatic activity of immobilized and free form of the enzyme should be determined with an appropriate procedure under varying pH, temperature conditions.

Table 1 Properties of the enzyme immobilized Fe_3O_4@SiO_2@p(GMA) magnetic nanoparticles with nanobrush polymer length 1–6 (Bayramoglu et al., 2017).

Fe_3O_4@SiO_2@ p(GMA)-GA	ATRP reaction time (h)	Grafting (%)	Average GMA units	Amount of loaded invertase (mg/g)	Retained enzyme activity (%)
1	2	20.3 ± 1.2	5.5	21.3 ± 0.7	89.4 ± 1.8
2	4	31.1 ± 1.4	7.7	29.6 ± 0.4	84.4 ± 2.3
3	6	47.2 ± 0.9	10.4	33.4 ± 1.3	76.9 ± 0.6
4	8	86.6 ± 2.3	15.0	38 ± 2.1	72.8 ± 2.7
5	10	102.5 ± 0.6	16.4	54.1 ± 1.5	67.4 ± 1.3
6	12	104.1 ± 1.8	16.5	54.3 ± 0.8	66.6 ± 2.6

3. Magnetic nanoparticles conjugate properties

In this section, physiochemical properties of poly(glycidyl methacrylate) grafted on $Fe_3O_4@SiO_2$-Br nanoparticles following experimental protocols in Section 2 will be discussed over previously reported examples in order to explain what type of final product should be expected in each step of the synthesis. The amination of hydroxyl groups of Fe_3O_4 nanoparticles using APTES and subsequent ATRP initiation (i.e., bromoacetylbromide) can be checked for covalent attachment and Br ended silica on magnetic nanoparticles were created. Br amount on magnetic particles surface was determined to be 0.218 mmol/g particles after bromoacetylbromide treatment. This result proved that ATPR application successfully initialized epoxy group formation from halogen atoms on particle surface in the presence of bipyridine and CuBr. Thus, GMA brushes were shown to have potentially to immobilize 0.218 mmol of enzyme per g of particles as functional epoxide group and they can be used for functionalizing grafted polymer on the nanoparticles. The grafted GMA molecules can be monitored as an increase in the mass of $Fe_3O_4@SiO_2$-Br nanoparticles at different time intervals and the reaction can be stopped at a particular reaction time to have specific polymer length. Grafting time is the main parameter for polymer length growth. Table 1 shows that 8 h of increase in grafting time (from 2.0 to 12.0 h reaction) results in about 80% rise in percent fractional weight gain on the $Fe_3O_4@SiO_2$-Br particles. Obviously, the 1 g of $Fe_3O_4@SiO_2$-Br particles could increase from 1.203 to 2.041 g by increasing grafting time, indicating to time-dependent growth of the polymers on the surface of particles. The grafting amount increases linearly up to 8.0 h of ATRP reaction, reaching a plateau. Similarly, immobilization percentage is not significantly improving after a 10 h of ATRP reaction treatment. After grafting with GMA, epoxy groups on the $Fe_3O_4@SiO_2@$pGMA-1-6 nanoparticles reached 1.38–6.97 mmol/g of nanoparticles.

The average GMA units could be increased from 5.5 to 16.5 repeating units by ATRP application time while the specific surface area of the $Fe_3O_4@SiO_2$ and $Fe_3O_4@SiO_2@$p(GMA)-3 particles decreased from 37.8 to 21.4 m^2/g particles as determined by the Brunauer–Emmett–Teller (BET) analysis, respectively. The reduction of specific surface area of the $Fe_3O_4@SiO_2$ upon the grafting with p(GMA) might be originating from less porosity of the grafted $Fe_3O_4@SiO_2$ nanoparticles. Therefore, BET analysis can be interpreted for successful polymerization on the surface of the nanoparticles. HMDA spacer arm molecules were fixed on p(GMA) brushes on magnetic nanoparticles via ring opening reaction under alkaline condition. The amino group amounts on magnetic particles were 2.54 and 13.24 mmol/g as determined by potentiometric titration method. The size of the particles were determined to be 300–450 μm in diameter.

The ATR-FTIR analysis of Fe_3O_4, Fe_3O_4@SiO_2, and Fe_3O_4@SiO_2@p(GMA) between 400 and 4000 cm^{-1} wavelength (Fig. 3) was used for GMA grafting characterization. The spectroscopic scan with Fe_3O_4 nanoparticles resulted in a broad band at around 3205 cm^{-1}, indicating to –OH group presence on the surface of the magnetic nanoparticles from silica shell (Fig. 3A(a)). Another FTIR band with Fe_3O_4 nanoparticles was a strong band around 594 cm^{-1}, assigned to Fe—O vibration related to Fe_3O_4. The analysis with Fe_3O_4@ SiO_2 nanoparticles produced 1051 cm^{-1} peak indicating to Si—O bond stretching of Si—O—Si bonds (Fig. 3A(b)). Fe—O vibration band shifted from 594 cm^{-1}in Fe_3O_4 to 664 cm^{-1}, indicating that Fe_3O_4 nanoparticles were successfully grafted with p(GMA) (Fig. 3A(c)).

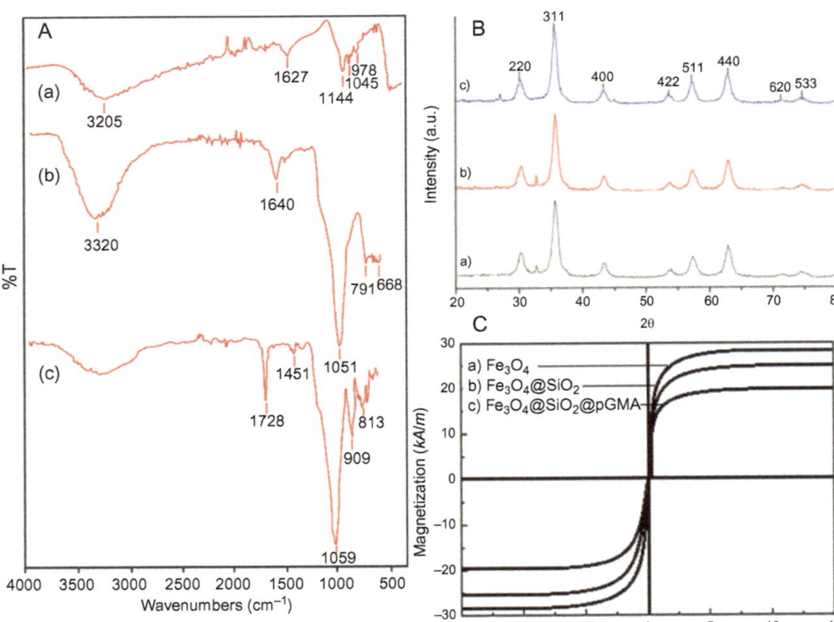

Fig. 3 (A) The ATR-FTIR spectra, (B) The XRD patterns and (C) VSM analysis for (a) Fe_3O_4; (b) Fe_3O_4@SiO_2; and (c) Fe_3O_4@SiO_2@pGMA-3 nanoparticles (Bayramoglu et al., 2017).

The epoxy vibration peaks of p(GMA) typically at 909 cm^{-1}, and the band at 1059 and 1728 cm^{-1} corresponding to ester group CAO vibration, and C@O group, respectively, was determined after epoxy functionalization.

X-ray diffraction (XRD) determined crystalline structure of nanoparticles (Fig. 3B). The XRD spectra of the Fe_3O_4@SiO_2 and Fe_3O_4@SiO_2@pGMA resulted in diffraction peaks typically observed with Fe_3O_4 nanoparticles. These results suggest that the inner core, Fe_3O_4 is well retained in Fe_3O_4 and Fe_3O_4@SiO_2@p(GMA) nanoparticles. The reflection peaks were compared to the X-ray diffraction data cards (ICDD number 88-0315) with eight diffraction peaks: (220), (311), (400), (422), (511), (440), (620) and (533) to find that all was correlated to desired synthesis. Fe_3O_4 grafted with silica or polymer grafted silica showed identical diffraction peaks (Fig. 3B(b) and (c)), which indicates to spinel structure and preservation of the magnetic core. The most important parameter after silanization is that magnetic core, thus magnetic force, should be maintained. The silica coated, and p(GMA) grafted Fe_3O_4 nanoparticles had the same spinel structures as magnetite. However, silica coating and p(GMA) grafting are not expected to result in detectable changes in the crystal structure after silica coating.

Vibrating sample magnetometer (VSM) is a standard technique used in the evaluation of magnetic properties of nanoparticles. The magnetic properties of the particles was studied by VSM analysis at 25 °C. Mass magnetization curves of the samples was obtained by applying magnetic field (Fig. 3C). The saturation magnetization was observed to be 28.4, 25.3 and 19.6 emu/g for Fe_3O_4, Fe_3O_4@SiO_2 and Fe_3O_4@SiO_2@p(GMA)-3, respectively. Magnetization curves exhibited zero remanence and coercivity, meaning that the particles were superparamagnetic. These results prove that magnetic nanoparticles can easily be removed efficiently from the solutions using an external permeant magnetic force. The synthesis procedures produced saturation magnetization strongly correlated to mass increase on Fe_3O_4. Modification with silica and grafting with p(GMA) resulted in lowering of magnetization properties probably due to lower magnetic moments from diamagnetic contribution from silica and p(GMA).

Direct visual examination of the particles can convincingly demonstrate that the synthesis was successful. Scanning electron microscopy (SEM) and atomic force microscopy (AFM) are commonly employed for examining morphology of the particles. The surface morphologies of chitosan particles before (Fig. 4A) and after (Fig. 4B) fibrous polymer grafting (at low and high magnifications) are investigated by SEM analysis. The irregular porous

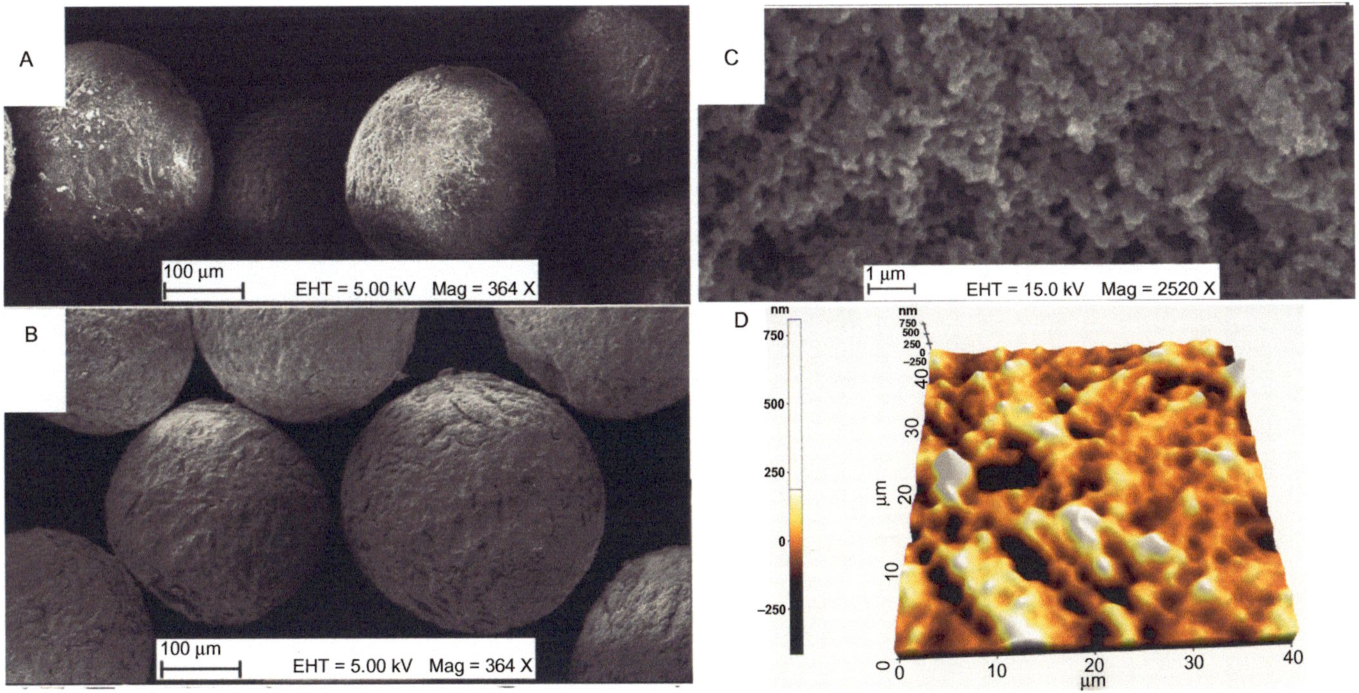

Fig. 4 SEM images of magnetic chitosan (A) and its p(GMA) grafted form at low (B) and high magnification (C). AFM image for a topographic analysis of p(GMA) grafted magnetic chitosan particles (D).

surface morphology of the magnetic chitosan beads disappeared and filled with grafted GMA polymer. This indicates that polymer grafting procedure was realized successfully. (Fig. 4C) is a close-up view to show the changing surface morphology, which can be confirmed by a detailed 3D AFM analysis (Fig. 4D). The surface is irregular with typical hills and valleys structure and the maximum height difference on the surface was determined to be about 500 nm. The diameter of the particles were measured as 75–150 μm by a molecular sieve method.

4. Stability, reusability and efficiency

In this section, the performance of nanofibrous polymer functionalized magnetic nanoparticles (Fe_3O_4@SiO_2@pGMA) will be discussed by using invertase as model enzyme. The kinetic parameters for enzymes have been the most important parameters for evaluating the usability of immobilized enzymes. Michaelis–Menten equation is the basis for estimating Michaelis constant (Km) and the maximum reaction velocity (Vmax) of free and immobilized forms of the enzyme, which are usually obtained by a modified equation Lineweaver–Burk plots (1/v versus 1/S). As an example, the kinetic parameters of fibrous magnetic chitosan beads will be discussed here. An apparent Km value of 39.4 mmol/L was obtained for immobilized invertase, which was very close to that obtained for the free enzyme (34.3 mmol/L). It indicated that immobilization did not significantly alter the affinity of invertase for its substrate. Since the magnetic nanoparticles have nanometer dimension size and extensive fibrous polymer matrix on the surface, hydrophilic properties on the nanoparticle surface should provide a comfortable microenvironment for invertase molecules. Thus, the enzyme would expand over the fibrous polymer surface with a good orientation for binding to the substrate with high affinity and readily available active sites. It is also worth to note that the grafting of magnetic particles with fibrous polymer could allow greater accessibility of sucrose to the active site of the enzyme immobilized, resulting from the decreased diffusion limitation. Due to the increase in hydrodynamic radius of the polymer-enzyme conjugate in the reaction medium, an extra-large surface area for enzymatic reaction could be provided opportunities to encounter its substrate. For free invertase, the Vmax value was determined.

4.1 Influence of enzyme concentration on stability

The influence of initial enzyme concentration on the immobilization amount of the Fe_3O_4@SiO_2@p(GMA)-GA-3 can be understood by studying a range of protein concentrations, for example, immobilization capacity for invertase were investigated from 0.5 to 4.0 mg/mL and related to chain length (Bayramoglu et al., 2017). The concentration of enzyme is the main parameter in determining immobilization capacity, testing in a range of concentration can indicate to an optimum concentration. Invertase was covalently bonded from amino groups of the enzyme (i.e., lysine residues) and aldehyde groups on nanoparticle surface. As expected, higher concentration of enzyme increased immobilization capacity on Fe_3O_4@SiO_2@p (GMA)-SA-GA-3 from 0.5 to 4.0 mg/mL. Enzyme immobilization at 2.0 mg/mL concentration was 36.1 mg/g; however, enzymatic activity decreased 76.9%. Adding spacer arm and glutaraldehyde activation increased enzymatic activity about 1.93-folds compared with Fe_3O_4@SiO_2@ p(GMA)-3 form due to the additional extra 11-carbon spacer arm (6-carbon from HMDA and 5-carbon from GA). Spacer arms locate the enzyme away from the surface of the particles and provide better substrate availability. High immobilization amount of protein decreases enzymatic activity probably due to the conformational changes in the active center, resulting from changing three-dimensional structure. Immobilization capacities for covalent method are generally higher than other methods, ranging between 0.1 and 63.0 mg/g. For examples, *Candida rugosa* lipase (EC 3.1.1.3) was covalently immobilized onto hybrid epoxy-silica polymer at 23.4 mg/g beads with immobilization capacity of 7.22 mg/g (Bayramoğlu & Arıca, 2008; Yuce-Dursun et al., 2016). Alcohol dehydrogenase (ADH) of *Saccharomyces cerevisiae* was immobilized on magnetic Fe_3O_4@SiO_2 nanoparticles via epoxy conjugation with 4.49 mg/g (Jiang et al., 2016). All these examples prove that high amount of protein can be immobilized on magnetic nanoparticles by covalent bond formation.

4.2 Influence of chain length on stability

The chain length of brush polymer is an influential parameter for the stability of magnetic nanoparticle enzyme conjugate. The relationship between average p(GMA) chain length density and immobilization efficiency were investigated with invertase enzyme (Bayramoglu et al., 2017). Invertase was immobilized on Fe_3O_4@SiO_2@pGMA-GA-1-6 nanoparticles through the glutaraldehyde activation in one-step reaction as explained in Section 2.3.1.

The p(GMA) polymer chain amount on Fe_3O_4@SiO_2@p(GMA)-GA was between 5.5 and 16.5 GMA units per chain when 2.0 mg/mL invertase concentration was used. A range of p(GMA) unit density was tried with Fe_3O_4@SiO_2@p(GMA)-GA-1-6 nanoparticles were used to relate chain length to enzymatic activity. Higher p(GMA) chain length was directly related to higher immobilization efficiency (Fig. 5). This means that higher number of free aldehyde groups on the particles results in more enzyme immobilization. The maximum immobilization amount on Fe_3O_4@SiO_2@p(GMA)-SA-GA-6 was achieved at 54.3 mg enzyme/g particles. However, this was the saturation density, resulted in lowering of enzymatic activity. The higher p(GMA) chain density (i.e., aldehyde density) over a threshold resulted in lower enzymatic activity, which might be resulting from the layers formed between fibrous polymer chains on the nanoparticles and immobilized enzyme molecules, leading to a sticky web like coating around the particle surface to restrict enzyme substrate interaction. Therefore, such layers might block the diffusion of the substrate molecules (i.e., sucrose) to the inner layer of the polymer enzyme brushes (Table 1). Similar observation for association between chain length and enzyme activity was reported in other studies as well. Lipase immobilization on alkyl silanized magnetic

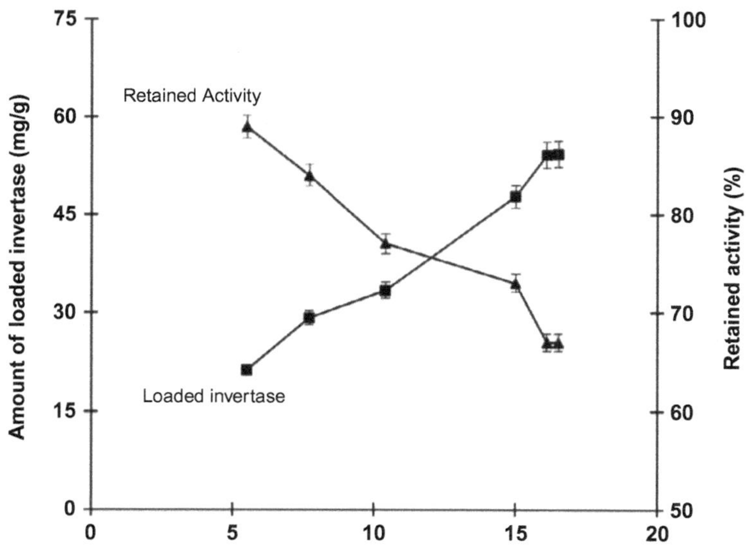

Fig. 5 Relationship between immobilization efficiency and p(GMA) average chain presence on Fe_3O_4@SiO_2@p(GMA)-GA- 1–6 nanoparticles (Bayramoglu et al., 2017).

nanoparticles is an example for increasing enzymatic activity along with alkyl chain length of the alkyltrimethoxy silanes (Wang et al., 2012). Mesoporous silica-lipase conjugates is another example for enzymatic activity increase with different carbon chain length (Zou et al., 2014).

4.3 Effect of pH and temperature

The protein conformation is labile to changes at the pH and temperature, and enzymes have optimum temperature and pH values. Therefore, the activity of enzymes are characterized after immobilization procedure at various pH and temperature. Invertase activity was determined at a pH range between 3.0 and 8.0 at 30 °C (Fig. 6A). A slight shift of optimum pH and wider activity range (from 5.0 to between 5.0 and 6.0) was observed for invertase activity with immobilization. Also, extreme pH activity was increased for immobilized form, for example, at pH 3.0, 38% of relative activity of free invertase increased to 59% after immobilization on fibrous magnetic beads. Similar activity improvements were observed for alkali pH, indicating to a general stabilizing activity of enzyme immobilization.

An improved activity of immobilization was also observed for higher temperatures. The optimum relative activity of immobilized form of invertase was obtained between 40.0 and 45.0 °C while free form had optimum at 35 °C (Fig. 6B). Over optimum temperature, both forms of enzyme had decreased activity, but immobilized form showed more stable activities (higher) at high temperatures (Bayramoğlu & Arıca, 2008).

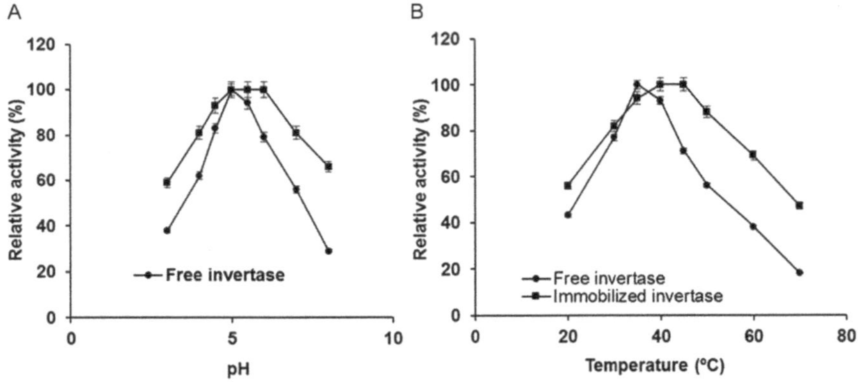

Fig. 6 The effect of (A) pH and (B) temperature on free and immobilized invertase on Fe_3O_4@SiO_2@pGMA-GA-3 nanoparticles.

4.4 Operational stability

Maintaining stability of enzyme is one of the challenges in continuous systems using enzymatic reaction. The immobilized form of invertase will be used as an example to explain stability change after immobilization. Invertase was investigated at 35 °C for 36h and monitored for stability parameters (Fig. 7A). A reactor of immobilized inverstase was prepared and continuously fed with 100mmol/L sucrose solution to find out that the enzyme activity reduced about 13% compared to initial activity in 36h. The first-order rate constant (k_{iop}) for operational inactivation was calculated for immobilized invertase as 6.94×10^{-5}/min for sucrose. Immobilization of invertase improved hydrolysis rates of sucrose significantly from 58% to 100% with the increase in the residence time from 0.024 to 0.081h. These results show that immobilization increased stability of invertase.

Temperature profiles of invertase showed a sharp inactivation for unbound enzyme while immobilization protected enzyme activity. Fig. 7D shows that free invertase activity was at 83% after 60min, 55 °C

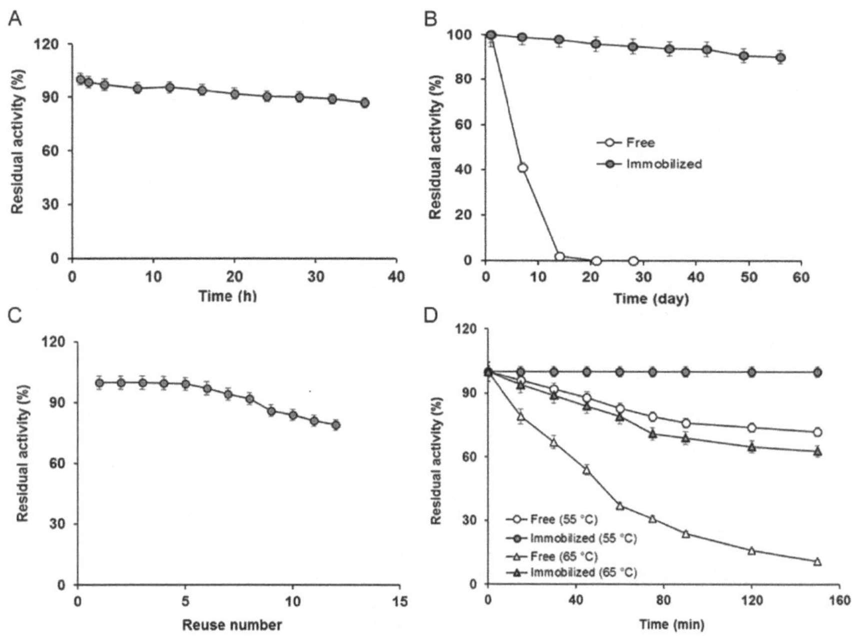

Fig. 7 (A) The operational stability and (B) repeated use of the immobilized invertase on Fe_3O_4@SiO_2@pGMA-SA-GA-3 nanoparticles (C) thermal stability (D) repeated use of the free and immobilized invertase on Fe_3O_4@SiO_2@pGMA-SA-GA-3 nanoparticles (Bayramoglu et al., 2017).

treatment, but immobilized form does not lose any activity during the same treatment. Enzyme inactivation rate constant and half-live time was determined according to the first-order inactivation kinetic model as 1.47×10^{-2} and 3.08×10^{-2} min, respectively, corresponding to about 2.1-fold more stability in immobilized form. Stability of the invertase were further investigated during repeated use to test the major advantage of enzyme immobilization on the magnetic support, which is easy separation and reusability. The residual activity for reusability of the immobilized enzyme at optimum temperature and pH for 30 min were assayed for a total of 12 cycles with magnetic separation in each cycle run and a wash step with acetate buffer (50 mmol/L, pH 5.5). The immobilized enzyme retained initial activity for five cycles and then a gradual decrease of 14% was observed at the end of ninth cycle (Fig. 6C). Stabilities of free and immobilized enzyme during storage conditions were investigated monitoring residual enzymatic activity (Fig. 6B). The samples of enzyme preparations were kept at 4 °C for 8 weeks and enzyme activity was measured. At the end of 6 weeks, immobilized enzyme was enzymatically active, but 2 weeks period was enough to observe activity decrease with free enzymes at the same conditions.

5. Conclusions

Enzyme immobilization is a useful technique for improving catalysis efficiency under production conditions. The fibrous polymer grafting on magnetic particles results in some invaluable advantages due to: (i) desired reactive chemical groups can easily be obtained via SI-ATRP method, (ii) being mild immobilization method for enzymes, (iii) providing high surface area and low volume ratio, (iv) easy separation from the reaction content with an external magnet. Immobilization might change the optimum pH and wider pH activity range. Immobilization might also raise the optimum temperature and improve thermal stability of the enzyme. The stability changes results from interactions of enzyme with hydrophilic fibrous polymer, providing conformational flexibility to immobilized enzyme. Additionally, fibrous polymer brushes provide a higher density of enzymes on the nanoparticle surface, leading to higher stability of protein conformations. Kinetic parameters for free and immobilized forms of enzyme can stay unchanged. The shelf life of immobilized enzyme might be prolonged several weeks with little loss in enzyme activity after tens of times reuse of the same magnetic particles. Thus, the immobilization of enzymes provides easy recovery and reuse of the enzyme.

Conflict of interest

All authors of this chapter have no conflict of interest.

References

Bayramoğlu, G., & Arıca, M. Y. (2008). Preparation of poly(glycidylmethacrylate–methylmethacrylate) magnetic beads: Application in lipase immobilization. *Journal of Molecular Catalysis B: Enzymatic*, *55*(1), 76–83.

Bayramoglu, G., Doz, T., Ozalp, V. C., & Arica, M. Y. (2017). Improvement stability and performance of invertase via immobilization on to silanized and polymer brush grafted magnetic nanoparticles. *Food Chemistry*, *221*, 1442–1450.

Bayramoglu, G., Tekinay, T., Ozalp, V. C., & Arica, M. Y. (2015). Fibrous polymer grafted magnetic chitosan beads with strong poly(cation-exchange) groups for single step purification of lysozyme. *Journal of Chromatography. B, Analytical Technologies in the Biomedical and Life Sciences*, *990*, 84–95.

Bradford, M. M. (1976). A rapid and sensitive method for the quantitation of microgram quantities of protein utilizing the principle of protein-dye binding. *Analytical Biochemistry*, *72*(1–2), 248–254.

Chen, M., Zeng, G. M., Xu, P., Lai, C., & Tang, L. (2017). How do enzymes 'meet' nanoparticles and nanomaterials? *Trends in Biochemical Sciences*, *42*(11), 914–930.

Fu, J., Reinhold, J., & Woodbury, N. W. (2011). Peptide-modified surfaces for enzyme immobilization. *PLoS One*, *6*(4), e18692. https://doi.org/10.1371/journal.pone.0018692.

Hartmann, M., & Kostrov, X. (2013). Immobilization of enzymes on porous silicas—Benefits and challenges. *Chemical Society Reviews*, *42*(15), 6277–6289.

Hong, T., Liu, W., Li, M., & Chen, C. (2019). Recent advances in the fabrication and application of nanomaterial-based enzymatic microsystems in chemical and biological sciences. *Analytica Chimica Acta*, *1067*, 31–47.

Jiang, X.-P., Lu, T.-T., Liu, C.-H., Ling, X.-M., Zhuang, M.-Y., Zhang, J.-X., et al. (2016). Immobilization of dehydrogenase onto epoxy-functionalized nanoparticles for synthesis of (R)-mandelic acid. *International Journal of Biological Macromolecules*, *88*, 9–17.

Johnson, R. D., Wang, Z. G., & Arnold, F. H. (1996). Surface site heterogeneity and lateral interactions in multipoint protein adsorption. *The Journal of Physical Chemistry*, *100*(12), 5134–5139.

Khoshnevisan, K., Bordbar, A.-K., Zare, D., Davoodi, D., Noruzi, M., Barkhi, M., et al. (2011). Immobilization of cellulase enzyme on superparamagnetic nanoparticles and determination of its activity and stability. *Chemical Engineering Journal*, *171*(2), 669–673.

Lee, B., Lopez-Ferrer, D., Kim, B. C., Na, H. B., Park, Y. I., Weitz, K. K., et al. (2011). Rapid and efficient protein digestion using trypsin-coated magnetic nanoparticles under pressure cycles. *Proteomics*, *11*(2), 309–318.

Monsan, P., & Combes, D. (1988). Enzyme stabilization by immobilization. *Methods in Enzymology*, *137*, 584–598.

Mu, X., Qiao, J., Qi, L., Dong, P., & Ma, H. (2014). Poly(2-vinyl-4,4-dimethylazlactone)-functionalized magnetic nanoparticles as carriers for enzyme immobilization and its application. *ACS Applied Materials & Interfaces*, *6*(23), 21346–21354. https://doi.org/10.1021/am5063025.

Prier, C. K., & Kosjek, B. (2019). Recent preparative applications of redox enzymes. *Current Opinion in Chemical Biology*, *49*, 105–112.

Qiu, J., Peng, H., & Liang, R. (2007). Ferrocene-modified Fe_3O_4@SiO_2 magnetic nanoparticles as building blocks for construction of reagentless enzyme-based biosensors. *Electrochemistry Communications*, *9*(11), 2734–2738.

Rao, S. V., Anderson, K. W., & Bachas, L. G. (1998). Oriented immobilization of proteins. *Mikrochimica Acta*, *128*(3–4), 127–143.

Singh, K., Mishra, A., Sharma, D., & Singh, K. (2019). Nanotechnology in enzyme immobilization: An overview on enzyme immobilization with nanoparticle matrix. *Current Nanoscience*, *15*(3), 234–241.
Spahn, C., & Minteer, S. (2008). Enzyme immobilization in biotechnology. *Recent Patents on Engineering*, *2*, 195–200.
Uygun, D. A., Öztürk, N., Akgöl, S., & Denizli, A. (2012). Novel magnetic nanoparticles for the hydrolysis of starch with Bacillus licheniformis α-amylase. *Journal of Applied Polymer Science*, *123*(5), 2574–2581.
Wang, J. Q., Meng, G., Tao, K., Feng, M., Zhao, X. B., Li, Z., et al. (2012). Immobilization of lipases on alkyl silane modified magnetic nanoparticles: Effect of alkyl chain length on enzyme activity. *PLoS One*, 7(8), e43478.
Yuce-Dursun, B., Cigil, A. B., Dongez, D., Kahraman, M. V., Ogan, A., & Demir, S. (2016). Preparation and characterization of sol–gel hybrid coating films for covalent immobilization of lipase enzyme. *Journal of Molecular Catalysis B: Enzymatic*, *127*, 18–25.
Zhang, Y. F., Ge, J., & Liu, Z. (2015). Enhanced activity of immobilized or chemically modified enzymes. *ACS Catalysis*, *5*(8), 4503–4513.
Zou, B., Hu, Y., Cui, F., Jiang, L., Yu, D., & Huang, H. (2014). Effect of surface modification of low cost mesoporous SiO_2 carriers on the properties of immobilized lipase. *Journal of Colloid and Interface Science*, *417*, 210–216.

CHAPTER SIX

Improvement in biochemical characteristics of cross-linked enzyme aggregates (CLEAs) with magnetic nanoparticles as support matrix

Nithyakalyani Doraiswamy[a], Mahalakshmi Sarathi[a], Gautam Pennathur[a,b,*]
[a]Centre for Biotechnology, Anna University, Chennai, Tamil Nadu, India
[b]AU-KBC Research Centre, Anna University, Chennai, Tamil Nadu, India
*Corresponding author: e-mail address: pgautam@annauniv.edu

This work is dedicated to Professor P. Balaram on the occasion of his 70th birthday.

Contents

Abstract

Recent developments in novel carriers for enzyme immobilization have led to improvement in the stability and cost-effectiveness of the biocatalysts for their enhanced suitability in the industrial applications. Cross-linked enzyme aggregates (CLEAs), a recent technique developed in the carrier-free type of enzyme immobilization is a simple and

Methods in Enzymology, Volume 630
ISSN 0076-6879
https://doi.org/10.1016/bs.mie.2019.10.019

straightforward method. Moreover, the magnetic property and the higher surface-to-volume ratio of the maghemite nanoparticles have also been utilized in the present immobilization technique as magnetic nanoparticle-supported CLEAs (Mgnp-CLEAs). The stability studies of the free and immobilized enzyme revealed the Mgnp-CLEAs to have enhanced enzyme stability with an increase in the reusability cycle. The physical characterization of the nanoparticles and immobilized enzymes by the Scanning Electron Microscopy (SEM), Fourier-Transform Infrared spectroscopy (FT-IR) and X-ray diffraction analysis (XRD) showed the successful immobilization of the enzyme for its improved stability.

1. Introduction

The versatility of the reactions carried out by enzymes makes them a promising biocatalyst for industrial applications. Nevertheless enzymes are sensitive to operational conditions for their catalytic activity (Alexandre, Hubert, & Peter, 2011). The main challenge in large-scale processes employing enzymes is that the enzyme activity decreases under industrial reaction conditions. These shortcomings are effectively solved by constraining the enzyme in a favorable micro-environment through immobilization, thus increasing its local concentration and also its robustness for its stability during large-scale applications (Chaubey, Parshad, Taneja, & Qazi, 2009).

Immobilization techniques are one of the best strategies to improve the biocatalyst owing to the enhancement of operational stability, reusability, prevention of sub-units dissociation, aggregation, autolysis and proteolysis. It can also alter the geometry of the active-site center thereby stabilizing the structural conformation for improved enzyme selectivity and successful catalytic reaction (Cui & Jia, 2013). Due to these advantages, immobilization of the biocatalysts has developed into a promising technique for the improvement of enzyme features that facilitates easier separation with minimalization of the downstream processing thus making the process cost-effective in the long run (Hakan, Fatma, & Aydan, 2012).

The methodologies for enzyme immobilization could be grouped as two types: physical or chemical, depending on the type of interactions involved during immobilization process. The physical method employs entrapment, adsorption and microencapsulation techniques which mainly involves physical forces for enzyme attachment such as van der Waals, hydrophobic interactions and hydrogen bonding in a reversible manner while the chemical method involves attachment of the enzyme to the matrix surface through chemical bonds such as covalent bonding, cross-linking, ionic binding and conjugation by affinity ligands (Dwevedi, 2016). An efficient process requires optimization of immobilization conditions to maintain and improve the catalytic activity of the enzyme. Moreover, these efficiencies are quantified by percentage of enzyme recovery, operational stability, selectivity and reduction in inhibition by substrates or products in the reaction medium. However the cost of immobilization also reflects on its applications at industrial level which is mainly dependent on enzyme source, expression level, isolation and purification of the enzyme, matrix and method of immobilization technique employed (Dwevedi, 2016).

There are numerous ways for the immobilization of an enzyme as either through entrapment onto different matrices or binding onto various supports. Recently, these methodologies have been modified by fabrication of supports and cross-linking of enzymes for diverse applications. CLEAs technique was introduced in 2000's for its remarkable features such as its simple preparation, highly concentrated activity per unit volume, low production cost for exclusion of carrier, capability for industrial scale-up and lack of requirement for purified enzymes along with enhanced stability against denaturing agents (Sachin, Shamraja, Asavari, & Gandhali, 2014). Successful CLEAs preparations for a wide range of enzymes such as nitrilases, lipases, esterases, aminoacylases, proteases, glycosidases and laccases has proved its versatility in its application for immobilization (Cui, Zhang, & Sun, 2012; Dwevedi, 2016).

Nanostructures have drawn attention for its high surface-to-volume ratio with excellent physicochemical characteristics thereby making them a promising support for the immobilization of the enzymes (Cao et al., 2016). Along with its biocompatible nature nanoparticles have become an integral part in the enzyme immobilization field for its varied applications (Leitgeb, Knez, & Vasic, 2016).

1.1 Magnetic nanoparticle (γ-Fe_2O_3) synthesis

The magnetic nanoparticles have drawn greater attention for its properties such as high surface area which enables higher enzyme loading with lowered mass-transfer resistance, non-toxic nature, superparamagnetic property and its ease of separation through magnetic decantation. This has led to their use as immobilizing matrices (Wang et al., 2012) as they have significantly proved to be useful supports for bioactive materials (Chaubey et al., 2009;

Ladole, Muley, Patil, Talib, & Parate, 2014) for its low toxicity and control over the catalysis. Earlier reports by Cui and Jia (2013) showed that the use of the magnetic nanoparticle as a matrix improves the overall enzyme activity and stability for its continuous operations.

In this study, maghemite nanoparticles (γ-Fe_2O_3) was preferred over magnetite nanoparticles (Fe_3O_4) as it acts as a better adsorbent with increased specificity for enzyme binding. Its strong magnetic behavior is due to its spinel structure with magnetically non-equivalent interpenetrating sub-lattices. And also other interesting properties such as its biocompatibility, biodegradability and low particle dimension with large surface area makes it a potential candidate for enzyme immobilization (Shalyda et al., 2016). However the bare magnetic cannot be used directly to immobilize the enzymes thus surface modification is necessary through amphiphiles (oleic acid, SDS) or silane derivatives (Wang et al., 2012).

Functionalization of the magnetic surface regulates enzymatic interaction, increases its biocompatible nature and has better dispersion in the aqueous environment. Improvement in the activity recovery and stability enhancement for the xylanase immobilized on functionalized magnetite nanoparticles as xylanase Mgnp-CLEAs has been reported by Shalyda et al. (2016). In this study, 3-aminopropyl triethoxysilane (APTES) functionalized maghemite nanoparticles have been used for the stabilization of the esterase-CLEAs.

1.2 Silanization with 3-aminopropyl triethoxysilane (APTES)

Surface modification of the magnetic nanoparticles through functionalization makes them more biocompatible and stable avoiding chemical corrosion and instability. It also gives additional stabilization, through chemical and physical interactions to these nanoparticles. The hydroxyl groups on the magnetic surface are the sites of modification for the amphiphiles or silane derivatives to create hydrophobic surface for enzyme attachment. The silane derivatives are usually preferred over the amphiphiles for surface modification as it provides an improved structural stability through chemical grafting and tuning of interfacial surface property (hydrophilicity and hydrophobicity). Different silane precursors can be used for its application in wide range of CLEAs. It allows for the attachment of additional functional groups such as NH_2, –SH and –COOH for further functionalization and protects the surface from oxidation with no loss of its magnetic property (Wang et al., 2012) (Fig. 1).

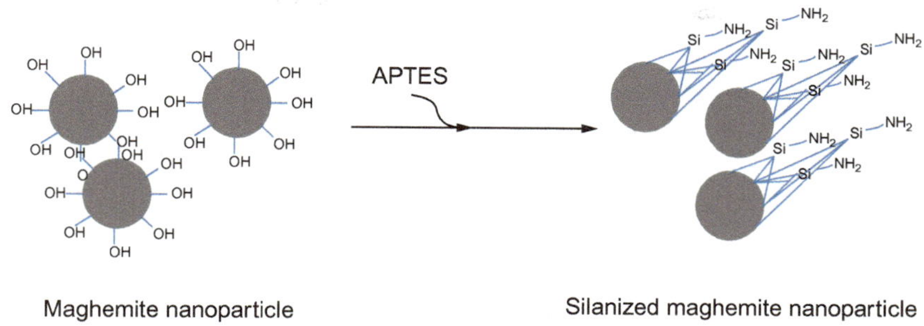

Fig. 1 Graphical illustration of silanization of the magnetic nanoparticles.

Amino-functionalization of the magnetic matrix allows for efficient cross-linking of the enzymes. Addition of magnetic nanoparticles was found to enhance the activity recovery of the enzymes (Cui & Jia, 2013). The magnetic-CLEAs were prepared by adsorbing the enzyme onto magnetic nanoparticles followed by cross-linking with glutaraldehyde. They shows ability over a wide temperature and pH range with good operational stability and magnetic response. Recent examples include cellulase immobilized on magnetic chitosan microspheres, *Candida rugosa* lipase immobilized on amino-functionalized magnetic beads for polycaprolactone (PCL) degradation and esterase immobilized on amino-functionalized magnetic beads. The Mgnp-CLEAs prepared with the amino-functionalized magnetite nanoparticles have shown to exhibit excellent operational stability and magnetic response (Cui et al., 2014) thus indicating its significant role in the supported CLEAs preparation (Lozano, Gallego, Diaz, & Torres, 2016).

1.3 Enzyme immobilization

Esterases belong to α/β hydrolase fold comprising the catalytic triad Ser-Asp-His residues (Sayali, Sadichha, & Surekha, 2013). They are capable of both esterification/transesterification and hydrolysis reactions based on the absence or presence of water in the reaction mixture respectively. They constitute the majority of enzymes used for industrial application in the synthesis reaction especially in synthesis of esters in an organic solvent (Pastorino, Piolo, Zilli, Converti, & Nicolini, 2004). The formation and hydrolysis of the ester bonds with high enantioselectivity along with regiospecificity is of special interest during its application (Sayali et al., 2013). The enantioselectivity of the enzyme is used for the resolution of enantiomeric compounds in the field of organic chemistry (Pastorino et al., 2004). Apart from these processes, they also catalyze aminolysis and transamidation reactions (Wang et al., 2012). As they act on a wide range of substrates, it has been effectively employed for plastic degradation, synthesis of chiral alcohols and deinking processes (Sayali et al., 2013).

Although the esterase has varied uses, its application in industries is limited by their high cost of production and poor performance due to operational instability which is effectively overcome by immobilization. Immobilization facilitates the stabilization and separation of the esterase for reuse, thus enhancing its potential as industrial biocatalysts (Susana et al., 2014). Though numerous methods are available for immobilization there is always a search for a simple technique for an efficient immobilization (Cao et al., 2016). CLEAs prepared via the physical aggregation of the enzyme molecules through alteration of electrostatic constant of the solution with appropriate precipitation agents or through a change in its hydration state. The precipitation is subsequently followed by cross-linking to form an insolubilize aggregate. CLEAs are highly preferred as the enzymes are precipitated in its native enzyme conformation leading to increased immobilization yield which is distorted during other immobilization techniques which decreases activity. And also it is desired for its mechanical stability with desired size range preparation. CLEAs provide a flexible immobilization platform where it allows the use of various precipitants, cross-linkers and a wide range of enzymes for the design of robust CLEAs (Hakan et al., 2012). As the CLEAs avoid the use of solid supports, the specific activity of the immobilized enzyme increases such that it can be applied in small-sized reactors in comparison to other immobilized preparations prepared using solid carriers (Alexandre et al., 2011).

Despite huge advantages of this technique, the drawback of its soft fragile nature and small particle size of the CLEAs ($<10\,\mu m$) makes it difficult for recovery from the reaction medium. The aggregation of CLEAs into clumps during centrifugation and filtration hampers distribution in aqueous medium leading to low catalytic efficiency. These drawbacks can be overcome by developing a strategy of introducing supports to the CLEAs to enhance the operational stability of biocatalyst at extreme environmental conditions with improved reusability and storage stability (Cui & Jia, 2013). Recent developments of nanotechnology have led to the use of nanoparticles especially with magnetic properties for immobilization of enzymes (Shalyda et al., 2016) where its semi-rigid nature is found to provide rigidity required for the enzymes at the industrial scale (Ladole et al., 2014) along with its ease of separation from the reaction mixture through magnetic decantation. Due to the ever-growing demand and widespread use of the esterase, in this study the stability of the esterase has been enhanced via efficient immobilization in its native conformation on the maghemite nanoparticles as Mgnp-CLEAs.

2. Material and methods

2.1 General equipment

1. *Safety equipment*: Laboratory coat, nitrile gloves, safety goggles, fume hood
2. Deionized water
3. Centrifuges
4. pH meter
5. Elisa microplate reader
6. Microcentrifuge tubes (2 mL)
7. Micropipettes (20–1000 μL)
8. Beakers (100–1000 mL)
9. Weighing balance
10. Incubator shaker
11. Sonicator
12. −20 °C freezer
13. Scanning electron microscope (SEM)
14. Fourier-transform infrared spectrometer (FT-IR)
15. X-ray diffractometer (XRD)

2.2 Synthesis of maghemite nanoparticle (γ-Fe_2O_3) for enzyme immobilization

1. Ferric chloride
2. Ferrous sulfate
3. 25% ammonia solution
4. Distilled water

The maghemite nanoparticles (γ-Fe_2O_3) were synthesized by chemical co-precipitation method of Sulistyaningsih, Santosa, Siswanta, and Rusdiarso (2017) with slight modifications.

1. Dissolve 6.0 g $FeCl_3 \bullet 6H_2O$ and 4.0 g $FeSO_4 \bullet 7H_2O$ in 25 mL distilled water and stir continuously until the iron salts dissolve completely.
2. Place the solution prepared in step 1 inside the fume hood and add 25% ammonia slowly. While doing so stir the mixture continuously at 80 °C and maintain the pH at 11.0 until a black precipitate is formed.
3. Allow the black precipitate to settle for 12 h.

4. Decant the black precipitate using an external magnet, and wash it several times with distilled water till the pH of the filtrate was neutral (pH 7.0).
5. Filter the black precipitate and dry it well in the oven.
6. Store the synthesized maghemite at room temperature for further studies.

2.2.1 Functionalization of maghemite nanoparticle (γ-Fe2O3) via APTES silanization

1. APTES
2. Maghemite nanoparticles
3. Ethanol
4. Methanol
5. Glycerol
6. Distilled water

The maghemite nanoparticles were functionalized with amino groups of APTES by the method of Talekar, Waingade, Gaikwad, Pati, and Nagavekar (2011).

1. In a 100 mL beaker, pipette out 2.5 mL ethanol, 25 μL distilled water and 100 μL APTES and shake well.
2. Weigh 20 mg maghemite nanoparticles and 1.5 mL glycerol to the solution prepared in the step 1 and heat it at 90 °C with continuous agitation for 6 h.
3. At the end of 6 h, stop the stirring and allow the functionalized magnetic nanoparticles to settle down.
4. Wash the settled functionalized magnetic nanoparticle with distilled water three times to remove the unreacted APTES.
5. Wash it with methanol to remove any remaining traces of water and air-dry to isolate the fine functionalized powder of maghemite nanoparticle.
6. Store the functionalized magnetic nanoparticles at room temperature for further studies.

2.2.2 Characterization of maghemite with SEM, FT-IR and XRD analysis

1. Characterize the surface morphology of maghemite nanoparticle using SEM (Hitachi Model S-3400 N electron microscope) at an operating voltage of 15 kV.
2. Characterize the functional groups of maghemite nanoparticle by FT-IR (JASCO FT-IR-4700 (Japan)) in the range of 4000–400 cm^{-1}, respectively.

3. Determine the crystalline structure of the maghemite nanoparticle using XRD, GE Inspection Technology Model 3003TT.

The SEM image as shown in the Fig. 2A shows the nanopowder of the maghemite nanoparticles as agglomerates of nanoparticles at the size range distribution of 700–1000 nm synthesized via co-precipitation methodology.

The FT-IR spectra of the maghemite nanoparticle in the Fig. 2B shows the characteristic peak for the magnetic nanoparticles around 580 cm^{-1} corresponding to the Fe—O bond stretch vibration (Asgari, Fakhari, & Berijani, 2014) indicating the formation of the maghemite. The bands in the range 1500–1650 cm^{-1} shows the presence of the amino groups (Kouassi, Irudayaraj, & McCarty, 2005). The bands around 3400 cm^{-1} shows the presence of hydroxyl groups corresponding to the water molecules associated on the particle surface (Natália, Marcelo, & Angela de Mello, 2017). The FT-IR spectra of the silanized magnetite nanoparticle in the Fig. 2B indicates the presence of the silanol, propyl and amino groups introduced by the APTES as the cross-linking sites on the surface of the maghemite nanoparticles. The Si—O—Si bond vibrations of the silanol group on the maghemite surface show characteristic bands at 1006 cm^{-1} and the propyl group of APTES absorbs at 2883 cm^{-1}. In addition to these bands, N—H stretching of the free amino groups of APTES shows the characteristic bands around 1600 cm^{-1} (Shalyda et al., 2016).

The XRD spectra in the Fig. 2C were used to further characterize the crystalline nature of the magnetic nanoparticles synthesized. It revealed the appearance of six peaks at $2\theta = 30.1°$, 35.5°, 43.1°, 53.4°, 57.0° and 62.6° marked by their indices ((210), (311), (400), (422), (511) and (440)), characteristic of the magnetic nanoparticles. Further, functionalization of the nanoparticles did not affect the position of the peaks characteristic of the magnetic nanoparticles (Asgari et al., 2014; Yuhan et al., 2018).

2.3 Immobilization of esterase via CLEAs

2.3.1 Preparation of CLEAs

1. Crude enzyme
2. 1 M tris-HCl buffer (pH 8.0)
3. Ammonium sulfate ($(NH_4)_2SO_4$)
4. Methanol
5. Trichloro acetic acid (TCA)
6. 2-Propanol
7. Ethanol
8. Acetone

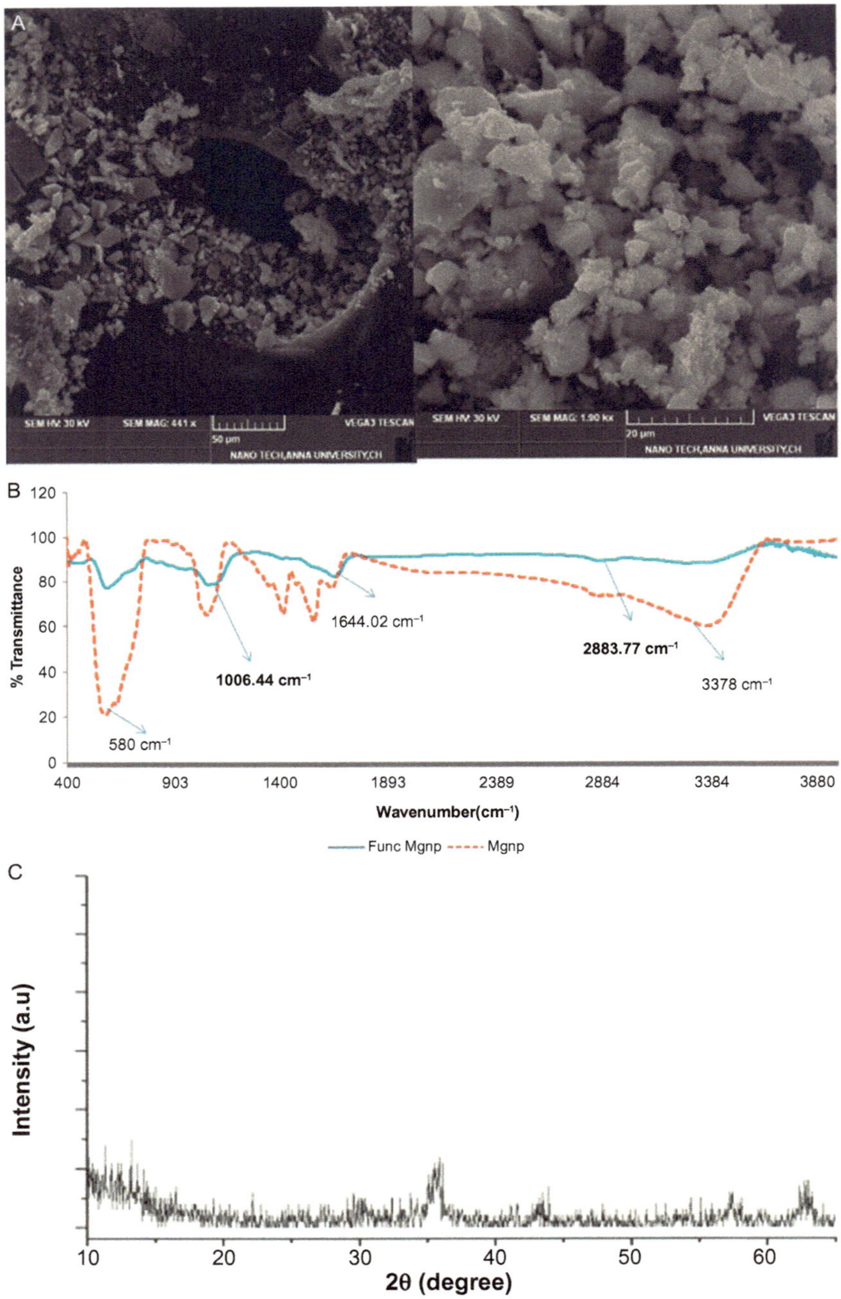

Fig. 2 Characterization of maghemite nanoparticle with (A) SEM of Mgnp, (B) FT-IR of bare and silanized Mgnp, (C) XRD of Mgnp.

9. Glutaraldehyde
10. Preparation of saturated ammonium sulfate (SAS)—It was prepared by adding 76.8 g of $(NH_4)_2SO_4$ salt in 100 mL of distilled water.

The CLEAs was prepared by the method of Cao, van Rantwijk, and Sheldon (2000).

1. Wear gloves and lab coat while taking 200 mL of the organic solvents in the 500 mL glass bottles in the fume hood and chill the solvents at −20 °C for 4 h.
2. In the fume hood, mix 90 mL chilled organic solvents with 10 mL crude enzyme solution (20 mg/mL in 10 mM tris-HCl buffer, pH 8.0) in the molar ratio of 9:1 at low temperature (4 °C) and store it overnight at 4 °C. For salt precipitation, mix 15 mL of SAS with 10 mL of crude enzyme with continuous stirring at 4 °C overnight.
3. Cross-link the precipitated enzyme at 30 °C and 150 rpm agitation with varying concentrations of glutaraldehyde (10, 25, 50, 75, 100, 125 and 150 mM) for varied time period intervals (3, 4, 5, 6 and 7 h) to optimize the parameters for CLEAs preparation.
4. At the end of the cross-linking time, quench the reaction with 50 mL of tris-HCl buffer (pH 8.0).
5. Thoroughly wash the CLEAs with tris-HCl buffer (pH 8.0) to remove the unreacted glutaraldehyde and free enzyme.
6. Store the CLEAs in tris-HCl buffer (pH 8.0) at 4 °C for further studies.

CLEAs helps in the stabilization of multimeric enzymes, which is a major limitation in other immobilization strategies (Dwevedi, 2016). The slightly distorted conformations of aggregates obtained during aggregation process is fixed by cross-linkers revealing the importance of the precipitating and cross-linking agents on the active conformation of the CLEAs (Lopez-Serrano, Cao, van Rantwijk, & Sheldon, 2002). The precipitants, cross-linker concentration and the time required for cross-linking are the major factors influencing the activity recovery of the CLEAs synthesized. The commonly employed salts and solvents used yields CLEAs with varying degrees of enzyme activity as it differs with the biochemical and structural properties of the enzyme. The Fig. 3A results showed saturated ammonium sulfate (SAS) to be the best precipitant at a concentration of 60% with 96% of the initial enzyme activity retained and precipitation by organic solvents as ineffective precipitants with <50% retention of initial activity. The SAS has been shown as the best precipitant of choice for CLEAs preparation (Rehman, Bhatti, Bilal, & Asgher, 2016). The organic solvents aggregate proteins through enhancement of the protein–protein interactions rather

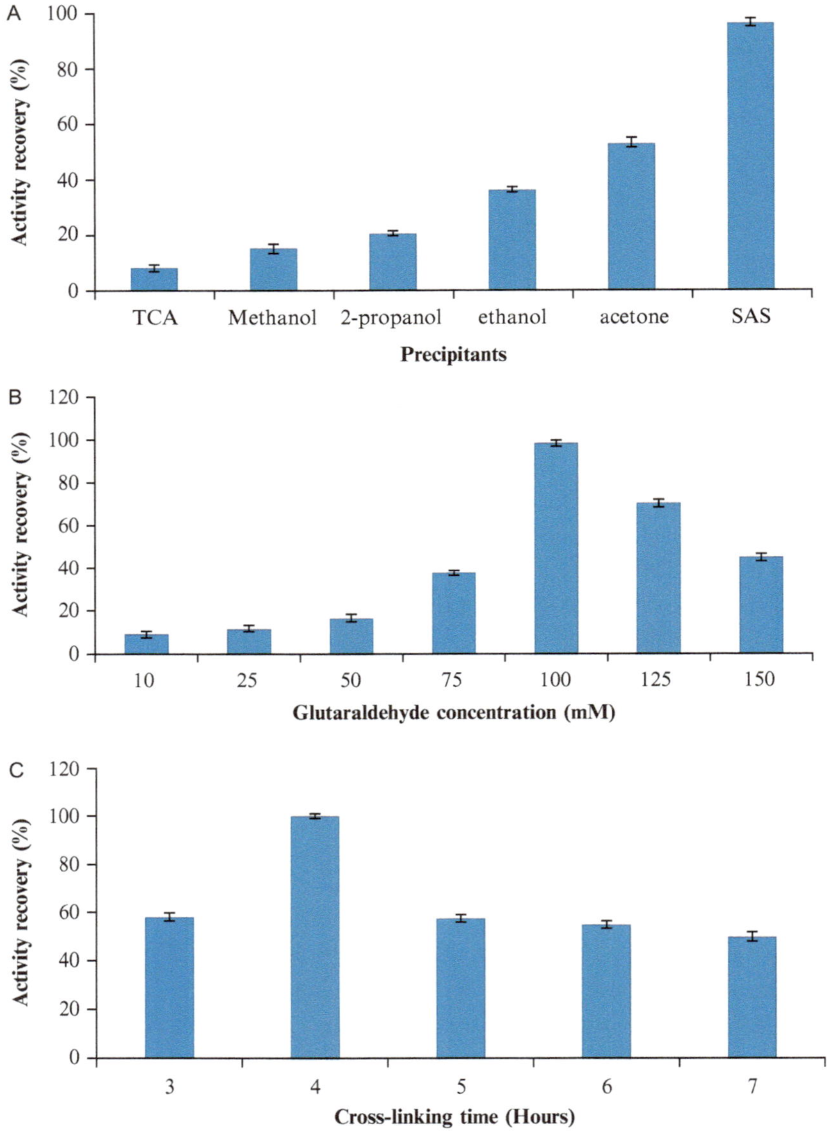

Fig. 3 Optimization of the variables for CLEAs preparation with (A) optimization of the precipitants, (B) cross-linker concentration and (C) cross-linking time.

than the protein–solvent interactions via reduction in the dielectric constant of the medium. The water in hydration spheres of the protein are replaced with the solvent molecules thereby bringing the protein molecules together. However the reduced activity recovery observed with solvents is probably due to change in the conformations or distortion of the active site with the penetration of solvent molecules (Shalyda et al., 2016).

The cross-linker influences the activity, stability and particle size of the CLEAs. The esterase activity increased with an increase in glutaraldehyde concentration as shown in the Fig. 3B and C, yielding the highest activity of 98.5% at a concentration of 100mM with 4h cross-linking time as reported in previous studies by Talekar et al. (2011), Shalyda et al. (2016) and Schoevaart et al. (2004). Nevertheless, increased concentration of cross-linker and cross-linking time beyond the optimal range caused a decrease in the activity which might be due to extensive cross-linking of enzymes thereby reducing its flexibility, giving rise to inactive CLEAs. Similarly, lower concentrations and lower time period duration for cross-linking also had lower activity recovery as the ineffective cross-linking produced CLEAs of increased flexibility resulting in leaching of enzyme into the medium (Shalyda et al., 2016).

2.3.2 Preparation of Mgnp-CLEAs

1. Crude enzyme
2. Functionalized maghemite nanoparticles
3. SAS
4. Glutaraldehyde

The maghemite nanoparticle (γ-Fe_2O_3) supported cross-linked enzyme aggregates (Mgnp-CLEAs) was prepared by the method of Talekar et al. (2011).

1. To 10mL of the crude enzyme (20mg/mL in 10mM tris-HCl buffer, pH 8.0) add 10mg of functionalized maghemite nanoparticles.
2. Incubate the mixture at 30°C and 150rpm agitation for 2h.
3. At the end of 2h, precipitate the enzyme adsorbed maghemite nanoparticles with 60% SAS at 4°C for overnight.
4. Cross-link the precipitated mixture with 100mM glutaraldehyde at 30°C and 150rpm agitation for 4h.
5. At the end of 4h, thoroughly wash the Mgnp-CLEAs with tris-HCl buffer (pH 8.0) to remove the free enzyme and unattached Mgnp.
6. Store the Mgnp-CLEAs in tris-HCl buffer (pH 8.0) at 4°C for further use.

The major limitation of the naked maghemite nanoparticles forming large aggregates was prevented by surface coating with APTES. The surface particle functionalization with organosilane such as APTES leads to efficient immobilization. Moreover decreased recovery of activity of the enzymes immobilized on non-silanized maghemite nanoparticles compared to the silanized counterpart has been observed due to lower number of enzyme molecules covalently bound to the non-silanized surface. The stability of the esterase immobilized on silanized maghemite nanoparticle is enhanced through an increase in the number of covalent bonds between the maghemite surface and enzyme molecule (Shalyda et al., 2016). Hence the esterase enzyme was immobilized with silanized maghemite nanoparticles to form Mgnp-CLEAs.

2.4 Characterization of CLEAs and Mgnp-CLEAs using SEM and FT-IR

1. The surface morphology of the CLEAs and Mgnp-CLEAs were characterized using SEM (Hitachi Model S-3400 N electron microscope) at an operating voltage of 15 kV.
2. The functional groups and amide bands involved in enzyme immobilization for CLEAs and Mgnp-CLEAs were characterized by FT-IR (JASCO FT-IR-4700 (Japan)) in the range of 4000–400 cm^{-1}, respectively.

The SEM image of the Esterase—CLEAs at 5500 × and 10,000 × magnification in the Fig. 4A revealed Type 1 "ball-like" aggregates morphology which shows the highly lipophilic nature of the esterase enzyme. The Mgnp-CLEAs at 5000 × magnification shows the enzyme molecules to be aggregated as clumps in Fig. 4B. The uniformity and small size (< 10 μm) of the CLEAs had increased in size to 40–60 μm with irregularities with the introduction of maghemite nanoparticles during the preparation of Mgnp-CLEAs. Similar transition of morphology from typical "ball-like" to amorphous morphology for CLEAs and Mgnp-CLEAs has been observed for xylanase Mgnp-CLEAs by Shalyda et al. (2016) (Table 1).

The immobilization of the esterase as CLEAs and Mgnp-CLEAs was analyzed using FT-IR and represented in Fig. 4C. The amide I and amide II bands of the C=O stretching, N—H bending and C—N stretching of the amino groups of the protein shows bands at 1650 and 1553.85 cm^{-1} band, respectively. The amide III band shows bands at 1238.76 and 1242.50 cm^{-1} which represent the N—H bending and C—N stretching, respectively. The absence of peaks around ~800 cm^{-1} confirms the loss of NH_2 groups of the free enzyme as they are the possible cross-linking sites used during the cross-linking as CLEAs. The amide bands confirm the presence of esterase immobilized on magnetic nanoparticles as Mgnp-CLEAs (Shalyda et al., 2016).

Fig. 4 Characterization of immobilized enzyme with (A) and (B) SEM of CLEAs and Mgnp-CLEAs, (C) FT-IR of CLEAs and Mgnp-CLEAs.

Table 1 Size distribution for the CLEAs and Mgnp-CLEAs.

	CLEAs	**Mgnp-CLEAs**
Size distribution (μm)	10	40–60

2.5 Enzyme assay

2.5.1 Standard graph using 4-nitrophenol

1. 4-nitrophenol
2. NaOH
3. Tris base
4. Glycine base
5. HCl
6. Acetonitrile
7. Sodium acetate
8. Glacial acetic acid
9. Potassium di-hydrogen orthophosphate
10. Preparation of 1 M sodium acetate buffer (50 mL): Dissolve 4.10 g of sodium acetate in 40 mL distilled water and adjust the pH of the solution to pH 5.0 with glacial acetic acid and make it upto 50 mL with distilled water.
11. Preparation of 1 M phosphate buffer (50 mL): Dissolve 6.80 g of potassium di-hydrogen orthophosphate in 40 mL distilled water and adjust the pH of the solution to pH 6.0 with 0.1 N sodium hydroxide and make it upto 50 mL with distilled water.
12. Preparation of 1 M tris-HCl buffer (50 mL): Dissolve 6.05 g of tris base in 40 mL distilled water and adjust the pH of the solution to pH (7.0–8.0) with 0.1 N hydrochloric acid and make it upto 50 mL with distilled water.
13. Preparation of 1 M glycine-NaOH buffer (50 mL): Dissolve 3.75 g of glycine base in 40 mL distilled water and adjust the pH of the solution to pH (9.0–10.0) with 0.1 N sodium hydroxide and make it upto 50 mL with distilled water.

Prepare 4-nitrophenol in buffers of varying pH (pH 5.0–10.0) and plot the respective standard graphs at different pH.

2.5.2 Esterase assay with 4-nitrophenyl valerate (p-NPV)

1. Crude enzyme
2. 1 M tris-HCl buffer (pH 8.0)
3. Deionized water
4. Preparation of 50 mM *p*-NPV: Mix 4 μL of *p*-nitrophenyl valerate with 996 μL of acetonitrile to make 1 mL of 50 mM *p*-NPV in 2 mL eppendorf tube.

The esterase activity was determined using *p*-nitrophenyl valerate (*p*-NPV) as the substrate following the method of Kulakova, Nakayama, Nishino, and Esaki (2004).

1. Prepare the standard enzyme assay mixture (200 μL) containing 1 M *p*-NPV, 10 mM tris-HCl buffer and enzyme by pipetting out 4 μL of 50 mM *p*-NPV, 10 μL of 1 M tris-HCl buffer (pH 8.0), 100 μL enzyme supernatant and 86 μL water in the 96-well microplate.
2. Incubate the enzyme mixture at 30 °C for 10 min for free enzyme along with the blank without the enzyme as control.
3. Measure the release of 4-nitrophenol at its absorption maximum of 410 nm in an ELISA reader MODEL 680 (Bio-Rad, Italy).
4. The enzyme activity is calculated from the standard graph where one unit of esterase activity (1 U) is defined as the amount of enzyme required to catalyze the release of 1 μmol of 4-nitrophenol per minute at the standard assay conditions (pH 8.0 and 30 °C).

2.5.3 Immobilization yield and activity recovery

The enzyme activity recovered during each step of immobilization process for the CLEAs and Mgnp-CLEAs was calculated using the following formula:

$$\text{Activity recovery (\%)} = \frac{\text{Total activity of esterase in CLEAs (U)}}{\text{Total activity of esterase used for CLEAs preparation (U)}} \times 100$$

The CLEAs and Mgnp-CLEAs had a recovery of almost 100% which implies that all of the immobilized enzymes on the magnetic nanoparticles were active which could be due to the intrinsic property of the nanoparticles as to reduce protein unfolding and its following deactivation upon immobilization thereby enhancing its stability (Ahmad & Sardar, 2015) (Table 2).

2.6 Stability of free enzyme and CLEAs

The performance of the CLEAs and Mgnp-CLEAs in comparison to the free enzyme was investigated by measuring its stability at different pH, temperature, organic solvents, along with operational stability where the two important criteria determining its stability during immobilization are rigidity retained at its active site and flexibility of the enzyme molecule (Susana et al., 2014).

Table 2 Activity recovery (%) for the immobilized enzyme preparations.

	CLEAs	Mgnp-CLEAs
Activity recovery (%)	99.8	100

2.6.1 Biochemical characterization of the immobilized enzyme

2.6.1.1 pH stability of free enzyme and CLEAs

1. Crude enzyme
2. Immobilized enzyme
3. 1 M sodium acetate buffer (pH 5.0)
4. 1 M tris-HCl buffer (pH 7.0–8.0)
5. 1 M phosphate buffer (pH 6.0)
6. 1 M glycine-NaOH buffer (pH 9.0–10.0)
7. *p*-NPV

The pH stability of the free and immobilized enzyme preparations was determined in the following steps:

1. Dissolve the free and immobilized enzymes in the respective pH buffers (sodium acetate (pH 5.0), phosphate (pH 6.0), tris-HCl (pH 7.0 and pH 8.0) and glycine-NaOH (pH 9.0 and pH 10.0) buffers).
2. Incubate the mixture for 5 h at 30 °C in the shaker.
3. At the end of 5 h, measure the enzyme activity of the free and immobilized enzyme at 410 nm using Elisa reader.
4. Then calculate the residual enzyme activity with the initial enzyme activity in the respective buffers as 100%.

The residual activity of cross-linked enzymes increased over a broad pH range compared to the free enzyme as shown in the Fig. 5. The stability of the nanoparticle-supported enzymes increased over a broader pH profile had been shown by Xiao et al. (2016) which implies that the maghemite nanoparticles enhanced the stability as compared to the free enzyme. This might be due to enzyme stabilization through covalent linkages between the enzyme molecule and silanized surface of the maghemite nanoparticle (Rehman et al., 2016; Seenuvasan et al., 2013).

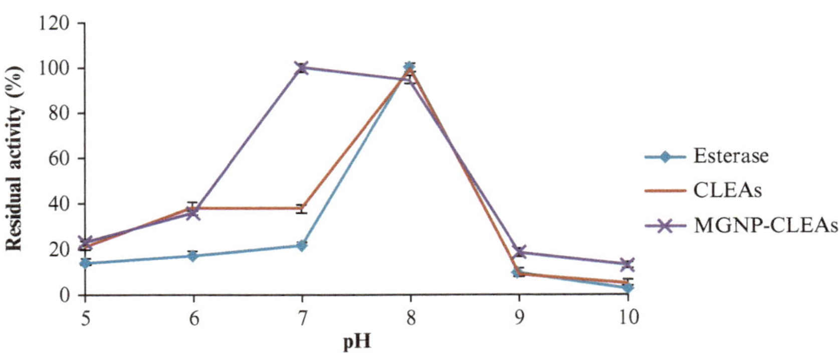

Fig. 5 The stability of the CLEAs and Mgnp-CLEAs along with free enzyme was investigated over the pH range of 5.0–10.0 at 30 °C for 5 h.

2.6.1.2 Temperature stability of free enzyme and CLEAs

1. Crude enzyme
2. Immobilized enzyme
3. Incubator shaker
4. *p*-NPV
5. 1 M tris-HCl buffer (pH 8.0)

The thermostability studies of the free and immobilized enzyme preparations were carried out using the following procedure:

1. Set the shaker incubator at temperature of 20–60 °C.
2. Incubate the free and immobilized enzyme with *p*-NPV at the respective temperatures for a period of 5 h.
3. At the end of 5 h, quantify the amount of 4-nitrophenol released by measuring its absorbance at 410 nm.
4. Then calculate the residual activity of the enzyme with the initial enzyme activity being taken as 100%.

Enhanced thermal stability is a prerequisite for the immobilized enzymes in an industrial scale at high temperatures (Rehman et al., 2016). The results show that the maghemite nanoparticle-supported CLEAs had a better residual activity than the CLEAs as shown in the Fig. 6. The enhanced thermostability could be due to reinforced rigidity of the enzyme molecules through formation of strong covalent bonds between the enzyme and nanoparticles (Kim, Grate, & Wang, 2006) along with the "vicinal effect" of the nanoparticles which favors the enzyme arrangement contributing to its stability at higher temperatures (Dong, Zhao, Huang, & Tan, 2010; Ibrahim, Al-Salamah, El-Toni, El-Tayeb, & Elbadawi, 2013; Muthulingam et al., 2013; Park et al., 2010; Sachin et al., 2012; Song, Sheng, & Zhang, 2012; Wang, Qi, Yu, Su, & He, 2010). The enhanced thermostability for

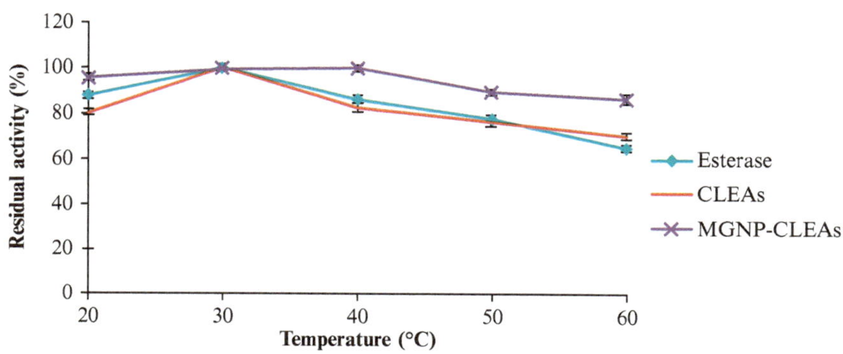

Fig. 6 The influence of temperature on the stability of the CLEAs and Mgnp-CLEAs was investigated in the range of 20–60 °C.

Mgnp-CLEAs could be due to cross-linking between the enzyme and nanoparticle which protects it from conformational distortion and heat-induced damage while in free enzymes increased temperature causes inactivation through unfolding of certain peptide segments which in turn results in changes in the structural conformation of enzyme molecule (Shalyda et al., 2016).

2.6.1.3 Organic solvents stability of free enzyme and CLEAs

1. Crude enzyme
2. Immobilized enzyme
3. Dimethyl formamide (DMF)
4. n-Hexane
5. n-Heptane
6. n-Octane
7. n-Decane
8. n-Undecane
9. *p*-NPV

The solvent stability of the free and immobilized enzyme was investigated by the following way;

1. Pipette out 0.3 mL of organic solvents to 0.7 mL of crude enzyme solution along with the control blank of the enzyme in buffer.
2. Incubate the enzyme-solvent mixture in the shaker at 30 °C and 150 rpm agitation for 24 h.
3. Calculate the residual enzyme activity with the enzyme activity in buffer taken as 100%.

The CLEAs and Mgnp-CLEAs were found to exhibit increased resistance to deactivation in the hydrophobic solvents as compared to the free enzyme which is inactivated in the solvents of higher log P value as shown in the Fig. 7. The free enzyme is deactivated in the presence of solvents as in general, the solvents strip off water molecules from the enzyme surface thus deactivating the protein (Sangeetha & Abraham, 2008) but the cross-linking increases the rigidity with reduction in its exposed surface thereby reducing its denaturation and inactivation. The smaller particle size of the CLEAs as shown in the SEM image is more susceptible to decreased activity recovery compared to Mgnp-CLEAs due to its direct contact with the solvent molecules while nanoparticle aggregated CLEAs with its large particle aggregate size shields the enzyme from direct contact with the solvent environment resulting in increased tolerance to the solvents (Shalyda et al., 2016).

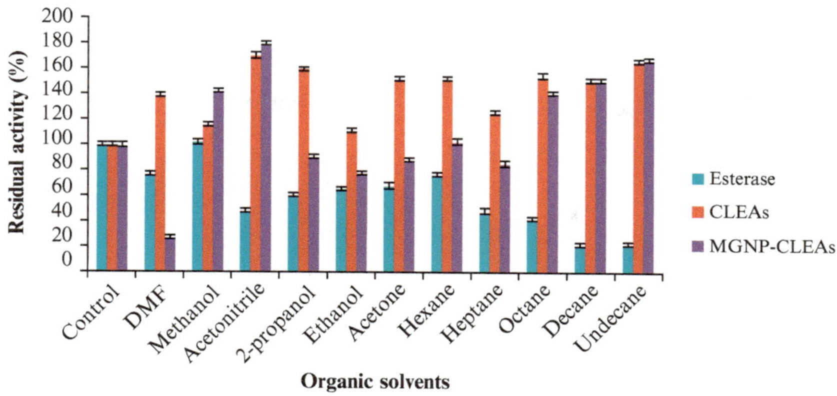

Fig. 7 The effect of the solvents on the CLEAs and Mgnp-CLEAs by incubation with different solvents of varying log *P* value for 24h at 30°C.

2.6.2 Operational stability

2.6.2.1 Reusability of CLEAs

1. Immobilized enzyme
2. *p*-NPV
3. 1M tris-HCl buffer (pH 8.0)

The reusability of the immobilized enzyme was determined in the following steps:

1. Determine the initial enzyme activity of the immobilized enzyme.
2. Measure the residual enzyme activity after its first usage.
3. Wash it thoroughly in buffer and air-dried.
4. Reuse the immobilized enzyme and repeat the steps 2 and 3.

The maghemite nanoparticle-supported CLEAs had better residual activity with 95% activity after 15 cycles of continuous use compared to the CLEAs as shown in the Fig. 8 which had retained only 85% enzyme activity. The decreased reusability of the CLEAs could be due to distortion of bonds leading to loss of enzyme upon its successive usage (Rehman et al., 2016). The addition of maghemite nanoparticles in the CLEAs enhanced the durability of the Mgnp-CLEAs during repetitive usage (Shalyda et al., 2016). Similar increased residual activity with Mgnp-CLEAs after repeated use has been showed by Baskar, Banu, Leuca, Gayathri, and Jeyashree (2015) and Sangeetha and Abraham (2008).

2.6.2.2 Storage stability of free enzyme and CLEAs

1. Free enzyme
2. Immobilized enzyme

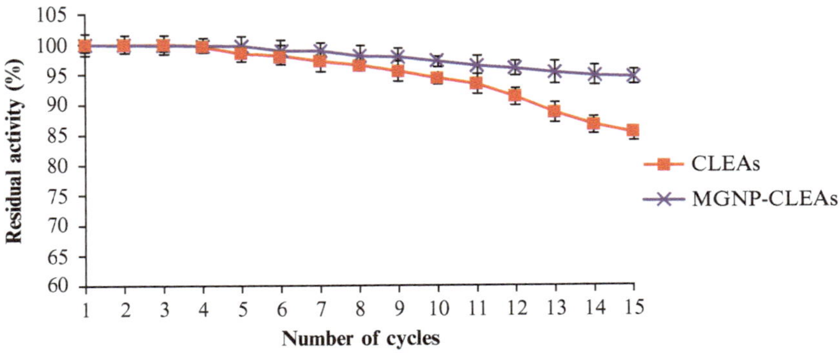

Fig. 8 The operational stability of the CLEAs and Mgnp-CLEAs were investigated.

3. 1 M tris-HCl buffer (pH 8.0)

The storage stability of the free enzyme and CLEAs was determined in the following steps:

1. Measure the enzyme activity for the free and immobilized enzyme stored at 4 °C.
2. Repeat step 1 at periodic time intervals of 15 days for a period of 120 days.
3. Measure the residual enzyme activity with the initial activity taken as 100%.

The stability of the enzyme during the storage process is an important variable influencing the cost of its application at the industries (Lee, Park, Yeo, & Kim, 2006). The immobilized enzyme had extended storage stability in comparison to the free enzyme due to the increased rigidity of the enzyme during immobilization process as shown in the Fig. 9. The Mgnp-CLEAs showed 53% residual activity in comparison to CLEAs with 31% activity after a period of 4 months which could be due to effective cross-linking with magnetic nanoparticles through strong hydrophobic interactions thus preventing enzyme leakage and dissociation (Sachin et al., 2012).

3. Concluding remarks

An improvement of the CLEAs technique has been explored in the study with the incorporation of silanized maghemite nanoparticles as support matrix during the preparation of cross-linked enzyme aggregates preparation as revealed with 100% activity recovery for the Mgnp-CLEAs. The major drawback in utilizing the nanoparticles as support matrix due to

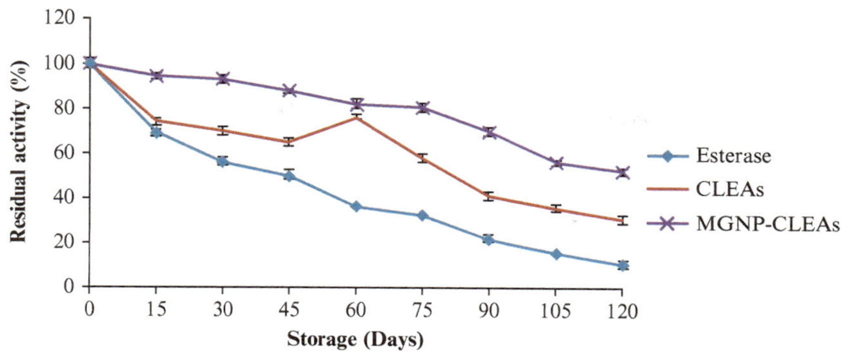

Fig. 9 Storage stability of free and immobilized enzyme for 120 days at 4 °C.

its high cost and not-so-simplified preparation methodology has been resolved with the co-precipitation methodology for its low cost and large-scale synthesis. The catalytic properties of the CLEAs technique had been combined with the unique properties of the maghemite nanoparticles as Mgnp-CLEAs where the stability and catalytic efficiency of the esterase was found to be improved at least by 20% along with an increased retention of activity during iterative cycles of reusability with 100% activity retained even after 15 cycles, thus providing an efficient biocatalyst preparation. The immobilized enzyme preparations characterized using SEM and FT-IR analysis implied successful immobilization of the enzymes with an increase in its size and amide I bands, respectively after immobilization on to the maghemite nanoparticle.

Acknowledgments

ND, MS thank University Grants Commission (UGC), Government of India for a research fellowship. PG thanks Department of Science and Technology (DST) and Department of Biotechnology (DBT), Government of India for continued support and funding. The authors also thank Taylor and Francis for permission to reproduce the figures.

References

Ahmad, R., & Sardar, M. (2015). Enzyme immobilization: An overview on nanoparticles as immobilization matrix. *Biochemistry and Analytical Biochemistry*, *4*, 1–8. https://doi.org/10.4172/2161-1009.1000178.

Alexandre, A., Hubert, C., & Peter, J. J. (2011). Laccase-based CLEAs: Chitosan as a novel cross-linking agent. *Enzyme Research*, *2011*, 1–10. https://doi.org/10.4061/2011/376015.

Asgari, S., Fakhari, Z., & Berijani, S. (2014). Synthesis and characterization of Fe_3O_4 magnetic nanoparticles coated with carboxymethyl chitosan grafted sodium methacrylate. *Journal of Nanostructures*, *4*, 55–63.

Baskar, G., Banu, N. A., Leuca, G. H., Gayathri, V., & Jeyashree, N. (2015). Magnetic immobilization and characterization of α-amylase as nanobiocatalyst for hydrolysis of sweet potato starch. *Biochemical Engineering Journal*, *102*, 18–23. https://doi.org/10.1016/j.bej.2015.02.020.

Cao, S. L., Huang, Y. M., Li, X. H., Xu, P., Wu, H., Li, N., et al. (2016). Preparation and characterization of immobilized lipase from *Pseudomonas cepacia* onto magnetic cellulose nanocrystals. *Scientific Reports*, *6*, 1–12.

Cao, L., van Rantwijk, F., & Sheldon, R. A. (2000). Cross-linked enzyme aggregates: a simple and effective method for the immobilization of penicillin acylase. *Organic Letters*, *2*, 1361–1364.

Chaubey, A., Parshad, R., Taneja, S. C., & Qazi, G. N. (2009). *Arthrobacter sp.* lipase immobilization on magnetic sol–gel composite supports for enantioselectivity improvement. *Process Biochemistry*, *44*, 154–160. https://doi.org/10.1016/j.procbio.2008.10.001.

Cui, J. D., Cui, L. L., Zhang, S. P., Zhang, Y. F., Su, Z. G., & Ma, G. H. (2014). Hybrid magnetic cross-linked enzyme aggregates of phenylalanine ammonia lyase from *Rhodotorula glutinis*. *PLoS One*, *9*, 1–8.

Cui, J. D., & Jia, S. R. (2013). Optimization protocols and improved strategies of cross-linked enzyme aggregates technology: Current development and future challenges. *Critical Reviews in Biotechnology*, *35*, 15–28.

Cui, J. D., Zhang, S., & Sun, L. M. (2012). Cross-linked enzyme aggregates of phenylalanine ammonia lyase: Novel biocatalysts for synthesis of L-phenylalanine. *Applied Biochemistry and Biotechnology*, *167*, 835. https://doi.org/10.1007/s12010-012-9738-0.

Dong, T., Zhao, L., Huang, Y., & Tan, X. (2010). Preparation of cross-linked aggregates of aminoacylase from *Aspergillus melleus* by using bovine serum albumin as an inert additive. *Bioresource Technology*, *101*, 6569–6571. https://doi.org/10.1016/j.biortech.2010.03.061.

Dwevedi, A. (2016). *Basics of enzyme immobilization (2nd chapter). In Enzyme immobilization. Advances in industry, agriculture, medicine and the environment* (p. 132). Switzerland: Springer International Publishing. ISBN: 978-3-319-41416-4 https://doi.org/10.1007/978-3-319-41418-8_2.

Hakan, A., Fatma, A., & Aydan, G. (2012). Highly biocompatible enzyme aggregates crosslinked by L-lysine. *Turkish Journal of Biochemistry*, *37*(1), 14–20.

Ibrahim, A. S. S., Al-Salamah, A. A., El-Toni, A. M., El-Tayeb, M. A., & Elbadawi, Y. B. (2013). Immobilization of cyclodextringlucanotransferase on aminopropyl-functionalized silica-coated superparamagnetic nanoparticles. *Electronic Journal of Biotechnology*, *16*, 1–6. https://doi.org/10.2225/vol16-issue6-fulltext-8.

Kim, J., Grate, J. W., & Wang, P. (2006). Nanostructures for enzyme stabilization. *Chemical Engineering Science*, *61*, 1017–1026.

Kouassi, G. K., Irudayaraj, J., & McCarty, G. (2005). Activity of glucose oxidase functionalized onto magnetic nanoparticles. *BioMagnetic Research and Technology*, *3*(1).

Kulakova, A., Nakayama, G. T., Nishino, T., & Esaki, N. (2004). Cold-active esterase from *Psychrobacter* sp. Ant 300: Gene cloning, characterization, and the effects of Gly/Pro substitution near the active site on its catalytic activity and stability. *Biochimica et Biophysica Acta*, *1696*, 59–65. https://doi.org/10.1016/j.bbapap.2003.09.008.

Ladole, M. R., Muley, A. B., Patil, I. D., Talib, M. L., & Parate, V. R. (2014). Immobilization of tropizyme-P on amino-functionalized magnetic nanoparticles for fruit juice clarification. *Journal of Biochemical Technology*, *5*, 838–845.

Lee, D. H., Park, C. H., Yeo, J. M., & Kim, S. W. (2006). Lipase immobilization on silica gel using a cross linking method. *Journal of Industrial and Engineering Chemistry*, *12*, 777–782.

Leitgeb, M., Knez, Z., & Vasic, K. (2016). Micro-carriers and nanocarriers for immobilization of enzymes. In: S. G. Stanciu (Ed.), *Micro and nanotechnologies for biotechnology*. InTechOpen publishershttps://doi.org/10.5772/63129. Book, Chapter 2.

Lopez-Serrano, P., Cao, L., van Rantwijk, F., & Sheldon, R. A. (2002). Cross-linked enzyme aggregates with enhanced activity: Application to lipases. *Biotechnology Letters*, *24*, 1379–1383.

Lozano, S. V., Gallego, F. L., Diaz, J. C. M., & Torres, E. F. (2016). CLEA in enzyme improvement—A review. *Biocatalysis*, *1*, 166–177. https://doi.org/10.1515/boca-2015-0012.

Muthulingam, S., Carlin, M., Sridhar, P., Nagarajan, B., Jayaraj, I., Madhava, K., et al. (2013). Immobilization of pectinase on co-precipitated magnetic nanoparticles for enhanced stability and activity. *Research Journal of Biotechnology*, *8*, 24–30.

Natália, C. C. L., Marcelo, B. M., & Angela de Mello, F. (2017). Characterization and chemical stability of hydrophilic and hydrophobic magnetic nanoparticles. *Materials Research*, *20*, 736–746.

Park, J. M., Kim, M., Park, J. Y., Lee, D. H., Lee, K. H., Min, J., et al. (2010). Immobilization of the cross-linked para-nitrobenzyl esterase of *Bacillus subtilis* aggregates onto magnetic beads. *Process Biochemistry*, *45*, 259–263. https://doi.org/10.1016/j.procbio.2009.09.015.

Pastorino, L., Piolo, F., Zilli, M., Converti, A., & Nicolini, C. (2004). Lipase catalyzed degradation of poly(ε-caprolactone). *Enzyme and Microbial Technology*, *35*, 321–326. https://doi.org/10.1016/j.enzmictec.2004.05.005.

Rehman, S., Bhatti, H. N., Bilal, M., & Asgher, M. (2016). Cross-linked enzyme aggregates (CLEAs) of *Pencilluim notatum* lipase enzyme with improved activity, stability and reusability characteristics. *International Journal of Biological Macromolecules*, *91*, 1161–1169. https://doi.org/10.1016/j.ijbiomac.2016.06.081.

Sachin, T., Shamraja, N., Asavari, J., & Gandhali, J. (2014). Pectin cross-linked enzyme aggregates (pectin-CLEAs) of glucoamylase. *RSC Advances*, *4*, 59444–59453.

Sachin, T., Vishal, G., Trupti, G., Pavan, R., Priyant, D., Shamraja, N., et al. (2012). Novel magnetic cross-linked enzyme aggregates (magnetic CLEAs) of alpha amylase. *Bioresource Technology*, *123*, 542–547.

Sangeetha, K., & Abraham, T. E. (2008). Preparation and characterization of cross-linked enzyme aggregates (CLEA) of subtilisin for controlled release applications. *International Journal of Biological Macromolecules*, *43*, 314–319.

Sayali, K., Sadichha, P., & Surekha, S. (2013). Microbial esterases: An overview. *International Journal of Current Microbiology and Applied Sciences*, *2*, 135–146.

Schoevaart, R., Wolbers, M. W., Golubovic, M., Ottens, M., Kieboom, A. P. G., van Rantwijk, F. V., et al. (2004). Preparation, optimization, and structures of cross-linked enzyme aggregates (CLEAS). *Biotechnology and Bioengineering*, *87*, 754–762.

Seenuvasan, M., Malar, C. G., Preethi, S., Balaji, N., Iyyappan, J., Kumar, M. A., et al. (2013). Immobilization of pectinase on co-precipitated magnetic nanoparticles for enhanced stability and activity. *Research Journal of Biotechnology*, *8*, 24–30.

Shalyda, M. S., Jamaliah, M. J., Roshanida, A. R., Ani, I., Abdul, M. A. M., et al. (2016). Silanizedmaghemite for cross-linked enzyme aggregates of recombinant xylanase from *Trichoderma reesei*. *Journal of Molecular Catalysis B: Enzymatic*, *133*, 65–76.

Song, C., Sheng, L., & Zhang, X. (2012). Preparation and characterization of a thermostable enzyme (Mn-SOD) immobilized on supermagnetic nanoparticles. *Applied Microbiology and Biotechnology*, *96*, 123–132. https://doi.org/10.1007/s00253-011-3835-9.

Sulistyaningsih, T., Santosa, S. J., Siswanta, D., & Rusdiarso, B. (2017). Synthesis and characterization of magnetites obtained from mechanically and sonochemically assisted co-precipitation and reverse co-precipitation methods. *International Journal of Materials, Mechanics and Manufacturing*, *5*, 1. https://doi.org/10.18178/ijmmm.2017.5.1.280.

Susana, V. L., Fernando, L. G., Rafael, V. D., Mateos-Díaz, J. C., Guisan, J. M., & Ernesto, F. T. (2014). Carrier-free immobilization of lipase from *Candida rugosa* with polyethyleneimines by carboxyl-activated cross-linking. *Biomacromolecules*, *15*, 1896–1903.

Talekar, S., Waingade, S., Gaikwad, V., Pati, S., & Nagavekar, N. (2011). Preparation and characterization of cross-linked enzyme aggregates (CLEAs) of *Bacillus amyloliquefaciens* alpha amylase. *Journal of Biochemical Technology*, *3*, 349–353.

Wang, J., Meng, G., Tao, K., Feng, M., Zhao, X., Li, Z., et al. (2012). Immobilization of lipases on alkyl silane modified magnetic nanoparticles: Effect of alkyl chain length on enzyme activity. *PLoS One*, *7*, e43478.

Wang, M., Qi, W., Yu, Q., Su, R., & He, Z. (2010). Cross-linking enzyme aggregates in the macropores of silica gel: A practical and efficient method for enzyme stabilization. *Biochemical Engineering Journal*, *52*, 168–174. https://doi.org/10.1016/j.bej.2010.08.003.

Xiao, A., Xu, C., Lin, Y., Ni, H., Zhu, Y., & Cai, H. (2016). Preparation and characterization of κ-carrageenase immobilized onto magnetic iron oxide nanoparticles. *Electronic Journal of Biotechnology*, *19*, 1–7.

Yuhan, C., Qihua, J., Lili, S., Qiang, L., Liping, Z., Qian, C., et al. (2018). Magnetic combined cross-linked enzyme aggregates of ketoreductase and alcohol dehydrogenase: An efficient and stable biocatalyst for asymmetric synthesis of (*R*)-3-quinuclidinol with regeneration of coenzymes in situ. *Catalysts*, *8*, 334.

Further reading

Bornscheuer, U. T. (2002). Microbial carboxyl esterases: Classification, properties and application in biocatalysis. *FEMS Microbiology Reviews*, *26*(1), 73–81.

Shuguang, W., Debing, Z., Lingyu, Y., & Fei, W. (2017). Preparation, activity and structure of cross-linked enzyme aggregates (CLEAs) with nanoparticle. *Enzyme and Microbial Technology*, *107*, 22–31.

Xiu, X., Jia, J. Q., Zhang, J. F., Zhou, Z. W., Jun, L., Na, W., et al. (2019). CALB immobilized onto magnetic nanoparticles for efficient kinetic resolution of racemic secondary alcohols: Long-term stability and reusability. *Molecules*, *24*(3), 490.

CHAPTER SEVEN

Magnetic bead-based semi-automated phage display panning strategy for the directed evolution of antibodies

Angela Chiew Wen Ch'ng[a], Zoltán Konthur[b], Theam Soon Lim[a,c,*]
[a]Institute for Research in Molecular Medicine, Universiti Sains Malaysia, Penang, Malaysia
[b]Max Planck Institute of Colloid and Interfaces, Department of Systems Biology, Potsdam, Germany
[c]Analytical Biochemistry Research Centre, Universiti Sains Malaysia, Penang, Malaysia
*Corresponding author: e-mail address: theamsoon@usm.my

Contents

Abstract

Directed evolution is a proven approach to fine tune or modify biomolecules for various applications ranging from research to industry. The process of evolution requires methods that are capable of not only generating genetic diversity but also to distinguish the variants of desired characteristics. One method that is synonymous with directed evolution of proteins is phage display. Here, we present a protocol describing the application of magnetic nanoparticles coupled with a processor to carry out the identification of monoclonal antibodies (mAbs) from a diverse antibody library via phage display. Target antigens are coupled to magnetic nanoparticles as the solid phase for the isolation of the binding mAbs via affinity. A gradual enrichment in clones would result in increasing ELISA readouts with increasing rounds of panning. During monoclonal level analysis, positivity can be deduced with comparison to background and controls. The biopanning process can also be adopted for the directed evolution of enzymes, scaffold proteins or even peptides.

Methods in Enzymology, Volume 630
ISSN 0076-6879
https://doi.org/10.1016/bs.mie.2019.10.023

1. Introduction

Directed evolution is a process to generate proteins that exhibit specific properties. Arnold first pioneered the concept of directed evolution of enzymes with the belief that biological selection may help in reprogramming of microbes to become a better chemical factory (Arnold, 2018). The main concept of protein evolution is random mutagenesis or DNA shuffling to improve the properties of proteins and enzymes in term of diversity and functionality in desired conditions (Könning & Kolmar, 2018). It usually involves error-prone PCR for random base substitution to create mutants, which contributes to the diversity of enzyme. The mutagenized enzyme pool will be screened to isolate an improved version of enzyme with either the same or evolved function.

Antibodies are proteins that can be reprogramed by directed evolution for improved characteristics just like enzymes. The rise of monoclonal antibodies (mAbs) in the medical arena in the past decades is greatly attributed to the improved understanding of immunology and value for diagnostics or therapy (Chen et al., 2016; Frenzel, Schirrmann, & Hust, 2016; Nixon, Sexton, & Ladner, 2014). In 2018, a total of 570 antibody therapeutics were reported to be undergoing clinical trials at various stages with 16 mAbs approved in the European Union (EU) and United States (US) (Kaplon & Reichert, 2019). With the advances in science and technology, the discovery and development of mAbs has become more robust and less complicated in comparison to the traditional hybridoma method for mAb generation. The gold standard method for the generation of human mAbs is phage display (dos Santos, Quintilio, Manieri, Tsuruta, & Moro, 2018; Frenzel et al., 2016).

A key element of directed evolution is the physical coupling of the phenotype and the corresponding genotype as part of the selection system (Tizei, Csibra, Torres, & Pinheiro, 2016). The introduction of phage display by George Smith has allowed for the directed evolution of proteins and peptides to be carried out seamlessly (Arap, 2005). This was later applied by Sir Gregory Winter for the evolution of antibodies on bacteriophage for antibody discovery purpose (Carter & Lazar, 2017) and has allowed for the generation of human antibodies in bacteria bypassing the need for animal immunization (Bornscheuer, Hauer, Jaeger, & Schwaneberg, 2019). Phage display is fundamentally an iterative process of selection carried out under different conditions to ultimately identify a specific antibody variant that satisfies the required characteristics. The process begins with the establishment of a collection of diverse antibody genes, commonly referred to as antibody library. There are different versions of antibody libraries based on the source of the antibody genes. Antibody libraries are generally classified as naïve, immune and synthetic antibody libraries (Lim et al., 2014). They mainly differ in their source of the antibody sequences, which are derived from healthy and infected donors or generated by chemical synthesis, respectively (Ch'ng, Choong, & Lim, 2016). The antibody library can be further evolved through random mutagenesis at the amino acids region responsible for target interaction, termed as Complementarity Determining Regions (CDR) by various molecular approaches, such as error-prone polymerase chain reaction (PCR) and chain recombination (Lim, Choong, & Lim, 2019). This aids in the isolation of mAbs that differ in terms of binding affinity, avidity and desired functionalities (Arnold, 2018).

Biopanning is the selection technique used to isolate single mAbs from a diverse phage display antibody library via affinity and is a repetitive process over 3–5 rounds of selection. During the selection rounds, different binding conditions and selection pressures can be introduced to enrich the best possible variant with the desired characteristic (Bazan, Całkosiński, & Gamian, 2012).

2. Magnetic nanoparticle-based panning protocol

The biopanning process in phage display involves the immobilization of a target antigen onto a solid phase prior to the introduction of antibody phage particles to bind via affinity. There are several different solid phases used for biopanning to date, which includes microtiter plate (MTP) wells, immunopins (Konthur & Walter, 2002), PVDF membranes from 2D-gels (Bugli et al., 2011; Liu, Huang, Sihlbom, Burlingame, & Marks, 2002; Liu & Marks, 2000), streptavidin mass spectrometry immunoassay (MSIA™) Disposable Automation Research Tips (Chin et al., 2016) or magnetic particles using a magnetic particle processor (Hallborn & Carlsson, 2002; Walter, Konthur, & Lehrach, 2001) as shown in Fig. 1. The traditional solid phase used is mainly MTPs, due to its ability to work with small volumes and enables simultaneous panning in a 96-well format. Even so, this is not ideal for large scale panning as the manual handling process using MTP allows for errors to be introduced thus increasing variation between samples (Ch'ng, Ahmad, Konthur, & Lim, 2019; Ch'ng, Hamidon, Konthur, & Lim, 2018).

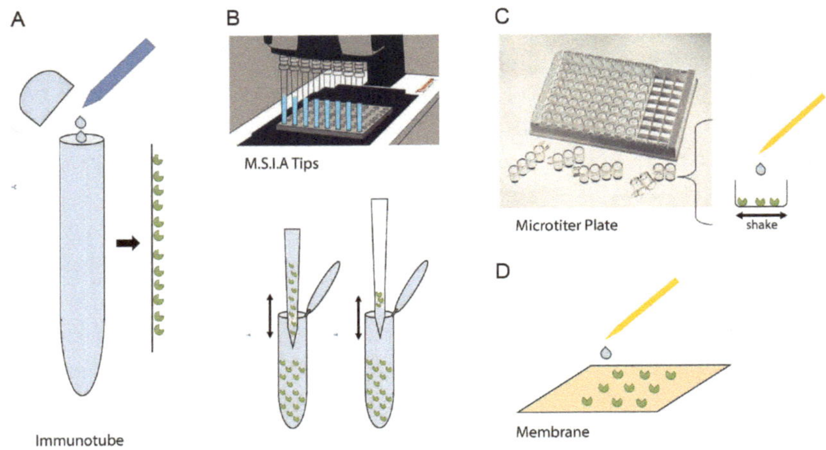

Fig. 1 The immobilization surface for phage display antibody selection other than magnetic beads surface. The decision on different immobilization surface such as (A) the maxi-binding immunotube surface, (B) affinity ligands in pipette tips, (C) microtiter plate (MTP) with hydrophobic surface, and (D) membrane filter paper that allow protein adhesion is strongly depend on the target that will be used for antibody selection.

The introduction of nano-sized particles with magnetic properties has allowed for easy separation and isolation of solid phase with liquid phase samples. Coupled with robotics in the form of a particle processor, the cumbersome workload associated with the multi-step panning process can be reduced tremendously (Konthur, Wilde, & Lim, 2010). The physical attachment of target antigens on the surface of magnetic nanoparticles can be done either using standard chemical conjugation methods or even direct immobilization using intermediate biomolecules like streptavidin (Shen, Rusling, & Dixit, 2017). In our proposed approach, the immobilization efficiency of the target antigen is improved with biotinylation of the target protein. This allows for a directed immobilization of the biotinylated antigen to the streptavidin conjugated magnetic particles (Welch et al., 2017). The nano-sized particles also allow for improved binding efficiency with a higher surface area to volume ratio for antigen presentation (Ch'ng et al., 2019). The use of a magnetic particle processor allows for majority of the steps in the panning process to be automated with minimal operator intervention. This holds well for reproducibility and a reduction of operator error between batches and samples.

2.1 General instruments

2.1.1 Loading of magnetic beads

1. DynaMag™-2 magnet (Invitrogen, USA)
2. Multi RS-60 Programmable rotator (Biosan, Latvia)
3. Mini-PROTEAN® Tetra Handcast Systems (Bio-Rad)

2.1.2 Semi-automated panning on magnetic particle processor

1. KingFisher™ Flex System (Thermo Fisher Scientific)
2. Multi RS-60 Programmable rotator (Biosan, Latvia)
3. Heraeus T6 Incubator (Thermo Scientific)

2.1.3 Packaging of phage particles

1. Thermo Scientific™ iEMS Incubator (Thermo Fisher Scientific)
2. PST-60HL-4 Plate Shaker-Thermostat (Biosan)
3. Heraeus T6 Incubator (Thermo Scientific)
4. Heraeus Megafuge 40R Refrigerated Centrifuge (Thermo Scientific)

2.1.4 Titration of phage particles

1. MaxQ™ 8000 Incubated/Refrigerated Stackable Shakers (Thermo Fisher Scientific)
2. Heraeus T6 Incubator (Thermo Scientific)

2.1.5 ELISA of polyclonal antibody phage

1. MTS 2/4 digital microtiter shaker (Ika)
2. PST-60HL-4 Plate Shaker-Thermostat (Biosan)
3. Heraeus Megafuge 40R Refrigerated Centrifuge (Thermo Scientific)

2.1.6 Monoclonal antibody selection

1. Thermo Scientific™ iEMS Incubator (Thermo Fisher Scientific)
2. PST-60HL-4 Plate Shaker-Thermostat (Biosan)
3. MTS 2/4 digital microtiter shaker (Ika)
4. Heraeus T6 Incubator (Thermo Scientific)
5. Heraeus Megafuge 40R Refrigerated Centrifuge (Thermo Scientific)

2.2 Materials and methods

2.2.1 Loading of magnetic beads

1. Dynabeads™ M-280 Streptavidin magnetic beads (Invitrogen Dynal AS)
2. Phosphate-buffered saline (PBS): 8 g/L NaCl, 0.2 g/L KCL, 1.44 g/L $Na_2HPO_4 \cdot 2H_2O$ and 0.24 g/L KH_2PO_4, pH 7.4

3. Phosphate-buffered saline Tween (PBST): PBS and 0.1% Tween-20. All beads must be washed three times with 1.5 mL of 0.1% (*v*/v) PBS Tween-20 (PBST) either on rotator or pipette aspiration at room temperature (RT) before use. Conjugate 1 mg of Dynabeads™ M-280 Streptavidin magnetic beads (Invitrogen) with 100–200 μg of biotinylated protein antigen or 1–2 μg of biotinylated peptide in 1 mL PBS (8 g/L NaCl, 0.2 g/L KCL, 1.44 g/L $Na_2HPO_4 \cdot 2H_2O$ and 0.24 g/L KH_2PO_4, pH 7.4) overnight (o/n) at 4 °C or 1 h at RT on a rotator. Then wash the conjugated beads three times with 1.5 mL of 0.1% (v/v) PBS Tween-20 (PBST) either on rotator or pipette aspiration at RT. Discard the leftover PBST with the help of a magnetic rack. Discard the wash solution and resuspend the magnetic beads with 200 μL PBS and store the *antigen-loaded bead stock* at 4 °C until use. The target protein should be validated via SDS PAGE and quantified before and after conjugation by using Bradford assay to estimate the binding capacity. Besides, the peptide conjugation will be different due to its size. Normally an ELISA will be used to determine the binding capacity with the use of Biotin HRP for detection.

2.2.2 Semi-automated panning on magnetic particle processor

1. *E. coli* TG1 genotype: *supE thi-1 Δ(lac-proAB) Δ(mcrB-hsdSM)5 (rK– mK–) [**F′** traD36 proAB lacIq ZΔM15]* (Stratagene)
2. KingFisher 96 KF microplate (PP) (Thermo Scientific™)
3. KingFisher Flex 96 tip comb for KF magnets (PP) (Thermo Scientific™)
4. Aera Seal breathable sealing film (Sigma-Aldrich)
5. Maxi-binding polystyrene (PS) Immunotube (SPL)
6. Parafilm
7. Dynabeads™ M-280 Streptavidin magnetic beads—unconjugated (Invitrogen)
8. Dynabeads™ M-280 Streptavidin magnetic beads—conjugated (Invitrogen)
9. PBST-Milk (PTM): 10 mg/mL stock solution in PBS (Fluka)
10. Phosphate-buffered Saline (PBS): 8 g/L NaCl, 0.2 g/L KCL, 1.44 g/L $Na_2HPO_4 \cdot 2H_2O$ and 0.24 g/L KH_2PO_4, pH 7.4
11. Phosphate-buffered saline Tween (PBST): PBS and 0.1% Tween-20
12. 2YT medium: 1.6% (*w*/*v*) tryptone, 1% (w/v) yeast extract, and 0.5% NaCl, pH 7.0
13. 10 × Amp/Glu solution: 100 μg/mL ampicillin, and 2% (w/v) glucose in 2YT medium

Any commercial magnetic particle processor can be used for this section. In this context, we will describe the use of the KingFisher Flex magnetic particle processor system. The system allows for 1–96 antigens to be screened simultaneously but we will focus on the process to screen for 12 targets (A1–A12) on the KingFisher 96 KF microplate (PP) (Thermo Fisher Scientific) as shown in Fig. 2; Table 1. The KingFisher 96 KF microplates (PP) and the KingFisher Flex 96 tip comb for KF magnets (PP) that will be used must be sterilized under UV before use to prevent contamination.

There are some preparations that need to be done before phage display panning is carried out according to Table 2.

First, prepare a fresh culture of *E. coli* TG1: *supE thi-1 Δ(lac-proAB) Δ(mcrB-hsdSM)5 (rK– mK–) [**F′** traD36 proAB lacIq ZΔM15]* (Stratagene) by using filter tips to prevent contaminations from pipette. A colony of

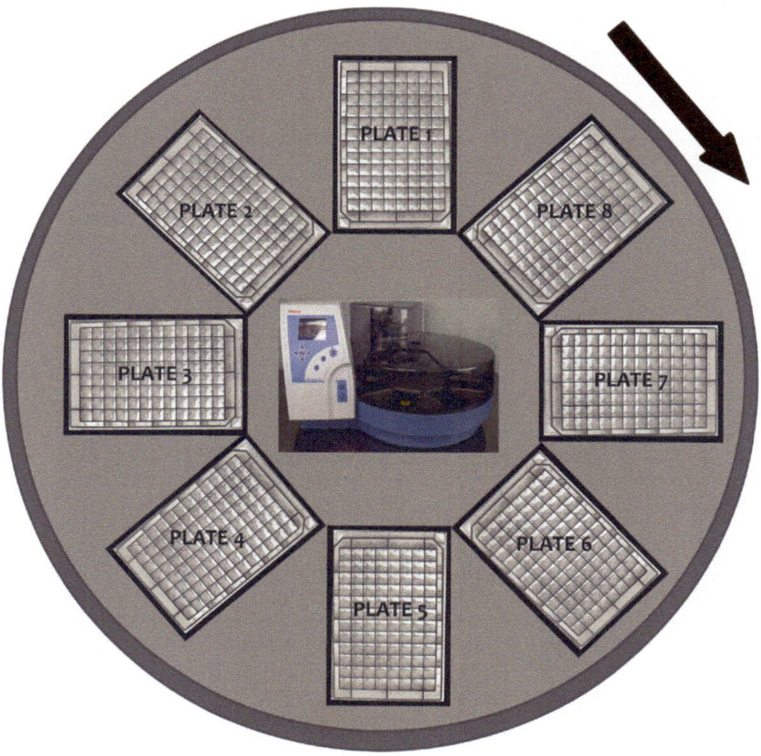

Fig. 2 The design of rotating plate platform of a KingFisher Flex magnetic particle processor. The magnetic head is at a fixed loading position. The beads are recovered from one plate to another new plate by rotating the plate holder table clockwise with the set program of Kingfisher Flex.

Table 1 Plate position on the platform of KingFisher Flex according to magnetic bead-based panning procedure.

Plate no.	Panning round 1	Panning round 2	Panning round 3	Panning round 4
1	Bead plate	Bead plate	Bead plate	Bead plate
2	Wash plate	Wash plate	Wash plate	Wash plate
3	Phage plate	Phage plate	Phage plate	Phage plate
4	Wash plate 1	Wash plate 1	Wash plate 1	Wash plate 1
5	Release plate	Wash plate 2	Wash plate 2	Wash plate 2
6	–	Release plate	Wash plate 3	Wash plate 3
7	–	–	Release plate	Wash plate 4
8	–	–	–	Release plate
Total time	~140 min	~150 min	~160 min	~170 min

The total time used for a complete panning round increased from round 1 to round 4 with the increasing of washing steps during the process.

Table 2 The detail procedure for round 4 magnetic bead-based panning for KingFisher Flex.

Plate no.	Plate name	Procedure	Volume (μL)	Time (min)
1	Bead plate	Blocking of antigen-conjugated magnetic beads with PTM	200	60
2	Wash plate	Wash of antigen-conjugated magnetic beads in PBST	200	10
3	Phage plate	Binding of antigen-conjugated magnetic beads with antibody phage library/antibody phage stock binding of the selection rounds	200	60
4	Wash plate 1	Wash 1 of antigen-conjugated magnetic beads in PBST	200	10
5	Wash plate 2	Wash 2 of antigen-conjugated magnetic beads in PBST	200	10
6	Wash plate 3	Wash 3 of antigen-conjugated magnetic beads in PBST	200	10
7	Wash plate 4	Wash 4 of antigen-conjugated magnetic beads in PBST	200	10
8	Release plate	Replaced by *E. coli* TG1 culture when it reached OD_{600} ~0.5 for infection	200	[a]
Total time				~170

[a]At this stage, the infection process takes place at 37 °C incubator. After Infection, 10 μL bead-bound phage solution will be collected for titration after removing the magnetic beads.

E. coli TG1 was picked and grown in 5 mL of 2YT o/n at 37 °C, 200 rpm on the day before phage display panning. At the same time, block a 5 mL maxi-binding polystyrene (PS) immunotube with 5 mL of PBST-Milk (PTM) o/n at 4 °C and seal with parafilm.

The phage display panning starts with the washing of the blocked immunotube with PBST. Then, 1×10^{10}–1×10^{11} phage particles are pre-incubated in 100 μL PTM and 20 μL unconjugated Dynabeads™ M-280 Streptavidin magnetic beads (Invitrogen) for 1–2 h at RT on a rotator. After that, the unconjugated magnetic beads were removed using a magnetic stand before transferring the beads to *plate no. 3*. The plates are arranged according to Table 1 and placed into KingFisher Flex machine as shown in Fig. 2 before starting the magnetic bead-based panning program.

1. Arranging *Plate no. 1 bead plate*. Fill the wells in positions A1–A12 of a KingFisher 96 KF microplate (PP) with 180 μL PTM. Add 20 μL of the corresponding *antigen-loaded bead stocks* for each antigen to the specified position, namely, beads with antigen 1 to position A1, beads of antigen 2 to position A2, and so on (total of 12 antigens in this case).
2. Arrange *Plate no. 3 phage plate*. Fill positions A1–A12 of a KingFisher 96 KF microplate (PP) with 200 μL of the antibody phage library solution (for round 1 panning) or 100 μL of the amplified phage solutions of the previous round with 100 μL PTM according to the same antigen after removing the magnetic beads with the help of a magnetic stand.
3. Prepare *Plate no. 2, 4, 5, 6 wash plate*. Fill positions A1–A12 of a KingFisher 96 KF microplate (PP) with 200 μL PBST.
4. Prepare *Plate no. 5, 6, 7 release plate*. Fill positions A1–A12 of a KingFisher 96 KF microplate (PP) with 200 μL phosphate-buffered saline (PBS): 8 g/L NaCl, 0.2 g/L KCL, 1.44 g/L $Na_2HPO_4 \cdot 2H_2O$ and 0.24 g/L KH_2PO_4, pH 7.4. The *release plate* will be replaced later by the *E. coli TG1 culture plate*.

During the magnetic bead-based panning program, the magnetic beads were allowed to be incubated in each plate up and down with the help of cupped magnetic pins. After the cycle done within each plate, it should then be moved from plate to plate according to the program with the help of the cupped magnetic pins as shown in Fig. 3. The inoculation of *E. coli* TG1 was done when there is about 105 min left in the program run (Table 1) according to round of panning. When the *E. coli* TG1 reaches OD_{600} ~0.5, replace *the release plate* with *E. coli TG1 culture plate*. The culture time may vary between laboratories. Thus, it is good to check the time needed for the TG1 culture $0D_{600}$ to reach 0.5 before panning started by using Kingfisher Flex.

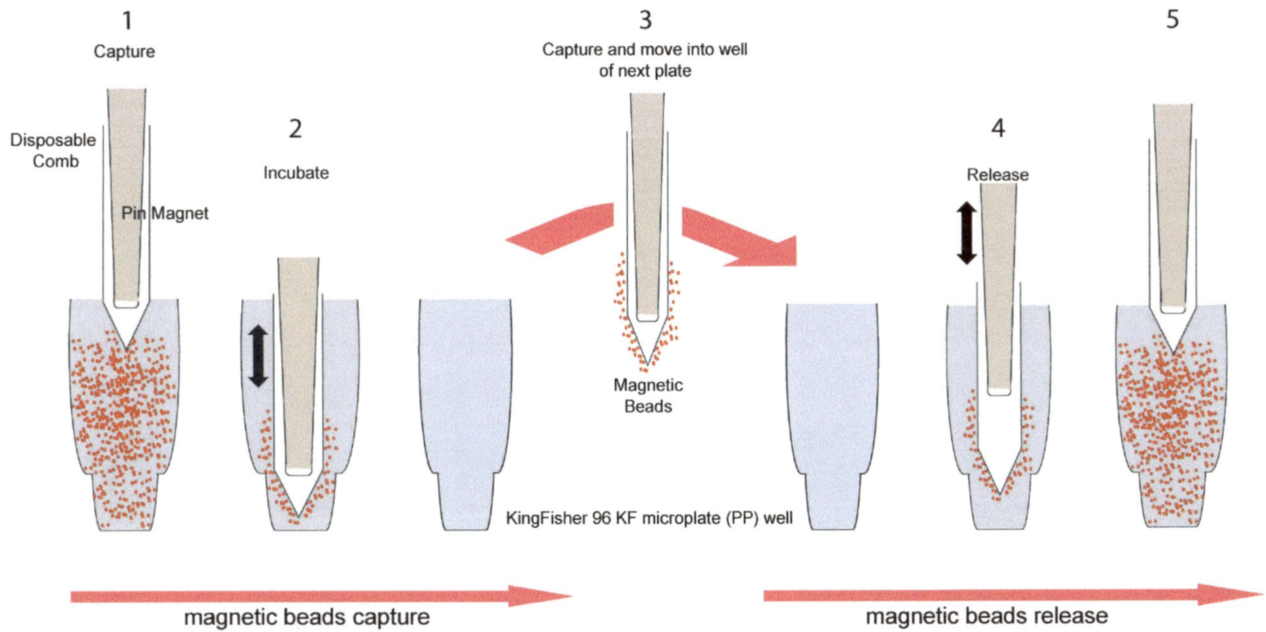

Fig. 3 The operating mode of KingFisher Flex magnetic particle processors. (1) The plastic comb cover pin-magnet moves into a solution containing suspended magnetic beads. (2) The incubation process by moving up and down the magnetic beads attracted to the cover and (3) the magnetic beads are captured and move to new solution. (4) Magnetic beads are released and suspend into solution by removing the pin-magnet from the comb. (5) Both 96 pin-magnet head and plastic comb are raised to the starting position to proceed to the next stage of the process.

When the OD_{600} reaches 0.5, place the culture on ice to slow the growth process if the program is not complete. In the last step of the machine program, the antigen-conjugated magnetic beads with the rescue phage will be released static in the *E. coli TG1 culture plate/selection stock plate.* The plate will be incubated static in incubator at 37 °C for 30 min after the magnetic bead-based panning program done on KingFisher Flex machine.

2.2.3 Packaging of phage particles

1. M13K07 Helper-phage (New England BioLabs)
2. Nunc™ 96-Well Filter Plates (Thermo Fisher Scientific)
3. 2YT-AG-2: 2YT medium containing 100 μg/mL ampicillin, 2% (*w/v*) glucose
4. 2YT-AKG: 2YT medium containing 100 μg/mL ampicillin, 60 mg/mL kanamycin, 0.1% (w/v) glucose
5. Glycerol solution: 80% (*v*/v) glycerol in distilled water, then autoclave

The rescue of the enriched phage particles is continued after the incubation of the released beads in the *culture plate/selection stock plate.* Add 20 μL 10 × Amp/Glu solution (100 μg/mL ampicillin, and 2% (w/v) glucose) into the *selection stock plate* after removing the beads from the plate. The plate is then sealed with breathable sealing film and incubated in MTP incubator shakers for 2 h at 37 °C at high speed, such as the iEMS™ Microplate Incubator/Shaker HT microplate shaker (Thermo Scientific). The OD_{600} of the cultures in the plate were considered 0.5 after 2 h of shaking incubation.

Before infection, 200 μL of 37 °C pre-warmed 2YT-AG medium (2YT medium containing 100 μg/mL ampicillin, 2% (w/v) glucose) is added to the culture and mixed thoroughly before transferring 200 μL into a 96-well filtration plate. Thus, there will be two plates, the *selection stock plate* and *phage production plate.* The *selection stock plate* is sealed again with breathable sealing film and continued with incubation in a microplate shaker o/n at 37 °C, 1400 rpm.

Add 10 μL M13K07 helper phage (stock concentration: 10^{11} pfu/mL) to the *phage production plate* and incubate static at 37 °C for 30 min. The bacterial culture is then filtered by centrifugation of 96-well filtration plate for 15 min at 3500 rpm (swing out rotor). Discard the supernatant with remaining M13K07 helper phage by using filter tips before resuspending the pelleted bacteria in 220 μL pre-warmed 2YT-AK (2YT medium containing 100 μg/mL ampicillin, 60 μg/mL kanamycin) and transfer to a fresh 96-well U-bottom PP microtiter plate. New, fresh plate is used for the packaging process to prevent contamination by the uninfected helper phage that sticking to the well of the plate as helper phage M13K07 had been recognized as sticky molecules in biopanning.

Finally, seal *phage production plate* with breathable sealing film and incubate in an iEMS™ Microplate Incubator/Shaker HT microplate shaker (Thermo Scientific) o/n at 30 °C, 1400 rpm.

The next day, mix 160 µL of autoclaved 80% glycerol solution to *selection stock plate* and keep as glycerol stock at −80 °C. At the same time, centrifuge the *phage production plate* for 15 min, 3500 rpm (swing out rotor) to retrieve the phage. Filtration plate is used to filter antibody phage particles to remove possible contaminating *E. coli* TG1 cells by centrifugation for 2–5 min, 2000 rpm. The filtrate is stored whereas the bacteria pellet is discarded. Then 100 µL of phage filtrate is used for the next round of selection whereas 10 µL of phage filtrate is used for phage titration.

2.2.4 Titration of phage particles

1. *E. coli* TG1 genotype: *supE thi-1 Δ(lac-proAB) Δ(mcrB-hsdSM)5 (rK– mK–) [F′ traD36 proAB lacIq ZΔM15]* (Stratagene)
2. 2YT-AG agar plates: 2YT medium containing 100 µg/mL ampicillin, 2% (*w/v*) glucose and 1.5% (w/v) agar-agar
3. 2YT-KG agar plates: 2YT medium containing 60 µg/mL kanamycin and 1.5% (w/v) agar-agar

A total of 100 µL o/n *E. coli* TG1 culture is inoculated in 10 mL 2YT and grow at 37 °C, 200 rpm until OD_{600} ~0.5. While waiting for the OD_{600} to reach ~0.5, 1:10 serial dilutions of phage suspension are prepared in PBS. Then, serially diluted phage is used to infect the 100 µL of *E. coli* TG1 for 30 min at 37 °C without shaking. Then, 10 µL droplets of each dilution series are spotted on a single 2YT-AG and 2YT-K agar plates to check on the rescue and enrichment of the antibody phage. At last, the plates are incubated top-down o/n at 37 °C after the droplets are dried. The whole process was done using filter tips to prevent contamination of TG1 and pipette.

2.2.5 ELISA of polyclonal antibody phage

1. KingFisher 96 KF microplate (PP) (Thermo Scientific™)
2. KingFisher Flex 96 tip comb for KF magnets (PP) (Thermo Scientific™)
3. PBST-Milk (PTM): 10 mg/mL stock solution in PBS (Sigma)
4. Anti-M13 Horseradish Peroxidase (HRP)-conjugated monoclonal antibody (GE)

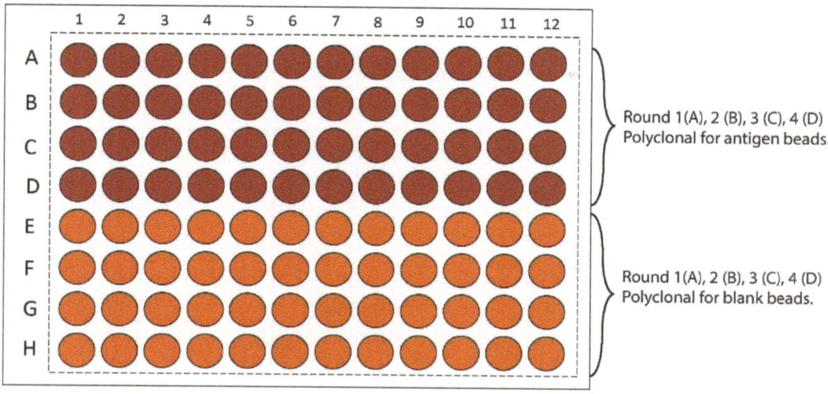

Fig. 4 The example polyclonal antibody phage ELISA by using magnetic bead for the selection of 12 different targets. The positions A–D indicate the targeted antigen selection for round 1 to round 4 whereas E–H of the same column indicated the reading of non-conjugated magnetic beads. All steps of the ELISA are performed in a 96-well magnetic particle processor.

5. ABTS developing solution (Roche): 50mg tablet ABTS in 5 mL of 10XABTS substrate buffer with 50mL of Ultrapure water as final volume. Store in dark.

The ELISA protocol in the KingFisher Flex consist of bead plate, wash plates, phage plate, antibody plate and substrate plate. The plate was labeled according to Fig. 4. Then, they are arranged according to Table 3; Fig. 4. The magnetic bead-based ELISA program is started after the plates are placed in their respective positions in the KingFisher Flex holder table.

1. Arrange *bead plate*. Fill each position of a KingFisher 96 KF microplate (PP) with 180μL PTM and add 20μL (roughly 10–20μg) of *antigen-loaded bead stock*. Then, add magnetic beads of antigen 1 to positions A1–D1, beads of antigen 2 to positions A2–D2, and so on.
2. Arrange *phage plate*. Fill each position of a KingFisher 96 KF microplate (PP) with 150μL PTM. Add 50μL of phage solution from the phage stock plates of the individual rounds to the plate. Add phage stocks of selection rounds 1–4 on antigen 1 to position A1–D1 and E1–H1, respectively. Add phage stocks of selection rounds 1–4 on antigen 2 to position A2–D2 and E2–H2, respectively, and so on.
3. Prepare three *wash plates* and fill KingFisher 96 KF microplate (PP) with 200μL PBST.
4. Prepare *antibody plate*. Fill KingFisher 96 KF microplate (PP) with 200μL mouse monoclonal anti-M13 HRP-conjugated-PTM solution (1:5000).

Table 3 The detail procedure for magnetic bead-based polyclonal antibody ELISA for KingFisher Flex.

Plate no.	Plate name	Procedure	Volume (μL)	Time (min)
1	Bead plate	Blocking of antigen-conjugated magnetic beads with PTM	200	60
2	Wash plate	Wash of antigen-conjugated magnetic beads in PBST	200	10
3	Phage plate	Binding of antigen-conjugated magnetic beads with antibody phage library/antibody phage stock of the selection rounds	200	60
4	Wash plate 1	Wash 1 of antigen-conjugated magnetic beads in PBST	200	20
5	M13 Antibody Plate	Incubating antigen-conjugated and control beads with anti-M13 HRP (1:5000) in PTM	200	60
6	Wash plate 2	Wash 3 of antigen-conjugated magnetic beads in PBST	200	10
7	Substrate plate	Incubating antigen-conjugated magnetic beads in ABTS	200	30[a]
8	–	–	–	–
Total time				170

[a]At this stage, the ABTS incubation take place in dark at microplate shaker at room temperature. The magnetic beads must be removed before measuring the absorbance in plate reader.

5. Prepare the *substrate plate*. Pipette 200 μL of ABTS substrate into each well of 96-well polystyrene microtiter plates and place plate in KingFisher Flex. Once beads are incubated in the substrate and color developed for 20 min, remove beads from the substrate manually.

The program should be set to move magnetic beads from plate to plate and incubate the beads in each plate (Fig. 3). Remove the *substrate plate* from the KingFisher Flex plate holder table and measure substrate absorbance at 405 nm in a plate reader machine.

2.2.6 Monoclonal antibody selection using magnetic nanoparticles

1. Nunc™ 96-Well Polystyrene Round Bottom Microwell Plates (Thermo Scientific)

2. *E. coli* TG1 genotype: *supE thi-1 Δ(lac-proAB) Δ(mcrB-hsdSM)5 (rK– mK–) [**F'** traD36 proAB lacIq ZΔM15]*
3. 2YT-AG-2: 2YT medium containing 100 μg/mL ampicillin, 2% (*w/v*) glucose
4. M13K07 Helper-phage (New England BioLabs)
5. 96-well Polystyrene Flat bottom Microplates (Greiner Bio-One)
6. PBST-Milk (PTM): 10 mg/mL stock solution in PBS (Fluka)
7. Anti-M13 Horseradish Peroxidase (HRP)-conjugated monoclonal antibody (GE)
8. Anti-his tag Antibody (Biolegend) or Anti-c-myc tag antibody (Roche)
9. ABTS developing solution: 50 mg tablet ABTS (Roche) in 5 mL of 10XABTS substrate buffer with 50 mL of Ultrapure water as final volume. Store in dark.

The monoclonal antibody selection can be done either using phage antibody or soluble antibody. Although the use of magnetic beads for monoclonal antibody ELISA has some clear advantageous in the reproducibility and reduced handling error, the sheer cost of carrying out monoclonal antibody level screening with magnetic beads will be high. This is because the screening scale per antigen can range between 96 and 384 clones. Therefore, biopanning of 12 antigens can result in a minimum or 1152 clones to a maximum of 4608 clones. Thus, 96-well polystyrene (PS) microtiter plate (MTP) was used for ELISA.

The production of phage monoclonal antibody fragments was carried out in 96-well polypropylene (PP) microtiter plate. A total of 96 individual clones were picked and cultured in each well as single antibody clones. Before colonies were inoculation, *E. coli* TG1 was inoculated from an o/n culture using filter tips in 5 mL of 2YT at 37 °C and 200 rpm until OD_{600} ~0.4–0.5. Before infection, prepare a 1:10 serial dilution by using filter tips with 10 μL of the desired panning round phage stock to 90 μL PBS. Then, add 200 μL of *E. coli* TG1 cell (OD_{600} ~0.4–0.5) to phage dilutions 10^{-5}—10^{-8} and incubate for 1 h at 37 °C before plating out on 2YT-AG agar plate. Once dried, incubate plates top-down at 37 °C o/n.

After that, pick 92 individual clones into 96-well U-bottom PP microtiter plate filled with 200 μL of pre-warmed 2YT-AG (*H5, H6, H11, and H12 are left empty for controls)—mother plate.* After colony picking, seal *mother plate* with breathable sealing film and incubate in a microplate shaker o/n at 37 °C and 1400 rpm.

Next day, inoculate each well of fresh 96-well U-bottom PP microtiter plate containing 180 μL 2YT-AG with 20 μL of the o/n culture (*daughter plate*) for 2 h at 37 °C and 1400 rpm before packaging with 10^9 M13K07 helper phage at 37 °C stationary for 1 h. The *mother plate* was stored as glycerol stock at −80 °C after adding 150 µL of 80% glycerol.

The bacterial culture in the *daughter plate* was filtered by centrifuge microtiter plate for 15 min at 3500 rpm (swing out rotor). The supernatant with the remaining M13K07 helper phage was discarded. The pelleted bacteria in the *daughter plate* were resuspended in 220 µL pre-warmed 2YT-AKG and transferred to a new and fresh 96-well U-bottom PP microtiter plate (*new daughter plate*). Seal *new daughter plate* with breathable sealing film and incubate in a microplate shaker o/n at 30 °C shaking at 1400 rpm.

The bacteria in the *daughter plate* were centrifuged down for 15 min at 3500 rpm with swing out rotor. The phage in the supernatant were transferred and filtered in a 96-well filtration plate before use for monoclonal antibody phage ELISA.

Two polystyrene (PS) ELISA plates are needed for 92 colonies of soluble monoclonal antibody fragments. The wells (A1–H6) are the antigen coated whereas the other half (A7–H12) are for background with either PTM or BSA coating as negative controls. The proposed layout is shown in Fig. 5.

The next day, discard coating solution and wash all wells three times for 5 min by completely filling them with PBST. Block all 96 wells with

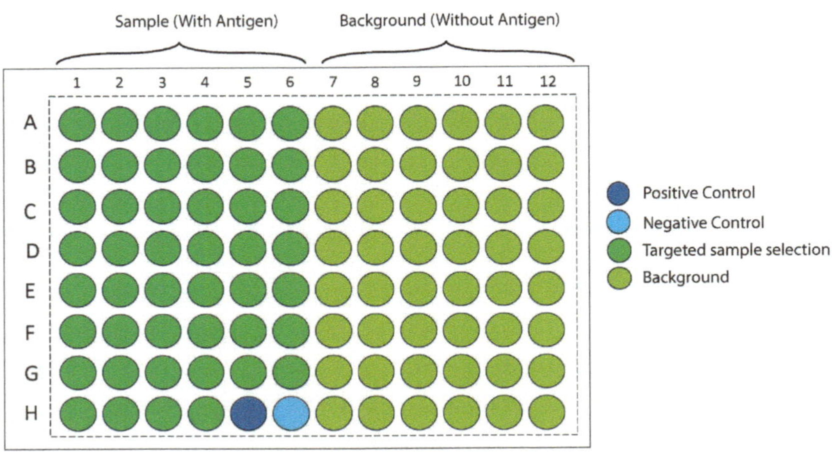

Fig. 5 The plate layout example for 46 monoclonal antibody phages in 96-well plate format. The plate was separated into half, the plate positions A1–H6 indicate the binding affinity between the target antigen coating and antibody phages, in which H5 is positive control and H6 is negative control. Besides, the other half of the plate position A7–H12 indicate the binding affinity between the background (blocking agent) and antibody phage particles. Thus, the monoclonal antibody phage of are added to each half of the assay plate.

300 μL of PTM/BSA for 1 h at RT, 600 rpm. Then, discard the blocking solution before washing all wells three times for 5 min with PBST. Then, binding was done in each well with 50 μL PTM/BSA and 50 μL phage antibody fragment solution for 1 h at RT, 600 rpm.

The phage antibody fragment solution was discarded by using pipette tips before washing wells three times for 5 min with PBST. Then, each well was incubated with 100 μL of mouse monoclonal anti-M13 HRP-conjugated-PTM solution (1:5000 in PTM) for 1 h at RT, 600 rpm. Then, wash the wells three times with PBST and two times with PBS after removing the anti-M13 HRP-conjugated-PTM solution (other antibody such as Anti-his tag or Anti-c-myc tag antibody were used as detecting antibody).

Finally, 100 μL of ABTS or TMB substrate was used to detect the binding phage for 5–30 min at RT in the dark. The plates were read at 405 nm extinction coefficient for ABTS in a plate reader.

3. Other phage display application of KingFisher

In principle, the described selection procedure applying magnetic bead-based panning for directed evolution is not confined to just antibody phage display libraries but can be used on all library-based display technologies (Galán et al., 2016). Here, we will only discuss a few phage display approaches as examples. These include peptide libraries (cyclic peptides), enzyme (variant) libraries, other scaffold-based binder molecules, (e.g., anticalins, SH2-domains, affibodies) or even complex cDNA libraries (Georgieva & Konthur, 2011).

In the context of peptide based evolution, mutagenesis of specific peptides can also be carried out using any molecular based method to achieve a diverse repertoire of peptides. This library can then be screened using the approach described above to isolate specific peptides with a desired characteristic. In terms of applications, this can be related to either mimotope identification or even epitope mapping of mAbs. Cyclic peptides binding to protein targets can also be identified using phage display. For these applications, a random peptide library is can be used in place of the antibody library as the repertoire source for panning. Here, with the target proteins are being mounted on the magnetic beads as mentioned described.

The proposed panning strategy can also be used for the evolution of enzymes. This is evident for enzymes like Sortase A (Schmohl, Bierlmeier, Gerth, Freund, & Schwarzer, 2017) (https://doi.org/10.1002/psc.2980) and transglutaminase engineering (Kuramoto, Yamasaki, Shimizu, Tatsukawa, & Hitomi, 2013) (https://doi.org/10.1016/j.abb.2013.07.010).

The evolution of enzymes can be two-pronged in the approach with the engineering of either the enzyme or the substrates. The engineering of site-selective enzymes like Sortase can be designed to either improve the characteristics on the enzyme itself or the identification of new substrate peptide motifs that reacts with the enzyme. These modifications are achieved by introducing mutations at specific sites on the reactive sites of the enzyme. These mutations will result in an enzyme with specific mutations being introduced at hotspot residues. This will result in alterations to the biological functions and characteristics in terms of the activity and specificity of the enzyme. Therefore, the proposed approach can also be used to screen this library of enzyme mutants for a specific function or activity. The approach will allow the screening of multiple mutants simultaneously which would reduce the work load required for conventional enzyme mutagenesis experiments.

4. Conclusion

The application of magnetic nanoparticles for panning experiments has a significant impact in the way directed evolution of biologics can be carried out. The ability to adapt the use of these nanoparticles in different conditions, samples and handling systems allows for easy automation and handling. Therefore, in addition to conventional panning strategies, magnetic nanoparticles offer an attractive approach for panning particularly with the use of automation. The design of the approach also allows for easy adaptation of the protocol to different biological molecules other than antibodies making it a robust alternative.

Acknowledgments

The authors would like to acknowledge financial support from Universiti Sains Malaysia Research University Individual Grant Scheme (Grant no: 1001/CIPPM/812173) and the Malaysian Ministry of Education through the Higher Institution Centre of Excellence (HICoE) Grant (Grant No. 311/CIPPM/4401005).

References

Arap, M. A. (2005). Phage display technology: Applications and innovations. *Genetics and Molecular Biology*, *28*, 1–9.

Arnold, F. H. (2018). Directed evolution: Bringing new chemistry to life. *Angewandte Chemie International Edition*, *57*(16), 4143–4148.

Bazan, J., Całkosiński, I., & Gamian, A. (2012). Phage display—A powerful technique for immunotherapy: 1. Introduction and potential of therapeutic applications. *Human Vaccines & Immunotherapeutics*, *8*(12), 1817–1828.

Bornscheuer, U. T., Hauer, B., Jaeger, K. E., & Schwaneberg, U. (2019). Directed evolution empowered redesign of natural proteins for the sustainable production of chemicals and pharmaceuticals. *Angewandte Chemie International Edition*, *58*(1), 36–40.

Bugli, F., Paroni Sterbini, F., Graffeo, R., Caridi, F., Iantomasi, R., Torelli, R., et al. (2011). Effective use of nitrocellulose-blotted antigens for phage display monoclonal antibody selection. *The New Microbiologica*, *34*(3), 281–286.

Carter, P. J., & Lazar, G. A. (2017). Next generation antibody drugs: Pursuit of the 'high-hanging fruit'. *Nature Reviews Drug Discovery*, *17*, 197–223.

Chen, Z., Zhang, L., Tang, A., Callahan, C., Pristatsky, P., Swoyer, R., et al. (2016). Discovery and characterization of phage display-derived human monoclonal antibodies against RSV F glycoprotein. *PLoS One*, *11*(6), e0156798.

Chin, C. F., Ler, L. W., Choong, Y. S., Ong, E. B. B., Ismail, A., Tye, G. J., et al. (2016). Application of streptavidin mass spectrometric immunoassay tips for immunoaffinity based antibody phage display panning. *Journal of Microbiological Methods*, *120*, 6–14.

Ch'ng, A. C. W., Ahmad, A., Konthur, Z., & Lim, T. S. (2019). A high-throughput magnetic nanoparticle-based semi-automated antibody phage display biopanning. In M. Steinitz (Ed.), *Human monoclonal antibodies: Methods and protocols* (pp. 377–400). New York, NY: Springer New York.

Ch'ng, A. C. W., Choong, Y. S., & Lim, T. S. (2016). Phage display-derived antibodies: Application of recombinant antibodies for diagnostics. In S. K. Saxena (Ed.), *Proof and concepts in rapid diagnostic tests and technologies* (pp. 107–135). London, UK: IntechOpen.

Ch'ng, A. C. W., Hamidon, N. H. B., Konthur, Z., & Lim, T. S. (2018). Magnetic nanoparticle-based semi-automated panning for high-throughput antibody selection. In M. Hust & T. S. Lim (Eds.), *Phage display: Methods and protocols* (pp. 301–319). New York, NY: Springer New York.

dos Santos, M. L., Quintilio, W., Manieri, T. M., Tsuruta, L. R., & Moro, A. M. (2018). Advances and challenges in therapeutic monoclonal antibodies drug development. *Brazilian Journal of Pharmaceutical Sciences*, *54*(spe), e01007.

Frenzel, A., Schirrmann, T., & Hust, M. (2016). Phage display-derived human antibodies in clinical development and therapy. *MAbs*, *8*(7), 1177–1194.

Galán, A., Comor, L., Horvatić, A., Kuleš, J., Guillemin, N., Mrljak, V., et al. (2016). Library-based display technologies: Where do we stand? *Molecular BioSystems*, *12*(8), 2342–2358.

Georgieva, Y., & Konthur, Z. (2011). Design and screening of M13 phage display cDNA libraries. *Molecules*, *16*(2), 1667–1681.

Hallborn, J., & Carlsson, R. (2002). Automated screening procedure for high-throughput generation of antibody fragments. *BioTechniques*, *33*, S30–S37.

Kaplon, H., & Reichert, J. M. (2019). Antibodies to watch in 2019. *MAbs*, *11*(2), 219–238.

Könning, D., & Kolmar, H. (2018). Beyond antibody engineering: Directed evolution of alternative binding scaffolds and enzymes using yeast surface display. *Microbial Cell Factories*, *17*(1), 32–49.

Konthur, Z., & Walter, G. (2002). Automation of phage display for high-throughput antibody development. *Targets*, *1*(1), 30–36.

Konthur, Z., Wilde, J., & Lim, T. S. (2010). Semi-automated magnetic bead-based antibody selection from phage display libraries. In R. Kontermann & S. Dübel (Eds.), *Antibody engineering* (pp. 267–287). Berlin, Heidelberg: Springer Berlin Heidelberg.

Kuramoto, K., Yamasaki, R., Shimizu, Y., Tatsukawa, H., & Hitomi, K. (2013). Phage-displayed peptide library screening for preferred human substrate peptide sequences for transglutaminase 7. *Archives of Biochemistry and Biophysics*, *537*(1), 138–143.

Lim, C. C., Choong, Y. S., & Lim, T. S. (2019). Cognizance of molecular methods for the generation of mutagenic phage display antibody libraries for affinity maturation. *International Journal of Molecular Sciences*, *20*(8), 1861–1893.

Lim, B. N., Tye, G. J., Choong, Y. S., Ong, E. B. B., Ismail, A., & Lim, T. S. (2014). Principles and application of antibody libraries for infectious diseases. *Biotechnology Letters*, *36*(12), 2381–2392.
Liu, B., Huang, L., Sihlbom, C., Burlingame, A., & Marks, J. D. (2002). Towards proteome-wide production of monoclonal antibody by phage display. 11Edited by J. Karn *Journal of Molecular Biology*, *315*(5), 1063–1073.
Liu, B., & Marks, J. D. (2000). Applying phage antibodies to proteomics: Selecting single chain Fv antibodies to antigens blotted on nitrocellulose. *Analytical Biochemistry*, *286*, 119–128.
Nixon, A. E., Sexton, D. J., & Ladner, R. C. (2014). Drugs derived from phage display: From candidate identification to clinical practice. *MAbs*, *6*(1), 73–85.
Schmohl, L., Bierlmeier, J., Gerth, F., Freund, C., & Schwarzer, D. (2017). Engineering sortase a by screening a second-generation library using phage display. *Journal of Peptide Science*, *23*(7–8), 631–635.
Shen, M., Rusling, J., & Dixit, C. K. (2017). Site-selective orientated immobilization of antibodies and conjugates for immunodiagnostics development. *Methods (San Diego, Calif.)*, *116*, 95–111.
Tizei, P. A. G., Csibra, E., Torres, L., & Pinheiro, V. B. (2016). Selection platforms for directed evolution in synthetic biology. *Biochemical Society Transactions*, *44*(4), 1165–1175.
Walter, G., Konthur, Z., & Lehrach, H. (2001). High-throughput screening of surface displayed gene products. *Combinatorial Chemistry & High Throughput Screening*, *4*(2), 193–205.
Welch, N. G., Scoble, J. A., Muir, B. W., Pigram, P. J., et al. (2017). Orientation and characterization of immobilized antibodies for improved immunoassays (Review). *Biointerphases*, *12*(2), 02D301.

Further reading

Cheng, F., Zhu, L., & Schwaneberg, U. (2015). Directed evolution 2.0: Improving and deciphering enzyme properties. *Chemical Communications*, *51*(48), 9760–9772.
Cherry, J. R., & Fidantsef, A. L. (2003). Directed evolution of industrial enzymes: An update. *Current Opinion in Biotechnology*, *14*(4), 438–443.
Cobb, R. E., Chao, R., & Zhao, H. (2013). Directed evolution: Past, present and future. *AIChE Journal*, *59*(5), 1432–1440, American Institute of Chemical Engineers.

CHAPTER EIGHT

On-bead enzyme-catalyzed signal amplification for the high-sensitive detection of disease biomarkers

Wenjiao Fan[a,b,c], Wei Ren[a,b,c,*], Liping Zhu[a,b,c], Chenghui Liu[a,b,c,*]
[a]Key Laboratory of Applied Surface and Colloid Chemistry, Ministry of Education, Xi'an, China
[b]Key Laboratory of Analytical Chemistry for Life Science of Shaanxi Province, Xi'an, China
[c]School of Chemistry & Chemical Engineering, Shaanxi Normal University, Xi'an, China
*Corresponding authors: e-mail address: wei.ren@snnu.edu.cn; liuch@snnu.edu.cn

Contents

Abstract

The high-sensitive and rapid detection of critical biomarkers, e.g., disease-related nucleic acids and proteins, is always desired. Compared with the routine homogenous detection strategies, the on-bead flow cytometry (FCM)-based assays have drawn a lot of interests owing to their unique advantages. On one hand, microbeads (MBs) are employed for the enrichment of fluorescent signals, allowing the size encoding for multiplexed detection of biomarkers. On the other hand, FCM enables the fast read-out of the total fluorescent signals enriched on the MBs and the decoding of MBs' size information. For an improved sensitivity and versatile application scenarios, the signal amplification on MBs is required. However, the enzyme-catalyzed on-bead reactions remain challenging owing to the critical reaction conditions on the MBs/solution interface. Toward the high-sensitive detection of target biomolecules in real-samples, a series

Methods in Enzymology, Volume 630
ISSN 0076-6879
https://doi.org/10.1016/bs.mie.2019.09.005

of on-bead enzyme-catalyzed signal amplification strategies have been developed. After careful optimization of the reaction conditions, the proposed sensors are proven to have ultra-high sensitivities to fulfill the requirement of real-sample detection.

1. Introduction

1.1 FCM beads-based assay

The flow cytometry (FCM) beads-based assay provides a powerful tool for the quantitative analysis of target biomolecules. Its application ranges from disease-associated biomolecule diagnosis to medical discovery (Gondhalekar et al., 2018). Compared with the homogenous detection strategies, the microbeads (MBs)-based, particularly magnetic beads-based assay possesses several distinct advantages. Firstly, the MBs offer appreciable binding capacity due to their high surface-to-volume ratio, which provides more reaction sites and consequently a higher sensitivity (Rödiger et al., 2014; Yu, Kim, Niessner, & Knopp, 2012). Secondly, the composition and surface modification of MBs are highly customizable. The MBs can be synthesized with a range of materials and even multiple compositions (Kim & Ligler, 2010). Meanwhile, the surface functionalization of the MBs is also variable (Taitt, Shriver-Lake, Anderson & Ligler, 2011). In this way, the target molecules can be specifically captured on the MBs' surface and easily isolated with the help of centrifuge or magnet. Thirdly, besides the composition and surface properties, the size of MBs is also tunable (Kim & Ligler, 2010). The size tunability permits the size-encoding for the simultaneous detection of multiplexed biomarkers (Qiu, Zhang, Liu, & Li, 2017; Zhu et al., 2018). Moreover, by integrating the size and fluorophore color for binary labeling, the encoding capacity of the MBs can be potentially expanded in an exponential manner (Akinfieva, Nabiev, & Sukhanova, 2013; Birtwell & Morgan, 2009; Guo et al., 2017).

FCM is widely recognized as a versatile technology for scientific research and clinical diagnosis (Darzynkiewicz, Huang, & Zhao, 2017; Robinson & Roederer, 2015; Wilkerson, 2012). Unlike the fluorescent analyzers to read out the spectral properties of the whole homogenous samples, FCM enables the interrogation of individual events one by one (Adan, Alizada, Kiraz, Baran, & Nalbant, 2017). As to the beads-based assay, the working principle of FCM depends on the simultaneous investigation of light scattering of the MBs and the fluorescence signal enriched on their surface (Adan et al., 2017).

More interestingly, a flow cytometer can be equipped with multiple lasers and detection channels, and once the beads-based reaction for target assay is finalized, the MBs are analyzed one after another by FCM in a high-efficient manner without any separation procedures, accessing high-throughput detection. Therefore, by using FCM technique that enables multiple optical parameters measurement of each individual MBs, the multiplexed detection of biomarkers can be achieved.

Although the integration of beads-based assay and FCM has allowed the high-throughput monitoring of life-associated biomarkers, the traditional on-bead reaction is generally in a one-to-one format, which means one target molecule binding on the surface of MBs can only capture one fluorescent probe. Thus, for a higher detection sensitivity of critical biomolecules, the on-bead signal amplification technique remains desired.

1.2 On-bead signal amplification methods

In general, to enlarge the number of fluorescent probes anchored on the MBs, one needs to expand the binding area of MBs or conduct signal amplification on the surface. For example, the Zr^{4+}-functionalized mesoporous SiO_2 beads (ZrMMs) have been synthesized for the detection of protein kinase (PKA) activity (Ren, Liu, Lian, & Li, 2013). PKA can specifically catalyze the substrate peptides' phosphorylation, after which the fluorophore-labeled phosphorylated peptide can be recognized and captured by the ZrMMs via Zr^{4+}-phosphate binding. In this way, the fluorescence signal enriched on the ZrMMs is in a positive relationship with the PKA activity. The enlarged surface area of the ZrMMs provides this sensor with a wide dynamic range. However, such strategies to expand the surface areas of MBs still rely on the one-to-one binding format. The sensitivity of beads-based sensors is to be largely improved if on-bead signal amplification can be realized.

Thanks to the development of molecular biotechnology, a series of enzyme-catalyzed signal amplification methods have been reported and successfully employed to the high-sensitive detection of biomolecules. These methods include polymerase chain reaction (PCR), loop-mediated isothermal amplification (LAMP), exponential amplification reaction (EXPAR), and rolling cycle amplification (RCA), etc., PCR, which relies on repetitive cycles of denaturation, hybridization, and polymerase extension of nucleic acid, is a classic signal amplification strategy that has been widely applied in biological studies. However, the PCR process focuses the analysis of nucleic acid and requires temperature cycling thus can hardly be applied to the on-bead assays in most cases (Mullis et al., 1986; Saiki et al., 1985). The LAMP strategy amplifies the target sequence with high efficiency but relies on a DNA polymerase and four specially designed primers to recognize six distinct sequences on the target DNA (Li, Li, Jia, & Yan, 2011; Notomi et al., 2000). However, its on-bead application is restricted by the complicated sequence design. Based on the integration of polymerase strand extension and single-strand nicking, EXPAR can achieve an amplification efficiency of $>10^6$ in isothermal conditions (Jia, Li, Liu, & Cheng, 2010; Ness, Van Ness, & Galas, 2003; Wang, Wang, Liu, Wang, & Li, 2017). Nevertheless, this strategy is designed for the amplification of short oligonucleotides and not fit for on-bead assays. RCA emerges to be the best choice for on-bead signal amplification. It depends on DNA polymerase catalysis to produce a lot of tandem repeats of monomers with circle DNA as a template (Fire & Xu, 1995). As a result of producing numbers of repeat sequences, RCA offers an exquisite strategy for signal amplification. Although it has been proven that RCA works well on the surface of MBs (Sato, Ishii, Sasaki, Sato, & Nilsson, 2013), each kind of target molecule demands specialized designed padlocks, leading to tedious work for achieving multiplexed detection of target biomarkers.

Compared to the enzyme-catalyzed reactions, enzyme-free amplification strategies seem to be more adaptable to the on-bead assay due to their much more simplified reaction process. For instance, hybridization chain reaction (HCR) is one of the most attractive strategies among the enzyme-free approaches (Dirks & Pierce, 2004; Yang, Liu, Ren, & Li, 2012). In an HCR system, two rational designed DNA hairpin probes can coexist in solution. With the introduction of trigger DNA/RNA, the two metastable hairpin probes will hybridize in a cascade manner. Finally, a long double-strand DNA whose sequence is composed of the two hairpin probes is formed. Based on this enzyme-free kinetics-controlled amplification method, a set of on-bead FCM-based strategies for the sensitive detection of target molecules have been proposed. Up to now, the combination of HCR and on-bead assay has been applied to the detection of DNA (Ren, Liu, Lian, & Li, 2013; Ren et al., 2013), miRNA (Oishi, 2015), and enzyme (Zhang, Liu, Sun, Tang, & Li, 2015). In these designs, the target molecules act as the anchors to link the initiator of HCR to the MBs' surface. Hence, HCR is to proceed on the MBs and the concentration of target molecules can be reflected by the amplified fluorescent signal. In another kind of design, the target miRNA molecules facilitate the click chemistry-mediated ligation chain reaction (CCLCR) between the Aza-dibenzocyclooctyne (Aza-DBCO) modified DNA probe immobilized on the surface of MBs and azide (N_3) modified DNA probe as the initiator of HCR (Qi, Qiu, Fan, Liu, & Li, 2017). The target miRNA links the two probes via hybridization to trigger the click chemistry ligation. After that, mild thermal cycling is applied to release the target miRNA for another round of ligation. By this means, the signal of the target miRNA is amplified by CCLCR and HCR in a twofold manner, resulting in higher sensitivity than that of routine HCR.

HCR promises in on-bead signal amplification, but the application is still hindered because the beads-based HCR is of lower efficiency than that in the homogeneous phase, and the design of complicated hairpin sequences is essential to conduct the HCR amplification. It is supposed that the best alternative for on-bead signal amplification is free of thermal cycling and sophisticated sequence design.

1.3 TdT-catalyzed DNA extension for on-bead signal amplification

Recently, a new kind of signal amplification strategy based on terminal deoxynucleotidyl transferase (TdT)-catalyzed DNA extension has been reported for the detection of nucleic acid and enzymes (Tjong, Yu, Hucknall, & Chilkoti, 2013). TdT, a sequence-independent enzyme that catalyzes the repetitive addition of mononucleotides to the 3′-OH terminal of a DNA by using dNTPs as fuel, shows an excellent performance in DNA extension and signal amplification. It is worth mentioning that the TdT-based extension can largely simplify the DNA probe design in biosensing systems because it starts from the 3′-OH end of DNA, and depends neither on the substrate DNA sequence nor a specially designed template (Idili et al., 2014). Taking benefit of beads-based FCM assay and TdT, the authors developed a series of on-bead signal amplification strategies for the high-sensitive detection of disease-related biomolecules such as miRNA and protein (Fig. 1).

1.3.1 Click chemistry-aided on-bead TdT-extension for miRNA analysis

The principle of the click chemistry-aided on-bead TdT-extension for miRNA analysis is illustrated in Fig. 2 (Fan et al., 2018). There are two steps in this strategy. In the first step, the binding probe (Probe A) is linked with the extension probe (Probe B) by target miRNA-templated click chemistry reaction. The click chemistry reaction between Aza-DBCO and N_3 has

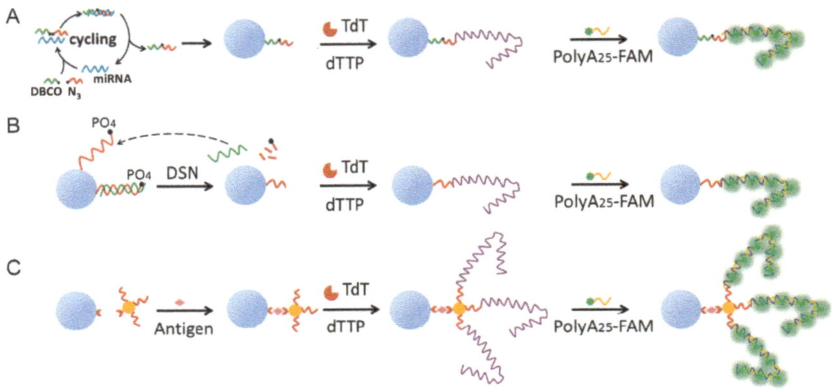

Fig. 1 (A) Principle of click chemistry-aided on-bead TdT-extension for miRNA analysis. (B) On-bead DSN-assisted TdT-extension for miRNA analysis. (C) Nanoparticle-mediated branched TdT-extension for protein analysis.

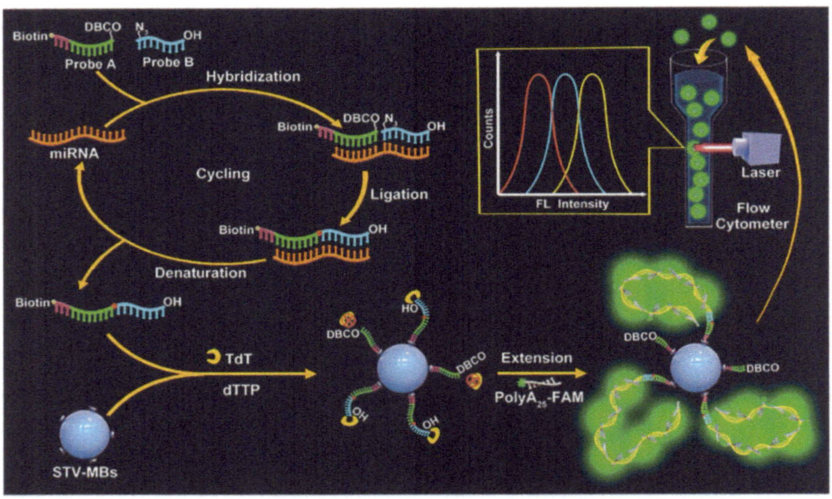

Fig. 2 Schematic illustration of the click chemistry-aided on-bead enzyme-catalyzed signal amplification for miRNA analysis (Fan et al., 2018).

been reported to be used for click nucleic acid ligation (CNAL) of two oligonucleotides (El-Sagheer & Brown, 2012; Kato & Oishi, 2014). It can only happen when Aza-DBCO group and N_3 group are brought into a close distance with the addition of target miRNA. In this consideration, the target miRNAs are introduced to template the CNAL reaction between

Aza-DBCO/biotin bi-modified Probe A and N_3 modified Probe B with high ligation efficiency and specificity. By introducing thermal cycling, each target miRNA can template a number of rounds of CNAL reaction to yield numerous ligation products as extension probes, forming the first-step signal amplification. In the second step, the extension probe is captured on the surface of streptavidin (STV) functionalized MBs via STV-biotin conjugation. After that, TdT recognizes the 3′-OH terminals of the ligated extension probes immobilized on the MBs to initiate the on-bead DNA polymerization, producing long poly(T) tails that can be up to several thousands of nucleotides. Each poly(T) tail can hybridize with plenty of fluorescein-labeled short poly(A) oligonucleotides to accumulate considerable fluorophores on the MBs, acting as the second step of signal amplification. In this way, the on-bead enzyme-catalyzed reaction is realized. Finally, by using a flow cytometer, the optical information of each individual fluorophore-anchored MB can be directly interrogated in multiple parameters without any separation or fluorophore elution procedures, which provides a powerful and easy-to-use analytical tool for the fluorescence signal readout of MBs. By integrating cycling CNAL, TdT-catalyzed on-bead terminal DNA polymerization, and FCM technology, the high-sensitive miRNA analysis is realized.

1.3.2 On-bead DSN-assisted TdT-extension for miRNA analysis

As shown in Fig. 3, the on-bead duplex-specific nuclease (DSN)-catalyzed target recycling is combined with TdT-mediated sequence-independent extension for signal amplification toward miRNA analysis (Qiu et al., 2017). In this design, the probe single-stranded DNA (ssDNA) molecules modified with biotin at 5′ terminal and phosphate group on the 3′ terminal are anchored on STV-modified MBs via STV-biotin interaction. At this point, TdT cannot initiate the terminal extension of the probe DNA because it recognizes 3′-OH only. When introducing target miRNA to the system, it hybridizes with the probe ssDNA on the surface of MBs to form the miRNA/DNA duplexes that can be recognized by DSN. DSN selectively cleave the probe DNA in the duplexes (Qiu, Zhang, Yu, Jiang, & Luo, 2015; Yin, Liu, & Ye, 2012), releasing 3′-OH terminal that can be recognized by TdT for sequence-independent extension. Once the probe DNA is not long enough to retain the miRNA/DNA duplex, the target miRNA will be released to participate in the next rounds of ssDNA cleavage to produce 3′-OH terminals constantly on the surface of MBs. Afterward, TdT-catalyzed extension will generate a long poly(T) tail from the 3′-OH ends of each cleaved ssDNA, which will hybridize with

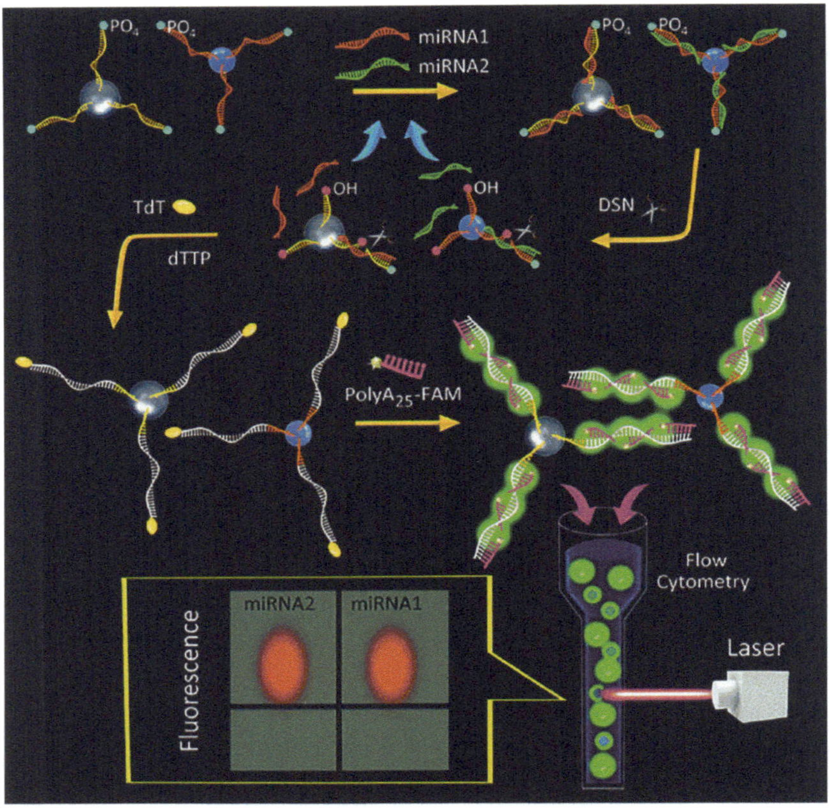

Fig. 3 Schematic illustration of on-bead DSN-assisted TdT-extension for miRNA analysis (Qiu et al., 2017).

numerous fluorescein-labeled short poly(A) oligonucleotides for fluorescence signal enrichment. The fluorophores enriched on the MBs can be analyzed by a flow cytometer.

Taking benefit of MBs' size-encoding, multiplexed analysis of distinct miRNAs can be also achieved. In general, the probe ssDNA molecules corresponding to different target miRNAs are immobilized on MBs of varied sizes. Accordingly, the target miRNAs will induce fluorescence enrichment on these MBs via DSN-catalyzed cleavage and TdT-extension reactions. Finally, the size information of the MBs is to be discriminated by FCM via light scattering, while the target miRNA's concentration is to be reflected by the fluorescence intensity. Therefore, by combining the DSN-catalyzed cleavage for target recycling, bead-size encoding for target recognition, TdT-catalyzed extension for signal amplification, and robust flow cytometer for signal readout, high sensitivity and multiplexed analysis of miRNA can be achieved.

1.3.3 Nanoparticle-mediated branched TdT extension for protein analysis

Unlike the nucleic acid that possesses the duplication mechanisms for signal amplification, the protein/antigen interaction obeys one-to-one binding, leaving the high-sensitive analysis challenging. To solve this issue, the TdT-catalyzed DNA extension is further conducted on the gold nanoparticles (AuNPs) that can be immobilized on MBs for a branched amplification toward protein/antigen detection (Zhu et al., 2018). Compared with MBs, nanoparticles have a much larger surface-to-volume ratio, making them ideal carriers for protein/antibody loading and nucleic acid amplification (Rosi & Mirkin, 2005; Esmaeili-bandboni et al., 2017). In the meantime, the concentration information of protein/antigen is converted to fluorescent signals that can be amplified via TdT-induced extension. The scheme of this work is illustrated in Fig. 4. Specially, the target-specific monoclonal antibodies (mAb2) and oligonucleotides with 5'-SH and 3′-OH on the two ends (ODN) are co-modified on AuNPs.

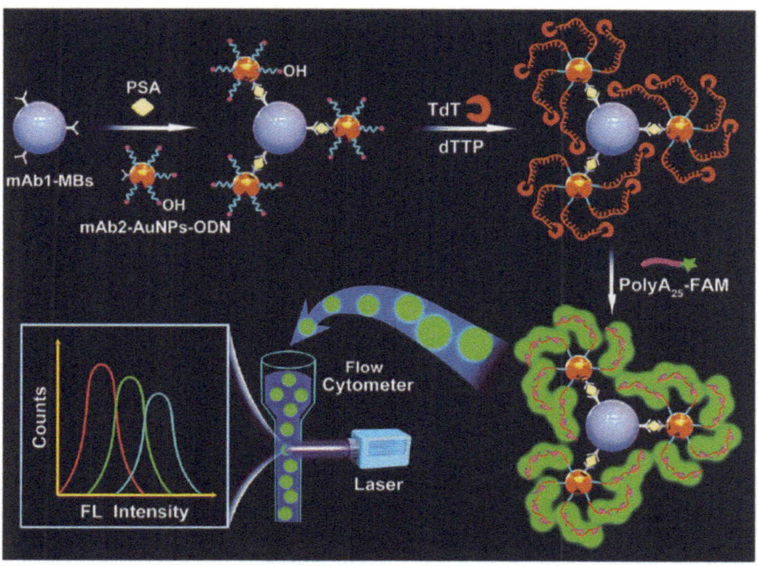

Fig. 4 Schematic illustration of AuNPs-mediated branched TdT extension for protein/antigen analysis (Zhu et al., 2018).

After the sandwich immunoreaction of target protein/antigen to conjugate mAb2-AuNPs to the surface of monoclonal antibodies (mAb1) modified MBs, TdT-extension happens on the surface of AuNPs to generate long poly(T) tails for fluorescein-labeled short poly(A) oligonucleotides accumulation on the MBs. As a result, on the surface of MBs, a great deal of long poly(T) tails are generated in branched manner because each MB carries a lot of AuNPs that act as carriers of TdT-catalyzed extension products. Finally, the flow cytometer is used for signal readout. Furthermore, based on the size-encoding principle as introduced in the last section, different antigen-specific antibodies can be labeled with MBs of various sizes and discriminated by FCM. Compared with the TdT-catalyzed DNA extension that directly happens on the surface of MBs, the AuNPs-mediated strategy not only realizes the signal amplification of proteins but also provides higher detection sensitivity owing to the branched amplification format.

2. Materials and methods

2.1 General equipment and reagents

Flow cytometric analysis was performed on a BD FACSCalibur flow cytometer. Dynabeads M-270 Streptavidin (STV-MBs, 2.8 μm in diameter) were purchased from Thermo Fisher Scientific (Invitrogen, Oslo, Norway). Terminal deoxynucleotidyl transferase (TdT) and 5 × TdT buffer (125 mM Tris-HCl, 1 M potassium cacodylate, 5 mM $CoCl_2$, 0.05% (v/v) Triton X-100, pH 7.2) were acquired from Thermo Fisher Scientific. Duplex-specific nuclease (DSN) and 10 × DSN reaction buffer (500 mM Tris-HCl, 50 mM $MgCl_2$, 10 mM DTT, pH 8.0) were purchased from Evrogen (Moscow, Russia). The PSA (prostate-specific antigen), biotinylated anti-PSA monoclonal antibody (anti-PSA-mAb1), anti-PSA monoclonal antibody (anti-PSA-mAb2), AFP (human alpha-fetoprotein antigen), biotinylated anti-AFP-mAb1, and anti-AFP-mAb2 used in AuNPs-mediated branched TdT extension for protein analysis were acquired from Linc-Bio Science (Shanghai, China). The CEA (Human carcinoembryonic antigen), biotinylated anti-CEA-mAb1 and anti-CEA-mAb2 were acquired from Biosynthesis Biotechnology (Beijing, China). All of the DNA and miRNA used were synthesized and purified by Takara Biotechnology (Dalian, China), and the nucleic acid sequences are listed below (5′–3′ direction):

(1) The DNA and miRNA sequences used in click chemistry-aided on-bead TdT-extension for miRNA analysis strategy.

ath-miR156a: UGACAGAAGAGAGUGAGCAC-2'me (methylation)
Probe A: biotin-CCCCCCCCCC(spacer)-GTGCTCACTC-DBCO
Probe B: N_3-TCTTCTGTCA
FAM-poly(A)25: AAAAAAAAAAAAAAAAAAAAAAAAA-FAM

(2) The sequences used in on-bead DSN-assisted TdT-extension for miRNA analysis method.
Let-7a: UGAGGUAGUAGGUUGUAUAGUU
miR21: UAGCUUAUCAGACUGAUGUUGA
anti-let-7a-P: CCCCCCCCCC(spacer)-AACTATACAACCTAC TACCTCA-PO_4
anti-miR21-P: CCCCCCCCCC(spacer)-TCAACATCAGTCTG ATAAGCTA-PO_4
FAM-poly(A)25: AAAAAAAAAAAAAAAAAAAAAAAAA-FAM

(3) The sequences used in nanoparticle-mediated branched TdT extension for protein analysis.
ODN: SH-TTTTTTCCCAGCACGAGACCAATT
FAM-poly(A)25: AAAAAAAAAAAAAAAAAAAAAAAAA-FAM

2.2 Preparation of the MBs

The stock vial was rotated with a mixer before taking out the STV-MBs to ensure the MBs are resuspended homogenously. Then the MBs were washed three times with 10 mM phosphate buffer (PBS, 0.3 M NaCl, pH 7.4) containing 0.1% Tween-20 to remove the preservative (*Notice*: 0.1% Tween-20 is an essential component in washing buffer and binding buffer to reduce non-specific binding).

2.3 Standard procedures of signal amplification

2.3.1 Click chemistry-aided on-bead TdT-extension for miRNA analysis

The standard click chemistry-aided on-bead TdT-extension for miRNA analysis consists of the following steps (Fan et al., 2018). Ath-miR156a was selected as a proof-of-concept target.

(1) Cycling CNAL reaction. The cycling CNAL was conducted in 10 mM PBS containing 0.01% Tween-20. *Critical Point*: The addition of Tween-20 is the key to restrain the non-specific CNAL. The STV sites on the surface of the MBs were full bound with biotinylated DNA based on calculation. In detail, 40 nM Probe A (3′-Aza-DBCO/5′-biotin), 40 nM Probe B (5′-N_3/3′-OH), and a series amount of target miRNA were mixed together in a total volume of 10 μL to proceed the CNAL reaction. *Notice*: For single-stranded oligonucleotides, the typical binding capacity for 100 μL original MBs slurry is 200 pmol based on the instruction of STV-MBs. The thermal cycling procedure (denaturation at 70 °C for 30 s and hybridization at 25 °C for 150 s) was set to perform target cycling as the first amplification (*Notice*: The amount of thermal cycle is another key experimental condition. The optimized cycle number is ~50).

(2) Immobilization of ligated probes on the MBs. The $\sim 1.2 \times 10^5$ (approximately 0.2 μL original MBs slurry) washed MBs was incubated with the mixture at room temperature for 30 min to ensure the maximum binding efficiency of the ligated probes to the MBs via STV-biotin conjugation.

(3) TdT-catalyzed DNA extension. After magnetic separation, the unbounded probes were removed. As a result, the purified MBs were added in 10 μL of 1× TdT buffer, containing two units (U) TdT (*Notice*: This is the optimized TdT dosage for the best resolution to discriminate the signals of blank control and samples in the proposed system) and 1 mM dTTP to conduct TdT template-free extension (*Critical points*: To avoid the inactivation of TdT (a) the operation should be performed on ice cube to guarantee the low temperature and the TdT should be restored at −20 °C other than is in use; (b) vortex or other violent vibration should be avoided when performing mix operation). The TdT-catalyzed extension was conducted at 37 °C for 1 h under gentle shaking.

(4) Enrichment of fluorescent probes on MBs. After magnetic purification (*Notice*: The magnetic purification is performed to stop the TdT induced-extension), 1 μM of FAM-poly(A)25 were introduced to hybridize with the poly(T) tails in 10 μL of 1× PBS. *Critical point*: In order to improve the hybridize stability of poly(A)25 and poly(T) produced by TdT-extension, 300 mM NaCl was added in the hybridization system. The hybridization was conducted at 4 °C for 1 h. Finally, the fluorophore-enriched MBs were inspected by FCM for the read-out of the fluorescence signal which reflects the target miRNA concentration.

2.3.2 On-bead DSN-assisted TdT-extension for miRNA analysis

The standard procedure of on-bead DSN-assisted TdT-extension for miRNA analysis contains the following steps (Qiu et al., 2017). Let-7a was employed as a proof-of-concept.

(1) Immobilization of let-7a-complement ssDNA (anti-let-7a-P) on MBs. According to the optimization, 20 nM anti-let-7a-P was chosen to incubate with ~1.2×10^6 M-270 STV-MBs (corresponding to 2 μL original MBs slurry) in 10 μL of Tris-HCl buffer (20 mM Tris-HCl containing 300 mM NaCl, pH 7.4) at room temperature for 0.5 h (*Notice*: The amount of anti-let-7a-P is optimized to avoid the background signal that can be observed at high anti-let-7a-P concentration/MB ratio). Afterward, the MBs were separated by a magnet and washed three times by using Tris-HCl buffer (containing 0.1% Tween-20). Then, the washed MBs were dispersed in 10 μL of 1 × DSN buffer for subsequent use.

(2) DSN-catalyzed target recycling reaction. The optimized temperature is 52 °C in the DSN cleavage reaction and the dosage of DSN is set as 0.5 U in cleavage according to optimization (*Notice*: The reaction temperature and the dosage of DSN are two key factors for the activity of DSN and the stability of miRNA and DNA hybridization product, respectively). Therefore, for each sample, 1 μL prepared anti-let-7a-P conjugated MBs (corresponding to 0.2 μL of the original MB slurry, ~1.2×10^5 beads) were incubated with 0.5 U DSN and different dosage of target let-7a in a total volume of 10 μL of 1 × DSN buffer (*Critical point*: The mixing of DSN with other substrates is performed on ice cube and DSN cannot be vortexed in case of inactivation). The DSN cleavage reaction was performed at 52 °C for 2 h under vigorous shaking.

(3) TdT-mediated signal amplification reaction. The MBs were magnetically separated and washed three times with Tris-HCl buffer. Practically, the MBs were mixed with 1 mM dTTP and 10 U TdT (*Notice*: This is the optimized TdT dosage for the maximal discrimination of the signals of blank control and samples in this protocol) in 10 μL of 1 × TdT buffer, and the mixture was incubated at 37 °C for 1 h under shaking to conduct sequence-independent TdT-extension.

(4) Enrichment of fluorescent probes to MBs. After magnetic isolation, 500 nM FAM-poly(A)25 were incubated with MBs in 10 μL of 50 mM phosphate buffer (containing 750 mM NaCl) at room temperature for 1 h. Finally, the fluorescence signal of MBs in each sample was monitored by FCM one by one.

For multiplexed detection, the simultaneous analysis of let-7a and miR21 was selected as an example. The anti-miR21-P was anchored on the Myone MBs (1 μm in diameter) according to the same protocols of anti-let-7a-P. At first, ~3×10^5 anti-let-7a-P functionalized M-270 STV-MBs and ~3×10^5

anti-miR21-P functionalized Myone MBs were mixed together. Then the sample contains both let-7a and miR21 were added to the mixture to perform the hybridization reaction of probe ssDNA and target miRNA. The subsequent DSN-catalyzed target recycling reaction, TdT-mediated signal amplification reaction and enrichment of fluorescent probes to MBs were conducted following the same procedures stated above.

2.3.3 Nanoparticle-mediated branched TdT extension for protein/antigen analysis

The AuNPs employed as carriers for branched DNA extension were synthesized according to the well-established protocol (Frens & Kolloid, 1973). Firstly, all of the glassware was cleaned with $K_2Cr_2O_7$-H_2SO_4 lotion (three parts $K_2Cr_2O_7$, one part H_2SO_4) and then rinsed with Millipore-filtered water thoroughly. The washed glassware was dried before use. For the synthesis of AuNPs, 100 mL 0.01% $HAuCl_4$ solution was heated in a flask with magnetic stir. As soon as the solution began to boil, 2 mL of 1% sodium citrate was pipetted in quickly. An obvious color change from pale yellow to brilliant red can be observed along with the formation of AuNPs. After the solution cools down to the room temperature slowly, the solution was filtered by a 0.22 μm nitrocellulose membrane and then kept at 4 °C for long-term storage. The average diameter of the AuNPs is ~16 nm.

The as-prepared AuNPs were then comodified with ODN and target antigen-specific antibody (mAb2). Typically, 1 mL 16 nm as-prepared AuNPs were taken out from the stoke vial, and the pH value was adjusted to 9.2–9.5 by dropping a certain amount of 0.1 M Na_2CO_3 solution. Then, according to the optimization result for the ratio of mAb2 to ODN, 10 μL of 1 $mg\,mL^{-1}$ mAb2 was added into the mixture and incubated for 20 min at room temperature to form mAb2-AuNPs conjugation. Afterward, 1 nmol alkylthiol-labeled ODNs were introduced into the mixture and incubated for 5 min at room temperature followed by 4 °C overnight incubation. After then, the mixture was dispersed in 10 mM PB (pH = 7.2) and stabilized by adding 0.15 M NaCl dropwise under a mild magnetic stir. The assembled mAb2-AuNPs-ODNs were purified by several rounds of centrifuge and stored in 1 mL PBS-BSA buffer at 4 °C for subsequent usage.

The standard nanoparticle-mediated branched TdT extension for protein/antigen analysis includes the following steps (Zhu et al., 2018). The detection of PSA is taken as an example.

(1) Immobilization of biotinylated capture antibody (anti-PSA-mAb1) on MBs. The maximal protein binding capacity of M-270 STV-MBs was adopted for anti-PSA-mAb1 incubation. In each sample, 0.02 μg of anti-PSA-mAb1 was incubated with $\sim 1.2 \times 10^5$ MBs (corresponding to 0.2 μL original slurry of MBs) at room temperature for 30 min under stirring (*Notice*: For biotinylated antibody, the typical binding capacity for 100 μL original MBs slurry is 10 μg according to the instruction of STV-MBs).

(2) Sandwich-type immunoreaction. After the magnetic separation to remove the unanchored anti-PSA-mAb1, the anti-PSA-mAb1-MBs conjugates were mixed with 5 μL of PSA-specific mAb2-AuNPs-ODN and different dilutions of PSA to conduct the sandwich-type immunoreaction for 2 h at room temperature under constant stirring.

(3) TdT-catalyzed extension. After the immunoreaction, another cycle of purification with 1 × PBS (containing 0.1% Tween-20) proceeded. As the optimized result, 1 U TdT and 1 mM dTTP were added to the washed MBs to perform on-bead TdT extension in 10 μL 1 × TdT buffer at 37 °C for 1 h.

(4) Enrichment of fluorescent probes on MBs. The washed MBs were incubated with 1 μM FAM-poly(A)25 at 4 °C for 1 h for the hybridization with poly(T) tails generated by the TdT-induced extension. In this way, the amplified signal is converted into the fluorescence on the MBs. Finally, the MBs of each sample were monitored by FCM.

The simultaneous detection of CEA and PSA was performed as the demonstration of the multiplexed detection of different antigens. At first, the anti-CEA mAb1-functionalized Myone MBs and anti-CEA mAb2-AuNPs-ODN were prepared according to the same protocols stated before. Then, $\sim 3 \times 10^5$ anti-PSA mAb1-functionalized M-270 STV-MBs and $\sim 3 \times 10^5$ anti-CEA mAb1-functionalized Myone MBs were mixed together. Afterward, the sample contains both PSA and CEA, which correspond to the recognized detection thresholds for clinical diagnosis, were incubated with the MBs to conduct sandwiched immunoreaction with anti-PSA mAb2-AuNPs-ODN and anti-CEA mAb2-AuNPs-ODN, respectively. Then the subsequent TdT-catalyzed extension and enrichment of fluorescent probes on MBs were conducted following the same procedures stated above.

2.4 Detection method

2.4.1 Analysis of a single type of biomarker target

The sample was diluted to 500 μL with 1 × PBS firstly and immediately subject to a BD FACSCalibur flow cytometer for analysis. For each sample, 10,000 MBs were collected for analysis by 488 nm laser excitation. The fluorescent signal was collected via the FL1 channel. According to the size information of M-270 STV-MBs, the voltage of FSC (forward light scattering) was set at E00 with 1.05 Amp Gain at Lin mode and SSC (side light scattering) was set at 270 V with 1.00 Amp Gain at Lin mode. For the click chemistry-aided on-bead TdT-extension for miRNA analysis, on-bead DSN-assisted TdT-extension for miRNA analysis, and nanoparticle-mediated branched TdT-extension for protein/antigen analysis the FL1 voltages were set at 280 V ~ 440 V to make sure the fluorescence signals of blank control and samples with serial dilutions of target can be clearly distinguished in the histogram of counts/fluorescence intensity. Then the concentration of target biomolecules can be analyzed quantitatively according to the mean fluorescence intensity (MFI) of the MBs.

2.4.2 Multiplexed analysis of target biomarkers

For the multiplexed detection, in order to differentiate the distinct size of MBs which were applied to encode the specific target, FSC and SSC were set at E-1 and 165 V, respectively, with Log mode. The two sorts of MBs were gated with G1 and G2, respectively, to make sure they can be clearly distinguished. The setting of FL1 voltage was the same as that of the single target analysis stated above.

3. Conclusion and perspective

In this chapter, a series of on-bead enzyme-catalyzed signal amplification methods are introduced for the high-sensitive analysis of biomarkers with FCM. These well-established strategies have several distinct advantages. First, due to the high surface-to-volume ratio and tunable size, the MBs are the perfect tool for the enrichment of signal and multiplexed analysis of biomarkers. Then, the TdT-catalyzed signal amplification is sequence-independent, thus it can be used for the sensitive detection of biomarkers without the complicated design of nucleic acid sequence. At last, the flow cytometer allows the multiparametric analysis of the MBs including fluorescence intensity and size information, accessing the multiplexed detection of biomarkers on MBs. In conclusion, the proposed FCM strategies based on beads-surface TdT-catalyzed signal amplification is promising in biological research and clinical diagnosis.

Acknowledgments

This work was supported by the National Natural Science Foundation of China (21622507, 21575086, 21904083), the Program for Changjiang Scholars and Innovative Research Team in University (IRT_15R43), and the Fundamental Research Funds for the Central Universities (GK201802016).

References

Adan, A., Alizada, G., Kiraz, Y., Baran, Y., & Nalbant, A. (2017). Flow cytometry: Basic principles and applications. *Critical Reviews in Biotechnology*, *37*, 163–176.

Akinfieva, O., Nabiev, I., & Sukhanova, A. (2013). New directions in quantum dot-based cytometry detection of cancer serum markers and tumor cells. *Critical Reviews in Oncology/Hematology*, *86*, 1–14.

Birtwell, S., & Morgan, H. (2009). Microparticle encoding technologies for high-throughput multiplexed suspension assays. *Integrative Biology*, *1*, 345–362.

Darzynkiewicz, Z., Huang, X., & Zhao, H. (2017). Analysis of cellular DNA content by flow cytometry. *Current Protocols in Immunology*, *119*, 571–572.

Dirks, R. M., & Pierce, N. A. (2004). Triggered amplification by hybridization chain reaction. *Proceedings of the National Academy of Sciences of the United States of America*, *101*, 15275–15278.

El-Sagheer, A., & Brown, T. (2012). Click nucleic acid ligation: Applications in biology and nanotechnology. *Accounts of Chemical Research*, *45*, 1258–1267.

Esmaeili-bandboni, A., Amini, S., Faridi-majidi, R., Bagheri, J., Mohammadnejad, J., & Sadroddiny, E. (2017). Cross-linking gold nanoparticles aggregation method based on localised surface plasmon resonance for quantitative detection of miR-155. *IET Nanobiotechnology*, *12*, 453–458.

Fan, W., Qi, Y., Qiu, L., He, P., Liu, C., & Li, Z. (2018). Click chemical ligation-initiated on-bead DNA polymerization for the sensitive flow cytometric detection of 3′-terminal 2′-O-methylated plant microRNA. *Analytical Chemistry*, *90*, 5390–5397.

Fire, A., & Xu, S.-Q. (1995). Rolling replication of short DNA circles. *Proceedings of the National Academy of Sciences of the United States of America*, *92*, 4641–4645.

Frens, G., & Kolloid, Z. (1973). Controlled nucleation for the regulation of the particle size in monodisperse gold suspensions. *Nature Physical Science*, *241*, 20–22.

Gondhalekar, C., Rajwa, B., Patsekin, V., Ragheb, K., Sturgis, J., & Robinson, J. P. (2018). Alternatives to current flow cytometry data analysis for clinical and research studies. *Methods*, *134–135*, 113–129.

Guo, Q., Bian, F., Liu, Y., Qu, X., Hu, X., & Sun, Q. (2017). Hybridization chain reactions on silica coated Qbeads for the colorimetric detection of multiplex microRNAs. *Chemical Communications*, *53*, 4954–4957.

Idili, A., Amodio, A., Vidonis, M., Feinberg-Somerson, J., Castronovo, M., & Ricci, F. (2014). Folding-upon-binding and signal-on electrochemical DNA sensor with high affinity and specificity. *Analytical Chemistry*, *86*, 9013–9019.

Jia, H., Li, Z., Liu, C., & Cheng, Y. (2010). Ultrasensitive detection of microRNAs by exponential isothermal amplification. *Angewandte Chemie International Edition*, *49*, 5498–5501.

Kato, D., & Oishi, M. (2014). Ultrasensitive detection of DNA and RNA based on enzyme-free click chemical ligation chain reaction on dispersed gold nanoparticles. *ACS Nano*, *8*, 9988–9997.

Kim, J. S., & Ligler, F. S. (2010). Utilization of microparticles in next-generation assays for microflow cytometers. *Analytical and Bioanalytical Chemistry*, *398*, 2373–2382.

Li, C., Li, Z., Jia, H., & Yan, J. (2011). One-step ultrasensitive detection of microRNAs with loop-mediated isothermal amplification (LAMP). *Chemical Communications*, *47*, 2595–2597.

Mullis, K., Faloona, F., Scharf, S., Saiki, R., Horn, G., & Erlich, H. (1992). Specific enzymatic amplification of DNA in vitro: The polymerase chain reaction. 1986. *Biotechnology*, *24*, 17–27.

Ness, J. V., Van Ness, L. K., & Galas, D. J. (2003). Isothermal reactions for the amplification of oligonucleotides. *Proceedings of the National Academy of Sciences of the United States of America*, *100*, 4504–4509.

Notomi, T., Okayama, H., Masubuchi, H., Yonekawa, T., Watanabe, K., Amino, N., et al. (2000). Loop-mediated isothermal amplification of DNA. *Nucleic Acids Research*, *28*, e63.

Oishi. (2015). Enzyme-free and isothermal detection of microRNA based on click-chemical ligation-assisted hybridization coupled with hybridization chain reaction signal amplification. *Analytical and Bioanalytical Chemistry*, *407*, 4165–4172.

Qi, Y., Qiu, L., Fan, W., Liu, C., & Li, Z. (2017). An enzyme-free flow cytometric bead assay for the sensitive detection of microRNAs based on click nucleic acid ligation-mediated signal amplification. *Analyst*, *142*, 2967–2973.

Qiu, L., Zhang, Y., Liu, C., & Li, Z. (2017). A versatile size-coded flow cytometric bead assay for simultaneous detection of multiple microRNAs coupled with a two-step cascading signal amplification. *Chemical Communications*, *53*, 2926–2929.

Qiu, X., Zhang, H., Yu, H., Jiang, T., & Luo, Y. (2015). Duplex-specific nuclease-mediated bioanalysis. *Trends in Biotechnology*, *33*, 180–188.

Ren, W., Liu, C., Lian, S., & Li, Z. (2013). Flow cytometry-assisted mix-and-read assay for ultrasensitive detection of protein kinase activity by use of Zr^{4+}-functionalized mesoporous SiO_2 microspheres. *Analytical Chemistry*, *85*, 10956–10961.

Ren, W., Liu, H., Yang, W., Fan, Y., Yang, L., Wang, Y., et al. (2013). A cytometric bead assay for sensitive DNA detection based on enzyme-free signal amplification of hybridization chain reaction. *Biosensors and Bioelectronics*, *49*, 380–386.

Robinson, J. P., & Roederer, M. (2015). Flow cytometry strikes gold. *History of Science*, *350*, 739–740.

Rödiger, S., Liebsch, C., Schmidt, C., Lehmann, W., Resch-Genger, U., Schedler, U., et al. (2014). Nucleic acid detection based on the use of MBs: A review. *Microchimica Acta*, *181*, 1151–1168.

Rosi, N., & Mirkin, C. (2005). Nanostructures in biodiagnostics. *Chemical Reviews*, *105*, 1547–1562.

Saiki, R. K., Scharf, S., Faloona, F., Mullis, K. B., Horn, G. T., Erlich, H. A., et al. (1985). Enzymatic amplification of ß-globin genomic sequences and restriction site analysis for diagnosis of sickle cell anemia. *Science*, *230*, 1350–1354.

Sato, K., Ishii, R., Sasaki, N., Sato, K., & Nilsson, M. (2013). Bead-based padlock rolling circle amplification for single DNA molecule counting. *Analytical Biochemistry*, *437*, 43–45.

Taitt, C. R., Shriver-Lake, L. C., Anderson, G. P., & Ligler, F. S. (2011). Surface modification and biomolecule immobilization on polymer spheres for biosensing applications. *Biomedical Nanotechnology*, *726*, 77–94.

Tjong, V., Yu, H., Hucknall, A., & Chilkoti, A. (2013). Direct fluorescence detection of RNA on microarrays by surface-initiated enzymatic polymerization. *Analytical Chemistry*, *85*, 426–433.

Wang, X., Wang, H., Liu, C., Wang, H., & Li, Z. (2017). A three-way junction structure-based isothermal exponential amplification strategy for sensitive detection of 3′-terminal 2'-O-methylated plant microRNA. *Chemical Communications*, *53*, 1124–1127.

Wilkerson, M. J. (2012). Principles and applications of flow cytometry and cell sorting in companion animal medicine. *Veterinary Medicine, Small Animal Clinician*, *42*, 53–71.

Yang, L., Liu, C., Ren, W., & Li, Z. (2012). Graphene surface-anchored fluorescence sensor for sensitive detection of microRNA coupled with enzyme-free signal amplification of hybridization chain reaction. *Applied Materials & Interfaces*, *4*, 6450–6453.

Yin, B., Liu, Y., & Ye, B. (2012). One-step, multiplexed fluorescence detection of microRNAs based on duplex-specific nuclease signal amplification. *Journal of the American Chemical Society*, *134*, 5064–5067.

Yu, H.-W., Kim, I. S., Niessner, R., & Knopp, D. (2012). Multiplex competitive microbead-based flow cytometric immunoassay using quantum dot fluorescent labels. *Analytica Chimica Acta*, *750*, 191–198.

Zhang, Y., Liu, C., Sun, S., Tang, Y., & Li, Z. (2015). Phosphorylation-induced hybridization chain reaction on beads: An ultrasensitive flow cytometric assay for the detection of T4 polynucleotide kinase activity. *Chemical Communications*, *51*, 5832–5835.

Zhu, L., Chen, D., Lu, X., Qi, Y., He, P., Liu, C., et al. (2018). An ultrasensitive flow cytometric immunoassay based on bead surface-initiated template-free DNA extension. *Chemical Science*, *9*, 6605–6613.

CHAPTER NINE

Organophosphonate functionalized Au/Si@Fe_3O_4: Versatile carrier for enzyme immobilization

Sunaina Kaul[a], Vishal Singh[a], Rajat Sandhir[b], Nitin Kumar Singhal[a,*]
[a]National Agri Food Biotechnology Institute, Food and Nutritional Biotechnology, Mohali, Punjab, India
[b]Department of Biochemistry, Panjab University, Chandigarh, India
*Corresponding author: e-mail address: nitin@nabi.res.in

Contents

Abstract

Magnetic nanoparticles with silica/gold (Au) shell were synthesized to be used as substrates for enzyme immobilization. Successful functionalization was followed by addition of phosphate linkers containing carboxyl end as free end for linking with enzyme. Effect of varied phospho-carboxyl chain lengths based linkers, 3-phosphonopropionic acid (3-PPA) and 16-phosphonopropionic acid (16-PHDA) on

Methods in Enzymology, Volume 630
ISSN 0076-6879
https://doi.org/10.1016/bs.mie.2019.10.020

kinetic parameters of enzyme were studied. Immobilization of upto 670 μg/mg of nanoparticles were observed. Enzyme bound with longer chain linker were observed to have K_m (11.8 mg/mL) and V_{max} (1.59 μmol/min/mg) which was found to be lower than 3-PPA. Reusability assay gave a minimum of 70% activity at 10th time of usage. Thus, an efficient, robust, cost effective and easily separable substrate for restraining the enzyme was prepared.

1. Introduction

Enzymes (biocatalysts) have been generally used to accelerate chemical reactions by providing features such as high activity, specificity and selectivity toward particular reactions, working best in ambient environment (Mateo, Palomo, Fernandez-Lorente, Guisan, & Fernandez-Lafuente, 2007). High temperature and pH limit the enzyme catalyzed reactions, as these conditions cause the enzyme degradation (Kishore, Kundu, & Kayastha, 2012). Therefore, their industrial application is hindered because of instability, inhibition, low yield or selectivity, narrow working range and lack of reusability (Rigoldi, Donini, Redaelli, Parisini, & Gautieri, 2018).

Advances in Biotechnology have solved the difficulties of activity enhancement and production of thermally stable enzymes by protein engineering (Han et al., 2019). Research area of protein or reaction engineering have paved the way to accomplish the effectivity of enzymes by directed evolution of microorganisms (enrichment culture technique), de novo protein design, use of non-established media, active-site imprinting and altering temperature (Kaul & Asano, 2012). These techniques however have not provided solutions for problems for example enzyme retrieval, enzyme constancy, enzyme selectivity, reduction of inhibition by the medium or products.

To overcome the limitations and to retain the efficacy of enzyme in the process of biocatalysis, immobilization of enzyme on different supports is emerging as a promising technique (Datta, Christena, & Rajaram, 2013). Immobilization involves restraining the enzyme to an inert matrix providing benefits such as ease in handling, effortlessness of separation of enzymes from the reaction mixture and recycle. Additionally, decrease in product cost with probable enhancement in thermal and pH stability were also rewarded by the immobilization process (Lei, Shin, Liu, & Ackerman, 2002). A study explained the stability of zinc-oxide nanoparticle bound β-Galactosidase against denaturants such as pH and temperature. It was also observed that galactose based competitive inhibition was also reduced in nanoparticle immobilized enzyme (Husain, Ansari, Alam, & Azam, 2011). However, earlier studies have also reported certain drawbacks of immobilization, including leaching of enzyme and low diffusional rate in reaction mixtures. For illustration, examination of the sustainability of geranyl propionate synthesis using lipase as biocatalyst was carried out, results showed linear decrease in residual activity of immobilized lipase with every next cycles of reuse, suggesting enzyme leaching from the support matrix after each cycle (Ferraz et al., 2015). In another study, successful immobilization of lipase onto the nano-fibrous membranes consist of repeating units of poly(vinyl al*co*hol-co-ethylene) (PVA-co-PE), via glutaraldehyde for the hydrolysis of *p*-nitrophenyl palmitate was carried out. An upsurge in K_m and a reduction in V_{max} was perceived due to slower substrate diffusion and diminished enzyme mobility. Nevertheless, immobilized lipase was discovered to maintain approximately 90% of its activity after incubation in a phosphate buffer system at 55 °C for 75 min, whereas free enzymes retain approximately 20% of their original activity (Zhu & Sun, 2012).

Nanotechnology as an upcoming field has provided improvement to the process of preventive enzyme on supports, offering increased effective surface area and surface to volume ratio. Increased surface area leads to a much higher loading density, and thus increase in bio-catalytic activity per unit volume or mass (Min & Yoo, 2014). Immobilization of enzymes on different nanosized supports such as spheres, fibers and tubes have been reported recently (Kim, Grate, & Wang, 2006; Martin & Kohli, 2003). Silica nanotubes as substrates for enzymes (Mitchell et al., 2002), encapsulation within phospholipid bilayers and binding on self-assembled monolayers (SAM) (Hung et al., 2006) has also been recently studied. Mesoporous nanomaterials have been also utilized for enzyme adsorption in pores owing to their unique characteristics of greater pore volume and tunable pore size (Ulu, Ozcan, Koytepe, & Ates, 2018). Organic groups or linkers, are responsible for covalent attachment of enzymes to the supports, prevent it from leaching (Yiu, Wright, & Botting, 2001). Magnetic nanoparticles (MNP) have come up as one of the promising substrate/support matrix for immobilization, providing separation of biocatalysts by use of external magnetic field, enhances the reusability of enzyme.

1.1 Magnetic nanoparticles

Iron oxide nanoparticles/MNP (diameters up to 30 nm) have been widely investigated as useful support matrix for enzyme immobilization as they provide a large surface area leading to augmented enzyme loading and thereby pre-eminent immobilization yield (Jiang et al., 2009). Biocompatibility and reusability in subsequent reactions, are also some of the key features of MNP immobilized enzyme (Bilal, Zhao, Rasheed, & Iqbal, 2018). These particles need to be kept in an inert atmosphere as they tend to oxidize easily (synthesis process carried in inert atmosphere). Oxidation, that further leads to loss in magnetism and dispensability, reducing the wide application of these particles in various applications (Li et al., 2013). Surface modification of magnetite with surfactants, polymers or noble metals is required to prevent these particles from oxidation (Moraes Silva, Tavallaie, Sandiford, Tilley, & Gooding, 2016; Williams, Gold, Holoman, Ehrman, & Wilson, 2006). For commercial and reaction feasibility, the polymer shell materials used should be biocompatible, biodegradable, non-toxic, and display low managing cost. Acrylamide, polysaccharide, chitosan, silica, etc., are generally, used for synthesis of shell of nanoparticles containing iron oxide cores (Bilal & Iqbal, 2019; Namdeo & Bajpai, 2009; Sun et al., 2006). In present study, functionalization of MNPs with Gold was carried out as it provides opportunities for surface modification and attachment of biological macromolecules (Majouga et al., 2015). While, silanol groups of silica allowed reaction with silanol coupling agent and provided increased charge density on MNPs (Ranjbakhsh, Bordbar, Abbasi, Khosropour, & Shams, 2012). Thus, the above said coatings were studied because of the distinctive chemical properties along with chemical inertness that prevent the particles from uncontrolled oxidation and irreversible aggregation. Techniques for immobilization of enzymes include physical (adsorption) and chemical (covalent) attachment with MNPs (Ahmed et al., 2018; Pan et al., 2009). Covalent immobilization causes more activity preservation of enzyme because of less leaching from the support (Hirsh et al., 2010). Functional groups containing spacer arms (organic chains) play an important role in linking surface of nanoparticles with enzymes and their degree of freedom.

1.2 Linkers

Linkers bind irreversibly with the functional groups of the side chains of lysine (ϵ-amino group), cysteine (thiol group) and aspartic and glutamic acids (carboxylic group), imidazole and phenolic groups which are not important for the overall catalytic activity of enzyme (Mohamad, Marzuki, Buang, Huyop, & Wahab, 2015). Nanoparticles are coated with aminosilane, carboxylic and epoxy silane coupling agent to introduce amino, carboxylic or epoxy groups on their surface.

Amino polymers such as chitosan (Li, Jiang, Huang, Ding, & Chen, 2008), polydopamine (Martín et al., 2014), and polyethyleneimine (PEI) (Wang, Xu, Zhang, & Wei, 2013) due to their remarkable biodegradability and biocompatibility structures are also used for surface functionalization. Aminosilane, mercaptosilane and glycidoxysilane can also be used as coupling agent while also working as silanization agent. Hydroxyl groups on MNPs displace alkoxy groups of silane thus forming covalent bond with metal oxide (Bini, Marques, Santos, Chaker, & Jafelicci, 2012).

In one of our study we immobilized enzyme on magnetic nanoparticles containing carboxylic functional group in the end of linker group (Singh et al., 2016). However, to study the effect of linker length and degree of freedom on catalytic activity of enzyme, we introduced carboxylic groups on the nanoparticles by binding with two different length linkers viz, 3-phosphonopropionic acid (3-PPA) and 16-phosphonohexadecanoic acid (16-PHDA) (Singh et al., 2018). In one of novel study, employed both viz protein engineering and modified enzyme nanoparticle immobilization to see the combined effect of the activity of enzyme (Narayan Patel et al., 2018). In another study, multiple enzyme was immobilized on nanoparticle to carry out multistep reaction with single catalyst (Kumari et al., 2018).

1.3 Enzymes

In our contemporary work, we have used industrially beneficial enzymes alpha amylase and xylanase to immobilized on nanoparticles surface. Alpha amylase is utilized in ethanol industry to break down starch into small fermentable sugars. Production of high fructose corn syrup is enabled by release of shorter chain oligosaccharides by the action of alpha amylase (Singh et al., 2016). Xylanase has major applications in pulp and paper industry, extraction of fruit juices, coffee and plant oil. To reduce the production cost of these commercially important enzymes it is essential to make them more stable and cheaper (Singh et al., 2018).

2. Material and methods

2.1 General equipment

- **I.** UV-visible spectrophotometer
- **II.** Zeta sizer
- **III.** Fourier transform Infrared spectroscopy (FTIR)
- **IV.** High resolution transmission electron microscope (HRTEM)

V. Fourier electron scanning electron microscope (FESEM)
VI. X-ray diffraction (XRD)
VII. Thermogravimetric analyzer
VIII. Round bottom flasks
IX. Condenser
X. Glass beakers
XI. Ultrasonicator
XII. pH meter
XIII. Stir/hot plate
XIV. Weighing balance

2.2 Synthesis and fabrication of magnetic nanoparticles

MNPs were synthesized by the method (Fig. 1) published by Lyon, Fleming, Stone, Schiffer, and Williams (2004). Steps followed were as given below:

I. $FeCl_2.7H_2O$ and $FeCl_3.6H_2O$ in the ratio of 1:2 mmol were mixed in 25 mL of 0.4 M HCl solution in round bottom flask till the solution turned transparent yellow-orange (solution 1).

II. 1000 mL closed round bottom flask containing 250 mL of 1.5 M NaOH solution on the other side was maintained at 80 °C and purged with nitrogen (solution 2).

Fig. 1 Reaction scheme for synthesis of magnetic nanoparticles.

III. Solution 1 it is poured into the solution 2 drop by drop to get instant black precipitates of magnetite. (Brownish precipitates are indicator of maghetite (oxidized) crystals).

IV. The mixture is stirred for 30 min and then allowed to cool to room temperature. Particles obtained were washed thoroughly (4–5 times) with MQ and then with 0.1 M TMAOH and later stored in it.

2.2.1 Synthesis of silica-coated magnetic nanoparticles

In 100 mL round bottom flask, the synthesized Fe_3O_4 (containing 250 mg magnetite) were added (Van Quy et al., 2013) into a flask comprising 12.5 mL of the 10% (w/v) solution of tetraethylorthosilicate (TEOS) in alcohol and 17.5 mL of absolute alcohol (molecular grade) at room temperature. (TEOS acts as the precursor of silicate ions).

I. The pH of the suspension was corrected to 9.0 with 0.1 M NaOH.

II. The flask was heated at 90 °C and agitated for 6 h under reflux. After cooling to room temperature (RT), the suspension was cleaned twice with absolute alcohol and six times with distilled water.

2.2.2 Synthesis of gold-coated magnetic nanoparticles

I. Washed Fe_3O_4 nanoparticles were dissolved in 250 mL of 0.1 M HNO_3 and the precipitate was recovered by the magnet decantation (Lyon et al., 2004).

II. The precipitate was dispersed in 250 mL of 0.01 M HNO_3 and heated with stirring to 90–100 °C on magnetic plate (the solution color transformed from dark brown–black to brown–red). The heating was stopped after 30 min, and the solution was kept at room temperature to cool down.

III. Gold shell around magnetic nanoparticles were synthesized using by 0.1 M TMAOH and were added to 0.1 M citric acid, 0.2 M $NH_2OH.HCl$ and 1% $HAuCl_4$ and stirred until the solution became purple in color.

2.3 Surface functionalization with organophosphorus linkers

I. 3-phosphonopropionic acid (3-PPA)

II. 16-phosphohexadecanoic acid (16-PHDA)

III. MES buffer (0.01 M)

IV. N-(3-dimethylaminopropyl)-N′-ethylcarbodiimide hydrochloride (EDC)

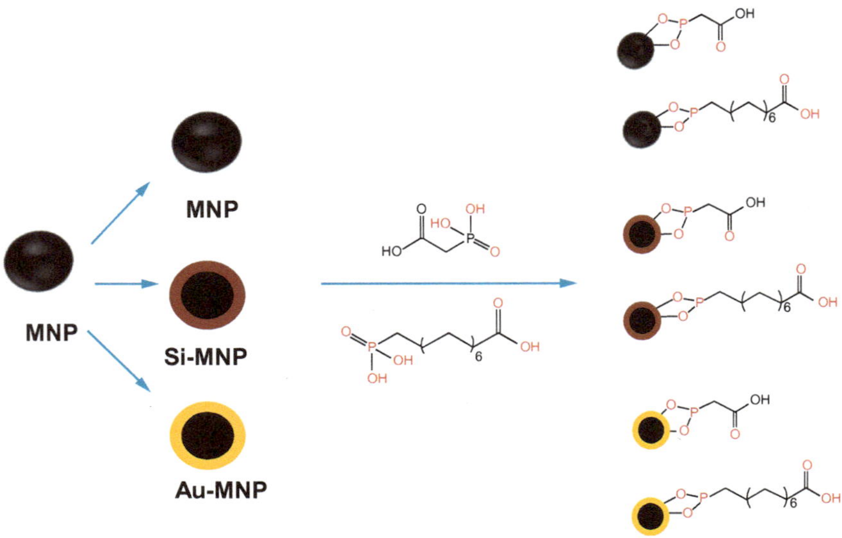

Fig. 2 Reaction mechanism for binding of phosphonates with hydroxyl groups of magnetic nanoparticles.

Freshly prepared nanoparticles were mixed with 3-PPA solution (29 mM 3-PPA in case of Fe_2O_3 and Fe_3O_4Si, whereas 44 mM of 3-PPA for Fe_3O_4Au) in 1:1 M ratio followed by ultrasonication for 30 min (Fig. 2). Nanoparticles dispersion with linkers were kept on a permanent magnet for 18 h (Sahoo, Sahu, & Pramanik, 2011). Unreacted 3-PPA was removed with the help of magnetic decantation and nanoparticles were washed with 0.01 M MES buffer (pH 5.30).

The nanoparticles were mixed with 16-PHDA linker maintaining the pH of suspension at 9. Ultrasonication of the suspension were carried out for 30 min, and then suspension separated using permanent magnet for 18 h. Separated nanoparticles were washed three times with 0.01 M PBS buffer and stored at 4 °C for further use.

2.4 Enzyme immobilization

I. The carboxylic acids groups of 3-PPA and 16-PHDA were activated by EDC and then coupled with α-amylase/xylanase by using 1:1 weight ratio of nanoparticles and enzyme in 0.1 M MES buffer (pH 5.30).

II. Reagents were stirred at intervals of 15 min for 2 h at room temperature, later kept on ice bath for another 30 min (Grabarek & Gergely, 1990).

2.5 Characterization of synthesized magnetic nanoparticles

The prepared synthesized magnetic nanoparticles were characterized by various techniques such as HRTEM (High Resolution Transmission Electron Microscopy), FESEM (Field Emission Scanning Electron Microscope), XRD (X-Ray Diffraction), FTIR (Fourier Transmission Infrared Spectroscopy), UV-visible absorption spectroscopy and TGA (Thermogravimetric analysis).

1. For HRTEM, a minimal amount around 2 μL of MNPs was placed onto the TEM grid.
2. Sample was allowed to dry at room temperature.
3. Visualization was followed in TEM.

To characterize the MNPs with FESEM, XRD, TGA and FTIR, MNP solution was washed with MQ and then allowed to dry in vacuum. Dried samples of MNPs were used for analysis.

2.6 Bicinchoninic acid (BCA) assay for amount of enzyme immobilized

I. BCA kit
II. BSA (bovine serum albumin) standard

The quantity of enzyme immobilized onto nanoparticles was determined in by using BCA method. The loading percentage of enzyme was calculated using the following equation:

$$\text{loading\%} = \frac{\text{Amount of enzyme used} - \text{amount of enzyme in supernatant}}{\text{Amount of enzyme used}} \times 100$$

2.7 Enzyme assay for free and immobilized enzyme

I. DNS reagent (1 g of DNS in 50 mL water + 30 g sodium potassium tartarate tetrahydrate + 20 mL of 2 N NaOH; made volume up to 100 mL)
II. Xylose/maltose
III. Xylan/potato starch

1 mL of free or immobilized enzyme with 1 mL of 1% (w/v) soluble starch/xylan solution made in (0.01 M) PBS buffer (pH 6.9) as substrate for 3 min. The reaction was stopped by adding 1 mL of DNS reagent followed by incubation of the mixture in a boiling water bath for 15 min. Amount of maltose formed in the mixture was determined spectrophotometrically at 540 nm.

One unit (IU) of amylase and xylanase activity was defined as the amount of enzyme which catalyzed production of 1.0 μmoL of maltose hydrolytically from soluble gelatinized starch and xylanase, respectively, per minute under the assay condition.

2.8 Calculations of kinetic parameters K_m and V_{max}

Kinetic parameters, K_m (substrate concentration at which V_{max} reaches half) and V_{max} (maximum velocity of reaction that can be achieved with standard assay conditions) were estimated for free and immobilized enzyme. Calculations were done using optimized enzyme dilution and varying the substrate (starch/xylan) concentration from 0.2% (w/v) to 2% (w/v) in reaction mixture. Lineweaver Burk plot was used for the estimation.

2.9 Reusability assay

Hydrolysis of 1% (w/v) starch solution in phosphate buffer (0.01 M) (pH 6.9), was carried out 10 times. After each cycle, the immobilized enzyme was retrieved from the reaction mixture by application of magnet and washed well 2–3 times with phosphate buffer to use for the next cycle of reaction. For the next cycle the immobilized enzyme was redissolved in phosphate buffer and added to the fresh substrate (starch/xylan) solution in phosphate buffer (0.01 M) (pH 6.9). The activity obtained in each cycle was compared to the activity obtained in first cycle with relative activity of 100%.

2.10 Comparison of activity parameters of free and immobilized enzymes

See Table 1.

3. Results and discussions

The functionalized MNPs with/without immobilized enzyme were synthesized as described above. Images obtained by TEM confirmed the successful formation of Fe_2O_3 nanoparticles with average diameter of 9 ± 2 nm.

Zeta potential values of the synthesized magnetic nanoparticles by DLS analysis were found to be as depicted in Table 2.

Table 1 Comparison of kinetic parameters and enzyme loadings on 3-PPA and 16-PHDA functionalized magnetic nanoparticles.

Nanoparticles with linkers	Free/immobilized xylanase	V_{max} (μmoles/min/mg)	K_m (mg/mL)	K_{cat} (sec^{-1})	μg enzyme/mg of nanoparticles	% Weight loss (TGA) due to functionalization
	Xylanase	2.53	9.41	1.23	100	NA
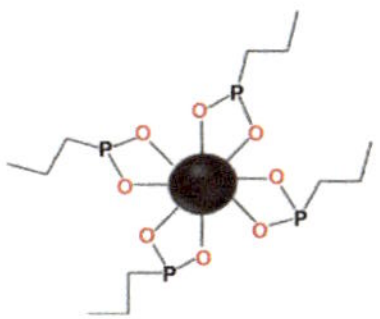	Fe_3O_4-3PPA-Xylanase	2.83	14.98	1.07	520	40
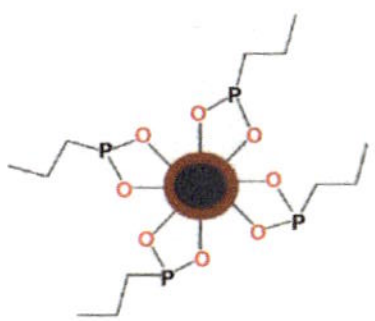	Fe_3O_4-Si-3PPA-Xylanase	2.13	13.5	1.04	680	NA
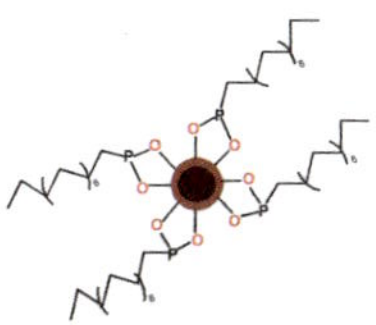	Fe_3O_4-Si-16PHDA-Xylanase	1.59	11.8	1.08	670	NA

Table 2 Zeta values of MNP and various functionalizations.

Types of nanoparticles	Zeta potential (mV)
Fe_3O_4	−39
Fe_2O_3-Au	−34
Fe_3O_4-Si	−44
Fe_3O_4-Si-3PPA	−41
Fe_3O_4-Si-16-PHDA	−43

Negative values for magnetic nanoparticles and silica-coated magnetic nanoparticles due to the presence of hydroxyl ions. Addition of Au lowered the negativity of MNP affirming the addition of Au coating. However, 3-PPA and 16-PHDA linker caused the surface potential to be more negative as compared to MNPs due to the presence of more ionizable carboxyl groups at the surface.

Functionalization of organic molecules on MNPs was confirmed by TGA analysis. Addition of 3-PPA caused weight loss of 25% upto 400 °C as compared to MNPs. Further, loss of 40% was obtained on enzyme immobilized on 3-PPA-MNPs.

Functional group modification by FTIR showed characteristic peaks of Fe-O, Fe-O-Si and Si-O-Si at 590 cm^{-1}, 1050–1250 cm^{-1} and 1070–1080 cm^{-1}. Peaks at 1100–1150 cm^{-1} specific for P-O and P-O-Fe stretching appeared on addition of 3-PPA and 16-PHDA.

Analysis by XPS, XRD and SEM also confirmed successful synthesis of MNPs.

On studying temperature and pH effects, immobilized xylanase showed stability at extreme conditions as compared to bare enzyme. Reusability profile confirmed recycling of the immobilized enzyme upto 10 times with retention of activity upto 70% in last batch. Silica-coated magnetite nanoparticles exhibited maximum comparative activity as associated to free and magnetite nanoparticles regardless of the linker. Better catalytic activity was observed in lengthier chain organophosphonate that allowed interaction of enzyme with the substrate more efficiently. Enzyme immobilized onto iron matrices through long alkyl chain length linker, i.e., 16-PHDA had less K_m value in comparison to the shorter 3-PPA linker. Less restriction of the enzyme and more distance of enzyme from the magnetite nanoparticles, allowed xylanase to interact with the substrate more effectively. Calculations of activation energy from the Arrhenius plot of all the above mentioned cases were carried out. It was observed that activation energy of 16-PHDA bound enzyme was lower than enzymes bound with 3-PPA. With reduced activation energy and K_m value, long chain alkyl length linker proved to be a better candidate for enzyme immobilization.

4. Conclusion

Magnetic nanoparticles as support for enzyme immobilization prove to be beneficial in terms of recyclability and green use of commercial enzyme. Reusability of the enzyme in a reaction for upto 10 times proves beneficial for cost cutting in industries. The alkyl chain length linker attachment to the nanoparticles provided functional group to bind covalently with enzyme lead to flexibility and reduction in stearic hindrance of the attached enzyme. All the above parameters caused better product formation. Linker applicability in higher yield of product formation was proven by low k_m values of PHDA immobilized enzyme than of PPA in our study.

Acknowledgments

The work was sponsored by NABI core grant and Department of Biotechnology, India.

References

Ahmed, I. N., Yang, X. L., Dubale, A. A., Li, R. F., Ma, Y. M., Wang, L. M., et al. (2018). Hydrolysis of cellulose using cellulase physically immobilized on highly stable zirconium based metal-organic frameworks. *Bioresource Technology*, *270*, 377–382. https://doi.org/10.1016/j.biortech.2018.09.077.

Bilal, M., & Iqbal, H. M. N. (2019). Chemical, physical, and biological coordination: An interplay between materials and enzymes as potential platforms for immobilization. *Coordination Chemistry Reviews*, *388*, 1–23. https://doi.org/10.1016/j.ccr.2019.02.024.

Bilal, M., Zhao, Y., Rasheed, T., & Iqbal, H. M. N. (2018). Magnetic nanoparticles as versatile carriers for enzymes immobilization: A review. *International Journal of Biological Macromolecules*, *120*, 2530–2544. https://doi.org/10.1016/j.ijbiomac.2018.09.025.

Bini, R. A., Marques, R. F. C., Santos, F. J., Chaker, J. A., & Jafelicci, M. (2012). Synthesis and functionalization of magnetite nanoparticles with different amino-functional alkoxysilanes. *Journal of Magnetism and Magnetic Materials*, *324*, 534–539. https://doi.org/10.1016/j.jmmm.2011.08.035.

Datta, S., Christena, L. R., & Rajaram, Y. R. S. (2013). Enzyme immobilization: An overview on techniques and support materials. *3 Biotech*, *3*(1), 1–9. https://doi.org/10.1007/s13205-012-0071-7.

Ferraz, L. I. R., Possebom, G., Alvez, E. V., Cansian, R. L., Paroul, N., de Oliveira, D., et al. (2015). Application of home-made lipase in the production of geranyl propionate by esterification of geraniol and propionic acid in solvent-free system. *Biocatalysis and Agricultural Biotechnology*, *4*, 44–48. https://doi.org/10.1016/j.bcab.2014.07.003.

Grabarek, Z., & Gergely, J. (1990). Zero-length crosslinking procedure with the use of active esters. *Analytical Biochemistry*, *185*, 131–135. https://doi.org/10.1016/0003-2697(90)90267-D.

Han, H., Ling, Z., Khan, A., Virk, A. K., Kulshrestha, S., & Li, X. (2019). Improvements of thermophilic enzymes: From genetic modifications to applications. *Bioresource Technology, 279*, 350–361. https://doi.org/10.1016/j.biortech.2019.01.087.

Hirsh, S. L., Bilek, M. M. M., Nosworthy, N. J., Kondyurin, A., Dos Remedios, C. G., & McKenzie, D. R. (2010). A comparison of covalent immobilization and physical adsorption of a cellulase enzyme mixture. *Langmuir, 26*, 14380–14388. https://doi.org/10.1021/la1019845.

Hung, S. W., Hwang, J. K., Tseng, F., Chang, J. M., Chen, C. C., & Chieng, C. C. (2006). Molecular dynamics simulation of the enhancement of cobra cardiotoxin and E6 protein binding on mixed self-assembled monolayer molecules. *Nanotechnology, 17*, S8–S13. https://doi.org/10.1088/0957-4484/17/4/002.

Husain, Q., Ansari, S. A., Alam, F., & Azam, A. (2011). Immobilization of *Aspergillus oryzae* β galactosidase on zinc oxide nanoparticles via simple adsorption mechanism. *International Journal of Biological Macromolecules, 49*, 37–43. https://doi.org/10.1016/j.ijbiomac.2011.03.011.

Jiang, Y., Guo, C., Xia, H., Mahmood, I., Liu, C., & Liu, H. (2009). Magnetic nanoparticles supported ionic liquids for lipase immobilization: Enzyme activity in catalyzing esterification. *Journal of Molecular Catalysis B: Enzymatic, 58*, 103–109. https://doi.org/10.1016/j.molcatb.2008.12.001.

Kaul, P., & Asano, Y. (2012). Strategies for discovery and improvement of enzyme function: State of the art and opportunities. *Microbial Biotechnology, 5*, 18–33. https://doi.org/10.1111/j.1751-7915.2011.00280.x.

Kim, J., Grate, J. W., & Wang, P. (2006). Nanostructures for enzyme stabilization. *Chemical Engineering Science, 61*, 1017–1026. https://doi.org/10.1016/j.ces.2005.05.067.

Kishore, D., Kundu, S., & Kayastha, A. M. (2012). Thermal, chemical and pH induced denaturation of a multimeric β-galactosidase reveals multiple unfolding pathways. *PLoS One, 7*, e50380. https://doi.org/10.1371/journal.pone.0050380.

Kumari, A., Kaila, P., Tiwari, P., Singh, V., Kaul, S., Singhal, N., et al. (2018). Multiple thermostable enzyme hydrolases on magnetic nanoparticles: An immobilized enzyme-mediated approach to saccharification through simultaneous xylanase, cellulase and amylolytic glucanotransferase action. *International Journal of Biological Macromolecules, 120*, 1650–1658. https://doi.org/10.1016/j.ijbiomac.2018.09.106.

Lei, C., Shin, Y., Liu, J., & Ackerman, E. J. (2002). Entrapping enzyme in a functionalized nanoporous support. *Journal of the American Chemical Society, 124*, 11242–11243. https://doi.org/10.1021/ja026855o.

Li, G. y., Jiang, Y. r., Huang, K. l., Ding, P., & Chen, J. (2008). Preparation and properties of magnetic Fe_3O_4-chitosan nanoparticles. *Journal of Alloys and Compounds, 466*, 451–456. https://doi.org/10.1016/j.jallcom.2007.11.100.

Li, X., Zhu, H., Feng, J., Zhang, J., Deng, X., Zhou, B., et al. (2013). One-pot polylol synthesis of graphene decorated with size- and density-tunable Fe_3O_4 nanoparticles for porcine pancreatic lipase immobilization. *Carbon, 60*, 488–497. https://doi.org/10.1016/j.carbon.2013.04.068.

Lyon, J. L., Fleming, D. A., Stone, M. B., Schiffer, P., & Williams, M. E. (2004). Synthesis of Fe oxide core/Au shell nanoparticles by iterative hydroxylamine seeding. *Nano Letters, 4*, 719–723. https://doi.org/10.1021/nl035253f.

Majouga, A., Sokolsky-Papkov, M., Kuznetsov, A., Lebedev, D., Efremova, M., Beloglazkina, E., et al. (2015). Enzyme-functionalized gold-coated magnetite nanoparticles as novel hybrid nanomaterials: Synthesis, purification and control of enzyme function by low-frequency magnetic field. *Colloids and Surfaces B, Biointerfaces, 125*, 104–109. https://doi.org/10.1016/j.colsurfb.2014.11.012.

Martin, C. R., & Kohli, P. (2003). The emerging field of nanotube biotechnology. *Nature Reviews. Drug Discovery, 2*, 29–37. https://doi.org/10.1038/nrd988.

Martín, M., Salazar, P., Villalonga, R., Campuzano, S., Pingarrón, J. M., & González-Mora, J. L. (2014). Preparation of core-shell Fe3O4@poly(dopamine) magnetic nanoparticles for biosensor construction. *Journal of Materials Chemistry B*, *2*, 739–746. https://doi.org/10.1039/c3tb21171a.

Mateo, C., Palomo, J. M., Fernandez-Lorente, G., Guisan, J. M., & Fernandez-Lafuente, R. (2007). Improvement of enzyme activity, stability and selectivity via immobilization techniques. *Enzyme and Microbial Technology*, *40*, 1451–1463. https://doi.org/10.1016/j.enzmictec.2007.01.018.

Min, K., & Yoo, Y. J. (2014). Recent progress in nanobiocatalysis for enzyme immobilization and its application. *Biotechnology and Bioprocess Engineering*, *19*, 553–567. https://doi.org/10.1007/s12257-014-0173-7.

Mitchell, D. T., Lee, S. B., Trofin, L., Li, N., Nevanen, T. K., Söderlund, H., et al. (2002). Smart nanotubes for bioseparations and biocatalysis. *Journal of the American Chemical Society*, *124*, 11864–11865. https://doi.org/10.1021/ja027247b.

Mohamad, N. R., Marzuki, N. H. C., Buang, N. A., Huyop, F., & Wahab, R. A. (2015). An overview of technologies for immobilization of enzymes and surface analysis techniques for immobilized enzymes. *Biotechnology and Biotechnological Equipment*, *29*, 205–220. https://doi.org/10.1080/13102818.2015.1008192.

Moraes Silva, S., Tavallaie, R., Sandiford, L., Tilley, R. D., & Gooding, J. J. (2016). Gold coated magnetic nanoparticles: From preparation to surface modification for analytical and biomedical applications. *Chemical Communications*, *52*, 7528–7540. https://doi.org/10.1039/c6cc03225g.

Namdeo, M., & Bajpai, S. K. (2009). Immobilization of α-amylase onto cellulose-coated magnetite (CCM) nanoparticles and preliminary starch degradation study. *Journal of Molecular Catalysis B: Enzymatic*, *59*, 134–139. https://doi.org/10.1016/j.molcatb.2009.02.005.

Narayan Patel, S., Singh, V., Sharma, M., Sangwan, R. S., Singhal, N. K., & Singh, S. P. (2018). Development of a thermo-stable and recyclable magnetic nanobiocatalyst for bioprocessing of fruit processing residues and D-allulose synthesis. *Bioresource Technology*, *247*, 633–639. https://doi.org/10.1016/j.biortech.2017.09.112.

Pan, C., Hu, B., Li, W., Sun, Y., Ye, H., & Zeng, X. (2009). Novel and efficient method for immobilization and stabilization of β-d-galactosidase by covalent attachment onto magnetic Fe_3O_4-chitosan nanoparticles. *Journal of Molecular Catalysis B: Enzymatic*, *61*, 208–215. https://doi.org/10.1016/j.molcatb.2009.07.003.

Ranjbakhsh, E., Bordbar, A. K., Abbasi, M., Khosropour, A. R., & Shams, E. (2012). Enhancement of stability and catalytic activity of immobilized lipase on silica-coated modified magnetite nanoparticles. *Chemical Engineering Journal*, *179*, 272–276. https://doi.org/10.1016/j.cej.2011.10.097.

Rigoldi, F., Donini, S., Redaelli, A., Parisini, E., & Gautieri, A. (2018). Review: Engineering of thermostable enzymes for industrial applications. *APL Bioengineering*, *2*, 011501. https://doi.org/10.1063/1.4997367.

Sahoo, B., Sahu, S. K., & Pramanik, P. (2011). A novel method for the immobilization of urease on phosphonate grafted iron oxide nanoparticle. *Journal of Molecular Catalysis B: Enzymatic*, *69*, 95–102. https://doi.org/10.1016/j.molcatb.2011.01.001.

Singh, V., Kaul, S., Singla, P., Kumar, V., Sandhir, R., Chung, J. H., et al. (2018). Xylanase immobilization on magnetite and magnetite core/shell nanocomposites using two different flexible alkyl length organophosphonates: Linker length and shell effect on enzyme catalytic activity. *International Journal of Biological Macromolecules*, *115*, 590–599. https://doi.org/10.1016/j.ijbiomac.2018.04.097.

Singh, V., Rakshit, K., Rathee, S., Angmo, S., Kaushal, S., Garg, P., et al. (2016). Metallic/bimetallic magnetic nanoparticle functionalization for immobilization of α-amylase for enhanced reusability in bio-catalytic processes. *Bioresource Technology*, *214*, 528–533. https://doi.org/10.1016/j.biortech.2016.05.002.

Sun, H., Hong, J., Meng, F., Gong, P., Yu, J., Xue, Y., et al. (2006). Novel core-shell structure polyacrylamide-coated magnetic nanoparticles synthesized via photochemical polymerization. *Surface and Coating Technology*, *201*, 250–254. https://doi.org/10.1016/j.surfcoat.2005.11.112.

Ulu, A., Ozcan, I., Koytepe, S., & Ates, B. (2018). Design of epoxy-functionalized Fe_3O_4@MCM-41 core–shell nanoparticles for enzyme immobilization. *International Journal of Biological Macromolecules*, *115*, 1122–1130. https://doi.org/10.1016/j.ijbiomac.2018.04.157.

Van Quy, D., Hieu, N. M., Tra, P. T., Nam, N. H., Hai, N. H., Thai Son, N., et al. (2013). Synthesis of silica-coated magnetic nanoparticles and application in the detection of pathogenic viruses. *Journal of Nanomaterials*, *2013*, 603940. https://doi.org/10.1155/2013/603940.

Wang, Y., Xu, F., Zhang, L., & Wei, X. (2013). One-pot solvothermal synthesis of Fe_3O_4-PEI composite and its further modification with Au nanoparticles. *Journal of Nanoparticle Research*, *15*, 1338. https://doi.org/10.1007/s11051-012-1338-y.

Williams, D. N., Gold, K. A., Holoman, T. R. P., Ehrman, S. H., & Wilson, O. C. (2006). Surface modification of magnetic nanoparticles using gum arabic. *Journal of Nanoparticle Research*, *8*, 749–753. https://doi.org/10.1007/s11051-006-9084-7.

Yiu, H. H. P., Wright, P. A., & Botting, N. P. (2001). Enzyme immobilisation using SBA-15 mesoporous molecular sieves with functionalised surfaces. *Journal of Molecular Catalysis B: Enzymatic*, *15*, 81–92. https://doi.org/10.1016/S1381-1177(01)00011-X.

Zhu, J., & Sun, G. (2012). Lipase immobilization on glutaraldehyde-activated nanofibrous membranes for improved enzyme stabilities and activities. *Reactive and Functional Polymers*, *72*, 839–845. https://doi.org/10.1016/j.reactfunctpolym.2012.08.001.

Further reading

Haiss, W., Thanh, N. T. K., Aveyard, J., & Fernig, D. G. (2015). Determination of size and concentration of gold nanoparticles from UV–Vis spectra. *Analytical Chemistry*, *79*, 4215–4221. https://doi.org/10.1021/ac0702084.

CHAPTER TEN

Bioelectrocatalysis at carbon nanotubes

Paolo Bollella*, Evgeny Katz
Department of Chemistry and Biomolecular Science, Clarkson University, Potsdam, NY, United States
*Corresponding author: e-mail address: pbollell@clarkson.edu

Contents

Abstract

This paper summarizes several examples of enzyme immobilization and bioelectrocatalysis at carbon nanotubes (CNTs). CNTs offer substantial improvements on the overall performance of amperometric enzyme electrodes mainly due to their unique structural, mechanical and electronic properties such as metallic, semi-conducting and superconducting electron transport. Unfortunately, their water insolubility restrains the kick-off in some particular fields. However, the chemical functionalization of CNTs, non-covalent and covalent, attracted a remarkable interest over the past several decades boosting the development of electrochemical biosensors and enzymatic fuel cells (EFCs) based on two different types of communications: mediated electron transfer (MET)-type, where the use of redox mediators, small electroactive molecules (freely diffusing or bound to side chains of flexible redox polymers), which are able to shuttle the

Methods in Enzymology, Volume 630
ISSN 0076-6879
https://doi.org/10.1016/bs.mie.2019.10.012

electrons between the enzyme active site and the electrode (second electron transfer generation system); direct electron transfer (DET)-type between the redox group of the enzyme and the electrode surface (third electron transfer generation system).

1. Introduction: Structural and electrochemical properties of carbon nanotubes

In the last 30 years, carbon nanotubes (CNTs), both single-walled carbon nanotubes (SWCNTs) and multi-walled carbon nanotubes (MWCNTs), raised a remarkable interest because of their unique structural, mechanical and electronic properties such as metallic, semi-conducting and superconducting electron transport (Gao, Dai, & Wallace, 2003; Wang, 2005). Furthermore, CNTs own a hollow core suitable for storing guest molecules and have the largest elastic modulus of any known material (Gong et al., 2005). Moreover, CNTs displayed also a high chemical and thermal stability, high elasticity, high tensile strength, which make them an excellent material to be used as nanowires in molecular electronics (Gao, Guo, Liu, & Huang, 2012; Pumera, 2009; Zhao, Gan, & Zhuang, 2002), as schematically displayed in Fig. 1.

Generally, CNTs can be defined as all sp^2 carbon atoms arranged in graphene sheets, which are rolled up to form a seamless hollow tube. They can be classified into two groups: SWCNTs, which consist of a single hollow tube with diameters between 0.4 and 2 nm, and MWCNTs, which are composed of multiple concentric nanotubes with a final diameter of 2–100 nm (Feng & Ji, 2011; Gooding, 2005).

In 1996, Britto et al. reported the first application of CNTs in electrochemistry, using a paste of CNTs tightly packed into a glass tube to study the oxidation of dopamine, which reveals unexpectedly a reversible two electron transfer process contrary to its usual irreversible behavior, as reported in Fig. 2 (labeled as A) (Britto, Santhanam, & Ajayan, 1996). Since the first application of CNTs in electrochemistry (1996), the number of papers increased slowly until 2003, growing then exponentially until 2009 (with 286 papers) and keeping constant about 320 papers in the last 10 years, as shown in Fig. 2 (Wildgoose, Banks, Leventis, & Compton, 2006). This particular trend can be probably divided into two different parts, the first until 2009 where CNTs were firstly discovered and heavily employed in novel electrode constructions (exponential trend), while in the second part (last 10 years) the use of CNTs raised a flat trend probably

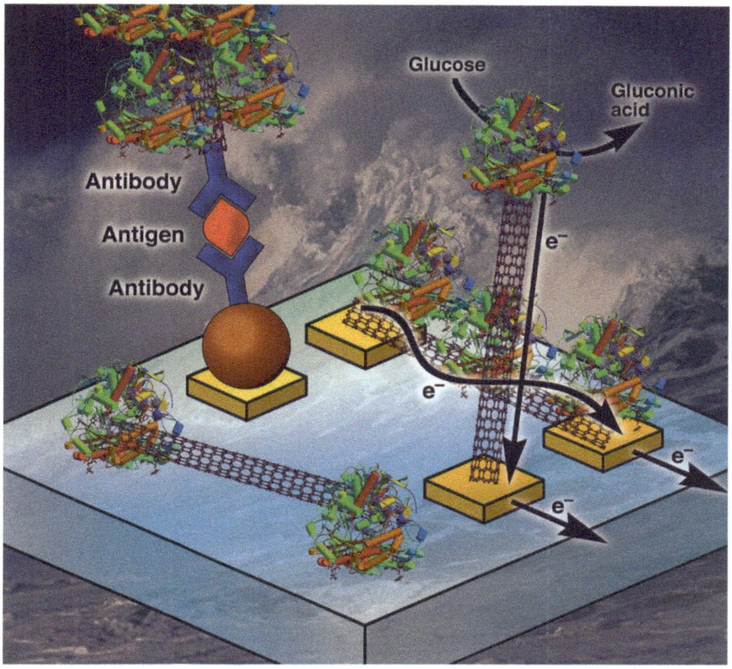

Fig. 1 Different wiring of redox enzymes using carbon nanotubes applied for various bioelectronic applications—An artistic vision. *The cartoon is copied from the cover picture of Katz, E. & Willner, I. (2004). Biomolecule-functionalized carbon nanotubes: Applications in nanobioelectronics. ChemPhysChem, 5(8), 1084–1104, with permission.*

due to the employment of different nanomaterials like graphene, porous nanomaterials and nanocomposites. In 2004, the direct electron transfer (DET) of hemoglobin has been reported for the first time mainly due to its commercial availability and its known and documented structure (Cai & Chen, 2004; Zhou, Hu, Zeng, & Rusling, 2002), which could be helpful to investigate the DET reaction of heme-based redox enzymes, as displayed in Fig. 2 (marked as B) (Gorton et al., 1999). In the same year, the reconstitution of the apo–flavoenzyme glucose oxidase (GOx) was successfully attempted exploiting CNTs vertically aligned onto gold electrode surface, as reported in Fig. 2 (labeled as C) (Patolsky, Weizmann, & Willner, 2004). Recently, CNTs have been widely used to realize single CNT-based field-effect transistor (FET), as shown in Fig. 2 (marked as D and E) (Allen, Kichambare, & Star, 2007; Hafizi et al., 2017; Martel, Schmidt, Shea, Hertel, & Avouris, 1998; Tans, Verschueren, & Dekker, 1998). In the last decade, several studies demonstrated that CNTs can

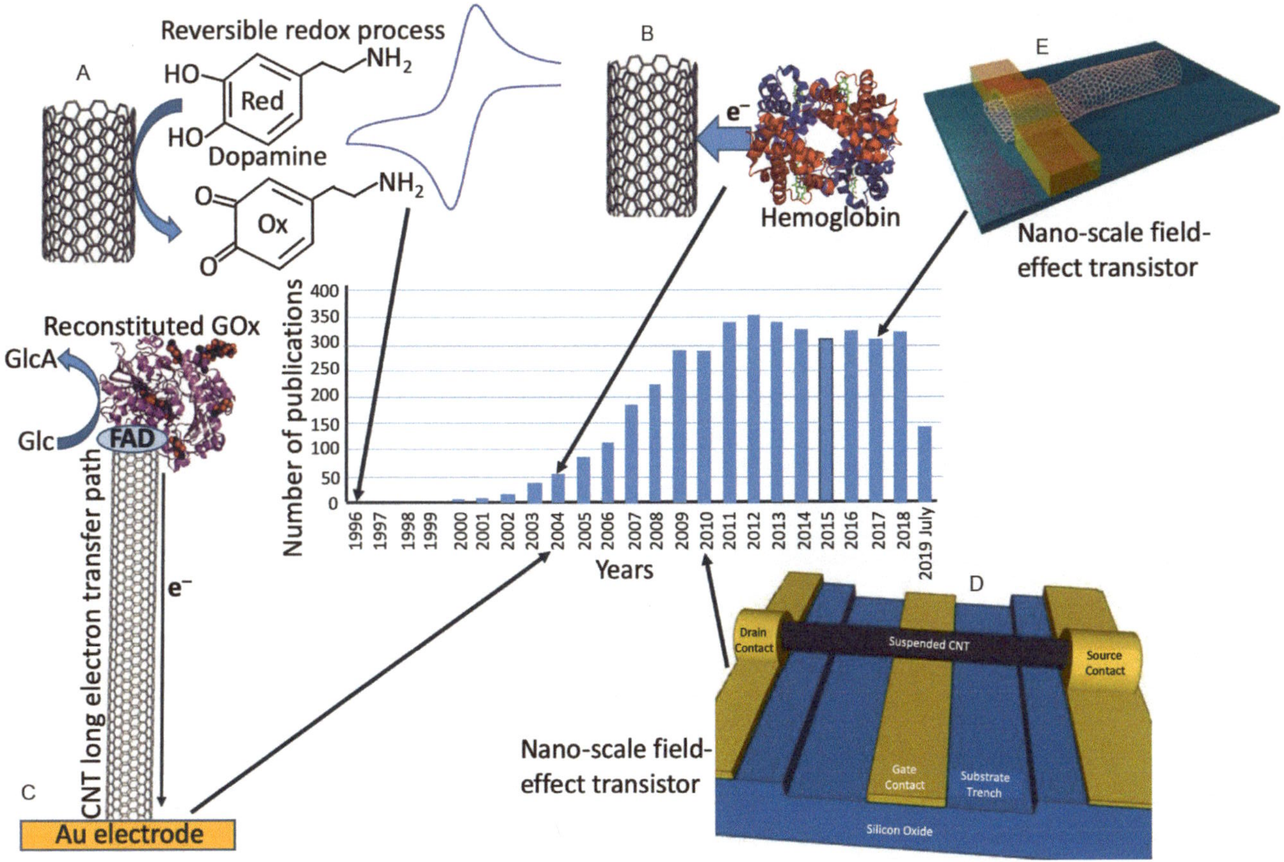

Fig. 2 Statistical data for number of publications during the last 30 years about carbon nanotube based electrochemical devices. (A) First application of CNTs based electrode for dopamine oxidation. (B) First direct electron transfer reaction at CNTs based electrode reported for hemoglobin. (C) First reconstitution of FAD glucose oxidase on aligned SWCNTs. (D and E) Field-effect transistor based on a single CNT. *The insets (D) and (E) of this figure were reproduced from Wikipedia public domain (https://upload.wikimedia.org/wikipedia/commons/4/4a/Wiki6.jpg) and Hafizi, R., Tersoff, J. & Perebeinos, V. (2017). Band structure and contact resistance of carbon nanotubes deformed by a metal contact. Physical Review Letters, 119(20), 207701, respectively, with permission.*

enhance the amount of biomolecules productively immobilized onto the electrode (surface coverage), and can promote the electron transfer reactions of proteins (including those where the redox center is deeply buried in the enzyme scaffold). In addition to enhance the electrochemical reactivity, CNT-modified electrodes have been shown to reduce surface fouling effects (Agüí, Yáñez-Sedeño, & Pingarrón, 2008; Wang & Lin, 2008).

2. Enzyme immobilization at carbon nanotubes

Different immobilization methods of enzymes onto CNTs have been explored to develop biosensors, biofuel cells and biocatalysts (Katz & Willner, 2004; Kim, Grate, & Wang, 2006; Laurent, Haddoub, & Flitsch, 2008). In particular, the immobilization method plays a crucial role in the preservation of the enzyme structure and retention of its native biological function (Wang, 2006). The immobilization methods should be distinguished in two classes, namely non-covalent and covalent linking, as summarized in Fig. 3 (Gao & Kyratzis, 2008; Huang et al., 2002).

Despite the low reliability of the first, the non-covalent methods are considered to be the most promising in terms of preservation of the immobilized enzyme conformational structure (Matsuura et al., 2006; Pedrosa et al., 2010). By considering the enzyme adsorption on CNTs, high surface loadings of enzymes or the crowding of the enzyme on the surface might help preventing enzyme inactivation due to surface spreading (Karajanagi, Vertegel, Kane, & Dordick, 2004). Moreover, Feng et al. demonstrated

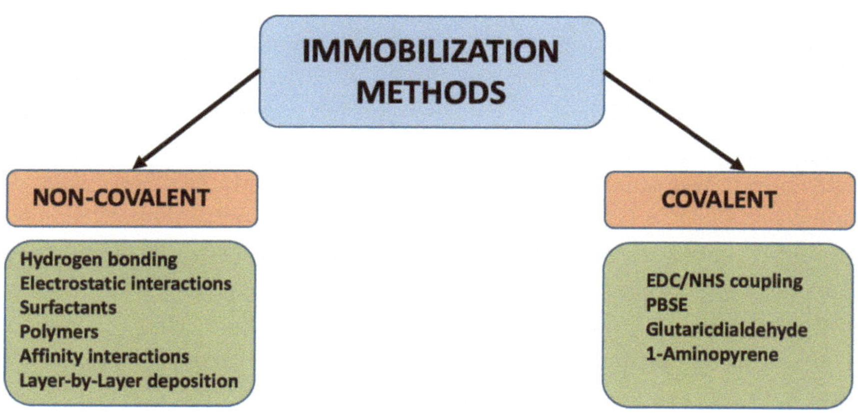

Fig. 3 Classification of enzyme immobilization methods on CNTs reported in literature.

through molecular dynamic simulations that the hydrogen bonds at the active site of the lipase were preserved through the lipase immobilization on the surface of the CNTs (Feng & Ji, 2011).

In general, the physical adsorption is driven by the interacting forces between the enzyme and CNTs such as hydrophobic interactions, π–π stacking interactions between the sidewalls of CNTs and the aromatic rings in the sidechains of the amino acids contained in the enzyme structure, and electrostatic interactions (Gao & Kyratzis, 2008; Yu, Yen, & Chen, 2010). Hydrogen bonding interactions between the carboxylated CNTs and enzymes have been also considered helpful for the enzyme adhesion onto CNTs (Asuri, Karajanagi, Yang, et al., 2006).

Enzymes can be also specifically adsorbed onto polymers coated or functionalized CNTs, assisted by surfactants and through layer-by-layer architecture (Bi, Zhou, & Zhang, 2009; Chen et al., 2003; McDonald et al., 2007). Polymer-coated CNTs can provide negatively and positively charged functional groups on the surface of CNT–polymer complexes (Yan et al., 2005). In the fabrication of biosensors, enzymes have been adsorbed onto nanotube composites through electrostatic interactions with charged polymers (Salvador-Morales et al., 2003). For example, many immobilization protocols have been reported using poly (sodium 4-styrenesulfonate) (PSS) (Zhao & Ju, 2006), chitosan (Ghica, Pauliukaite, Fatibello-Filho, & Brett, 2009), room temperature ionic liquids (RTILs), etc. (Polo-Luque, Simonet, & Valcárcel, 2013). Furthermore, enzymes can be specifically adsorbed on the surface of CNTs wrapped with single strand DNA (ssDNA) (Taft et al., 2004), where the sequence matches with a complementary ssDNA labeled with redox enzyme (e.g., alcohol dehydrogenase, etc.), forming a duplex DNA. CNTs can be also functionalized through biotinylation, displaying enzyme adsorption through avidin–biotin interaction (Yim, Liu, Lu, Kane, & Dordick, 2005). Similarly, CNTs can be also functionalized by immobilizing the enzyme cofactors (e.g., flavin adenine dinucleotide, FAD, nicotinamide adenine dinucleotide, NAD^+, pyrroloquinoline quinone, PQQ) (Patolsky et al., 2004; Yan, Yehezkeli, & Willner, 2007; Zayats, Katz, Baron, & Willner, 2005), which would interact through affinity with the apo-proteins (proteins without cofactors) forming the enzyme complex (e.g., PQQ-glucose dehydrogenase, FAD-alcohol dehydrogenase, etc.) (Raitman, Patolsky, Katz, & Willner, 2002).

In order to increase the amount of immobilized enzymes, different layer-by-layer approaches have been proposed based on the electrostatic interactions between polymer layers (e.g., polydiallyldimethylammonium chloride, PDDA, polyethyleneimine, PEI, etc.) and enzymes, or ssDNA layers and enzyme layers where both van der Waals and π–π attractions are likely involved (Asuri, Karajanagi, Kane, & Dordick, 2007; Nepal, Balasubramanian, Simonian, & Davis, 2008; Yu et al., 2006).

However, durability and leaching represent the main drawbacks of the non-covalent immobilization methods (Chen & Collier, 2005; Kurppa et al., 2007). Conversely, covalent immobilization of enzymes can result in high stability and reusability (Asuri, Karajanagi, Sellitto, et al., 2006; Huang et al., 2002; Jiang et al., 2004).

In this regard, the covalent immobilization of enzymes onto CNTs is generally carried out through free carboxy groups generated on the surface of the sidewalls by using an oxidative mixture 3:1 H_2SO_4:HNO_3 (Datsyuk et al., 2008; Zhang et al., 2003). The carboxyl groups are further activated by using 1-ethyl-3-(3-dimethylaminopropyl)carbodiimide (EDC) crosslinking reaction, performed in two steps, as shown in Fig. 4, namely: (1) active ester formation where EDC reacts with carboxylic acid groups to form an active *o*-acylisourea intermediate that is easily displaced by nucleophilic attack from *N*-hydroxysuccinimide (NHS) to create dry-stable (amino-reactive) intermediate; (2) the dry-stable (amino-reactive) intermediate can be easily displaced by nucleophilic attack from primary amino groups available on the surface of the enzyme scaffold (Gao & Kyratzis, 2008).

Other covalent immobilization methods are frequently employing linking molecules like 1-pyrenebutanoic acid succinimidyl ester (PBSE), which interacts through π–π stacking with the sidewalls of CNTs and with the enzymes through the high reactive succinimidyl ester group (Kim, Kang, Bahk, Yoo, & Lim, 2009; Yang et al., 2008); aminopyrene, which introduces amino groups onto the sidewalls of CNTs (Pang, Liu, Hu, Zhang, & Chen, 2010), performing the coupling by using glutaraldehyde as a linker (Chauhan, Narang, & Pundir, 2011).

3. Bioelectrocatalysis at carbon nanotubes

Bioelectrocatalysis can be defined as a coupled reaction between the enzymatic catalytic reaction and the electrode reaction (Karyakin, 2012). Bioelectrocatalytic reactions are usually distinguished in three groups:

(1) First generation electron transfer systems, where oxidation/reduction at the electrode surface of substrate/product of the enzymatic reaction is observed;

(2) Second generation electron transfer systems, where redox mediators are able to shuttle the electrons between the enzyme active site and the electrode (mediated electron transfer, MET);

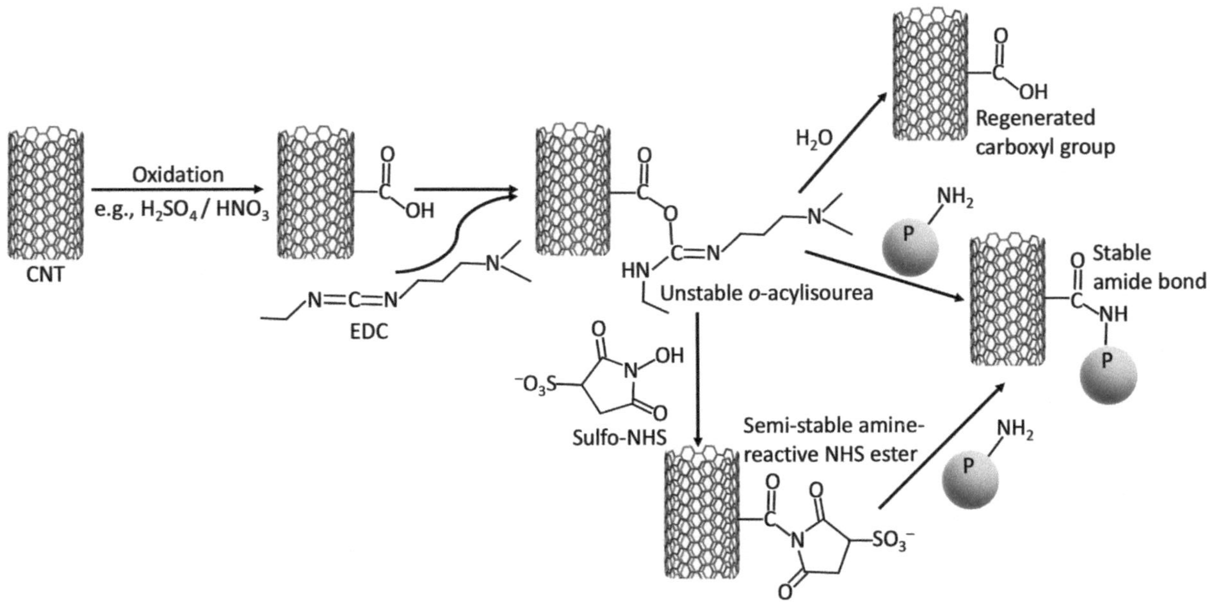

Fig. 4 Covalent conjugation of proteins to carboxylated CNTs using EDC in the presence or absence of sulfo-NHS.

(3) Third generation electron transfer systems, where direct electron transfer (DET)-type between the redox group of the enzyme and the electrode surface occurs (Bollella, Gorton, & Antiochia, 2018; Bollella & Gorton, 2018).

In all cases, the electron transfer kinetics is largely affected by the electrode materials and surface properties, raising a higher interest on the study of the electron transfer processes at various substrates (Bollella, Ludwig, & Gorton, 2017). Such investigations are both of fundamental and practical interest. Different carbon-based materials such as glassy carbon (GC), graphite, fullerene and diamond possess different electronic and structural properties and exhibit distinct electrochemical properties from each other (Antiochia, Tortolini, Tasca, Gorton, & Bollella, 2018; Bollella, Fusco, et al., 2017; Meredith et al., 2011). CNTs exist in three different configurations: "zig-zag" and "armchair," both with metallic conductivity, and "chiral," which are semiconducting, Fig. 5.

The highly conducting CNTs are particularly useful for electron transfer between redox enzymes and electrode supports, while semiconducting CNTs are used in semiconductor devices, for example, in field-effect transistors. In general, CNTs exhibit two kinds of carbon atoms: those at the tube ends, which behave like the edge plane of highly orientated pyrolytic graphite (HOPG) and possess favorable electrochemical properties; and those at the sidewalls of the nanotubes, which are similar to the basal plane of HOPG showing very slow electron transfer kinetics. The observed electrocatalytic activities of the CNTs can be ascribed to the edge-plane-like component of the CNTs according to results reported in the literature. Thus, from a mechanistic point of view, the CNTs are similar to the HOPG in their electrocatalytic activities, being crucial for the development of new electroanalytical methods, especially for biological analysis (Baughman, Zakhidov, & De Heer, 2002; Dresselhaus, Dresselhaus, & Jorio, 2004; Wilder, Venema, Rinzler, Smalley, & Dekker, 1998).

4. Homogeneous dispersion of carbon nanotubes by using room temperature ionic liquids (RTILs)

In this regard, we presented a new group of cholinium-amino acids room temperature ionic liquids (ChAARTILs) that can be used as an efficient immobilization matrix for electrochemical biosensor development. As mentioned above, the ideal immobilization strategy should be

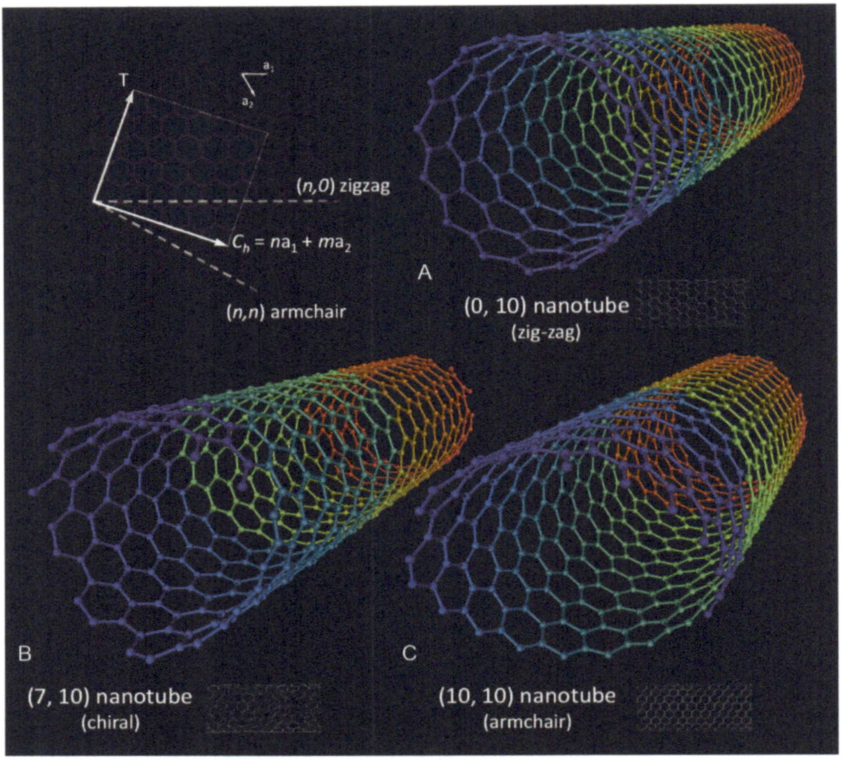

Fig. 5 Types of single-walled carbon nanotubes (SWCNTs): (A) zig-zag (metallic conductivity), (B) chiral (semiconducting) and (C) armchair (metallic conductivity). *The image was reproduced from Wikipedia public domain (http://en.wikipedia.org/wiki/Image:Types_of_Carbon_Nanotubes.png).*

able to ensure the highest enzyme loading and a tight enzymatic immobilization, preserving enzyme native structure and biological activity (Zappi et al., 2018). In particular, ChAARTILs present different side chains on the amino acids giving rise to van der Waals, π–π stacking and hydrogen bonding interactions. All these interactions can affect the nanomaterial organization onto the electrode surface (Zappi et al., 2017, 2019). Therefore, we evaluated the possibility to use the ChAARTILs as a novel immobilization matrix for the development of enzymatic-based biosensors, which ensured a homogeneous dispersion of MWCNTs bundles supporting the DET of *Trametes versicolor* laccase, as shown in Fig. 6A.

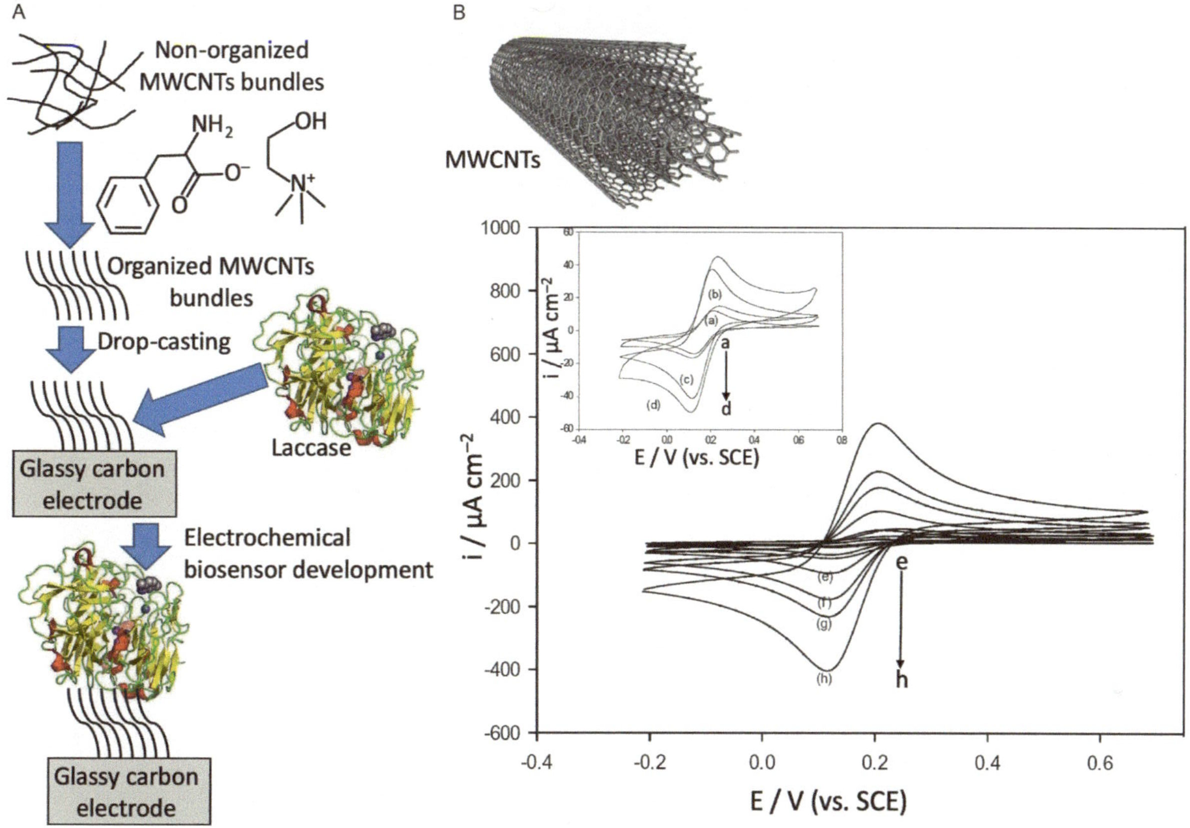

Fig. 6 (A) Reorganization of CNTs bundles promoting the efficient electron transfer from *Trametes versicolor* laccase; (B) CVs recorded in 1 mM $K_3Fe\ (CN)_6$ in 50 mM PBS buffer, pH 7, at $v = 50\,mV\,s^{-1}$ with different electrodes: (a) GCE; (b) GCE/MWCNTs/[Ch][Ala]; (c) GCE/MWCNTs; (d) GCE/MWCNTs/[Ch][Gly]; (e) GCE/MWCNTs/[BMIM][NTf_2]; (f) GCE/MWCNTs/[Ch][Ser]; (g) GCE/MWCNTs/[Ch][Cys]; (h) GCE/MWCNTs [Ch][Phe]. The cholinium-amino acids based ionic liquids are formed by cholinium [Ch] cation and different amino acids like glycine [Gly], alanine [Ala], serine [Ser], cysteine [Cys] and phenylalanine [Phe], that work as anions. *A part of this figure was reproduced from Zappi, D., Masci, G., Sadun, C., Tortolini, C., Antonelli, M.L., & Bollella, P. (2018). Evaluation of new cholinium-amino acids based room temperature ionic liquids (RTILs) as immobilization matrix for electrochemical biosensor development: Proof-of-concept with Trametes versicolor laccase. Microchemical Journal, 141, 346–352 with permission.*

4.1 Experimental realization

4.1.1 Chemicals and reagents

The five choline-amino acid ionic liquids, choline-glycine [Ch][Gly], choline-alanine [Ch][Ala], choline-serine [Ch][Ser], choline-cysteine [Ch][Cys] and choline-phenylalanine [Ch][Phe], (see structures in Fig. 1A SM) were synthesized as reported by Masci et al. (De Santis et al., 2015). Choline hydroxide and all amino acids (AAs) were purchased from Sigma Aldrich (St. Louis, MO, USA). 1-butyl-3-methylimidazolium bis(trifluoromethanesulfonyl)imide (Bmim NTF_2) was obtained from IOLITEC Ionic Liquids Technologies GmbH Germany.

Multi-walled carbon nanotubes, obtained by chemical vapor deposition method MWCNT, outer diameter (O.D.) × inner diameter (I.D.) × length (L): 10 nm (O.D) × 4.5 nm (I.D) × 4 μm (L), were supplied from Sigma Aldrich (Sigma Aldrich, SouthWest NanoTechnologies, Inc., 2501 Technology Pl, Norman, OK, USA). Potassium hexacyanoferrate (III) ($K_3[Fe(CN)_6]$), acetic acid (CH_3COOH), sodium dihydrogen phosphate (NaH_2PO_4), sodium hydrogen phosphate (Na_2HPO_4), sodium acetate (CH_3COONa), glutaraldehyde solution (GA, Grade I, 25% in H_2O) were obtained from Sigma Aldrich (Sigma Chemical Company, St. Louis, MO, USA).

Fungal laccase from *Trametes versicolor* (TvL, EC 1.10.3.2, activity: 30.6 U mg^{-1}) was supplied from Sigma Aldrich and stored at −18 °C.

4.1.2 Equipment

Cyclic voltammograms (CVs) were recorded by using a μ-Autolab Type III potentiostat (equipped with GPES 4.9, Autolab, Utrecht, The Netherlands). CVs were performed in a three-electrode electrochemical cell containing a saturated calomel electrode (SCE, Amel, Milano, Italy), a graphite rod counter electrode and a modified glassy carbon (GC) electrode as the working electrode. The modified GC electrode, an SCE reference electrode, and a platinum wire counter electrode were fitted into a wall-jet cell. The electrochemical system was equipped with a flow system consisting of a peristaltic pump (Gilson, Villier-le-Bel, France) and a six-port valve electrical injector (Cheminert® Injection Valves for FIA Model C22 Sample Injector—1/4–28 fitting, Seattle, WA, USA) with a 50 μL injection loop was used for the injection of analytes into the carrier flow (constant flow rate of 0.5 mL min^{-1}).

4.1.3 Electrode modification procedure

MWCNTs were homogeneously dispersed into RTILs and deposited onto glassy carbon electrodes (GCE), as follows:

(1) Glassy-carbon electrodes (GCEs; Bioanalytical Systems Inc., West Lafayette, IN, USA, i.d. 3 mm; $A_{geometric} = 0.073\,cm^2$) were polished with aqueous alumina (Al_2O_3) slurry (Struers, Copenhagen, Denmark), ultrasonicated, and dried under a N_2 stream;

(2) MWCNTs were purified and activated by stirring in 2 M nitric acid solution for 20 h (Pauliukaite, Doherty, Murnaghan, & Brett, 2011). The solid product was collected on a filter paper and washed several times with deionized water up to neutrality. Activated MWCNTs were dried in an oven at ~80 °C for 24 h (Pauliukaite, Murnaghan, Doherty, & Brett, 2009). This procedure ensured the complete removal of transition metal ions, used as catalysts in the production of carbon precursors. The oxidative process produces carboxylic groups on the MWCNT surface;

(3) 1 mg of activated MWCNTs-COOH was re-suspended in 100 mM aqueous solution of each ChAARTILs;

(4) 4 μL of MWCNTs-COOH/ChAARTILs suspension were drop-cast onto the GC electrode and let it dry overnight at room temperature (Pauliukaite, Doherty, Murnaghan, & Brett, 2008). The best ChAARTILs modified electrode was selected on the basis of the electroactive surface area (A_{EA}), heterogeneous electron transfer rate constant (k^0, $cm\,s^{-1}$) and roughness factor (ρ), calculated from cyclic voltammetry (CV) measurements carried out in 1 mM $Fe(CN)_6^{3-/4-}$ (50 mM phosphate-buffer saline (PBS buffer) pH 7.4) (Lavagnini, Antiochia, & Magno, 2004);

(5) Enzyme immobilization was performed by mixing 5 μL of *Trametes versicolor* laccase with MWCNTs-COOH/ChAARTILs suspension and further cross-linked with 2 μL of glutaraldehyde (2.5%, v/v in distilled water) (Pauliukaite et al., 2008). The modified electrodes were let to dry overnight at room temperature and stored in 50 mM acetate buffer, pH 5, at 4 °C.

4.2 Dispersion of carbon nanotubes in cholinium-amino acids based RTILs

Since MWCNTs tend to form bundles with a random organization onto the electrode surface, RTILs can play a key role in the organization through hydrogen bonding and π–π interactions (Rajesh, Majumder, Mizuseki, & Kawazoe, 2009). MWCNTs can interact with both cations cholinium and imidazolium forming fine bundles that can physically cross-link through cation–π or π–π interaction, respectively (Roman, Dino, Nakanishi, & Kasai, 2006; Wang, Chu, & Li, 2008). The cholinium-amino acids based ionic liquids are formed by cholinium [Ch] cation and different amino acids like glycine [Gly], alanine [Ala], serine [Ser], cysteine [Cys] and phenylalanine [Phe], that work as anions. Moreover, the different amino acid side chains may influence the MWCNTs bundles organization, as shown in Fig. 6B. In particular, [Ch][Gly] and [Ch][Ala] did not show any particular effect on A_{EA}, probably because of their simple side chains (H, CH_3). With respect to the k^0 values, the [Ch][Gly] and [Ch][Ala] showed an opposite behavior by comparison with the MWCNTs modified GCE, [Ch][Ser] and [Ch][Cys] showed a remarkable increase of A_{EA} and k^0 due to their side chains polar functional groups (OH, SH). Finally, [Ch][Phe] exhibits the best performance in terms of A_{EA} (3.432 cm^2) and k^0 (4.71×10^{-3} $cm\,s^{-1}$) due to the homogeneous distribution of MWCNTs bundles onto the electrode surface, promoted by π–π interactions (Rajesh et al., 2009).

In order to prove our hypothesis, the modified electrode with MWCNTs-[Ch][Phe] was further characterized by scanning electron microscopy (SEM), as shown in Fig. 7A–D. From the SEM picture reported in Fig. 7A, it is possible to see a homogeneous distribution of MWCNTs bundles which could be attributed to cation–π and π–π interactions between RTIL ions and MWCNTs surface. Moreover, to validate our hypothesis, the elemental maps of the modified electrode were performed and are reported in Fig. 7B–D. It is possible to see the homogeneous distributions of carbon, nitrogen, and oxygen, confirming a homogeneous distribution of MWCNTs onto the electrode surface, which could be beneficial for any possible electrochemical biosensor development.

The modified electrode platform was loaded approximately with 4.5 μg of TvL per μg of MWCNTs. After biosensor performance evaluation, the modified electrode exhibited an apparent Michaelis-Menten constant (K_M^{app}) of 0.92 mM and a steady-state current density (j_{max}) of 118 $\mu A\,cm^{-2}$. Afterward, the lifetime of the GCE/MWCNTs/[Ch][Phe]/TvL biosensor has been evaluated by injecting 0.2 mM gallic acid over 20 days showing a decrease of 10% after the first 10 days, reaching 33% after 20 days (67% of retained activity).

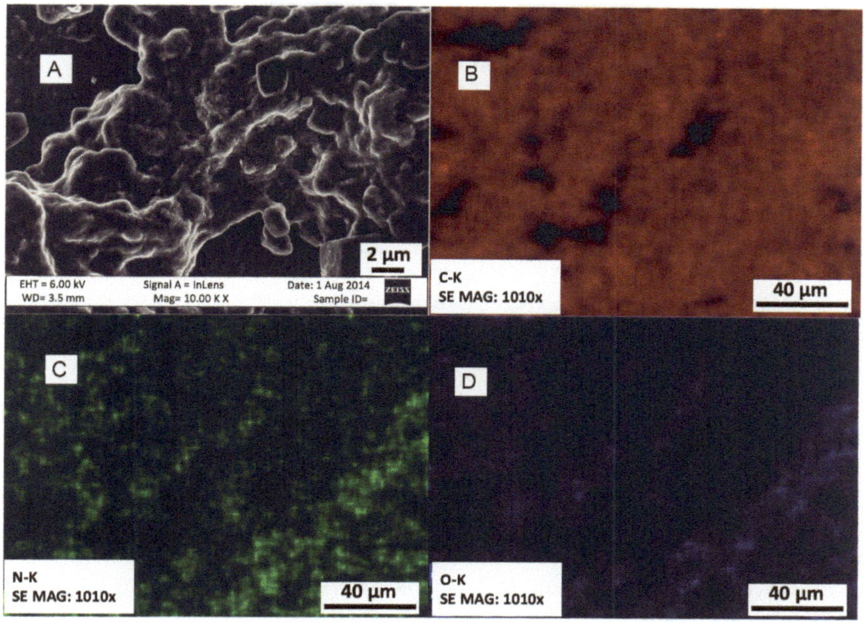

Fig. 7 A SEM picture of GCE/MWCNTs/[Ch][Phe] (A) and elemental map analysis of carbon (B), nitrogen (C) and oxygen (D). *This figure was reproduced from Zappi, D., Masci, G., Sadun, C., Tortolini, C., Antonelli, M.L., & Bollella, P. (2018). Evaluation of new cholinium-amino acids based room temperature ionic liquids (RTILs) as immobilization matrix for electrochemical biosensor development: Proof-of-concept with Trametes versicolor laccase. Microchemical Journal, 141, 346–352 with permission.*

5. Rational grafting of carbon nanotubes for enzyme immobilization

Considering another non-covalent immobilization method to preserve the enzyme structure, we aimed at obtaining an efficient DET reaction pathway between fructose dehydrogenase (FDH) and a SWCNTs/GCE further modified through diazonium coupling of an aromatic compound, namely 2-aminoanthracene (Bollella, Hibino, Kano, Gorton, & Antiochia, 2018a). The aromatic anthracenyl groups available at the electrode surface were interacting with the hydrophobic pocket within the enzyme, as shown by the enzymatic sequence reported in Fig. 8A (red and green highlights denote cysteine and methionine, respectively, and residues that coordinate

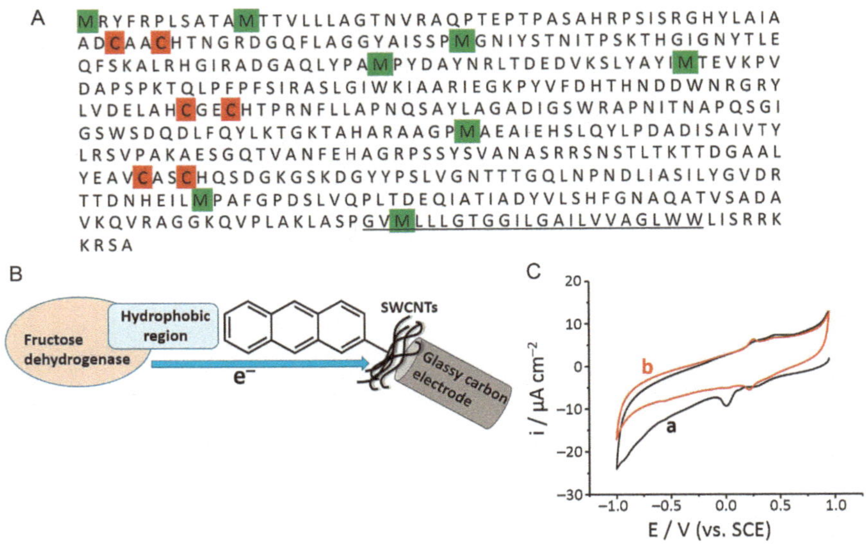

Fig. 8 (A) Enzymatic sequence of the subunit II (Cyt_{FDH}) and interaction with the anthracenyl moieties at the electrode surface. Red and green highlights denote cysteine and methionine, respectively, and residues that coordinate the heme sites and hydrophobic motif are underlined; (B) DET of fructose dehydrogenase through anthracenyl groups interacting with hydrophobic portion of FDH; (C) CVs of SWCNT/GCE in 1 mM 2-amino-anthracene solution with 5 mM $NaNO_2$ in a 0.5 M HCl solution scanning the electrode from +1000 to −1000 mV vs Ag|$AgCl_{sat}$ at 100 mV s^{-1} for two cycles (scan 1: a, black curve; scan 2: b, red curve), purged prior to the experiment with Ar for 20 min to achieve anaerobicity. *A part of this figure was reproduced from Bollella, P., Hibino, Y., Kano, K., Gorton, L., Antiochia, R. (2018b). Enhanced direct electron transfer of fructose dehydrogenase rationally immobilized on a 2-aminoanthracene diazonium cation grafted single-walled carbon nanotube based electrode. ACS Catalysis, 8(11), 10279–10289 with permission.*

the heme *c* sites and hydrophobic motif are underlined) (Bollella, Hibino, Kano, Gorton, & Antiochia, 2018b). This region is located on subunit II of FDH, leading to an improvement in the DET reaction of the enzyme, probably shortening the distance between the deeply buried prosthetic groups (i.e., heme) and the electrode surface (Bollella, Hibino, Conejo-Valverde, Soto-Cruz, et al., 2019), as shown in Fig. 8B.

5.1 Chemicals, reagents and equipment

2-Aminoanthracene (2-ANT), sodium nitrite ($NaNO_2$), hydrochloric acid (HCl), D-(−)-fructose, sodium acetate (NaAc), sodium hydroxide (NaOH), and single-walled carbon nanotubes (SWCNTs) were obtained from Sigma Aldrich (St. Louis, MO, USA). *Gluconobacter japonicus* FDH (EC 1.1.99.11) was extracted, purified, and solubilized in 50 mM PBS buffer, pH 6, containing 0.1 mM 2-mercaptoethanol and 0.1%, v/v Triton X-100. The enzymatic volumetric activity was 420 ± 30 U/mL at pH 4.5, as evaluated by using ferricyanide.

Cyclic voltammetry (CV) experiments were performed by using an Autolab potentiostat (Model PGSTAT30, Metrohm Autolab BV Ecochemie, Utrecht, The Netherlands) equipped with GPES, version 4.9. The experiments were carried out using a three-electrode electrochemical cell equipped with an Ag|AgCl (saturated KCl) as reference electrode, a Pt wire as a counter electrode, and a modified GCE as a working electrode. All potential values reported in this work are referenced to $Ag|AgCl_{sat}$ ($E = +199$ mV vs standard hydrogen electrode (SHE)). Square wave voltammetry (SWV) experiments were carried out using a PalmSens potentiostat (Model Emstat^2, Palm Instruments BV, Utrecht, The Netherlands) equipped with PSTrace, version 4.5. All measurements were performed under temperature control by using a thermostat ($T \pm 0.01$ °C, LAUDA RM6, Delran, NJ, USA). The rotating disk electrode (RDE) experiments were carried out using a 616A Electrode Rotator (EC&CG Princeton, GammaData Instruments AB, Uppsala, Sweden).

The amperometric experiments were performed by using a flow-injection setup equipped with a peristaltic pump (Gilson, Villier-le-Bel, France), a six-port valve electrical injector (Rheodyne, Cotati, CA, USA), and a 50 μL loop mounted onto the injector. The potential was applied by using an analogic potentiostat (Zäta Elektronik, Höör, Sweden) equipped with a strip chart recorder (Kipp & Zonen, Utrecht, The Netherlands) for signal recording. The enzyme-modified electrode and the reference and counter electrodes were placed into a wall-jet cell.

Secondary structure modifications occurring at FDH immobilized onto a bare indium tin oxide (ITO) electrode (cat. CEC007, Präzisions Glas & Optik GmbH, Iserlohn, Germany) and modified with 2-aminoanthracene diazonium salt were investigated by performing circular dichroism (CD) measurements using a CD spectrometer (J-815 Circular Dichroic Spectrometer, JASCO, Easton, MD, USA). These measurements were performed by placing the modified electrodes into a 1 cm cuvette containing 50 mM NaAc buffer at pH 4.5.

5.2 Grafting of carbon nanotubes: In situ reaction and enzyme immobilization

In particular, the in situ grafting of single-walled carbon nanotubes was performed according to the following procedure:

(1) Glassy-carbon electrodes (GCEs; Bioanalytical Systems Inc., West Lafayette, IN, USA, i.d. 3 mm; $A_{geometric} = 0.073\,cm^2$) were polished with aqueous alumina (Al_2O_3) slurry (Struers, Copenhagen, Denmark), ultrasonicated, and dried under a N_2 stream;

(2) 3 μL portion of SWCNTs aqueous suspension (30 mg/mL), previously ultrasonicated for 1 week, was drop-cast onto the electrodes and allowed to dry at room temperature (Ortiz, Ludwig, & Gorton, 2014);

(3) The SWCNTs/GCE was immersed into an ice-cold 1 mM solution of 2-aminoanthracene (2-ANT) diazonium cation generated in situ and purged with argon for 20 min to achieve anaerobicity. The in situ generated 2-aminoanthracene diazonium cation solution was prepared by mixing a 0.5 mM HCl/ethanol (50:50 v/v) solution containing 1 mM 2-aminoanthracene and 5 mM $NaNO_2$ for 5 min, with stirring (Liu, Chockalingham, Khor, Gui, & Gooding, 2010; Liu, Liu, Böcking, Eggers, & Gooding, 2005);

(4) The aryl diazonium cation was electrochemically reduced by scanning the potential between +1000 and −1000 mV at 100 mV s^{-1} for two cycles, as shown in Fig. 8C (Gooding, 2008; Liu, Böcking, & Gooding, 2007);

(5) The electrode was thoroughly rinsed with Milli-Q water and dried under a stream of N_2 and 3 μL of FDH solution was drop-cast onto the modified electrode surface. The modified electrode was rinsed thoroughly with 50 mM NaAc buffer pH 4.5 and stored in the same buffer overnight at 4 °C.

To investigate the role of the anthracenyl groups on the catalytic current density, several CV experiments have been performed in the absence (50 mM NaAc buffer at pH 4.5, non-turnover condition) and in the presence of 10 mM D-(−)-fructose as substrate (turnover condition). It has been proven in several papers that the orientation of the enzyme is a key issue to obtain an efficient DET reaction, especially considering the access to a particular part of the enzyme, e.g., a hydrophobic region. The appropriate immobilization strategy resulting in the optimized orientation of enzyme molecules at an electrode surface is highly important and beneficial both for the development of biosensors and EFCs (Blanford, Heath, & Armstrong, 2007; Giroud & Minteer, 2013).

Fig. 9A shows the CV for FDH/GC modified electrode under non-turnover conditions, where it was not possible to observe any redox wave

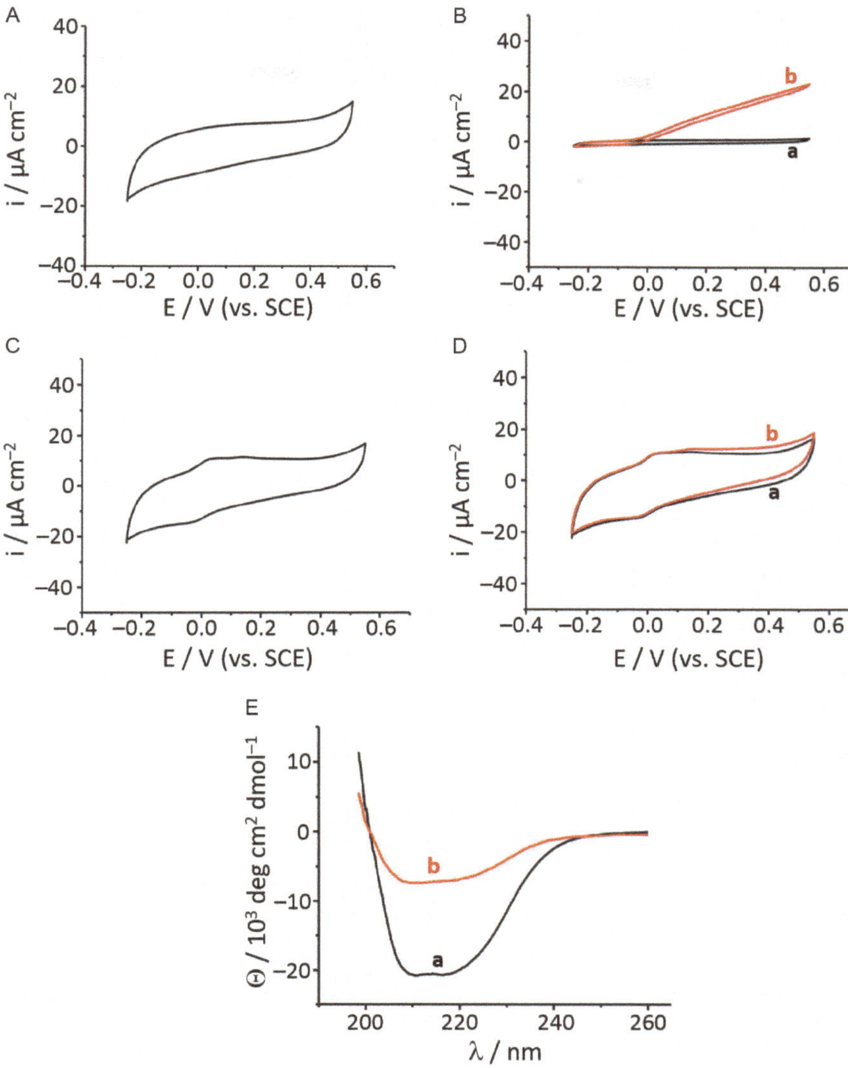

Fig. 9 (A) CV of FDH/GCE in 50mM NaAc buffer, pH 4.5, without the substrate at 10mVs^{-1}. (B) CVs of FDH/GCE in 50mM NaAc buffer, pH 4.5, in the absence (a, black line) and in the presence of 10mM D-(−)-fructose at 10mVs^{-1} (b, red line). (C) CV of FDH/2-ANT/GCE in 50mM NaAc buffer, pH 4.5, without the substrate at 10mVs^{-1}. (D) CVs of FDH/2-ANT/GCE in 50mM NaAc buffer, pH 4.5, in the absence (a, black line) and in the presence of 10mM D-(−)-fructose at 10mVs^{-1} (b, red line). All measurements were carried out at 25°C. (E) circular dichroism (CD) spectra of a FDH/ITO electrode (a, black line) and FDH/2-ANT/ITO electrode (b, red line) obtained in 10mM NaAc at pH 4.5. *This figure was reproduced from Bollella, P., Hibino, Y., Kano, K., Gorton, L., Antiochia, R. (2018b). Enhanced direct electron transfer of fructose dehydrogenase rationally immobilized on a 2-aminoanthracene diazonium cation grafted single-walled carbon nanotube based electrode. ACS Catalysis, 8(11), 10279–10289 with permission.*

related to a DET reaction of any heme *c*, probably due to FDH molecules randomly oriented onto the electrode surface and the low roughness of GCE not allowing a high enzyme loading. Under turnover conditions, the FDH/GC modified electrode showed a slight electrocatalytic wave starting at −0.058 V rising up to 18 μA cm^{-2} at +0.4 V, as shown in Fig. 9B.

Afterward, the GCE was modified through electrodeposition of the 2-aminoanthracene diazonium cation in order to highlight the effect of the anthracenyl group on the orientation of FDH. The CV of the FDH/2-ANT/GCE in non-turnover conditions is depicted in Fig. 9C, showing a couple of redox waves with a formal potential of $E^{\circ\prime} = -0.002$ V and another wave in the forward scan at a potential of E = +0.140 V, values close to those for heme c_1 ($E^{\circ\prime} = -0.010$ V) and heme c_3 ($E^{\circ\prime} = +0.150$ V), respectively, as reported in the literature (Bollella, Hibino, Kano, Gorton, & Antiochia, 2018c). The results are ascribable to the interaction between the anthracenyl groups exposed on the electrode surface and the hydrophobic region of FDH located on the subunit II of the enzyme. Moreover, when the modified electrode was tested in the presence of the substrate, it unexpectedly showed a low electrocatalytic wave starting at +0.028 V rising up to 13.3 μA cm^{-2} at +0.4 V, as shown in Fig. 9D. Despite the great results obtained under non-turnover conditions for FDH/2-ANT/GCE, the electrocatalytic wave showed lower current density values in comparison to those of the FDH/GCE probably because there might be some denaturation process occurring to FDH due to the strong interaction between the enzyme and the anthracenyl groups or there might be no space for the substrate to approach the catalytic site due to the compact arrangement of the enzyme on the smooth electrode (Hibino, Kawai, Kitazumi, Shirai, & Kano, 2017). The denaturation process was demonstrated by using circular dichroism experiments to verify possible changes occurring at the secondary structure of the enzyme immobilized on different modified electrodes. As shown in Fig. 9E, it was possible to observe a circular dichroism (CD) signal for a random coiled secondary structure immobilized onto an unmodified electrode (black line), while for an anthracenyl-modified electrode it was not possible to observe any clear CD signal (red line), meaning that a clear denaturation process has occurred, certainly due to the strong interaction between the enzyme and the electrode on a smooth electrode (Yao, Liang, Tian, & Zou, 1985).

In order to enhance the roughness of the electrode surface, the following experiments were performed on modifying the GCEs with SWCNTs. Fig. 10A shows the CV for an FDH/SWCNT/GCE under non-turnover

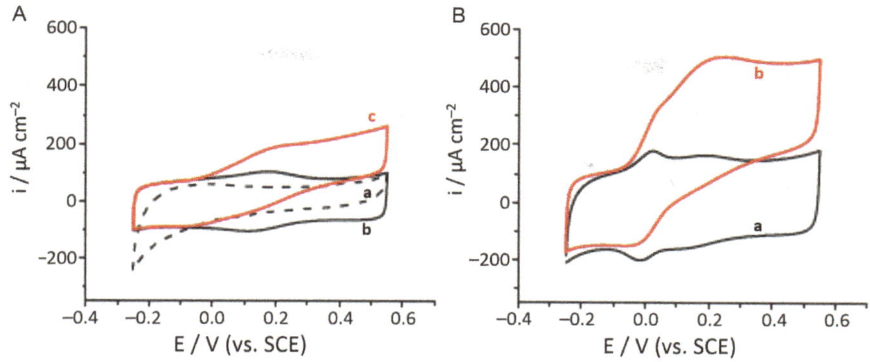

Fig. 10 (A) CVs of SWCNT/GCE in 50 mM NaAc buffer pH 4.5 (dashed black curve, a) and FDH/SWCNT/GCE in 50 mM NaAc buffer, pH 4.5, in the absence (solid black curve, b) and in the presence of 10 mM D-(−)-fructose (solid red curve, c) at 10 mV s^{-1}. (B) CVs of FDH/2-ANT/SWCNT/GCE in 50 mM NaAc buffer, pH 4.5, in the absence (solid black curve, a) and in the presence of 10 mM D-(−)-fructose (solid red curve, b) at 10 mV s^{-1}. All measurements were carried out at 25 °C. *This figure was reproduced from Bollella, P., Hibino, Y., Kano, K., Gorton, L., Antiochia, R. (2018b). Enhanced direct electron transfer of fructose dehydrogenase rationally immobilized on a 2-aminoanthracene diazonium cation grafted single-walled carbon nanotube based electrode. ACS Catalysis, 8(11), 10279–10289 with permission.*

conditions (black curve), where it is possible to observe a couple of redox waves with a formal potential of $E^{o\prime}=+0.142$ V, close to the $E^{o\prime}$ value of heme c_3 ($E^{o\prime}=+0.150$ V). This result can be ascribed to the random orientation of FDH onto the electrode surface and to the position of heme c_3, which is probably close to the edge of the enzyme. After the addition of 10 mM D-(−)-fructose, the FDH/SWCNT/GCE showed a large electrocatalytic wave starting at −0.052 V rising up to 223 μA cm^{-2} at +0.4 V, as shown in Fig. 8A (red curve). The large electrocatalytic wave is probably due to the high enzyme loading directly related to the enhanced roughness of the electrode surface achieved with the deposition of the SWCNTs. From these results it is possible to observe that a couple of redox peaks is visible in both curves (absence and presence of substrate); therefore the heme c_3 seems not to be involved in the electron transfer process during the catalysis of D-(−)-fructose oxidation.

Finally, the CV of an FDH/2-ANT/SWCNT/GCE in 50 mM NaAc at pH 4.5 is depicted in Fig. 10B, showing two couples of redox waves with $E^{o\prime}=-0.006$ V and $E^{o\prime}=+0.169$ V, respectively. In both cases, the $E^{o\prime}$ values of the redox waves are in agreement with the $E^{o\prime}$ values of heme c_1 ($E=-0.010$ V). The electrode was tested in the presence of substrate, showing a larger electrocatalytic wave in comparison to that of the SWCNT/GCE starting at −0.095 V rising up to 485 μA cm^{-2} at +0.4 V, as shown in Fig. 10B. In the overall electrocatalytic wave, it is possible to clearly see that it is divided into three parts: (i) the first sigmoidal catalytic wave starting in correspondence to heme c_1 of FDH (−0.1 to +0.050 V), (ii) the second sigmoidal catalytic wave starting in correspondence to heme c_2 of FDH not visible in the non-turnover signal (+0.050 to +0.250 V), and (iii) the mass-transfer-limited current due to the limited diffusion of the substrate to the electrode (+0.250 to +0.550 V). This result could be explained by the enhanced roughness of the electrode surface improving the enzyme loading and the correct orientation of FDH due to the interaction between the anthracenyl group and the hydrophobic part of subunit II. In particular, the interaction occurs between the aromatic rings of the anthracenyl group and the aromatic rings available on the side chain of the amino acids present in the hydrophobic domain of subunit II (π–π interaction) (Kizling & Bilewicz, 2018; Léger & Bertrand, 2008; So et al., 2016).

The modified electrode platform was loaded approximately with 0.06 μg of FDH per μg of SWCNTs. After biosensor performance evaluation, the modified electrode exhibited a K_M^{app} of 7.9 mM and a steady-state current density (j_{max}) of 592 μA cm^{-2} for FDH/2-ANT/SWCNT/GCE. Afterward, the lifetime of FDH/2-ANT/SWCNT/GCE biosensor has been evaluated by injecting 10 mM D-(−)-fructose over 60 days showing a decrease of 10% on biosensor signal (90% of retained activity).

6. Electrodeposition of gold-carbon nanotubes composites for enhanced bioelectrocatalysis

Herein, we report on another example of non-covalent immobilization method exploiting the electrostatic interaction between the electrode surface, which was positively charged, and the enzyme at working pH (negatively charged, pI 4.6), namely lactate oxidase (LOx) (Bollella, Sharma, Cass, & Antiochia, 2019a). The electrode, notably Au microneedles electrode, was modified with gold-nanotube nanocomposite, which was electrodeposited following the protocol reported in the next section. Next, poly-methylene blue (pMB) was electropolymerized onto the so modified electrode surface (Bollella, Sharma, Cass, & Antiochia, 2019b), being able to promote an efficient mediated electron transfer (MET) of LOX, which is oxidizing L-lactate to pyruvate, as shown in Fig. 11A.

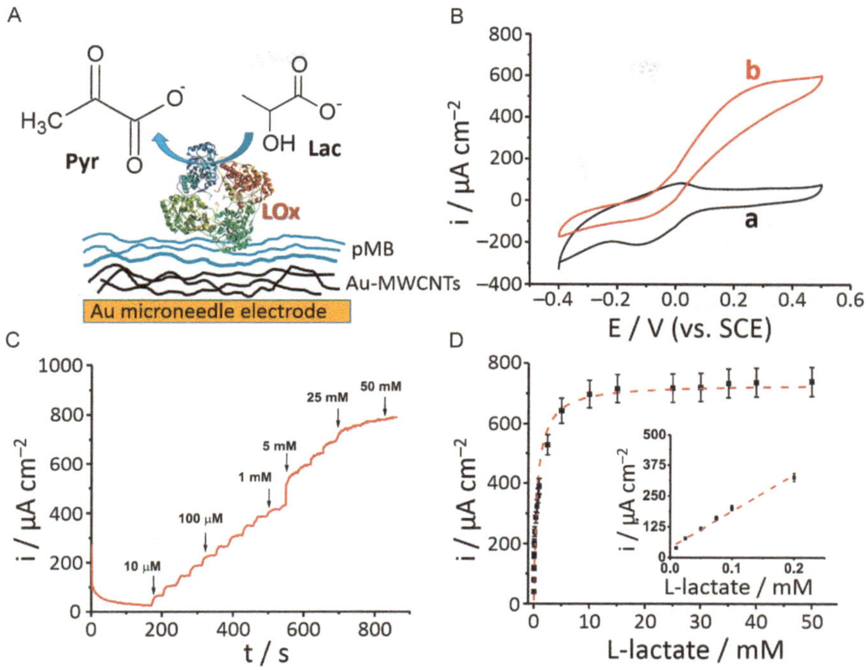

Fig. 11 (A) Schematic representation of the modified electrode with Au-MWCNTs electrodeposited composite, electropolymerized methylene blue (pMB) and lactate oxidase (LOx), which catalizes the oxidation of L-lactate to pyruvate. (B) Cyclic voltammograms of Au microneedle/Au-MWCNTs/pMB/LOx in absence (black) and in presence (red) of 10 mM sodium L-lactate in 50 mM PBS buffer, pH 7.4, with 137 mM NaCl at $\nu = 5\,mV\,s^{-1}$. (C) chronoamperometric and (D) calibration curves for Au microneedles/Au-MWCNTs/pMB/LOx based biosensor in 50 mM PBS buffer with 137 mM NaCl. Inset: low micromolar range. Experimental conditions: applied potential +0.15 V vs SCE, adding time = 40 s, under stirring = 400 rpm and 25 °C. *A part of this figure was reproduced from Bollella, P., Sharma, S., Cass, A.E.G., & Antiochia, R. (2019a). Microneedle-based biosensor for minimally-invasive lactate detection. Biosensors & Bioelectronics, 123, 152–159 with permission.*

6.1 Electrodeposition of gold-carbon nanotubes composites and enzyme immobilization

6.1.1 Chemicals and reagents

3-(N-morpholino)propanesulfonic acid (MOPS), 4-(2-hydroxyethyl)piperazine-1-ethanesulfonic acid (HEPES), ammonium chloride (NH_4Cl), boric acid (H_3BO_3), calcium chloride ($CaCl_2$), methylene blue (MB), multi-walled carbon nanotubes (MWCNTs), potassium chloride (KCl), potassium nitrate (KNO_3), sodium acetate (CH_3COONa), sodium tetraborate, sodium chloride (NaCl), sodium L-lactate, sodium phosphate dibasic (Na_2HPO_4), sodium phosphate monobasic (NaH_2PO_4), sulfuric acid (H_2SO_4) and tris(hydroxymethyl)aminomethane (TRIS) were obtained from Sigma Aldrich (St. Louis, MO, USA).

Lactate oxidase (LOx) from *Aerococcus viridans* was obtained from Creative Enzyme Ltd. The LOx (activity 106U/mL) was dissolved in a phosphate buffer at pH 6.5.

6.1.2 Equipment

Scanning electron microscopy (SEM) measurements were performed with high-resolution field emission scanning electron microscopy (HR FESEM, Zeiss Auriga Microscopy, Jena, Germany). All samples were prepared according to the modification protocol, as reported in Section 6.1.3, using gold plates (25 Å∼25 Å∼1 mm, ALS Co. Ltd., Tokyo, Japan) instead of gold electrodes and Au microneedles electrodes.

Cyclic voltammetry and amperometric experiments were performed by using a PGSAT204N potentiostat (Eco Chemie, The Netherlands) controlled by Nova 2.1 software (Eco Chemie, The Netherlands) with a conventional three-electrodes electrochemical cell equipped with a modified gold planar electrode or Au microneedles as working electrodes, a saturated calomel electrode (SCE, +244mV vs NHE, Cat. 303/SCG/12, AMEL, Milano, Italy) and a glassy carbon rod electrode (d=2mm, Cat. 6.1241.020, Metrohm, Herisau, Switzerland) were used as reference and counter electrodes, respectively. The temperature-controlled experiments were carried out by using a thermostatic bath (T ± 0.01 °C, LAUDA RM6, Delran, NJ, USA).

6.1.3 Electrode modification procedure

The Au microneedle electrodes were modified according to the following procedure:

(1) electrochemical cleaning was performed in 0.5 M H_2SO_4 by cycling 20 times between −0.3 and +1.7 V vs SCE (saturated calomel electrode used as the reference) at a scan rate of 300 mV s^{-1} (Bollella, Sharma, Cass, Tasca, & Antiochia, 2019);

(2) Au microneedle electrodes were modified by electrodeposition of Au-MWCNTs sweeping the potential between +0.8 and 0 V vs SCE for 25 scans at 50 mV s^{-1} in a suspension containing 5 mg/mL of MWCNTs (10 mM $HAuCl_4$ in 2.5 M NH_4Cl) (Dai & Compton, 2006);

(3) Au microneedle electrodes/Au-MWCNTs were further modified by electrodeposition of pMB sweeping the potential between −0.4 and +1.2 vs SCE in 0.25 mM methylene blue (MB) solution (0.02 M borate buffer pH 9.14, supporting electrolyte 0.1 M KNO_3, sweep rate $50\,mV\,s^{-1}$) (Karyakin, Karyakina, & Schmidt, 1999);

(4) 5 μL of LOx were drop-cast onto the modified electrode and left to dry at room temperature. Finally, the electrode was gently rinsed with 50 mM PBS buffer (pH 7.4) and stored at 4 °C overnight.

Fig. 11A shows schematically the structure and electrocatalytic operation of the modified electrode produced according to the procedure detailed above. Fig. 11B shows a comparative CVs of the Au-MWCNTs/pMB/LOx microneedle electrode in the absence and in the presence of lactate, where it is possible to note a very good electrocatalytic current with an onset potential of −90 mV, close to the redox potential of the mediator pMB ($E^{0'} = -95\,mV$ vs SCE). Such low potential for lactate oxidation reflects the efficient electron donor–acceptor interactions between the carbon nanotubes and the mediator, which supports the shuttling of the electrons between the redox center of the enzyme and the electrode surface. This platform enabled the continuous monitoring of lactate in artificial interstitial fluid (ISF) spiked with lactate, Fig. 11C, with a limit of detection (LOD) of 3 μM, definitely lower compared to results reported in the literature (Abrar, Dong, Lee, & Kim, 2016; Thomas, Lähdesmäki, & Parviz, 2012), and with a very good sensitivity of $178.4\,mA\,cm^{-2}\,mM^{-1}$, as shown in the calibration curve shown in Fig. 11D (Tasca, Tortolini, Bollella, & Antiochia, 2019).

Then, SEM images were used to evaluate the physical appearance and the surface characteristics of the gold microneedles electrode before and after the electrodeposition of Au-MWCNTs, obtained by cycling the potential between +0.8 and 0 V for 25 cycles. Fig. 12 shows the SEM images relative to the bare Au microneedles electrode (Fig. 12A) and to the Au-MWCNTs microneedles electrode at two different magnifications (Fig. 12B and C). The optimal surface coverage obtained is probably due to the initial reduction of Au^{3+}–Au^{0} with the formation of Au nanoparticles onto the electrode surface which allow the entrapment of the MWCNTs (Mazzei et al., 2015).

This process is further facilitated by the formation of stacking interactions between the nanotubes allowing a good coating. The Au-MWCNTs microneedles electrode was successively modified by electrodeposition of the redox mediator methylene blue, by cycling the potential values

Fig. 12 SEM images of: a bare Au microneedle electrode (A); a Au microneedle/Au-MWCNTs electrode at two different magnifications (B and C); a Au microneedle/Au-MWCNTs/pMB electrode (D). *This figure was reproduced from Bollella, P., Sharma, S., Cass, A.E.G., & Antiochia, R. (2019a). Microneedle-based biosensor for minimally-invasive lactate detection. Biosensors & Bioelectronics, 123, 152–159 with permission.*

from −0.4 to ±1.2 V vs SCE for 25 cycles in borate buffer pH 9.1 (Karyakin et al., 1999). Fig. 12D shows the SEM image of the Au microneedle/Au-MWCNTs/polyMB modified electrode, where it is possible to see the polymer growth onto the Au microneedle/Au-MWCNTs modified electrode.

The modified electrode platform was loaded approximately with 0.6 μg of LOx per μg of MWCNTs. After biosensor performance evaluation, the modified electrode exhibited a K_M^{app} of 0.64 mM and a steady-state current density (j_{max}) of 733 μA cm^{-2} for FDH/2-ANT/SWCNT/GCE. Afterwards, the lifetime of FDH/2-ANT/SWCNT/GCE biosensor has been evaluated by injecting 0.2 mM sodium-L-lactate over 30 days showing a decrease of 10% on biosensor signal (90% of retained activity).

7. Conclusions

The carbon nanotubes (CNTs) bioelectrocatalytic systems reviewed in this paper showed different approaches to enzyme immobilization at modified CNTs by using mainly non-covalent linking methods. In all aforementioned cases, it was possible to observe great bioelectrocatalytic effects mainly due to preserved enzyme structure and optimized enzyme orientation at the electrode surface. Nowadays, many CNTs modified electrodes have been successfully used for in vitro and in vivo operating biosensors and bioelectronic devices. Therefore, it should be clear that the realization of device prototypes based on CNTs modified electrodes for multiple medical applications still lie in the future.

Acknowledgments

PB would like to acknowledge the following funding: The Swedish Research Council (Vetenskapsrådet project 2014-5908), the European Commission (project "Bioenergy" FP7-PEOPLE-2013-ITN-607793), and Erasmus+ Project Unipharma-Graduates. All the aforementioned funds supported the realization of the experimental works overviewed in this paper. *Gluconobacter japonicus* FDH was kindly donated by Dr. Yuya Hibino and Prof. Kenji Kano. Au microneedles electrodes were kindly provided by Dr. Sanjiv Sharma and Prof. Anthony E.G. Cass. PB and EK would like to acknowledge Human Frontier Science Program for the financial support (project grant RGP0002/2018).

References

Abrar, M. A., Dong, Y., Lee, P. K., & Kim, W. S. (2016). Bendable electro-chemical lactate sensor printed with silver nano-particles. *Scientific Reports*, *6*, 30565.

Agüí, L., Yáñez-Sedeño, P., & Pingarrón, J. M. (2008). Role of carbon nanotubes in electroanalytical chemistry: A review. *Analytica Chimica Acta*, *622*(1–2), 11–47.

Allen, B. L., Kichambare, P. D., & Star, A. (2007). Carbon nanotube field-effect-transistor-based biosensors. *Advanced Materials*, *19*(11), 1439–1451.

Antiochia, R., Tortolini, C., Tasca, F., Gorton, L., & Bollella, P. (2018). Graphene and 2D-like nanomaterials: Different biofunctionalization pathways for electrochemical biosensor development. In A. Tiwari (Ed.), *Graphene bioelectronics* (pp. 1–35). Amsterdam (The Netherlands): Elsevier BV.

Asuri, P., Karajanagi, S. S., Kane, R. S., & Dordick, J. S. (2007). Polymer–nanotube–enzyme composites as active antifouling films. *Small*, *3*(1), 50–53.

Asuri, P., Karajanagi, S. S., Sellitto, E., Kim, D. Y., Kane, R. S., & Dordick, J. S. (2006). Water-soluble carbon nanotube-enzyme conjugates as functional biocatalytic formulations. *Biotechnology and Bioengineering*, *95*(5), 804–811.

Asuri, P., Karajanagi, S. S., Yang, H., Yim, T.-J., Kane, R. S., & Dordick, J. S. (2006). Increasing protein stability through control of the nanoscale environment. *Langmuir*, *22*(13), 5833–5836.

Baughman, R. H., Zakhidov, A. A., & De Heer, W. A. (2002). Carbon nanotubes—The route toward applications. *Science*, *297*(5582), 787–792.

Bi, S., Zhou, H., & Zhang, S. (2009). Multilayers enzyme-coated carbon nanotubes as biolabel for ultrasensitive chemiluminescence immunoassay of cancer biomarker. *Biosensors & Bioelectronics*, *24*(10), 2961–2966.

Blanford, C. F., Heath, R. S., & Armstrong, F. A. (2007). A stable electrode for high-potential, electrocatalytic O_2 reduction based on rational attachment of a blue copper oxidase to a graphite surface. *Chemical Communications*, *17*, 1710–1712.

Bollella, P., Fusco, G., Tortolini, C., Sanzò, G., Favero, G., Gorton, L., et al. (2017). Beyond graphene: Electrochemical sensors and biosensors for biomarkers detection. *Biosensors & Bioelectronics*, *89*, 152–166.

Bollella, P., & Gorton, L. (2018). Enzyme based amperometric biosensors. *Current Opinion in Electrochemistry*, *10*, 157–173.

Bollella, P., Gorton, L., & Antiochia, R. (2018). Direct electron transfer of dehydrogenases for development of 3rd generation biosensors and enzymatic fuel cells. *Sensors*, *18*(5), 1319.

Bollella, P., Hibino, Y., Conejo-Valverde, P., Soto-Cruz, J., Bergueiro, J., Calderón, M., et al. (2019). The influence of the shape of Au nanoparticles on the catalytic current of fructose dehydrogenase. *Analytical and Bioanalytical Chemistry*, [in press]. https://doi.org/10.1007/s00216-019-01944-6.

Bollella, P., Hibino, Y., Kano, K., Gorton, L., & Antiochia, R. (2018a). The influence of pH and divalent/monovalent cations on the internal electron transfer (IET), enzymatic activity, and structure of fructose dehydrogenase. *Analytical and Bioanalytical Chemistry*, *410*(14), 3253–3264.

Bollella, P., Hibino, Y., Kano, K., Gorton, L., & Antiochia, R. (2018b). Enhanced direct electron transfer of fructose dehydrogenase rationally immobilized on a 2-aminoanthracene diazonium cation grafted single-walled carbon nanotube based electrode. *ACS Catalysis*, *8*(11), 10279–10289.

Bollella, P., Hibino, Y., Kano, K., Gorton, L., & Antiochia, R. (2018c). Highly sensitive membraneless fructose biosensor based on fructose dehydrogenase immobilized onto aryl thiol modified highly porous gold electrode: Characterization and application in food samples. *Analytical Chemistry*, *90*(20), 12131–12136.

Bollella, P., Ludwig, R., & Gorton, L. (2017). Cellobiose dehydrogenase: Insights on the nanostructuration of electrodes for improved development of biosensors and biofuel cells. *Applied Materials Today*, *9*, 319–332.

Bollella, P., Sharma, S., Cass, A. E. G., & Antiochia, R. (2019a). Microneedle-based biosensor for minimally-invasive lactate detection. *Biosensors & Bioelectronics*, *123*, 152–159.

Bollella, P., Sharma, S., Cass, A. E. G., & Antiochia, R. (2019b). Minimally-invasive microneedle-based biosensor array for simultaneous lactate and glucose monitoring in artificial interstitial fluid. *Electroanalysis*, *31*(2), 374–382.

Bollella, P., Sharma, S., Cass, A. E. G., Tasca, F., & Antiochia, R. (2019). Minimally invasive glucose monitoring using a highly porous gold microneedles-based biosensor: Characterization and application in artificial interstitial fluid. *Catalysts*, *9*(7), 580.

Britto, P., Santhanam, K., & Ajayan, P. (1996). Carbon nanotube electrode for oxidation of dopamine. *Bioelectrochemistry and Bioenergetics*, *41*(1), 121–125.

Cai, C., & Chen, J. (2004). Direct electron transfer and bioelectrocatalysis of hemoglobin at a carbon nanotube electrode. *Analytical Biochemistry*, *325*(2), 285–292.

Chauhan, N., Narang, J., & Pundir, C. (2011). Fabrication of multiwalled carbon nanotubes/polyaniline modified Au electrode for ascorbic acid determination. *Analyst*, *136*(9), 1938–1945.

Chen, R. J., Bangsaruntip, S., Drouvalakis, K. A., Kam, N. W. S., Shim, M., Li, Y., et al. (2003). Noncovalent functionalization of carbon nanotubes for highly specific electronic biosensors. *PNAS*, *100*(9), 4984–4989.

Chen, J., & Collier, C. P. (2005). Noncovalent functionalization of single-walled carbon nanotubes with water-soluble porphyrins. *The Journal of Physical Chemistry. B*, *109*(16), 7605–7609.

Dai, X., & Compton, R. G. (2006). Direct electrodeposition of gold nanoparticles onto indium tin oxide film coated glass: Application to the detection of arsenic (III). *Analytical Sciences*, *22*(4), 567–570.

Datsyuk, V., Kalyva, M., Papagelis, K., Parthenios, J., Tasis, D., Siokou, A., et al. (2008). Chemical oxidation of multiwalled carbon nanotubes. *Carbon*, *46*(6), 833–840.

De Santis, S., Masci, G., Casciotta, F., Caminiti, R., Scarpellini, E., Campetella, M., et al. (2015). Cholinium-amino acid based ionic liquids: A new method of synthesis and physico-chemical characterization. *Physical Chemistry Chemical Physics*, *17*(32), 20687–20698.

Dresselhaus, M., Dresselhaus, G., & Jorio, A. (2004). Unusual properties and structure of carbon nanotubes. *Annual Review of Materials Research*, *34*, 247–278.

Feng, W., & Ji, P. (2011). Enzymes immobilized on carbon nanotubes. *Biotechnology Advances*, *29*(6), 889–895.

Gao, M., Dai, L., & Wallace, G. G. (2003). Biosensors based on aligned carbon nanotubes coated with inherently conducting polymers. *Electroanalysis*, *15*(13), 1089–1094.

Gao, C., Guo, Z., Liu, J.-H., & Huang, X.-J. (2012). The new age of carbon nanotubes: An updated review of functionalized carbon nanotubes in electrochemical sensors. *Nanoscale*, *4*(6), 1948–1963.

Gao, Y., & Kyratzis, I. (2008). Covalent immobilization of proteins on carbon nanotubes using the cross-linker 1-ethyl-3-(3-dimethylaminopropyl) carbodiimide—A critical assessment. *Bioconjugate Chemistry*, *19*(10), 1945–1950.

Ghica, M. E., Pauliukaite, R., Fatibello-Filho, O., & Brett, C. M. (2009). Application of functionalised carbon nanotubes immobilised into chitosan films in amperometric enzyme biosensors. *Sensors and Actuators, B: Chemical*, *142*(1), 308–315.

Giroud, F., & Minteer, S. D. (2013). Anthracene-modified pyrenes immobilized on carbon nanotubes for direct electroreduction of O_2 by laccase. *Electrochemistry Communications*, *34*, 157–160.

Gong, K., Yan, Y., Zhang, M., Su, L., Xiong, S., & Mao, L. (2005). Electrochemistry and electroanalytical applications of carbon nanotubes: A review. *Analytical Sciences*, *21*(12), 1383–1393.

Gooding, J. J. (2005). Nanostructuring electrodes with carbon nanotubes: A review on electrochemistry and applications for sensing. *Electrochimica Acta*, *50*(15), 3049–3060.

Gooding, J. J. (2008). Advances in interfacial design for electrochemical biosensors and sensors: Aryl diazonium salts for modifying carbon and metal electrodes. *Electroanalysis, 20*(6), 573–582.
Gorton, L., Lindgren, A., Larsson, T., Munteanu, F., Ruzgas, T., & Gazaryan, I. (1999). Direct electron transfer between heme-containing enzymes and electrodes as basis for third generation biosensors. *Analytica Chimica Acta, 400*(1–3), 91–108.
Hafizi, R., Tersoff, J., & Perebeinos, V. (2017). Band structure and contact resistance of carbon nanotubes deformed by a metal contact. *Physical Review Letters, 119*(20) 207701.
Hibino, Y., Kawai, S., Kitazumi, Y., Shirai, O., & Kano, K. (2017). Construction of a protein-engineered variant of D-fructose dehydrogenase for direct electron transfer-type bioelectrocatalysis. *Electrochemistry Communications, 77*, 112–115.
Huang, W., Taylor, S., Fu, K., Lin, Y., Zhang, D., Hanks, T. W., et al. (2002). Attaching proteins to carbon nanotubes via diimide-activated amidation. *Nano Letters, 2*(4), 311–314.
Jiang, K., Schadler, L. S., Siegel, R. W., Zhang, X., Zhang, H., & Terrones, M. J. (2004). Protein immobilization on carbon nanotubes via a two-step process of diimide-activated amidation. *Journal of Materials Chemistry, 14*(1), 37–39.
Karajanagi, S. S., Vertegel, A. A., Kane, R. S., & Dordick, J. S. (2004). Structure and function of enzymes adsorbed onto single-walled carbon nanotubes. *Langmuir, 20*(26), 11594–11599.
Karyakin, A. A. (2012). Principles of direct (mediator free) bioelectrocatalysis. *Bioelectrochemistry, 88*, 70–75.
Karyakin, A. A., Karyakina, E. E., & Schmidt, H. L. (1999). Electropolymerized azines: A new group of electroactive polymers. *Electroanalysis, 11*(3), 149–155.
Katz, E., & Willner, I. (2004). Biomolecule-functionalized carbon nanotubes: Applications in nanobioelectronics. *ChemPhysChem, 5*(8), 1084–1104.
Kim, J., Grate, J. W., & Wang, P. (2006). Nanostructures for enzyme stabilization. *Chemical Engineering Science, 61*(3), 1017–1026.
Kim, B. J., Kang, B. K., Bahk, Y. Y., Yoo, K. H., & Lim, K. J. (2009). Immobilization of horseradish peroxidase on multi-walled carbon nanotubes and its enzymatic stability. *Current Applied Physics, 9*(4), e263–e265.
Kizling, M., & Bilewicz, R. (2018). Fructose dehydrogenase electron transfer pathway in bioelectrocatalytic reactions. *ChemElectroChem, 5*(1), 166–174.
Kurppa, K., Jiang, H., Szilvay, G. R., Nasibulin, A. G., Kauppinen, E. I., & Linder, M. B. (2007). Controlled hybrid nanostructures through protein-mediated noncovalent functionalization of carbon nanotubes. *Angewandte Chemie, 46*(34), 6446–6449.
Laurent, N., Haddoub, R., & Flitsch, S. L. (2008). Enzyme catalysis on solid surfaces. *Trends in Biotechnology, 26*(6), 328–337.
Lavagnini, I., Antiochia, R., & Magno, F. (2004). An extended method for the practical evaluation of the standard rate constant from cyclic voltammetric data. *Electroanalysis, 16*(6), 505–506.
Léger, C., & Bertrand, P. (2008). Direct electrochemistry of redox enzymes as a tool for mechanistic studies. *Chemical Reviews, 108*(7), 2379–2438.
Liu, G., Böcking, T., & Gooding, J. J. (2007). Diazonium salts: Stable monolayers on gold electrodes for sensing applications. *Journal of Electroanalytical Chemistry, 600*(2), 335–344.
Liu, G., Chockalingham, M., Khor, S. M., Gui, A. L., & Gooding, J. J. (2010). A comparative study of the modification of gold and glassy carbon surfaces with mixed layers of in situ generated aryl diazonium compounds. *Electroanalysis, 22*(9), 918–926.
Liu, G., Liu, J., Böcking, T., Eggers, P. K., & Gooding, J. J. (2005). The modification of glassy carbon and gold electrodes with aryl diazonium salt: The impact of the electrode materials on the rate of heterogeneous electron transfer. *Chemical Physics, 319*(1–3), 136–146.

Martel, R., Schmidt, T., Shea, H., Hertel, T., & Avouris, P. (1998). Single-and multi-wall carbon nanotube field-effect transistors. *Applied Physics Letters*, *73*(17), 2447–2449.

Matsuura, K., Saito, T., Okazaki, T., Ohshima, S., Yumura, M., & Iijima, S. (2006). Selectivity of water-soluble proteins in single-walled carbon nanotube dispersions. *Chemical Physics Letters*, *429*(4–6), 497–502.

Mazzei, F., Favero, G., Bollella, P., Tortolini, C., Mannina, L., Conti, M. E., et al. (2015). Recent trends in electrochemical nanobiosensors for environmental analysis. *International Journal of Environment and Health*, 7(3), 267–291.

McDonald, T. J., Svedruzic, D., Kim, Y.-H., Blackburn, J. L., Zhang, S., King, P. W., et al. (2007). Wiring-up hydrogenase with single-walled carbon nanotubes. *Nano Letters*, 7(11), 3528–3534.

Meredith, M. T., Minson, M., Hickey, D., Artyushkova, K., Glatzhofer, D. T., & Minteer, S. D. (2011). Anthracene-modified multi-walled carbon nanotubes as direct electron transfer scaffolds for enzymatic oxygen reduction. *ACS Catalysis*, *1*(12), 1683–1690.

Nepal, D., Balasubramanian, S., Simonian, A. L., & Davis, V. A. (2008). Strong antimicrobial coatings: Single-walled carbon nanotubes armored with biopolymers. *Nano Letters*, *8*(7), 1896–1901.

Ortiz, R., Ludwig, R., & Gorton, L. (2014). Highly efficient membraneless glucose bioanode based on *Corynascus thermophilus* cellobiose dehydrogenase on aryl diazonium-activated single-walled carbon nanotubes. *ChemElectroChem*, *1*(11), 1948–1956.

Pang, H., Liu, J., Hu, D., Zhang, X., & Chen, J. (2010). Immobilization of laccase onto 1-aminopyrene functionalized carbon nanotubes and their electrocatalytic activity for oxygen reduction. *Electrochimica Acta*, *55*(22), 6611–6616.

Patolsky, F., Weizmann, Y., & Willner, I. (2004). Long-range electrical contacting of redox enzymes by SWCNT connectors. *Angewandte Chemie*, *43*(16), 2113–2117.

Pauliukaite, R., Doherty, A. P., Murnaghan, K. D., & Brett, C. M. (2008). Application of some room temperature ionic liquids in the development of biosensors at carbon film electrodes. *Electroanalysis*, *20*(5), 485–490.

Pauliukaite, R., Doherty, A. P., Murnaghan, K. D., & Brett, C. M. (2011). Application of room temperature ionic liquids to the development of electrochemical lipase biosensing systems for water-insoluble analytes. *Journal of Electroanalytical Chemistry*, *656*(1–2), 96–101.

Pauliukaite, R., Murnaghan, K. D., Doherty, A. P., & Brett, C. M. (2009). A strategy for immobilisation of carbon nanotubes homogenised in room temperature ionic liquids on carbon electrodes. *Journal of Electroanalytical Chemistry*, *633*(1), 106–112.

Pedrosa, V. A., Paliwal, S., Balasubramanian, S., Nepal, D., Davis, V., Wild, J., et al. (2010). Enhanced stability of enzyme organophosphate hydrolase interfaced on the carbon nanotubes. *Colloids and Surfaces, B: Biointerfaces*, 77(1), 69–74.

Polo-Luque, M., Simonet, B., & Valcárcel, M. (2013). Functionalization and dispersion of carbon nanotubes in ionic liquids. *TrAC, Trends in Analytical Chemistry*, *47*, 99–110.

Pumera, M. (2009). The electrochemistry of carbon nanotubes: Fundamentals and applications. *Chemistry—A European Journal*, *15*(20), 4970–4978.

Raitman, O. A., Patolsky, F., Katz, E., & Willner, I. (2002). Electrical contacting of glucose dehydrogenase by the reconstitution of a pyrroloquinoline quinone-functionalized polyaniline film associated with an Au-electrode: An in situ electrochemical SPR study. *Chemical Communications*, *17*, 1936–1937.

Rajesh, C., Majumder, C., Mizuseki, H., & Kawazoe, Y. (2009). A theoretical study on the interaction of aromatic amino acids with graphene and single walled carbon nanotube. *The Journal of Chemical Physics*, *130*(12) 124911.

Roman, T., Dino, W., Nakanishi, H., & Kasai, H. (2006). Amino acid adsorption on single-walled carbon nanotubes. *European Physical Journal D: Atomic, Molecular, Optical and Plasma Physics*, *38*(1), 117–120.

Salvador-Morales, C., Franklin, E., Chambers, G., Fonseca, A., Nagy, J. B., Blau, W. J., et al. (2003). Characterization of an interaction between functionalized carbon nanotubes and an enzyme. *Journal of Nanoscience and Nanotechnology*, *3*(3), 209–213.

So, K., Hamamoto, R., Takeuchi, R., Kitazumi, Y., Shirai, O., Endo, R., et al. (2016). Bioelectrochemical analysis of thermodynamics of the catalytic cycle and kinetics of the oxidative inactivation of oxygen-tolerant [NiFe]-hydrogenase. *Journal of Electroanalytical Chemistry*, *766*, 152–161.

Taft, B. J., Lazareck, A. D., Withey, G. D., Yin, A., Xu, J., & Kelley, S. O. (2004). Site-specific assembly of DNA and appended cargo on arrayed carbon nanotubes. *Journal of the American Chemical Society*, *126*(40), 12750–12751.

Tans, S. J., Verschueren, A. R., & Dekker, C. (1998). Room-temperature transistor based on a single carbon nanotube. *Nature*, *393*, 49–52.

Tasca, F., Tortolini, C., Bollella, P., & Antiochia, R. (2019). Microneedles-based electrochemical devices for transdermal biosensing: A review. *Current Opinion in Electrochemistry*, *16*, 42–49.

Thomas, N., Lähdesmäki, I., & Parviz, B. A. (2012). A contact lens with an integrated lactate sensor. *Sensors and Actuators, B: Chemical*, *162*(1), 128–134.

Wang, J. (2005). Carbon-nanotube based electrochemical biosensors: A review. *Electroanalysis*, *17*(1), 7–14.

Wang, P. (2006). Nanoscale biocatalyst systems. *Current Opinion in Biotechnology*, *17*(6), 574–579.

Wang, J., Chu, H., & Li, Y. (2008). Why single-walled carbon nanotubes can be dispersed in imidazolium-based ionic liquids. *ACS Nano*, *2*(12), 2540–2546.

Wang, J., & Lin, Y. (2008). Functionalized carbon nanotubes and nanofibers for biosensing applications. *TrAC, Trends in Analytical Chemistry*, *27*(7), 619–626.

Wilder, J. W., Venema, L. C., Rinzler, A. G., Smalley, R. E., & Dekker, C. (1998). Electronic structure of atomically resolved carbon nanotubes. *Nature*, *391*, 59–62.

Wildgoose, G. G., Banks, C. E., Leventis, H. C., & Compton, R. G. (2006). Chemically modified carbon nanotubes for use in electroanalysis. *Microchimica Acta*, *152*(3–4), 187–214.

Yan, Y. M., Yehezkeli, O., & Willner, I. (2007). Integrated, electrically contacted NAD (P) +–dependent enzyme–carbon nanotube electrodes for biosensors and biofuel cell applications. *Chemistry—A European Journal*, *13*(36), 10168–10175.

Yan, Y., Zheng, W., Zhang, M., Wang, L., Su, L., & Mao, L. (2005). Bioelectrochemically functional nanohybrids through co-assembling of proteins and surfactants onto carbon nanotubes: Facilitated electron transfer of assembled proteins with enhanced faradic response. *Langmuir*, *21*(14), 6560–6566.

Yang, J., Pang, F., Zhang, R., Xu, Y., He, P., & Fang, Y. (2008). Electrochemistry and electrocatalysis of hemoglobin on 1-pyrenebutanoic acid succinimidyl ester/multiwalled carbon nanotube and Au nanoparticle modified electrode. *Electroanalysis*, *20*(19), 2134–2140.

Yao, Q., Liang, S., Tian, M., & Zou, C. (1985). Changes in circular dichroism and exposure of buried thiol groups during denaturation of rabbit muscle creatine kinase. *Scientia Sinica. Series B*, *28*(5), 484–493.

Yim, T.-J., Liu, J., Lu, Y., Kane, R. S., & Dordick, J. S. (2005). Highly active and stable DNAzyme–carbon nanotube hybrids. *Journal of the American Chemical Society*, *127*(35), 12200–12201.

Yu, X., Munge, B., Patel, V., Jensen, G., Bhirde, A., Gong, J. D., et al. (2006). Carbon nanotube amplification strategies for highly sensitive immunodetection of cancer biomarkers. *Journal of the American Chemical Society*, *128*(34), 11199–11205.

Yu, C.-M., Yen, M.-J., & Chen, L.-C. (2010). A bioanode based on MWCNT/protein-assisted co-immobilization of glucose oxidase and 2, 5-dihydroxybenzaldehyde for glucose fuel cells. *Biosensors & Bioelectronics*, *25*(11), 2515–2521.

Zappi, D., Caminiti, R., Ingo, G., Sadun, C., Tortolini, C., & Antonelli, M. L. (2017). Biologically friendly room temperature ionic liquids and nanomaterials for the development of innovative enzymatic biosensors. *Talanta, 175*, 566–572.

Zappi, D., Gabriele, S., Gontrani, L., Dini, D., Sadun, C., Marini, F., et al. (2019). Biologically friendly room temperature ionic liquids and nanomaterials for the development of innovative enzymatic biosensors: Part II. *Talanta, 194*, 26–31.

Zappi, D., Masci, G., Sadun, C., Tortolini, C., Antonelli, M. L., & Bollella, P. (2018). Evaluation of new cholinium-amino acids based room temperature ionic liquids (RTILs) as immobilization matrix for electrochemical biosensor development: Proof-of-concept with Trametes versicolor laccase. *Microchemical Journal, 141*, 346–352.

Zayats, M., Katz, E., Baron, R., & Willner, I. (2005). Reconstitution of apo-glucose dehydrogenase on pyrroloquinoline quinone-functionalized Au nanoparticles yields an electrically contacted biocatalyst. *Journal of the American Chemical Society, 127*(35), 12400–12406.

Zhang, J., Zou, H., Qing, Q., Yang, Y., Li, Q., Liu, Z., et al. (2003). Effect of chemical oxidation on the structure of single-walled carbon nanotubes. *The Journal of Physical Chemistry. B, 107*(16), 3712–3718.

Zhao, Q., Gan, Z., & Zhuang, Q. (2002). Electrochemical sensors based on carbon nanotubes. *Electroanalysis, 14*(23), 1609–1613.

Zhao, H., & Ju, H. (2006). Multilayer membranes for glucose biosensing via layer-by-layer assembly of multiwall carbon nanotubes and glucose oxidase. *Analytical Biochemistry, 350*(1), 138–144.

Zhou, Y., Hu, N., Zeng, Y., & Rusling, J. F. (2002). Heme protein – clay films: Direct electrochemistry and electrochemical catalysis. *Langmuir, 18*(1), 211–219.

CHAPTER ELEVEN

Molecular wiring of glucose oxidase enzyme with Mn polypyridine complex on MWCNT modified electrode surface and its bio-electrocatalytic oxidation and glucose sensing

Natarajan Saravanan[a], Annamalai Senthil Kumar[a,b,*]
[a]Nano and Bioelectrochemistry Research Laboratory, Department of Chemistry, School of Advanced Sciences, Vellore Institute of Technology, Vellore, India
[b]Carbon Dioxide Research and Green Technology Centre, Vellore Institute of Technology, Vellore, India
*Corresponding author: e-mail address: askumarchem@yahoo.com

Contents

Abstract

A simple method for molecular wiring of glucose oxidase (GOx) enzyme with a low cost Mn polypyridine complex, $Mn(phen)_2Cl_2$, carboxylic acid functionalized multiwalled

Methods in Enzymology, Volume 630
ISSN 0076-6879
https://doi.org/10.1016/bs.mie.2019.10.024

carbon nanotube (f-MWCNT) and Nafion (Nf), which is useful for glucose oxidation and sensing application in pH 7 phosphate buffer solution, has been demonstrated. In the typical preparation, f-MWCNT, Mn(phen)$_2$Cl$_2$, Nafion and GOx solution/suspension were successfully drop-casted as layer-by-layer on a cleaned glassy carbon electrode and potential cycled using cylic voltametric (CV) technique. In this preparation procedure, the Mn(phen)$_2$Cl$_2$ complex is in-situ converted as a dimer complex, Mn$_2$(phen)$_2$(O)(Cl$_2$). A cooperative interaction based on π-π, covalent, ionic, hydrophilic and hydrophobic are operated in the bioelectrode for molecular wiring and electron-transfer shutting reaction. The modified electrode is designated as GCE/f-MWCNT@Mn$_2$(phen)$_2$(O)(Cl$_2$)-Nf@GOx. CV response of the bioelectrode showed a defined redox peak current signal at an apparent standard electrode potential, $E^{\circ\prime} = 0.55$ V vs Ag/AgCl. Upon exposure of glucose, the modified electrode showed a current linearity in a range, 0—6 mM with a current sensitivity value, 349.4 μA mM^{-1} cm^{-2} by CV and a current linearity in a window, 50–550 μM with a current sensitivity, 316.8 μA mM^{-1} cm^{-2} at applied biased potential, 0.65 V vs Ag/AgCl by amperometric i–t methods. Obtained glucose oxidation current sensitivity values are relatively higher than Os-complex based transducer systems.

1. Introduction

Search for an efficient transducer system that can shuttle electron between redox active enzyme and underlying electrode material is continued research interest in interdisciplinary fields of biology and electrochemistry (Schuhmann, Ohara, Schmidt, & Heller, 1991; Yates, Fascione, & Parkin, 2018). Bioelectrocatalysis of glucose has received an enormous level of attention because of its practical applications in monitoring blood glucose levels of diabetic patients, implantable biofuel cell, biofuel cell applications, etc. (Conzuelo, Ruff, & Schuhmann, 2018; Mano, 2019; Yoo & Lee, 2010). Structurally, Glucose oxidase enzyme (GOx) is a holoenzyme consisting of two identical 80 kDa subunits at the active site that contain a cofactor flavin adenine dinucleotide (FAD) (Scheme 1) (Bankar, Bule, Singhal, & Ananthanarayan, 2009). GOx catalyzes the oxidation of glucose into D-glucono-δ-lactone at its first hydroxyl group using atomic oxygen (O_2) as the electron acceptor with the synchronous generation of hydrogen peroxide (H_2O_2), according to a ping-pong mechanism (Mano, 2019; Nakamura & Ogura, 1968). The overall reaction is given below:

GOx (FAD) + β-D-glucose → GOx (FADH$_2$) + D-Glucono-δ-lactone

GOx (FADH$_2$) + O_2 → GOx (FAD) + H_2O_2

β − D-Glucose + GOx (FAD) + O_2 → D-glucono-δ-lactone + GOx (FADH$_2$) + H_2O_2

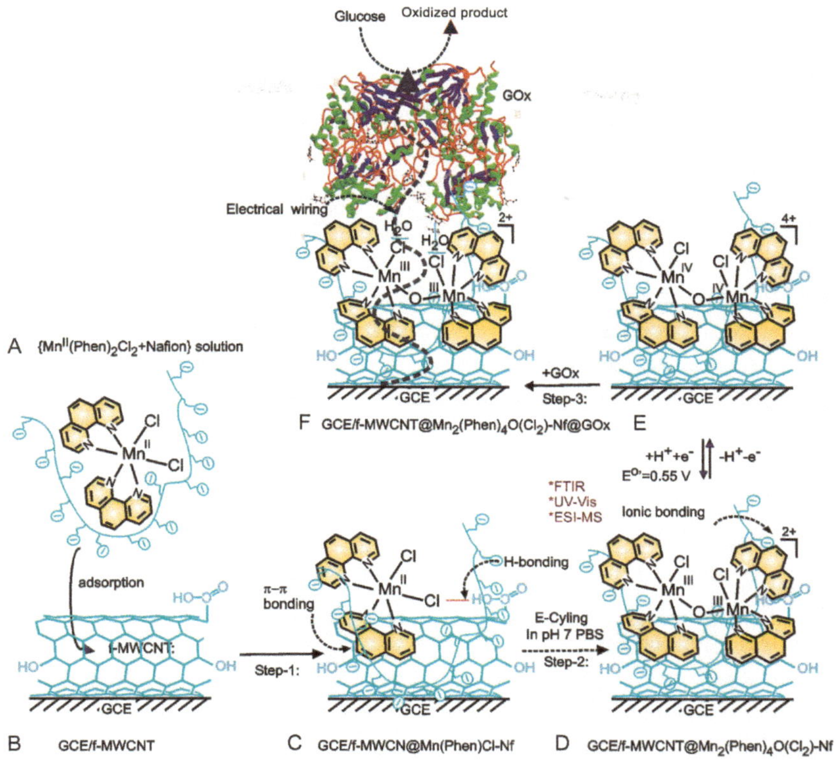

Scheme 1 Schematic representation of layer-by-layer bio-nanoarchitectures built with carbon nanotubes (A); Adsorption on a f-MWCNT modified GCE (B); Immobilization of the Mn-complex by various interactions, (C) and its in-situ conversion to the binuclear complex, $[Mn(phen)_4O(Cl)_2]^{2+}$ (D); Its redox step elevation to high-valent complex (E); Electrical Wiring (F) in the bio-electro-catalytic glucose oxidation reaction. The cartoon of the carbon nanomaterial is MWCNT. For clarity sake only outer layer is shown. *Reprinted from Ref. Saravanan, N., Mayuri, P., & Senthil Kumar, A. (2018). Improved electrical wiring of glucose oxidase enzyme with an in-situ immobilized Mn(1,10-Phenanthroline)$_2$Cl$_2$-complex/multiwalled carbon nanotube-modified electrode displaying superior performance to Os-complex for high-current sensitivity Bioelectrocatalytic and biofuel cell applications. ACS Applied Bio Materials, 1(5), 1758–1767. doi:10.1021/acsabm.8b00584. Copyright 2018 American Chemical Society.*

1.1 Basics of bioelectrocatalysis and sensing of glucose

Taking advantage of the electron-transfer activity, GOx and its associated chemical reagents like H_2O_2 or O_2 is linked with electrochemical reaction and performed quantitative bioelectrocatalytic oxidation and sensing of glucose in aqueous solution (first generation glucose biosensor) (Chen et al., 2013). In this connection, in early 1960, Clark type dissolved oxygen sensor that is made of Pt based chemically modified electrode facilitating the oxygen reduction reaction has been used as a quantitative tool for the glucose sensing in real sample (Clark & Lyons, 2006; Peteu, Emerson, & Mark Worden, 1996). In that case, a decrement in the dissolved oxygen current signal against increase in the glucose concentration was noticed as a calibration graph. Similarly, amperometric H_2O_2 sensing based glucose monitoring system for commercial use was introduced in 1975 by Yellow Springs Instrument Company analyser, Model 23A YSI analyser (Mano, 2019). Indeed, the main challenge with the 1st generation sensor is high operating potential and interference effect of coexisting biochemicals such as ascorbic acid, uric acid, dopamine and certain drugs at the H_2O_2 detection potential (Mano, 2019). As an improvement, in 1980s, transducer concept was introduced as second generation bioelectrocatalysis system (Borgmann, Hartwich, Schulte, & Schuhmann, 2005; Chaubey, & Malhotra, 2002; Mano, 2019;). The role of transducer is shuttle the electron-reaction between the GOx enzyme and underlying electrode surface (Scheme 1F), which is defined as molecular wiring (Alves, Fiorito, Córdoba de Torresi, & Torresi, 2006).

In this connection, variety of molecular mediators such as ferrocene and its derivatives, ferricyanide, quinone, tetrathialfulvalene (TTF), tetracyanoquinodimethane (TCNQ), thionine, methylene blue, and methyl viologen, transition-metal complexes like$[Ru(bpy)_3]^{2+}$ (bpy= 2,2′-bipyridine) and Os(bpy)-complexes with pendent polypyridine and phenothiazine phenoxazine compounds have been electrically connected with GOx enzyme for bioelectrocatalytic and sensing applications (Chaubey & Malhotra, 2002; Mano, 2019; Menger, 1985). The major challenge with the second generation biosensor system is achieving the proper molecular wiring and electron-shuttling reaction successfully. Note that the redox-active FAD moiety in GOx is deeply buried within the protein shell of the GOx enzyme and hence it is difficult to attain the electron transfer communication effectively (Mano, 2019; Milton & Minteer, 2017). To solve this problem, third generation biosensor was introduced (Das, Das, Chinnadayyala, Singha, & Goswami, 2016). In this case, direct electron-transfer (DET) reaction between the GOx enzyme and underlying electrode surface is achieved without any addition reagent and transducer (Mano, 2019). Carbon nanomaterials such as multiwalled carbon nanotube (MWCNT), single walled carbon nanotube (SWCNT), graphene oxide (GO) and graphene, etc., have been referred as the best choice for this purpose (Dinesh, Shalini Devi, & Krishnan, 2019; Guo & Li, 2010;

Tominaga, Nomura, & Taniguchi, 2008; Wang, Zhou, Zhang, Boey, & Zhang, 2009; Zhang, Feng, & Tachikawa, 2007; Zhang, Liao, Chu, Liu, & Zhang, 2014). Nevertheless, limited number of enzymes and electrode materials has been identified for this purpose. In some case, rather to show the DET, the immobilized GOx carbon modified electrodes behave as a dissolved oxygen sensor (Dinesh et al., 2019; Guo & Li, 2010). Overall, in consideration with efficiency and practical applicability, the second generation glucose bio-electrocatalytic system has been referred as an ideal reaction.

1.2 Molecular wiring $Mn(Phen)_2Cl_2$ with GOx and carboxylic acid functionalized MWCNT

Development of new molecular wiring system which is simple to prepare, low cost and high efficiency, is a demanding research interest in glucose bioelecrocatalytic system. Among various transducers reported, ferrocene derivatives and Os(bpy)-based metal complexes have been referred often for this purpose (Borgmann et al., 2005; Ohara, Rajagopalan, & Heller, 1993; Tran et al., 2011). In general, preparation of ferrocene and $Os(bpy)_3$-based complexes are completed and require skilled synthetic chemist. In addition, Os-based precursor salt, $[OsCl_6]^{2-}$ is expensive, and interference due to dissolved oxygen is a serious concern in the electrochemical studies of the complex. In this context, our group has introduced $Mn(phen)_2Cl_2$ complex, where phen = 1,10 Phenonthroline ligand, based transducer for the GOx molecular wiring (Saravanan, Mayuri, & Senthil Kumar, 2018). In the typical preparation, $Mn(Phen)_2Cl_2$ complex immobilized carboxylic acid functionalized-MWCNT (HOOC-MWCNT, f-MWCNT, f = carboxylic acid fuctionalized) chemically modified glassy carbon electrode, designated as GCE/f-MWCNT@$Mn(phen)_2Cl_2$ is coupled with dilution solution of GOx and perfluoro cation exchanging polymer, Nafion (Nf) for molecular wiring and electron-transfer reaction (GCE/f-MWCNT@$Mn(phen)_2Cl_2$-Nf@GOx) (Saravanan et al., 2018).

The present system offers followings four ways of molecular wiring with GOx enzyme (Scheme 1): (i) Noncovalent immobilization as the π–π interaction between π-extended benzene rings of Phen ligand in the $Mn(phen)_2Cl_2$ complex and sp^2 carbon site of MWCNT, (ii) Covalent immobilization as the carboxylic acid functional group interacts with amino functional group of GOx enzyme and there by formation of amide linkage, (iii) Ionic interaction as the cationic sits of Nafion interact with various ionic sites (–COO–) of the proteins, (iv) Hydrogen bonding and hydrophilic interaction as the Mn-OH_2 site created from the Mn(phen)$_2$$Cl_2$ complex linked with N and O terminals of GOx and HCOOH–MWCNT and (v) Hydrophobic interaction as basal plane of the HOOC–MWCNT and hydrophobic site of GOx cooperate with each other. This work provides a systematic study of preparation, electrochemical characterization, bioelectrocatalytic oxidation and sensing of GCE/MWCNT@Mn (phen)$_2$$Cl_2$-Nf@GOx.

2. Materials and methods

2.1 General safety

Please use safety glasses, latex, or nitrile gloves, lab coat and protective mask while handling the chemicals.

2.2 Equipment

1. Beakers
2. Round bottom flask
3. Ultrasonic bath
4. Magnetic stirrer and pellet
5. G4 sintered crucible
6. Pipette (3 mL)
7. Filter paper
8. Suction flask

2.3 Chemicals and material

1. $MnCl_2.4H_2O$ were purchased form SD fine chemicals (India)
2. 1,10-Phenanthroline (Sigma Aldrich)
3. Buffers and Reagents were purchased from Merck
4. Distilled water
5. A supporting electrolyte, pH 7 phosphate buffer solution (PBS) of 0.1 M ionic strength

2.4 Synthesis of [Mn(phen)$_2$$Cl_2$] complex

1. Preparation of ethanolic solution of 1,10-phenanthroline (0.984 g, 6.3 mmol in 15 mL) and $MnCl_2.4H_2O$ (0.5 g, 2.527 mmol 15 mL).
2. Mix gently the above mixture and stirred it in room temperature (25 ± 2 °C) for 30 ± 2 min.

3. The yellow precipitate obtained was filtered off, washed with small amounts of cold methanol and dried under vacuum.
4. The complex is soluble in water, methanol and ethanol, but insoluble in chloroform and ethyl acetate.

2.5 Preparation of the GCE/f-MWCNT@Mn(phen)$_2$Cl$_2$-Nf@GOx bioelectrode

2.5.1 Chemicals and reagents

1. Glassy carbon (Basi, Japan, 3 mm diameter, 0.0707 cm^2) as a working electrode
2. Ag/AgCl (3 M KCl solution) as a reference electrode
3. Platinum disc (Basi, Japan, 3 mm diameter, 0.0707 cm^2) as an auxiliary electrode
4. Electrode polishing kit & 0.5 μm alumina powder Aldrich (Basi, Japan)
5. Autolab Type III, potentiostat/galvanostat, Metrohm/Autolab, Netherlands
6. Carboxylic acid functionalized MWCNT (~90% purity on carbon basis, size 7–15 nm × 0.5–10 μm, Aldrich, USA)
7. Glucose Oxidase from *Aspergillus niger* Type VII, lyophilized powder, ≥100,000 units/g solid Aldrich (USA)
8. 5% Nafion -lower aliphatic alcohol solution, Aldrich, USA.

2.5.2 Preparation procedure

1. As a surface pretreatment step, in first, glassy carbon electrode (GCE) was cleaned both mechanically and electrochemically by polishing with 0.5 micron alumina powder.
2. Performing cyclic voltammetry (CV) for 10 number cycles in a potential window 0.2—1 V vs Ag/AgCl at a scan rate (v), 50 mV s^{-1} in pH 7 PBS.

Preparation of stock solutions

a. 1 mg of f-MWCNT dispersed 500 μL ethanol and sonication for 30 min at room temperature (25 ± 2 °C).
b. 1 mg of Mn(phen)$_2$Cl$_2$ dispersed in 500 μL of ethanol.
c. 10 mg of GOx dispersed in 1 mL of pH 7 PBS.
d. 5% Nafion diluted in 500 μL of ethanol.
e. 5 μL aliquot of each of the f-MWCNT, Mn(Phen)$_2$Cl$_2$, Nafion and GOx solutions were successively drop casted layer-by-layer on a cleaned GC electrode surface and air dried at room temperature for 5 ± 1 min individually. The modified electrode is tentatively designated as GCE/f-MWCNT@Mn(phen)$_2$Cl$_2$-Nf@GOx.

f. As a conditioning step, the above bioelectrode was subjected to CV in a window, −0.1 —1.0 V vs Ag/AgCl for 10 cycles at $v = 50\,mV\,s^{-1}$ in pH 7 PBS. During this treatment, $Mn(phen)_2Cl_2$ complex in the GCE/f-MWCNT@$Mn(phen)_2Cl_2$-Nf is in-sitully converted to $[Mn_2(phen)_4(O)Cl_2]^{2+}$ complex, which was confirmed by ESI-Mass spectroscopic characterization technique [m/z peak at 459.04]. The modified electrode has been designated as GCE/f-MWCNT@$[Mn_2(phen)_4(O)Cl_2]^{2+}$-Nf@GOx. This way of preparation procedure allows for an efficient electrical wiring of the Mn-complex with GOx and underlying electrode as illustrated in Scheme 1.

3. Characterization of the f-MWCNT@$[Mn_2(phen)_4(O)Cl_2]^{2+}$-Nf@GOx

The electrochemical and physicochemical characterizations of the modified electrode GCE/f-MWCNT@$[Mn_2(phen)_4(O)Cl_2]^{2+}$-Nf@GOx was done using following techniques;

3.1 Instrumentation

1. Autolab Type III, potentiostat/galvanostat, Metrohm/Autolab, Netherlands for electrochemical studies.
2. Transmission electron microscopy (TEM) analysis, FEI TECNAI G2 20-TWIN (Netherland) instrument.
3. FT-IR measurements JASCO 4100 Spectrophotometer, Japan.
4. UV–vis analysis, Ocean Optics instrument (JAZ-EL200-XR1).
5. Raman spectroscopic analysis, AZILTRON, PRO 532 (USA) instrument used with a 532 nm laser excitation source.
6. Electrospray ionization mass analysis was carried with Q-Tof Mass Spectrometer Micromass instrument (Waters), and the data was processed using the Mass Lynx NT software system.

3.2 Sampling procedure and observation

1. The electrochemical characterization of the GCE/f-MWCNT-@$[Mn_2(phen)_4(O)Cl_2]^{2+}$-Nf@GOx by CV showed a well-defined redox peak at an apparent standard electrode potential, $E^{o\prime} = 0.55\,V$ vs Ag/AgCl in pH 7 PBS in a potential window, −0.2 to −1 V vs Ag/AgCl at a scan rate (v) of $50\,mV\,s^{-1}$ in pH 7 PBS with a surface excess (Γ_{Mn}) value of $5.6 \times 10^{-9}\,mol\,cm^{-2}$.

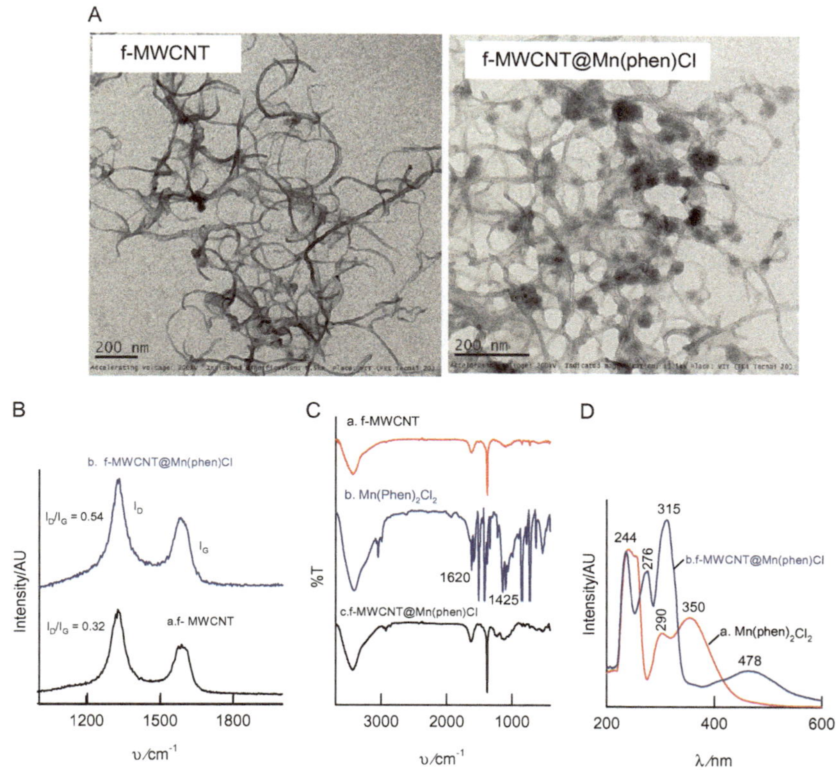

Fig. 1 Comparative (A) TEM images of f-MWCNT and f-MWCNT@Mn(phen)Cl. (B) Raman spectra of f-MWCNT (curve a) and f-MWCNT@Mn(phen)Cl (curve b). (C) FTIR spectra data of f-MWCNT (curve a), Mn(phen)2Cl2 (curve b), and f-MWCNT@ Mn(phen)Cl (curve c). (D) UV–vis spectra of Mn(phen)2Cl2 (curve a) and f-MWCNT@Mn(phen) Cl (curve b) modified bioelectrode. For UV–vis, an ethanolic extract was used for the analysis. *Reprinted from Ref. Saravanan, N., Mayuri, P., & Senthil Kumar, A. (2018). Improved electrical wiring of glucose oxidase enzyme with an in-situ immobilized Mn(1,10-Phenanthroline)$_2$Cl$_2$-complex/multiwalled carbon nanotube-modified electrode displaying superior performance to Os-complex for high-current sensitivity Bioelectrocatalytic and biofuel cell applications. ACS Applied Bio Materials, 1(5), 1758–1767. doi:10.1021/acsabm.8b00584. Copyright 2018 American Chemical Society.*

2. For TEM analysis, an ethanol extract of MWCNT@$[Mn_2(phen)_4(O)Cl_2]^{2+}$ (without nafion-polymer) obtained by sonication of the modified layer from GCE working electrode is mixed with 500 μL of ethanol and then casted on a 3 mm diameter Cu grid. An agglomerated carbon nanotube-composite like observation was noticed (Fig. 1A).

3. Raman analysis was performed by preparing the chemically modified electrode on gold screen printed electrode (Au-SPE) as a working electrode. A decrease in the intensity ratio of D and G-bands, I_D/I_G from 0.54 to 0.32 of f-MWCNT after $Mn(phen)_2Cl_2$ immobilization, which is due to π-generation on the modified electrode (Fig. 1B).
4. FTIR analysis, the modified layer scratched from GCE working electrode using a doctor's needle (0.5 mm × 3.5 cm) from the electrode surface and the samples were measured as 2 wt% KBr pellets, and averaged to obtain a good signal-to-noise ratio. The changes occurring to the chemical functionalities. Characteristic peak for C–N stretching vibration of phen-ligand of the complex was noticed (Fig. 1C).
5. UV–vis and ESI-Mass spectroscopic analysis, an ethanol extract of $MWCNT@[Mn_2(phen)_4(O)Cl_2]^{2+}$ layer obtained by sonication of modified layer with 500 μL of ethanol, filtered using micro syringe was used. Characteristic peaks at λ_{max}, 244, 290 and 350 nm due to $\pi-\pi^*$ and metal-to-ligand charge transitions of the Mn-complex (Fig. 1D).

4. Bioelectrocatalytic application of GOx modified electrode

4.1 Sampling procedure

1. 15 mL of cylindrical vial with 10 mL of pH 7 PBS and 5 mm Teflon stir pellet.
2. Cyclic voltammetry (non-stirred condition) or amperometric (under stirred condition) techniques.
3. Experiments carried out in room temperature and at normal dissolved oxygen condition.

Bioelectrocatalytic responses of the GCE/f-$MWCNT@Mn_2(phen)_4(O)(Cl_2)$-Nf@GOx was tested by performing CV experiment without and with 500 μM of glucose dissolved in 10 mL of pH 7 PBS at a scan rate = 2 mV s^{-1}. A well-defined oxidation current signal starting at 0.4 V vs Ag/AgCl with a peak maximum, i.e., at anodic peak potential, E_{pa} = 0.55 V vs Ag/AgCl, where the A1/C1 redox peak exist was noticed. No such oxidation was noticed when GOx was not immobilized on the electrode surface. CV response of the bioanode with increasing concentration of glucose (0–6 mM) at v = 10 mV s^{-1} as in Fig. 2. This observation highlights the molecular wiring, electron-transfer shuttling and bioelectrocatalytic glucose oxidation reaction. The term "molecular wiring" is denoted here to represent the electron-shutting path from

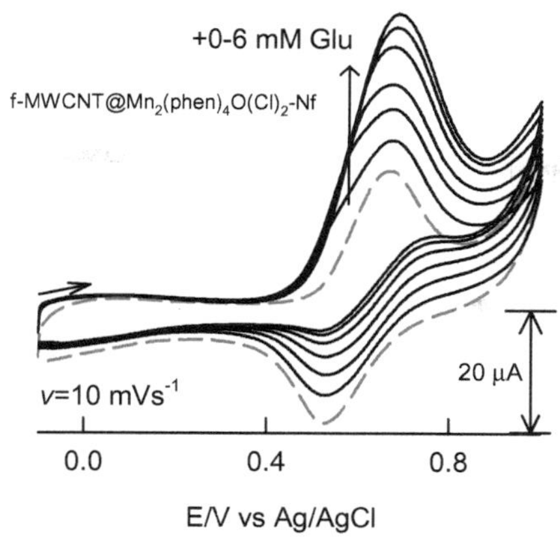

Fig. 2 CV response of the optimal bioanode with increasing concentration of glucose (0–6 mM) at v = 10 mV s^{-1} in pH 7 PBS. *Reprinted from Ref. Saravanan, N., Mayuri, P., & Senthil Kumar, A. (2018). Improved electrical wiring of glucose oxidase enzyme with an in-situ immobilized Mn(1,10-Phenanthroline)$_2$Cl$_2$-complex/multiwalled carbon nanotube-modified electrode displaying superior performance to Os-complex for high-current sensitivity Bioelectrocatalytic and biofuel cell applications. ACS Applied Bio Materials, 1(5), 1758–1767. doi:10.1021/acsabm.8b00584. Copyright 2018 American Chemical Society.*

the electrode surface to solution as GCE/f-MWCNT → Mn (Phen)$_2$Cl$_2$ → GOx → Glucose (Scheme 1F). Note that performing several control experiments with omission of individual component/s of the bioelectrode resulted to absence of any bioelectrocatalytic oxidation current signal of glucose. This observation confirms the molecular wiring between the GCE/f-MWCNT, Mn(Phen)$_2$Cl$_2$ and GOx on the bioelectrode surface. Stability was tested by performing 20 continuous CV cycles of the bioanode with 2 mM of glucose at v = 50 mV s^{-1} showing stable voltammetric response (data not enclosed). Amperometric i–t technique was adopted for the bioelectrocatalytic sensing of glucose at a fixed applied potential, E_{app} = 0.65 V vs Ag/AgCl (optimal) under a hydrodynamic condition as in Fig. 3A curve a. For this experiment high concentration stock solution of glucose (10 mM) was successively spiked (50 μL) in to a stirred 10 mL pH 7 PBS electrolyte, so that it reaches desired final concentration of glucose (50 μM). A systematic increase in the amperometric i–t current signal with increase in the glucose spikes was noticed. Plot of amperometric current signal against glucose concentration is linear in a window 50–550 μM with a current sensitivity value 22.4 μA mM^{-1} (316.8 μA mM^{-1} cm^{-2}) (Fig. 3B). As a limitation to this system, ascorbic acid showed significance interference and such a real sample can be avoided for the quantitative analysis.

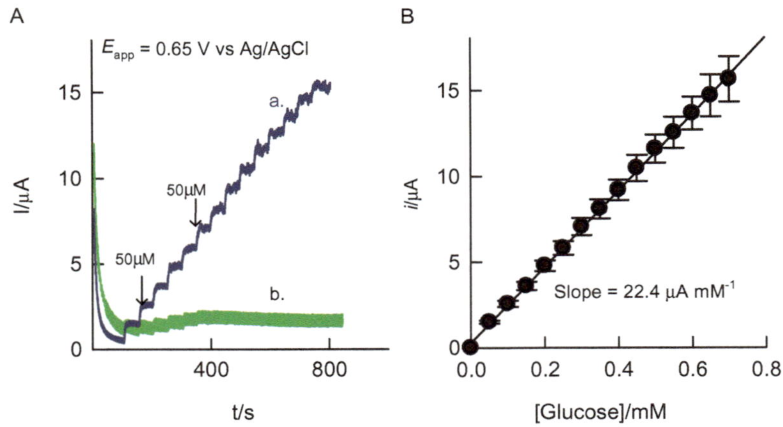

Fig. 3 (A) Comparative amperometric i–t responses of GCE/f-MWCNT@$Mn_2(phen)_4(O)(Cl)_2$-Nf@GOx (curve a) and GCE/f-MWCNT-Nf@GOx (curve b) for successive detection of 50 μM of glucose at an applied potential of +0.65 V vs Ag/AgCl in pH 7 PBS. (B) Typical calibration response of glucose bio-electro-catalytic current vs glucose concentration. *Reprinted from Ref. Saravanan, N., Mayuri, P., & Senthil Kumar, A. (2018). Improved electrical wiring of glucose oxidase enzyme with an in-situ immobilized Mn(1,10-Phenanthroline)$_2$Cl$_2$-complex/multiwalled carbon nanotube-modified electrode displaying superior performance to Os-complex for high-current sensitivity Bioelectrocatalytic and biofuel cell applications. ACS Applied Bio Materials, 1(5), 1758–1767. doi:10.1021/acsabm.8b00584. Copyright 2018 American Chemical Society.*

5. Conclusion

Bioeletrocatalytic oxidation and sensing of glucose using $Mn(phen)_2Cl_2$ complex molecularly wired MWCNT/Nafion modified glassy carbon electrode has been studied. A systematic procedure for the preparation of $Mn(phen)_2Cl_2$ and GOx chemically modified electrode and bioelectrocatalytic oxidation and sensing of glucose by CV and amperometric i–t methods have been demonstrated.

Acknowledgments

The authors acknowledge the DST-SERB, Scheme (DST-SERB-EMR/2016/002818) for financial support of this work. N.S. acknowledges the DST-SERB, national postdoctoral fellowship (PDF/2015/000253) program.

References

Alves, W. A., Fiorito, P. A., Córdoba de Torresi, S. I., & Torresi, R. M. (2006). Design of molecular wires based on supramolecular structures for application in glucose biosensors. *Biosensors and Bioelectronics*, *22*(2), 298–305. https://doi.org/10.1016/j.bios.2006.01.009.

Bankar, S. B., Bule, M. V., Singhal, R. S., & Ananthanarayan, L. (2009). Glucose oxidase—An overview. *Biotechnology Advances*, *27*(4), 489–501. https://doi.org/10.1016/j.biotechadv.2009.04.003.
Borgmann, S., Hartwich, G., Schulte, A., & Schuhmann, W. (2005). Amperometric enzyme sensors based on direct and mediated electron transfer. *Perspectives in Bioanalysis*, *1*, 599–655. https://doi.org/10.1016/S1871-0069(05)01017-7.
Chaubey, A., & Malhotra, B. D. (2002). Mediated biosensors. *Biosensors and Bioelectronics*, *17*(6–7), 441–456. https://doi.org/10.1016/S0956-5663(01)00313-X.
Chen, C., Xie, Q., Yang, D., Xiao, H., Fu, Y., Tan, Y., et al. (2013). Recent advances in electrochemical glucose biosensors: A review. *RSC Advances*, *3*(14), 4473. https://doi.org/10.1039/c2ra22351a.
Clark, L. C., & Lyons, C. (2006). Electrode systems for continuous monitoring in cardiovascular surgery. *Annals of the New York Academy of Sciences*, 102(1), 29–45. https://doi.org/10.1111/j.1749-6632.1962.tb13623.x.
Conzuelo, F., Ruff, A., & Schuhmann, W. (2018). Self-powered bioelectrochemical devices. *Current Opinion in Electrochemistry*, *12*, 156–163. https://doi.org/10.1016/j.coelec.2018.05.010.
Das, P., Das, M., Chinnadayyala, S. R., Singha, I. M., & Goswami, P. (2016). Recent advances on developing 3rd generation enzyme electrode for biosensor applications. *Biosensors and Bioelectronics*, *79*, 386–397. https://doi.org/10.1016/j.bios.2015.12.055.
Dinesh, B., Shalini Devi, K. S., & Krishnan, U. M. (2019). Achieving a stable high surface excess of glucose oxidase on pristine multiwalled carbon nanotubes for glucose quantification. *ACS Applied Bio Materials*, *2*(4), 1740–1750. https://doi.org/10.1021/acsabm.9b00145.
Guo, C. X., & Li, C. M. (2010). Direct electron transfer of glucose oxidase and biosensing of glucose on hollow sphere-nanostructured conducting polymer/metal oxide composite. *Physical Chemistry Chemical Physics*, *12*(38). 12153https://doi.org/10.1039/c0cp00378f.
Mano, N. (2019). Engineering glucose oxidase for bioelectrochemical applications. *Bioelectrochemistry*, *128*, 218–240. https://doi.org/10.1016/j.bioelechem.2019.04.015.
Menger, F. M. (1985). On the source of intramolecular and enzymatic reactivity. *Accounts of Chemical Research*, *18*(5), 128–134. https://doi.org/10.1021/ar00113a001.
Milton, R. D., & Minteer, S. D. (2017). Direct enzymatic bioelectrocatalysis: Differentiating between myth and reality. *Journal of the Royal Society Interface*, *14*(131), 20170253. https://doi.org/10.1098/rsif.2017.0253.
Nakamura, S., & Ogura, Y. (1968). Action mechanism of glucose oxidase of Aspergillus Niger. *Journal of Biochemistry*, *63*(3), 308–316.
Ohara, T. J., Rajagopalan, R., & Heller, A. (1993). Glucose electrodes based on cross-linked bis(2,2′-bipyridine)chloroosmium(+/2+) complexed poly(1-vinylimidazole) films. *Analytical Chemistry*, *65*(23), 3512–3517. https://doi.org/10.1021/ac00071a031.
Peteu, S. F., Emerson, D., & Mark Worden, R. (1996). A Clark-type oxidase enzyme-based amperometric microbiosensor for sensing glucose, galactose, or choline. *Biosensors and Bioelectronics*, *11*(10), 1059–1071. https://doi.org/10.1016/0956-5663(96)87665-2.
Saravanan, N., Mayuri, P., & Senthil Kumar, A. (2018). Improved electrical wiring of glucose oxidase enzyme with an *in-situ* immobilized Mn(1,10-Phenanthroline)$_2$Cl$_2$-complex/multiwalled carbon nanotube-modified electrode displaying superior performance to Os-complex for high-current sensitivity Bioelectrocatalytic and biofuel cell applications. *ACS Applied Bio Materials*, *1*(5), 1758–1767. https://doi.org/10.1021/acsabm.8b00584.
Schuhmann, W., Ohara, T. J., Schmidt, H. L., & Heller, A. (1991). Electron transfer between glucose oxidase and electrodes via redox mediators bound with flexible chains to the enzyme surface. *Journal of the American Chemical Society*, *113*(4), 1394–1397. https://doi.org/10.1021/ja00004a048.

Tominaga, M., Nomura, S., & Taniguchi, I. (2008). Bioelectrocatalytic current based on direct heterogeneous electron transfer reaction of glucose oxidase adsorbed onto multi-walled carbon nanotubes synthesized on platinum electrode surfaces. *Electrochemistry Communications*, *10*(6), 888–890. https://doi.org/10.1016/j.elecom.2008.04.011.

Tran, T. O., Lammert, E. G., Chen, J., Merchant, S. A., Brunski, D. B., Keay, J. C., et al. (2011). Incorporation of single-walled carbon nanotubes into ferrocene-modified linear Polyethylenimine redox polymer films. *Langmuir*, *27*(10), 6201–6210. https://doi.org/10.1021/la104999f.

Wang, Z., Zhou, X., Zhang, J., Boey, F., & Zhang, H. (2009). Direct electrochemical reduction of single-layer graphene oxide and subsequent functionalization with glucose oxidase. *The Journal of Physical Chemistry C*, *113*(32), 14071–14075. https://doi.org/10.1021/jp906348x.

Yates, N. D. J., Fascione, M. A., & Parkin, A. (2018). Methodologies for "wiring" redox proteins/enzymes to electrode surfaces. *Chemistry—a European Journal*, *24*(47), 12164–12182. https://doi.org/10.1002/chem.201800750.

Yoo, E.-H., & Lee, S.-Y. (2010). Glucose biosensors: An overview of use in clinical practice. *Sensors*, *10*(5), 4558–4576. https://doi.org/10.3390/s100504558.

Zhang, J., Feng, M., & Tachikawa, H. (2007). Layer-by-layer fabrication and direct electrochemistry of glucose oxidase on single wall carbon nanotubes. *Biosensors and Bioelectronics*, *22*(12), 3036–3041. https://doi.org/10.1016/j.bios.2007.01.009.

Zhang, X., Liao, Q., Chu, M., Liu, S., & Zhang, Y. (2014). Structure effect on graphene-modified enzyme electrode glucose sensors. *Biosensors and Bioelectronics*, *52*, 281–287. https://doi.org/10.1016/j.bios.2013.07.022.

CHAPTER TWELVE

Use of functionalized carbon nanotubes for the development of robust nanobiocatalysts

Michaela Patila[a], Nikolaos Chalmpes[b], Evangelia Dounousi[c], Haralambos Stamatis[a,*], Dimitrios Gournis[b,*]

[a]Laboratory of Biotechnology, Department of Biological Applications and Technology, University of Ioannina, Ioannina, Greece
[b]Department of Materials Science & Engineering, University of Ioannina, Ioannina, Greece
[c]Department of Nephrology, Medical School, University of Ioannina, Ioannina, Greece
*Corresponding authors: e-mail address: hstamati@uoi.gr; dgourni@uoi.gr

Contents

Abstract

This chapter deals with the use of functionalized carbon nanotubes (fCNTs) as supports for the development of nanobiocatalytic systems through the immobilization of enzymes. The surface characteristics, properties and production of carbon nanotubes

Methods in Enzymology, Volume 630
ISSN 0076-6879
https://doi.org/10.1016/bs.mie.2019.10.015

are described, while an analysis in their biological applications is also presented. The results presenting within the text are giving insights to the effect of carbon nanotubes on the catalytic and structural characteristics of different proteins, such as cytochrome *c* from equine heart (cyt *c*) and laccase from *Trametes versicolor* (TvL), either when they are used as additives in the reaction medium or as supports for protein immobilization. A variety of biochemical and spectroscopic techniques is applied to investigate the interactions between the protein biomolecules and carbon nanotubes. The results showed that the presence of fCNTs enables cyt *c* to maintain both its secondary structure and heme microenvironment. Non-covalent and covalent immobilization approaches are also described, while the immobilized biocatalysts are characterized with respect to their catalytic and structural characteristics. Immobilized TvL was found to exhibit higher catalytic activity when non-specific binding was used as immobilization procedure (up to $0.85\,U\,\mu g^{-1}$), compared to covalent immobilization (up to $0.7\,U\,\mu g^{-1}$), while the increase of the alkyl chain of the functionalized CNTs also seems to affect the catalytic efficiency of the immobilized enzymes. The nanobiocatalytic systems that are presented here demonstrated exceptional stability (up to 31% of their initial activity is maintained after 24 h incubation at 60 °C) and reusability (up to 58% remaining activity after 8 successive catalytic cycles) compared to the native enzymes, leading to robust biocatalytic systems appropriate for various applications of biotechnological and industrial interest.

Abbreviations

ABTS	2,2′-azino-bis(3-ethylbenzothiazoline-6-sulphonic acid)
CCVD	catalytic chemical vapor deposition
CD	circular dichroism
CNTs	carbon nanotubes
Cyt *c*	cytochrome *c*
DC	direct current
DMF	*N*,*N*-dimethylformamide
EDC	1-ethyl-3-(3-dimethylaminopropyl) carbodiimide hydrochloride
fCNTs	functionalized multi-walled carbon nanotubes
FTIR	Fourier-transform infrared
HEPES	4-(2-hydroxyethyl)piperazine-1-ethanesulfonic acid
KBr	potassium bromide
MRE	mean residue ellipticity
MWCNTs	multi-walled carbon nanotubes
NHS	N-hydroxysuccinimide
PE	plasma-enhanced
SWCNTs	single-walled carbon nanotubes
THF	tetrahydrofuran
TG	thermo-gravimetric analysis
TvL	laccase from *Trametes versicolor*
UV-vis	ultraviolet-visible

1. Introduction

1.1 Carbon nanotubes: Structure and properties

The rapid evolution of nanotechnology has led to the synthesis of nanomaterials for the development of novel nanobiocatalytic systems and their use in various applications of biotechnological and industrial interest. Great scientific attention has been paid to carbon-based nanostructured materials, which due to their particular morphological and mechanical properties, find wide application in various fields. Among these nanomaterials, carbon nanotubes (CNTs) present excellent properties as nanosupports for the immobilization of biomolecules and their subsequent use in biocatalytic processes.

Carbon nanotubes, an allotrope of carbon first discovered by Ijima in 1991 (Iijima, 1991), are thin graphite sheets of condensed benzene rings rolled up into a cylindrical shape (Mehra, Mishra, & Jain, 2014). The structure of CNTs could be correlated to graphite sheets that have been rolled-up to form one or more tubes. They can be classified in single-walled carbon nanotubes, which formation is occurred by rolling a single sheet of graphite -known as graphene- and multi-walled carbon nanotubes, consisted of several concentric tubes of graphene fitted one inside the other (Erol et al., 2018; Sharma, Kumar Mehra, Jain, & Jain, 2016). In addition, in the case of SWCNTs, the way the graphite sheet is rolled up during its creation process is very important for the final electronic properties, thus CNTs can be divided into chiral, zig-zag and armchair carbon nanotubes (Dresselhaus, Dresselhaus, & Saito, 1995; Schnorr & Swager, 2011). Typically, CNTs have their tips closed by graphitic layers, while they can also have open ends on one or both sides, if template methods are used for their growth (Kyotani & Tomita, 1995; Poudel & Li, 2018). Each nanotube is composed by millions of carbon atoms, and its length can be tens of micrometers long with very small outer diameters, from 3–30 nm, in the case of MWCNTs, to 0.8–2 nm in the case of SWCNTs. Nanotubes are considered as one-dimensional nanomaterials, as they generally present a large length-to-diameter ratio of about 1000 (De Volder, Tawfick, Baughman, & Hart, 2013; Saifuddin, Raziah, & Junizah, 2013).

Due to their outstanding mechanical and physical properties CNTs have attracted the interest of the scientific community during the last decades leading them as one of the most promising candidates in the fields of aerospace, electronic industries and textile (Baughman, Zakhidov, & De Heer, 2002; Wu et al., 2004). As a matter of fact, this nanomaterial possesses high chemical and thermal stability, mechanical strength, flexibility, and electrical and thermal conductivity (De Zhang, Shen, Phang, & Liu, 2004; Iijima, 2002; Sahoo, Chitturi, Agarwal, Jiang, & Katiyar, 2014). Moreover, CNTs have large surface area, low density, which is one-sixth of that of stainless steel, they are the strongest materials ever discovered, while their lightness and high thermal and electrical conductivity make them ideal supports for a wide range of applications (He et al., 2013; Kaushik & Majumder, 2015). However, in spite of the perceptual properties they present, CNTs show some significant disadvantages that limit their use in biotechnological and biomedical applications. Their hydrophobic surface prevents their solubility in water, leading to high agglomeration yields (He et al., 2013; Nagaraju, Reddy, & Reddy, 2015). Moreover, pristine carbon nanotubes exhibit some toxicity and show low biodegradation rates, hence they are unsuitable for aqueous-based bio-applications, by limiting the effective interactions between biomolecules and CNTs (Kim, Lee, et al., 2017).

To overcome these problems, different techniques have been developed for the modification of CNTs. Functionalization improves the aqueous dispersability of CNTs, and thus reduces their aggregation behavior, enhances the internalization of CNTs within the cells, and enables the conjugation with biomolecules due to the attachment of different functional groups and chemical moieties onto their surface (Sharma et al., 2016). Main approaches for the functionalization of CNTs can be generally classified in three categories: covalent, non-covalent, and endohedral functionalization (Erol et al., 2018).

1.2 Applications of carbon nanotubes in enzymatic procedures

Carbon nanotubes have rapidly become one of the most studied materials, as they are ideal supports for the design of novel biomaterials, due to their exceptional structural features and physical properties. In various biological applications, their use requires the immobilization of a biomolecule onto their surface, either by physical adsorption or covalent binding. Through biomolecule attachment, CNTs can be applied in a variety of research fields, form sensing to biocatalysis.

CNTs, due to their excellent electrical conductivity, large specific surface area and superior mechanical strength, are ideal materials for the design and synthesis of biosensors for the detection of different kind of analytes and biological compounds (Zhu, 2017). Moreover, owing to their extraordinary low detection limit and their ability to promote electron transfer, CNTs are often preferred as working electrodes (Galano, 2010; Kim, Park, et al., 2017). In general, the composition of CNTs-based biosensors includes two main parts; a biological sensitive element, e.g., an enzyme, that specifically recognizes the analyte, and a transducer, which converts the biological signal to other detectable physical signals (Sireesha, Jagadeesh Babu, Kranthi Kiran, & Ramakrishna, 2018). Immobilization of enzymes on CNTs is the key step for the preparation of CNT-based enzymatic biosensors. Enzymes are attached to the CNTs-electrode by means of physical adsorption or covalent binding. Depending on the analyte, different enzymes have been immobilized on CNTs-electrodes, such as glucose oxidase (Boussema et al., 2018), cholesterol oxidase (Zhu et al., 2013) and horseradish peroxidase (Moyo, Okonkwo, & Agyei, 2014).

In a similar manner, CNTs are also used in the design of enzymatic biofuel cells. Enzymatic biofuel cells (EBFCs) are devices that directly transform chemical to electric energy through enzyme-based catalysis (Willner, Yan, Willner, & Tel-Vered, 2009). The anode of a biofuel cell usually consists of an oxidoreductase, e.g., glucose oxidase, which oxidizes the fuel to generate electrons, while the cathode consists of an oxidase, e.g., laccase, which uses these electrons to catalyze the reaction of an oxidant, generating water. Carbon nanotubes, due to their aspect ratio, high electronic conductivity, and the fact that they share similar dimensions with enzymes, have been considered as ideal materials for the design of enzymatic biofuel cells, promoting the direct electron transfer of the immobilized enzymes, thus enhancing their capacity (Ben Tahar et al., 2018; Cosnier, Holzinger, & Le Goff, 2014).

Nanobiocatalysis is an emerging research field which refers to the development of novel nanobiocatalytic systems through enzyme immobilization on nanostructured materials, using conventional approaches, such as covalent bonding and physical adsorption (Kim, Grate, & Wang, 2008). An ideal immobilization support should be able to prevent enzymes from aggregation or denaturation, but maintain at the same time their native conformation (Pavlidis, Patila, Bornscheuer, et al., 2014). The large surface area and cylindrical shape of CNTs allows them to be functionalized with a variety of organic compounds providing them with different functional groups on their surface for the efficient and oriented immobilization of enzymes, resulting in the development of robust nanobiocatalytic systems (Patila et al., 2013; Pavlidis, Vorhaben, Tsoufis, et al., 2012). Several studies have been conducted concerning the use of immobilized enzymes on CNTs in biotechnological applications with industrial and environmental interest, such as phenol degradation, dye decolorization, waste water treatment etc. Zhang & Cai have reported the immobilization of horseradish peroxidase on Fe_3O_4/nanotubes composites for phenol degradation (Zhang & Cai, 2019). The phenol removal rate reached up to 82.8% due to the activity of the immobilized enzyme. In another study, immobilized laccase on functionalized CNTs was used for dye decolorization, where it was found that the decolorization of Reactive Black 5 by immobilized laccase on CNTs was efficient, and reached up to 84.26% decolorization yields after 24 h (Othman, González-Domínguez, Sanromán, Correa-Duarte, & Moldes, 2016). Lipases immobilized on CNTs have also been widely used for synthetic bioprocesses (Ke, Li, Huang, Xu, & Yan, 2014; Markiton, Boncel, Janas, & Chrobok, 2017). For instance, covalently immobilized lipase on different functionalized CNTs was used for the enantioselective synthesis of Solketal esters (Zniszczoł et al., 2016). The immobilized lipase exhibited transesterification yields (up to 54%) higher than the native enzyme form, while the enantioselectivity of the biocatalyst depended on the kind of functionalization of CNTs.

2. Synthesis of CNTs

2.1 Methods for CNTs production

The CNTs synthesis defines the characteristics of the product such as dimensions, geometry, chirality, purity etc. The majority of the processes are conducted in a vacuum or with process gasses. The three main techniques currently used are: (i) electric arc discharge, (ii) laser ablation of carbon, and (iii) catalytic chemical vapor deposition (CCVD). Other synthesis techniques have also been utilized including the condensed-phase electrolysis (Sen, Suzuki, Kataura, & Achiba, 2001), the flame pyrolysis (Sun et al., 2013) and the plasma torch (Smiljanic, Stansfield, Dodelet, Serventi, & Désilets, 2002).

The electric arc discharge method is the technique responsible for the discovery of carbon nanotubes in 1991 by Iijima and his co-workers (Iijima, 1991). In this method, two vertical graphite electrodes are placed in a reactor and a direct current (DC) is run between them and an arc-induced plasma is generated that evaporates the carbon atoms in the graphite. The main disadvantage of the arc discharge method is the requirement of high energies and temperatures for sublimation of solid targets, however, single wall carbon nanotubes can also be produced by this process (Guo, Nikolaev, Rinzler, et al., 1995).

Laser ablation was developed by Richard Smalley and co-workers at Rice University (Guo, Nikolaev, Thess, et al., 1995); in this method carbon nanotubes are created when a graphite target is vaporized by a pulsed laser in a high-temperature reactor while an inert gas, e.g., argon or helium, is bled into the chamber. Along with the electric arc discharge method, the laser ablation technique is considered to be a high-temperature and single wall carbon nanotube's synthesis method (Arepalli, 2004).

Catalytic chemical vapor deposition (CCVD), either thermal (Steiner III et al., 2009) or plasma enhanced (PE), is considered to be the most commonly used technique for the CNTs production. Huge amounts of CNTs can be produced, for example, by catalytic decomposition of acetylene over iron/cobalt catalysts, which are supported on zeolite or silica (Triantafyllidis et al., 2008). This method involves the heating of a catalytic material at high-temperatures (500–1000 °C) into a tube furnace while a precursor carbon gas penetrates the furnace. There is a variety of carbon sources that can be used in this process such as ethylene, acetylene, methane, carbon monoxide. The catalyst has a fundamental role in CCVD due to its significant impact on CNTs growth. A variety of transition metals nanoparticles, e.g., nickel, iron, and cobalt have been extensively used as catalysts (Tsoufis et al., 2007).

2.2 Functionalization of CNTs

Although CNTs are widely applied in numerous scientific fields, their stable chemical structure and high surface energy prevents them from forming complexes with other molecules. Due to strong van der Waals forces and π–π stacking interactions developed between the tubes, CNTs tend to agglomerate, thus exhibiting extremely low solubility in common solvents, such as water, and preventing the interactions between them and other molecules, e.g., enzymes (Mallakpour & Soltanian, 2016). To address this problem, CNTs are functionalized by attaching groups with active functionalities, such as carboxyl, hydroxyl, and sulfonyl groups, onto the carbon walls or at the end of the tubes. Different strategies are applied to modify CNTs, namely non-covalent exohedral functionalization with surfactants or polymers, endohedral functionalization with metals, nanoparticles, salts or even fullerenes, defect-group functionalization and covalent sidewall functionalization (Jeon, Chang, Kumar Ashok, & Baek, 2011).

Non-covalent functionalization of CNTs is based on the development of different adsorption forces, such as electrostatic and hydrophobic interactions, π–π stacking, van der Waals forces or hydrogen bonds. This kind of functionalization does not usually interrupt the graphitic structure of CNTs sidewalls, preserving the properties of the nanomaterials (Meng, Fu, & Lu, 2009). However, as non-covalent functionalization is governed by weak forces, the loading of the chemical compound might be low. CNTs can be non-covalently functionalized by surfactants, polymers, aromatic compounds and metals.

In the case of covalent functionalization, a chemical bond between functional groups of an agent and the carbon skeleton of CNTs is formed. Covalent functionalization can be performed either at the sidewalls of CNTs or at the end caps of the tubes. The most common method widely applied is chemical oxidation, by means of strong acid treatment, which adds carboxyl groups in the graphitic sidewalls of the nanotubes. In spite of the fact that covalent functionalization disrupts the surface conjugated π-network of CNTs, altering some of their pristine properties (e.g., electrical conductivity), oxidation of CNTs offers the possibility of further modification of the nanomaterials with amines, diamines or amino-acids (Pavlidis, Tsoufis, Enotiadis, Gournis, & Stamatis, 2010). For instance, carboxyl groups on the surface of CNTs may react in a first step with thionyl chloride, while in a second step the addition of an organic compound, containing free amine groups, is performed (Wang, Iqbal, & Malhotra, 2005). By this method, it is possible to produce different CNTs-derivatives with variable alkyl chains or terminal functional groups (Patila et al., 2013), enhancing their potential for targeted attachment of biomolecules and their following use in bio-applications.

2.2.1 General safety

Nitrile gloves, protective mask

2.2.2 Equipment

1. Beakers
2. Spherical bottle
3. Volumetric cylinders
4. Magnetic stirrer
5. Reflux accessories
6. Centrifuge
7. Vacuum desiccators

8. Pipette Pasteur
9. Ultrasonic bath

2.2.3 Chemicals and materials

1. MWCNTs (95% pure) from Sigma Aldrich Co., MO, USA.
2. Sulfuric acid (95–97%) from Merck KGaA, Darmstadt, Germany.
3. Nitric acid (65%) from Merck KGaA, Darmstadt, Germany.
4. 1-hexamethylenediamine (99%) from Sigma Aldrich Co., MO, USA.
5. Thionyl chloride (99%) from Merck KGaA, Darmstadt, Germany.
6. Chloroform from Merck KGaA, Darmstadt, Germany.
7. Anhydrous tetrahydrofuran (99%) from Merck KGaA, Darmstadt, Germany.
8. Anhydrous toluene (99.9%) from Merck KGaA, Darmstadt, Germany.
9. *N,N*-dimethyl formamide from Riedel-de-Haën, Seelze, Germany.
10. 11-aminoundecanoic acid (99%) from Sigma Aldrich Co., MO, USA.
11. Distilled and double distilled water.

2.2.4 Methodology

Covalent functionalization of MWCNTs is achieved through oxidation with sulfuric and nitric acid for the attachment of carboxyl groups on the sidewalls of nanotubes. In a next step, oxidized CNTs react with thionyl chloride, followed by addition of hexamethylenediamine or aminoundecanoic acid to obtain amine- and carboxyl- terminated functionalized CNTs, respectively (Fig. 1). The functionalized MWCNTs derivatives will be referred as fCNTs, from now on.

2.2.5 Protocol

2.2.5.1 Covalent functionalization of MWCNTs

1. 100 mg of pristine MWCNTs are added in a mixture consisting of concentrated H_2SO_4 (60 mL) and HNO_3 (20 mL).
2. The admixture is sonicated for 3 h.
3. The resulting suspension is centrifuged and washed six times with double distilled water.
4. The solid sample is left to dry at 50 °C in vacuum overnight.

2.2.5.2 Modification of functionalized MWCNTs-COOH with diamines

1. 50 mg of MWCNTs-COOH are suspended in a mixture of 20:1 v/v of thionyl chloride and *N,N*-dimethyl formamide.
2. The suspension is refluxed at 70 °C for 24 h.

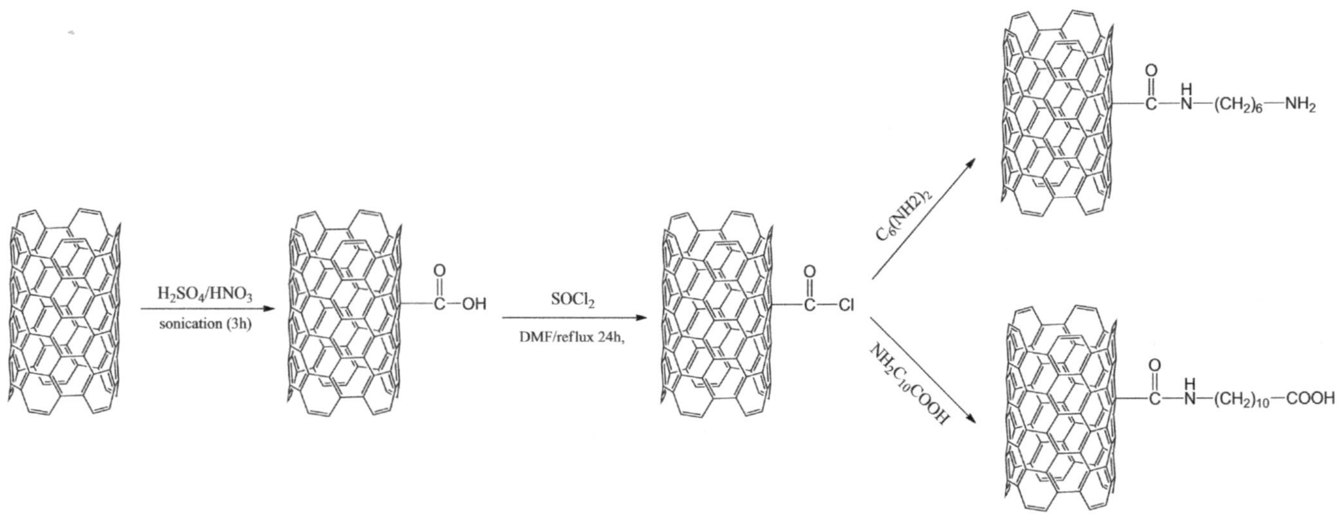

Fig. 1 Schematic representation of covalent functionalization of CNTs and further modification with a diamine (up) or an amino-acid (down).

3. The resulting nanomaterial is centrifuged and washed with anhydrous toluene.
4. The sample is dried under vacuum at 50 °C overnight.
5. The modified MWCNTs derivatives react with an excess quantity (50%) of 1,6 hexamethylenediamine using chloroform as solvent, at 80 °C for 3 days.
6. The removal of excess amine is occurred via washings with chloroform and finally with anhydrous tetrahydrofuran.
7. The amine-terminated MWCNTs are left to dry at room temperature in vacuum and then collected as black powder and stored until used.

2.2.5.3 Modification of MWCNTs with amino-acids

1. A suspension of 165 mg of MWCNTs-COOH in a mixture consisting of 20:1 v/v of thionyl chloride (126 mL) and *N,N*-dimethyl formamide (6 mL) is refluxed at 70 °C for 24 h.
2. The alcyl-chlorinated MWCNTs are centrifuged.
3. The sample is dried under vacuum at 50 °C for 1 day.
4. The modified MWCNTs are dispersed in chloroform.
5. 50 mg of 11-aminoundecanoic acid are dissolved in 40 mL chloroform, 40 mL ethanol and 20 mL acetic acid and are added in the above dispersion.
6. The mixture is refluxed at 80 °C for 3 days.
7. The excess amino-acid is removed from the solution via centrifugations and washings with tetrahydrofuran and chloroform.
8. The sample is left to dry at room temperature.

TIP 1: For the maximum solubility of 11-aminoundecanoic acid the temperature is increased at 50 °C.

TIP 2: The following diamines and amino-acids can also be employed for the surface modification of MWCNTs: $C_nH_{2n}(NH_2)_2$ (n=2, 4, 6), and $H_2N(CH_2)_n$ COOH (n=2, 4, 10).

3. Interactions between CNTs and proteins

3.1 Effect of functionalized CNTs on cytochrome *c* structure

The understanding of the interactions between CNTs and enzymes is of crucial importance in order to develop efficient nanobiocatalytic systems. Enzymes can interact with CNTs via different ways; (1) π–π stacking between the aromatic amino-acids present on the surface of the enzymes and the polyaromatic sidewalls of CNTs, (2) hydrophobic interactions between hydrophobic amino-acids on enzymes' surface and the hydrophobic walls of CNTs, and (3) electrostatic interactions between charged amino-acids of the enzyme and charged functional groups on CNTs surface (Feng & Ji, 2011).

CNTs can strongly interact with enzymes and alter their structure, thus affecting their catalytic activity, either through occupation of their active sites inhibiting their activities, or through modification of their conformational state, namely secondary and/or tertiary structure (Ganazzoli & Raffaini, 2019). Investigating the structure and function of immobilized enzymes on CNTs is crucial for obtaining a better understanding of nanomaterials–enzyme interactions, as well as for designing functional nanobiocatalysts. For instance, our group has previously demonstrated that functionalized CNTs with different alkyl chains and terminal active groups increase the activity of cytochrome *c*, due to their high hydrophobicity, while they preserve its structure (Patila et al., 2013). On the other hand, fCNTs have been reported to cause a slight decrease of the α-helical content with an equal increase of the β-sheet content of lipase, resulting in a more rigid structure of the enzyme, and thus enhanced stability (Pavlidis, Vorhaben, Gournis, et al., 2012; Verma, Naebe, Barrow, & Puri, 2013). It is obvious, that the details of the CNTs–enzyme interactions in a molecular level are still challenging due to the complexity the biomolecules present (Saifuddin et al., 2013). In this direction, many spectroscopic techniques have been used to provide insights into the interactions-structure correlation. UV-visible spectroscopy (UV-vis), fluorescence spectroscopy and circular dichroism spectroscopy (CD) can be applied for the detection of the effect of CNTs on the structural characteristics of enzymes.

UV-vis spectroscopy can provide information about alterations in the microenvironment of a protein. Hence, the extent of retention of the conformational state of cytochrome *c* (cyt *c*) in the presence of fCNTs is measured by UV-vis. Cyt *c* is a small protein found in mitochondria and has a heme moiety attached to its peptide backbone. Due to the heme iron, cyt *c* exhibits a characteristic spectrum in the UV-vis region. The absorbance spectrum of cyt *c* consists of three major peaks, which differ according to the iron oxidative state; (a) a high, sharp main Soret band (350–490 nm), which is attributed to π–π^* transitions and reflects the oxidative state of heme iron, as well as the spin of iron, (b) a smaller, broader Q band (520–550), which gives information about the coordination of the heme iron with its ligands, and (c) a weaker transfer charge band (695 nm), which corresponds to the

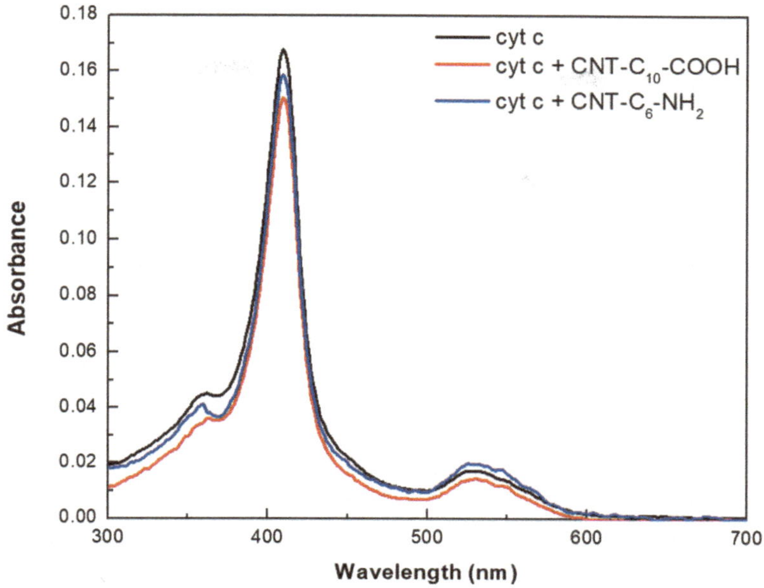

Fig. 2 UV-vis spectra of cyt *c* in the presence of 50 μg mL^{-1} fCNTs.

stability of the Fe-Met80 bond (Prasad, Maiti, Mazumdar, & Mitra, 2002). Changes to the above peaks entail changes in heme microenvironment (Miller, Dunn, Valentine, & Zink, 1996). The effect of fCNTs on the heme microenvironment of cyt *c* was investigated by monitoring the absorbance in the UV-vis region (300–700 nm). As seen in Fig. 2, two peaks appeared in the spectrum of cyt *c*; a big, sharp Soret band at 409 nm and a weaker, broad Q band at 537 nm, arising from electronic transitions of the porphyrin chromophore and corresponding to the low-spin six-coordinated ferric heme (Ray et al., 2011). Functionalized CNTs did not seem to remarkably alter the spectrum of cyt *c* as no shift of the maximum absorbance was observed, indicating that cyt *c* preserves its tertiary structure in the presence of functionalized CNTs.

Circular dichroism (CD) spectroscopy is a valuable tool, especially for recording changes in the conformation state of a protein. The chirality of natural amino-acids and their arrangement in the protein secondary structure result in characteristic circular dichroism spectra (Greenfield, 2006; Kelly, Jess, & Price, 2005). CD can be used to study how the secondary structure of a protein changes as a function of the concentration of agents, for example, carbon nanotubes. The far-UV CD spectrum of proteins can reveal important information about their secondary composition.

CD spectra can be readily used to estimate the fraction of a molecule that is in the α-helix, the β-sheet, the β-turn, or some other (e.g., random coil) conformation (Pelton & McLean, 2000; Whitmore & Wallace, 2008). Each secondary element exhibits a characteristic peak in the far-UV CD spectrum. In a typical CD spectrum of a protein in the far-UV range (190–260 nm), two negative peaks at 208 and 222 nm correspond to α-helixes, while β-sheets are indicated by a single, broad, negative peak at 212 nm, and random coils are indicated by a sharp negative peak at 195 nm (Pelton & McLean, 2000). In our work, CD spectroscopy was used for the detection of changes in the structure of cyt *c* upon addition of fCNTs. As seen from Fig. 3, cyt *c* in the absence of nanomaterials shows a typical α-helical profile, with two negative maxima at 222 and 208 nm. When functionalized CNTs are added in the protein solution, no significant changes in the cyt *c* spectra are observed. This result indicates that fCNTs do not disorder the conformational state of cyt *c* (Patila et al., 2013). CNT-C_6-NH_2 seems to alter in a higher extent the structure of cyt *c*, compared to CNT-C_{10}-COOH. The amine-functionalized CNTs have a smaller alkyl chain than CNT-C_{10}-COOH, which makes them more hydrophilic, and thus it is expected to interact with the water which is essential for the correct folding of cyt *c*, altering its conformational state.

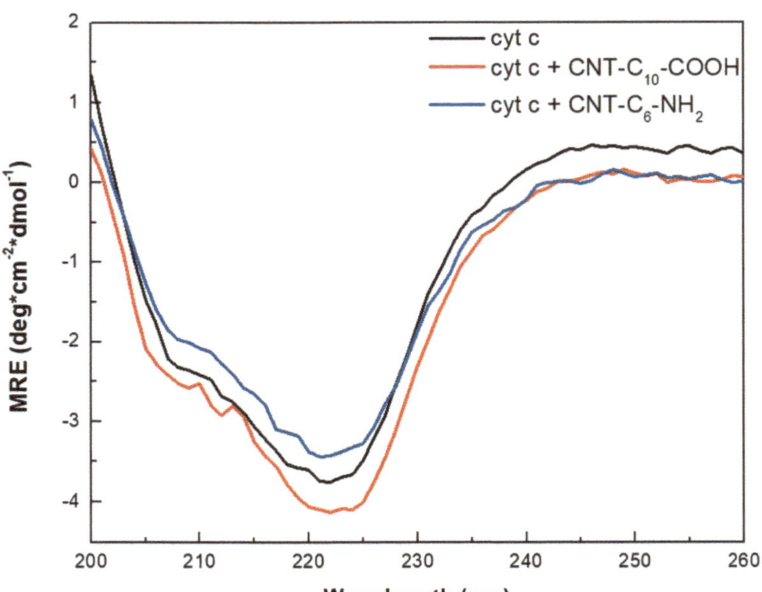

Fig. 3 Far-UV CD spectra of cyt *c* in the presence of 5 μg mL^{-1} fCNTs.

3.1.1 General safety

Nitrile gloves, protective mask

3.1.2 Equipment

1. Eppendorf tubes (1.5 mL)
2. Micropipettes (0.1–1 mL)
3. Double beam UV-1601 spectrophotometer (Shimadzu, Japan)
4. J-1500 CD spectrometer (Jasco, Japan)
5. Ultrasonic bath

3.1.3 Chemicals and materials

1. CNTs functionalized with terminal amine or carboxyl groups (fCNTs).
2. Cytochrome *c* from equine heart (cyt *c*) from Sigma Aldrich Co., MO, USA.
3. Phosphate buffer (0.5 mM, pH 7.0).
4. Phosphate buffer (0.05 mM, pH 7.0).

3.1.4 Methodology

UV-vis spectroscopy is applied to investigate alterations in the heme microenvironment of cyt *c*, while CD spectroscopy is applied to observe changes in the secondary structure of the protein.

3.1.5 Protocol

3.1.5.1 UV-Vis measurements

1. Suspend cyt *c* in 1 mL of 0.5 mM phosphate buffer at pH 7.0 (to reach a final concentration of 10 μg mL^{-1}), containing 50 μg mL^{-1} of fCNTs.
2. Collect the UV-vis spectra of the samples in the region of 300–600 nm, using a 1.0-cm path length quartz cuvette, at a medium scan rate, with a step resolution of 1.0 nm/data point.
3. Accumulate three scans for each sample and record the average spectrum.
4. Plot the samples on the same panel to identify even minor spectral changes due to the presence of fCNTs.

3.1.5.2 CD measurements

1. Suspend cyt *c* in 1 mL of 0.05 mM phosphate buffer at pH 7.0 (to reach a final concentration of 10 μg mL^{-1}), containing a nanomaterial concentration of 5 μg mL^{-1}.
2. Collect the CD spectra of the samples in the far-UV region (200–250 nm), using a 1.0-cm path length quartz cuvette, at a scan rate

of 20 nm min^{-1} with a step resolution of 1.0 nm/data point. The bandwidth and sensitivity are set at 1 nm and 50 millidegrees, respectively.
3. Accumulate three scans for each sample and record the average spectrum for each sample, in order the ratio of signal-to-noise is achieved.
4. Plot the samples on the same panel to identify even minor spectral changes due to the presence of fCNTs.

TIP 1: Before sample measurement, a blank measurement is always required to be recorded, to test if the nanomaterials contribute to the observed spectra.

TIP 2: The absorbance of the samples should be <1.0 in the cuvette.

TIP 3: Samples under above conditions give good quality spectra, but more scans may be required for some samples.

TIP 4: Before usage, the nanomaterial solution should be sonicated for approximately 5 min, till full dispersability is achieved.

TIP 5: For the determination of the transfer band at 695 nm in the UV-vis spectra, a higher (10-fold) concentration of cyt *c* must be used.

TIP 6: For CD measurements the buffer used should be of low salt concentration to avoid undesired absorbance.

TIP 7: CD can also be recorded at the near-UV and Soret regions (280–350 nm and 350–450 nm, respectively) which correspond to the aromatic region and heme region, respectively (Ahluwalia, Nayeem, & Deep, 2011), but a higher protein concentration (up to 10-fold) should be used.

TIP 8: CD results can be further analyzed to give some information about the secondary element composition of a protein. Through the determination of mean residue ellipticity (MRE) and the use of different algorithms, the secondary elements of the protein can be calculated (Pavlidis, Vorhaben, Gournis, et al., 2012).

TIP 9: Fluorescence spectroscopy can also be used for recording changes in the tertiary structure of proteins, upon nanomaterials addition. Phenylalanine, tyrosine and tryptophan residues present in a protein account for the fluorescent properties of the biomolecule.

3.2 Immobilization of enzymes on functionalized carbon nanotubes

Enzyme immobilization is a widely applied technology which results in the improved stability against denaturant factors such as organic solvents, high-temperature and pressure, extreme pH values and reaction time (Kim, Grate, & Wang, 2006). Especially in industrial applications, these advantages are critical for the development of robust biocatalysts with high stability, recovery and reusability, reducing this way the operational costs. Nanomaterials can serve as excellent supports for enzyme immobilization, stabilizing the biocatalysts against non-biological conditions. For instance, immobilized enzymes on magnetic nanomaterials show improved thermal stability and they can be easily isolated and reused by applying an external magnetic field (Orfanakis et al., 2018).

Carbon nanotubes are ideal supports for enzyme immobilization compared to other nanomaterials, as they are more stable under harsh conditions and provide high enzyme loading, due to their high surface area. Also, they are able to reduce the diffusion limitations and create a biocompatible microenvironment that promotes the retention of the catalytic properties of the enzymes. Immobilization on fCNTs can be carried out via two different methods: physical adsorption and covalent binding. Non-specific binding through physical adsorption is an easy process that can be conducted under mild conditions, and may increase the immobilization efficiency, leaving undisrupted the surface of the nanocarrier (Patila, Orfanakis, Polydera, Pavlidis, & Stamatis, 2017). Physical adsorption is governed by weak interactions between the enzyme and CNTs, such as van der Waals interactions, hydrophobic and electrostatic interactions, as well as π–π stacking interactions and hydrogen bonding (Patila et al., 2017; Pavlidis, Patila, Polydera, et al., 2014; Zhang, Yuwen, & Peng, 2013). Immobilization of enzymes through non-covalent attachment is based on a simple mixing of the enzyme solution with the fCNTs dispersion and incubation for a predominant time. Although physical adsorption is often preferred due to the easy preparation of the nanobiocatalyst and the fact that it does not significantly alter the conformation state of the enzyme, the protein can be gradually detached from the nanosupport.

Covalent immobilization addresses this drawback, by the formation of a strong covalent bond between functional groups of the nanomaterial and functional groups of the amino-acids in the surface of the enzyme, leading to a robust and stable nanobiocatalyst (Stavyiannoudaki, Vamvakaki, & Chaniotakis, 2009). Chemical modification of CNTs by adding various functional groups to their surface facilitates the formation of bonds between the nanomaterials and the enzymes. This approach requires the presence of a chemical agent acting as a bridge between the terminal functional groups of the modified nanomaterial and the side chains of the amino-acids found on the surface of the enzyme.

One of the most common approaches applied to carboxyl-functionalized CNTs is based on the use of a carbodiimide, such as 1-ethyl-3-(3-

dimethylaminopropyl) carbodiimide (1-ethyl-3-(3-dimethylaminopropyl) carbodiimide, EDC), and N-hydroxysuccinimide (NHS) (Gao & Kyratzis, 2008). The coupling reaction which takes place consists of two consecutive steps (Fig. 4). In a first step, EDC reacts with the terminal carboxyl group of fCNTs, to form an *O*-acylisourea intermediate. This intermediated is very unstable and sensitive to hydrolysis. For this reason, the addition of NHS in a second step is important, as it stabilizes the *O*-acylisourea intermediate by converting it to a semi-stable NHS-ester. This product can further react with the free amine groups of lysine (mainly) residues on the enzyme surface, increasing the coupling efficiency. Covalent immobilization using EDC/NHS has been applied to immobilize a number of enzymes and proteins, indicatively the case of tyrosinase (Subrizi et al., 2014), lipase (Raghavendra, Basak, Manocha, Shah, & Madamwar, 2013), and organophosphorus hydrolase (Pedrosa et al., 2010).

An alternative approach to covalent immobilization is the cross-linking of the terminal amine groups present on the fCNTs and the free amine groups present on the enzyme's surface. The most common compound used for this purpose is glutaraldehyde. During the cross-linking process, the free amine groups of fCNTs are activated by glutaraldehyde and then the free keto- groups are cross-linked with the amine groups of the enzyme via the formation of a Schiff's base (Fig. 5). Covalent immobilization using glutaraldehyde, although not particularly preferred, has been applied for immobilization of glucose oxidase (Kim, Lee, et al., 2017) and lipase on ethylenediamine-functionalized CNTs (Verma et al., 2013).

3.2.1 General safety

Nitrile gloves, protective mask

3.2.2 Equipment

1. Falcon tubes (15 mL)
2. Micropipettes (0.1–1 mL)
3. Incubator with stirring
4. Ultrasonic bath
5. Centrifuge

3.2.3 Chemicals and materials

1. CNTs functionalized with terminal amine or carboxyl groups (fCNTs).
2. Laccase from *Trametes versicolor* (TvL) from Sigma Aldrich Co., MO, USA.

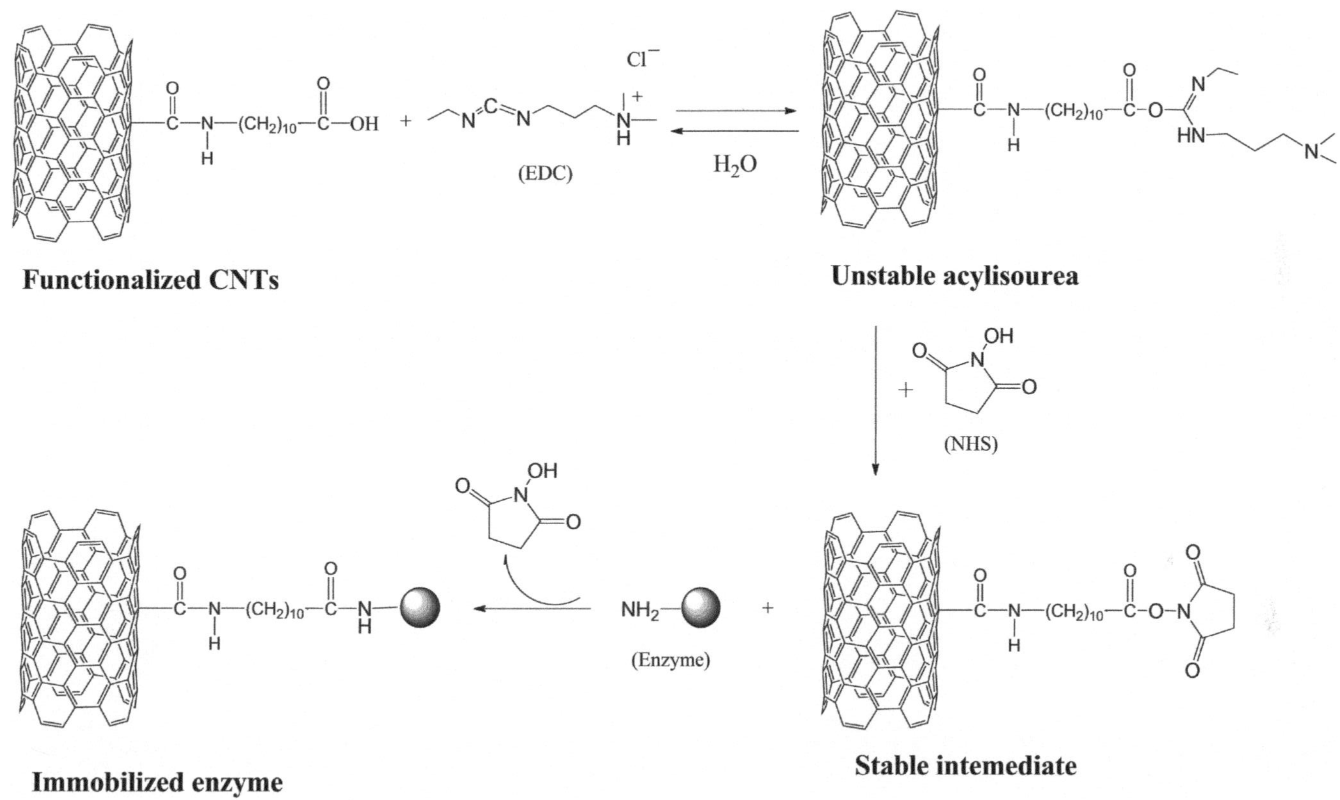

Fig. 4 Covalent immobilization of enzymes on fCNTs bearing free carboxyl groups, through the carbodiimide chemistry.

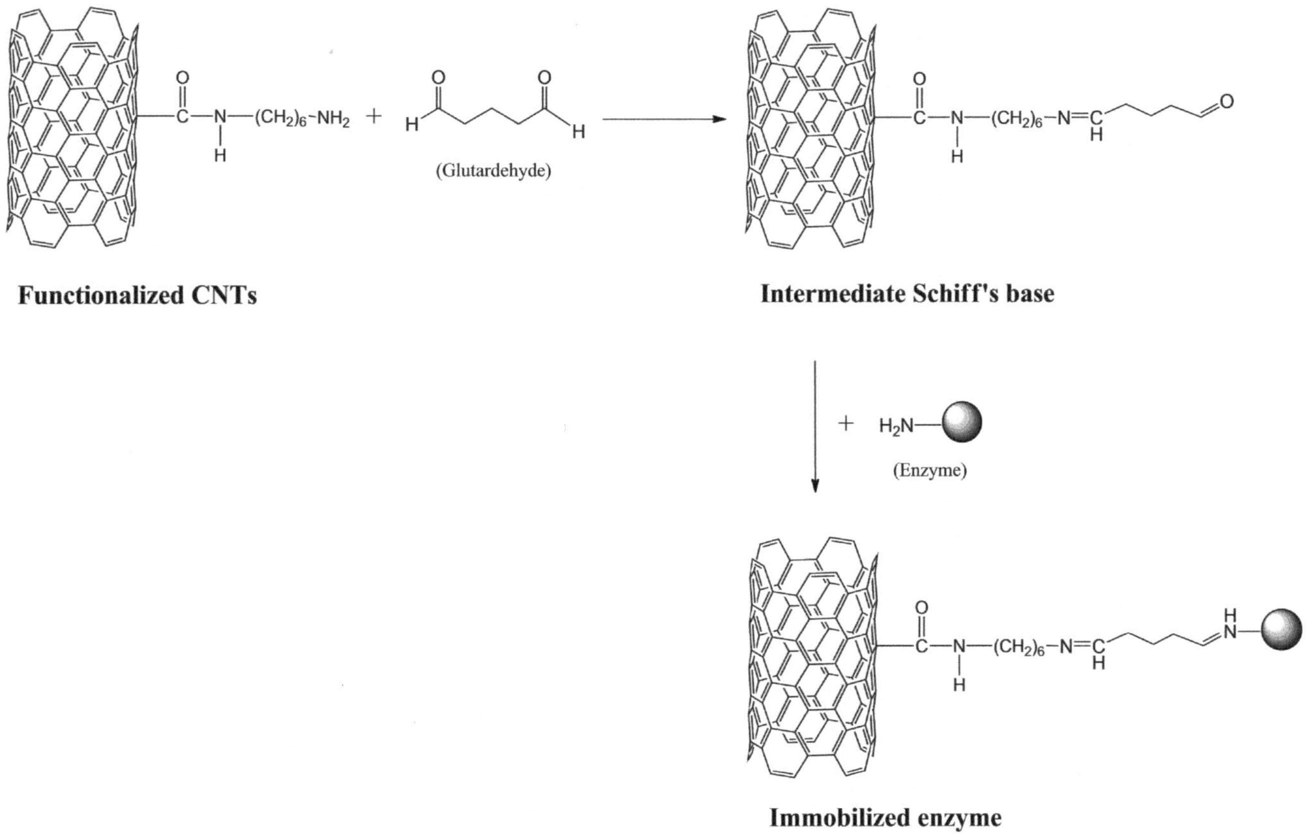

Fig. 5 Covalent immobilization of enzymes on fCNTs bearing free amine groups, through cross-linking with glutaraldehyde.

3. Cytochrome *c* from equine heart (cyt *c*) from Sigma Aldrich Co., MO, USA.
4. 1-ethyl-3-[3-dimethylaminopropyl] carbodiimide hydrochloride (EDC) from Sigma Aldrich Co., MO, USA.
5. N-hydroxysuccinimide (NHS) from Sigma Aldrich Co., MO, USA.
6. Glutaraldehyde solution 25% for electron microscopy from Merck KGaA, Darmstadt, Germany.
7. Acetate buffer (0.1 M, pH 4.58).
8. Phosphate buffer (0.05 M, pH 7.0).
9. 4-(2-hydroxyethyl)piperazine-1-ethanesulfonic acid (HEPES) buffer (0.1 M, pH 7.0 and pH 4.58).
10. HPLC water.

3.2.4 Methodology

Non-covalent immobilization of cyt *c* and TvL is performed by simple mixing of the nanomaterials with the enzymes. Covalent immobilization of cyt *c* and TvL is performed using carbodiimide chemistry in the case of carboxyl-functionalized CNTs, while covalent immobilization of cyt *c* and TvL on amine-functionalized CNTs is achieved using glutaraldehyde as the cross-linking agent.

3.2.5 Protocol

3.2.5.1 Non-covalent immobilization of cyt *c* or TvL on fCNTs

1. 3 mg of fCNTs are weighted in a falcon tube.
2. 5 mL of phosphate or acetate buffer (for cyt *c* or TvL immobilization, respectively) are added to fCNTs and the solution is left at an ultrasonic bath for 30 min.
3. 3 mg of cyt *c* or TvL are suspended in 1 mL phosphate or acetate buffer, respectively.
4. The protein preparation is transferred to the fCNTs dispersion and the mixture is incubated for 1 h at 30 °C.
5. The immobilized nanobiocatalyst is separated by centrifugation at 6000 rpm, followed by washing with phosphate or acetate buffer, thrice, in order to remove loosely bound protein (collect 1 mL of the supernatant for the determination of the immobilization yield).
6. The immobilized protein is dried over silica gel and stored at 4 °C until used.

3.2.5.2 Covalent immobilization of cyt *c* or TvL on fCNTs with terminal carboxyl groups

1. 3 mg of fCNTs are weighted in a falcon tube.
2. 5 mL of HPLC water are added to fCNTs and the solution is left at an ultrasonic bath for 30 min.
3. 1 mL of HEPES buffer and 1.2 mL of a 10 mg mL^{-1} EDC aqueous solution are added to the fCNTs dispersion.
4. Under fast stirring, 2.3 mL of a 50 mg mL^{-1} NHS aqueous solution are added quickly, and the mixture is incubated for 30 min at 30 °C.
5. The EDC/NHS excess is removed from the solution by centrifugation at 6000 rpm, followed by washing with HEPES buffer, thrice.
6. The activated fCNTs are re-dispersed in 5 mL of HEPES buffer.
7. 3 mg of cyt *c* or TvL are suspended in 1 mL HEPES buffer.
8. The protein preparation is transferred to the fCNTs dispersion and the mixture is incubated for 1 h at 30 °C.
9. The immobilized nanobiocatalyst is separated by centrifugation at 6000 rpm, followed by washing with HEPES buffer, thrice, in order to remove loosely bound protein (collect 1 mL of the supernatant for the determination of the immobilization yield).
10. The immobilized protein is dried over silica gel and stored at 4 °C until used.

3.2.5.3 Covalent immobilization of cyt *c* or TvL on CNTs with terminal amine groups

1. 3 mg of fCNTs are weighted in a falcon tube.
2. 9.24 mL phosphate or acetate buffer (for cyt *c* or TvL immobilization, respectively) are added to fCNTs and the solution is left at an ultrasonic bath for 30 min.
3. 1.76 mL of glutaraldehyde is added to the fCNTs dispersion and the mixture is incubated for 30 min at 30 °C.
4. The glutaraldehyde excess is removed from the solution by centrifugation at 6000 rpm, followed by washing with phosphate or acetate buffer (for cyt *c* or TvL immobilization, respectively), thrice.
5. The activated fCNTs are re-dispersed in 5 mL with phosphate or acetate buffer (for cyt *c* or TvL immobilization, respectively).
6. 3 mg of cyt *c* or TvL are suspended in 1 mL phosphate or acetate buffer, respectively.
7. The protein preparation is transferred to the fCNTs dispersion and the mixture is incubated for 1 h at 30 °C.

8. The immobilized nanobiocatalyst is separated by centrifugation at 6000 rpm, followed by washing with phosphate or acetate buffer, thrice, in order to remove loosely bound protein (collect 1 mL of the supernatant for the determination of the immobilization yield).
9. The immobilized protein is dried over silica gel and stored at 4 °C until used.

TIP 1: The fCNTs should be well-dispersed during the immobilization procedure, in order to achieve the highest surface area available for high enzyme loading.

TIP 2: Glutaraldehyde, due to its high toxicity, should be used in a fume hood.

4. Characterization of CNTs-enzyme conjugates

4.1 Determination of immobilization yield

The determination of the immobilization yield of an immobilization procedure is a crucial step for evaluation of the effective attachment of enzymes on nanosupports. The efficient immobilization of enzymes on fCNTs depends on different parameters. For instance, in case of physical adsorption, the specific surface area of CNTs is an important factor that influences the adsorption amount of enzyme (Zhang et al., 2013). Moreover, the amount of adsorbed enzyme is susceptible to the immobilization parameters such as pH, temperature, ionic strength, and incubation time. Respectively, in the case of covalent immobilization, the key parameters that affect the enzyme loading are the number of functional groups on the surface of fCNTs available for the formation of a covalent bond and the cross-linking agents that bind the enzyme on the nanomaterials.

The determination of the immobilization yield usually involves a quantification method, where the un-reacted enzyme concentration is determined. Bradford assay has been widely applied for protein quantification in a solution (Bradford, 1976). The amount of a protein bound on a nanosupport can be estimated by calculating the difference between the amount of the protein in the supernatant after immobilization and the initial protein amount used. Except for protein quantification methods, thermogravimetric analysis (TGA) can also be used for estimating the amount of the enzyme attached onto the nanomaterials. The method is based on the appearance of exothermal peaks due to decomposition and combustion of the sample (Patila, Pavlidis, et al., 2016).

Table 1 Immobilization yield (%) of cyt *c* and TvL on fCNTs.

	Immobilization yield (%)	
Nanomaterial	**Cyt *c***	**TvL**
CNT-C_6-NH_2	25	68
CNT-C_{10}-COOH	60	83

The immobilization efficiency of covalently immobilized cyt *c* and TvL onto functionalized CNTs is presented in Table 1. Both proteins are able to efficiently bind to the nanomaterials. However, the immobilization yield strongly depends on the functionalization of CNTs, as well as on the protein preparation. Comparing the two different protein preparations, higher immobilization yields are observed in the case of TvL. This result could render from the different surface characteristics these proteins present and allow them to interact and bind in different ways to the nanosupports. The functionalization of CNTs also plays a critical role for the successful grafting of the proteins on the nanomaterials. Both cyt *c* and TvL show higher immobilization efficiency when carboxyl-functionalized CNTs are used. CNT-C_{10}-COOH has been functionalized by using a longer organic molecule (aminoundecanoic acid) than CNT-C_6-NH_2 (hexamethylenediamine), which is expected to increase the distance between the protein and the nanomaterials' surface and thus to reduce the undesired interactions upon immobilization, leading to the formation of stable covalent bonds (Patila et al., 2018; Patila, Kouloumpis, et al., 2016).

4.1.1 General safety

Nitrile gloves, protective mask

4.1.2 Equipment

1. Eppendorf tubes (1.5 mL)
2. Micropipettes (2–20 μL, 20–200 μL, 100–1000 μL)
3. Pipette tips
4. Double beam UV-1601 spectrophotometer (Shimadzu, Japan)

4.1.3 Chemicals and materials

1. Bradford reagent (Coomassie Brilliant Blue G-250, from Sigma).
2. Milli-Q SP ultrapure water.

4.1.4 Methodology

The amount of immobilized cyt *c* and TvL is determined by calculating the protein concentration in the supernatant after the immobilization procedure using the Bradford method (Bradford, 1976). Loading was estimated as the difference between the amount of protein in the supernatant after immobilization and the amount of initial protein.

4.1.5 Protocol

1. 0.2 mL of ultrapure water and suitable amount of the protein solution are added in 0.8 mL of Bradford reagent, and are mixed gently.
2. The samples are incubated at room temperature, in dark, for approximately 15 min.
3. The absorbance of the samples is measured at 595 nm.
4. The observed absorptions are quantified by applying a standard albumin curve (1–20 $\mu g\,mL^{-1}$ in the measuring solution). The % immobilization yield is expressed as:

$$\text{Immobilization yield (\%)} = \frac{\text{Amount of initial enzyme} - \text{Amount of unbound enzyme}}{\text{Amount of initial enzyme}} \times 100$$

4.2 Determination of free and immobilized enzyme activity

The determination of the catalytic activity of an immobilized biocatalyst, in comparison with the native protein form, provides insights on the catalytic performance of the enzyme after the attachment on a support. The activity of the immobilized enzyme can be affected by mass transfer limitations, due to low substrate accessibility to its active site. Moreover, pH, temperature, immobilization time and protocol, surface coverage, and surface curvatures and chemistries can substantially impact the activity of enzymes (Arsalan & Younus, 2018; Guzik, Hupert-Kocurek, & Wojcieszynska, 2014).

The type of immobilization procedure also affects the catalytic characteristics of the resulted nanobiocatalyst. Non-specific binding prevails among the immobilization procedures, as the final biocatalyst usually preserves its molecular structure and catalytic activity. On the contrary, covalent immobilization may affect the structure of the bound enzyme, decreasing its catalytic efficiency in a higher extent. The specific activity of non-covalently and covalently immobilized TvL onto fCNTs is presented in Table 2. Immobilized TvL on fCNTs by non-specific binding exhibits

Table 2 Activity of immobilized and TvL on fCNTs.

	Activity ($U\,\mu g^{-1}$ immobilized TvL)	
Nanomaterial	**Non-covalent**	**Covalent**
CNT-C_6-NH_2	0.31	0.25
CNT-C_{10}-COOH	0.85	0.70

higher catalytic activity compared to the covalently immobilized enzyme, indicating that the formation of a stable chemical bond may cause changes in the native conformational state of the enzyme during the immobilization procedure, which in turn may lead to a lower catalytic activity. The functionalization of CNTs also plays a critical role in the catalytic activity of TvL. Both in covalent and non-covalent procedure, TvL exhibits higher catalytic activity when it is immobilized on CNT-C_{10}-COOH, rather than on CNT-C_6-NH_2. When a larger alkyl chain is attached on CNTs, the distance between the enzyme and the nanomaterials increases, preventing strong interactions with the surface of CNTs, as well as enabling the substrate diffusion to the catalytic site of the enzyme, resulting in enhanced activity of the immobilized enzyme (Patila et al., 2018; Patila, Pavlidis, et al., 2016).

4.2.1 General safety

Nitrile gloves, protective mask

4.2.2 Equipment

1. 96-well plates
2. Micropipettes (2–20 μL, 20–200 μL)
3. Pipette tips
4. Elisa reader
5. Ultrasonic bath

4.2.3 Chemicals and materials

1. Laccase from *Trametes versicolor* (TvL) from Sigma Aldrich Co., MO, USA.
2. 2,2′-azino-bis(3-ethylbenzothiazoline-6-sulphonic acid)(ABTS), from Sigma.
3. Sodium acetate buffer (0.1 M, pH 4.58).

4.2.4 Methodology

The activity of immobilized TvL is determined by ABTS oxidation. Oxidation of ABTS results in the production of a cation radical ($ABTS^{+}$) which results in the formation of a dark blue-green color and can be followed at the visible light region. The enzymatic activity depends on the concentration of the produced cation radical and can be determined by measuring the change in the absorbance at 415nm (Wolfenden & Willson, 1982).

4.2.5 Protocol

1. A suitable amount of free or immobilized TvL (dispersed in acetate buffer) is added in 1mM of ABTS dissolved in sodium acetate buffer (0.1M, pH 4.58).
2. The oxidation of ABTS is measured by following the change in the absorbance at 415nm.
3. To evaluate the results, a graph of kinetic data in terms of absorbance at 415nm vs time is created and the initial slope of the graph is used to calculate the initial rates of the reaction.

TIP 1: Use an ultrasonic bath for the well dispersion of the immobilized enzyme before use.

TIP 2: In order to compare different enzyme preparations, the specific activity of the enzyme (Units/mg protein) should be calculated. One Unit is defined as the amount of TvL that oxidizes 1 μmol of ABTS substrate per min.

4.3 Characterization of nanobiocatalysts by Fourier-transform infrared (FTIR) spectroscopy

FTIR is one of the classical methods for analyzing the structure of proteins through which information about their secondary structure can be extracted (Hammes, 2015). Infrared spectra do not contain information about the three-dimensional structure of a protein, but they can be analyzed and used to detect changes in its secondary structure. This analysis results from the characteristic spectrum of proteins arising from the vibrations of the polypeptide chain bonds. The peptide bond shows nine peaks in the infrared absorption spectrum called Amide peaks (I-VII, A and B), due to different vibrational conditions and bonds. The most commonly used regions for analysis of the protein structure are Amide*I*(1600–1700 cm^{-1}) which corresponds to stretching vibrations of C=O, Amide II (1500–1600 cm^{-1}), due to stretching vibrations of N—H and bending vibrations of C—N, and Amide III (1200–1350 cm^{-1}), arising from stretching vibrations of N—H and bending vibrations of C—N (Venyaminov & Kalnin, 1990).

Our group has already reported the use of FTIR spectroscopy to confirm the successful immobilization of cyt *c* on fCNTs (Patila et al., 2018). Both support and protein have a characteristic spectrum rendering from their surface functionalities. The spectrum of fCNTs showed all the characteristic peaks resulted from the oxidation procedure (Georgakilas et al., 2008). The absorption peak observed at $1630\,cm^{-1}$ corresponds to the C=O stretching vibrations of the carboxyl groups of fCNTs. The peak presented at $1383\,cm^{-1}$ arises from modulation vibrations of the hydroxyl groups, while stretching vibrations of C—O bonds and bending vibrations of C=O bonds were reflected to the presence of two bands at 1060 and $815\,cm^{-1}$, respectively. The spectrum of immobilized cyt *c* showed the characteristic peaks corresponding to Amide I, Amide II and Amide III bands; a peak at $1650\,cm^{-1}$ associated with the C=O stretching vibration and directly related to the backbone conformation, a peak at $1550\,cm^{-1}$ which renders from the N—H vibration and the C—N stretching vibration, and a peak at $1412\,cm^{-1}$ resulting from a mixture of several coordinate displacements.

4.3.1 General safety

Nitrile gloves, protective mask

4.3.2 Equipment

1. Hand press (PIKE technologies, USA).
2. FTIR-8400 infrared spectrometer (Jasco, Japan).

4.3.3 Chemicals and materials

1. Potassium bromide (KBr) from Sigma Aldrich Co., MO, USA.
2. Immobilized cyt *c* on fCNTs-derivatives.

4.3.4 Methodology

FTIR was used to confirm the successful immobilization of cyt *c* on fCNTs, due to the presence of the Amide bands in the recorded spectra.

4.3.5 Protocol

1. Prepare the pellets for the recording of FTIR spectra, by mixing an amount of KBr with the immobilized enzyme.
2. Collect the FTIR spectra of the samples in the 400–4000 cm^{-1} region, using a $4.0\,cm^{-1}$ resolution and $1.0\,cm^{-1}$ interval.
3. Accumulate 32 scans to obtain a good signal-to-noise ratio.
4. Plot the samples on the same panel to identify the presence of the characteristics peaks of nanomaterials and protein.

TIP 1: Data analysis and band assignment can provide further information about the secondary element composition of a protein (Kong & Yu, 2007). The deconvolution of the Amide I band, followed by curve fitting with a Gaussian function, results in the determination of the secondary structural components of the protein (Patila, Pavlidis, et al., 2016; Serefoglou et al., 2008; Tzialla et al., 2010).

TIP 2: Similarities between the secondary structure of an enzyme before and after immobilization can be determined by comparing their 2s-derivative spectra, and calculate the simple correlation coefficient, *r* (Prestrelski, Tedeschi, Arakawa, & Carpenter, 1993).

5. Stability and reusability of nanobiocatalysts

5.1 Determination of the stability of immobilized enzymes

One of the main benefits of the usage of immobilized enzymes, is the enhanced stability they exhibit, under both storage and operational conditions, e.g., toward denaturation by heat or organic solvents. The enhanced stability improves enzyme performance which leads to higher catalytic productivities for economic viability. Depending on the immobilization procedure and the support used, enzymes may preserve their activity in a high extent (Pavlidis, Patila, Bornscheuer, et al., 2014; Pavlidis, Patila, Polydera, et al., 2014). For instance, carbon-based nanomaterials such as graphene oxide and carbon nanotubes have been shown to protect the conformational state of proteins against thermal and organic solvent denaturation, leading to high stability preservation (Patila, Kouloumpis, et al., 2016).

In our study, functionalized CNTs with carboxyl and amine groups exhibited high potential in TvL stabilization (Fig. 6). The stability of covalently immobilized TvL on CNT-C_{10}-COOH and CNT-C_6-NH_2 was significantly higher compared to free protein. For instance, after 24 h of incubation, free TvL is almost deactivated, while immobilized TvL on CNT-C_{10}-COOH and CNT-C_6-NH_2 retains up to 17% and 31%, respectively of its initial activity. The results indicate that fCNTs preserve TvL from thermal denaturation, maintaining the stability of the protein. The increased stability of the immobilized TvL could be attributed to a reduction of the protein structure mobility due to the covalent attachment on fCNTs, preventing the enzyme form denaturation (Taqieddin & Amiji, 2004). It has been previously reported by our group that CNTs-derivatives protect proteins like cytochrome *c* from denaturing conditions (Patila et al., 2018).

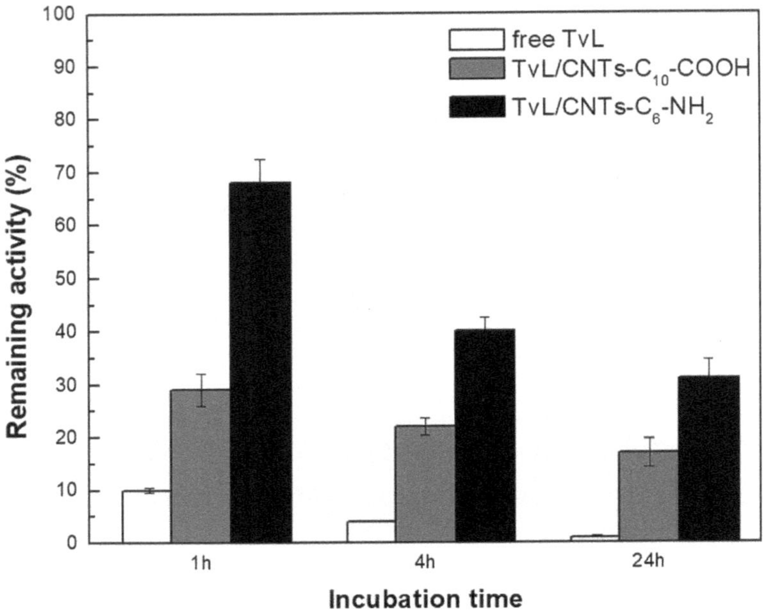

Fig. 6 Stability of free and covalently immobilized TvL on fCNTs.

5.1.1 General safety

Nitrile gloves, protective mask

5.1.2 Equipment

1. 96-well plates
2. Micropipettes (2–20 μL, 20–200 μL, 100–1000 μL)
3. Pipette tips
4. Eppendorf tubes (1.5 mL)
5. Thermoshaker
6. Elisa Reader
7. Ultrasonic bath

5.1.3 Chemicals and materials

1. Laccase from *T. versicolor* (TvL) from Sigma Aldrich Co., MO, USA.
2. 2,2′-azino-bis(3-ethylbenzothiazoline-6-sulphonic acid) (ABTS), from Sigma.
3. Sodium acetate buffer (0.1 M, pH 4.58).
4. Immobilized TvL on fCNTs.

5.1.4 Methodology

The thermal stability of immobilized TvL at different temperatures is determined by calculating the remaining oxidation activity of laccase through the ABTS oxidation (Patila, Kouloumpis, et al., 2016).

5.1.5 Protocol

1. Enzyme solutions are prepared with a final concentration of 50 $\mu g\,mL^{-1}$ for free and immobilized TvL on fCNTs, in acetate buffer (0.1 M, pH 4.58).
2. The enzyme solutions are incubated at 60 °C, up to 24 h.
3. A specimen at predetermined time intervals is sampled in order to determine the remaining oxidation activity of laccase through the ABTS oxidation as described before.

TIP: Before taking a sample from the immobilized enzyme solution, sonication for approximately 1 min is recommended to achieve high dispersability.

5.2 Determination of the reusability of immobilized TvL

A crucial step for the effective use of enzymes in industrial applications is their ability of repeatable use in several reaction cycles. As soluble enzymes are unable to be isolated from the reaction media and further used, immobilization of proteins on nanosupports overcomes this problem, offering the potential of stabilization, easy separation and reuse of the biocatalysts. The choice of the carrier is of high importance, as each supporting material affects enzyme stabilization and conformational state in a distinctive way. The reusability of the immobilized enzymes can be determined by using different substrates, depending on the enzyme catalytic characteristics and the application of interest. The experimental procedure is simple and the biocatalyst can be recovered either with centrifugation or by applying an external magnetic field, depending on the properties of the support.

The operational stability of immobilized TvL on fCNTs was investigated through ABTS oxidation. Laccase covalently immobilized on fCNTs was found to be able to oxidize ABTS up to eight reaction cycles (Fig. 7). Immobilized TvL on CNT-C_{10}-COOH and CNT-C_6-NH_2 retains 82% and 67%, respectively, of its initial activity after five catalytic cycles, indicating that immobilization on fCNT results in stable and reusable nanobiocatalysts. By comparing the two supports it can be observed that immobilized TvL on CNT-C_{10}-COOH is more stable than immobilized TvL on CNT-C_6-NH_2. This result could be attributed to the different immobilization procedures and the cross-linker used for the covalent attachment of TvL on fCNTs,

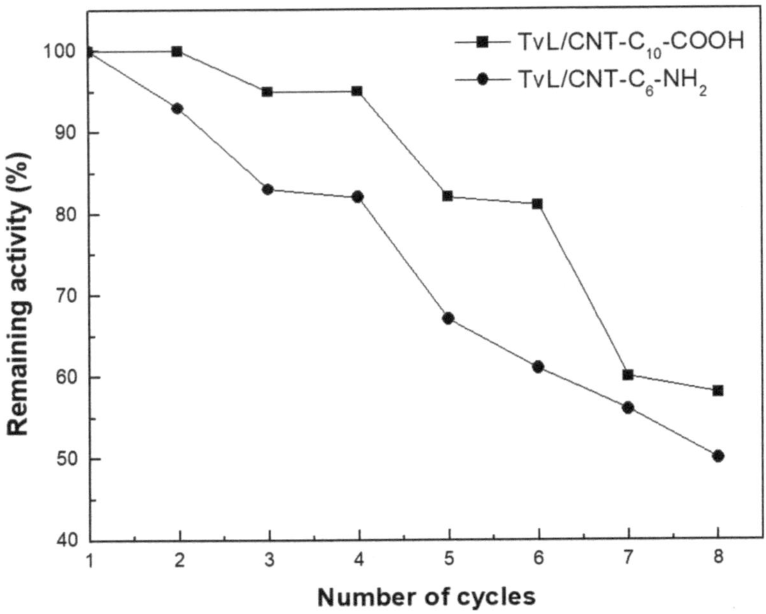

Fig. 7 Reusability of covalently immobilized TvL on fCNTs.

indicating that a zero-length cross-linker such as EDC could be more efficient for such applications compared with a lengthy linker, such as glutaraldehyde (Raghavendra et al., 2013).

5.2.1 General safety

Nitrile gloves, protective mask

5.2.2 Equipment

1. Micropipettes (2–20 μL, 20–200 μL, 100–1000 μL)
2. Pipette tips
3. Eppendorf tubes (1.5 mL)
4. Double beam UV-1601 spectrophotometer (Shimadzu, Japan)
5. Ultrasonic bath
6. Centrifuge for Eppendorf tubes

5.2.3 Chemicals and materials

1. 2,2′-azino-bis(3-ethylbenzothiazoline-6-sulphonic acid) (ABTS), from Sigma.
2. Sodium acetate buffer (0.1 M, pH 4.58).
3. Immobilized TvL on fCNTs.

5.2.4 Methodology

The reusability of TvL can be determined by measuring the remaining activity of the immobilized enzyme after several catalytic cycles, with respect to its initial activity.

5.2.5 Protocol

1. 1 mL of 1 mM ABTS is added in sodium acetate buffer (0.1 M, pH 4.58) containing 0.5 mg immobilized TvL pre-weighted in an Eppendorf tube.
2. The sample is put in an ultrasonic bath to fully disperse the immobilized enzyme.
3. The reaction medium is transferred to a cuvette and the oxidation of ABTS is measured by following the change in the absorbance at 415 nm.
4. The reaction medium is transferred back to an Eppendorf tube.
5. The immobilized TvL is separated by centrifugation at 12,000 rpm, after each reaction cycle.
6. The precipitate immobilized TvL is washed three times with acetate buffer (0.1 M, pH 4.58).
7. The process (steps 1–6) is repeated until there is no enzymatic activity.
8. A graph of kinetic data in terms of absorbance at 415 nm vs time is created and the initial slope of the graph is used to calculate the initial rates each reaction cycle.

TIP: During centrifugation and washing steps at the end of each catalytic cycle, be sure that the immobilized biocatalyst is well precipitated, in order to maintain the same enzyme concentration.

6. Conclusion

In this chapter, we present the use of functionalized carbon nanotubes for the development of robust nanobiocatalysts. Different routes for the functionalization of CNTs are presented, resulting in the production of CNTs-derivatives with variable alkyl chain lengths and different terminal functional groups. The interactions developed between functionalized CNTs and cytochrome *c*, a small heme protein with biological interest, are also commented, and how these affect the structural characteristics of the protein. Covalent immobilization of cyt *c* and laccase from *T. versicolor* on amine- and carboxyl- functionalized CNTs is also presented. It is interesting to note that immobilization efficiency depends not only on the nature of the protein, but also on the structural characteristics of functionalized CNTs. It is also worth noting that as the length of the alkyl chain of the organic molecule attached to CNTs increases, the immobilization performance also increases, possibly due the increased space between the nanomaterials and the biomolecules, reducing that way the undesired interactions between them. Immobilized TvL on fCNTs demonstrated high thermal stability compared to the native enzyme, as well as good operational stability for the repeated oxidation of ABTS, indicating that functionalized CNTs are efficient supports for the development of robust nanobiocatalysts.

Acknowledgments

This research has been co-financed by the European Regional Development Fund of the European Union and Greek national funds through the Operational Program Competitiveness, Entrepreneurship and Innovation, under the call "Aquaculture"—"Industrial Materials"—"Open Innovation In Culture" (project: Biomalga, project code: T6YBΠ-00033). Nikolaos Chalmpes gratefully acknowledges the IKY Foundation for the financial support: "This research is co-financed by Greece and the European Union (European Social Fund- ESF) through the Operational Program 'Human Resources Development, Education and Lifelong Learning' in the context of the project 'Strengthening Human Resources Research Potential via Doctorate Research' (MIS-5000432), implemented by the State Scholarships Foundation (IKY)." We acknowledge support for this work by the project MIS 5002772, implemented under the Action "Reinforcement of the Research and Innovation Infrastructure," funded by the Operational Program "Competitiveness, Entrepreneurship and Innovation" (NSRF 2014-2020) and co-financed by Greece and the European Union (European Regional Development Fund).

References

Ahluwalia, U., Nayeem, S. M., & Deep, S. (2011). The non-native conformations of cytochrome *c* in sodium dodecyl sulfate and their modulation by ATP. *European Biophysics Journal*, *40*, 259–271.

Arepalli, S. (2004). Laser ablation process for single-walled carbon nanotube production. *Journal of Nanoscience and Nanotechnology*, *4*, 317–325.

Arsalan, A., & Younus, H. (2018). Enzymes and nanoparticles: Modulation of enzymatic activity via nanoparticles. *International Journal of Biological Macromolecules*, *118*, 1833–1847.

Baughman, R. H., Zakhidov, A. A., & De Heer, W. A. (2002). Carbon nanotubes—The route toward applications. *Science*, *297*, 787–792.

Ben Tahar, A., Romdhane, A., Lalaoui, N., Reverdy-Bruas, N., Le Goff, A., Holzinger, M., et al. (2018). Carbon nanotube-based flexible biocathode for enzymatic biofuel cells by spray coating. *Journal of Power Sources*, *408*, 1–6.

Boussema, F., Gross, A. J., Hmida, F., Ayed, B., Majdoub, H., Cosnier, S., et al. (2018). Dawson-type polyoxometalate nanoclusters confined in a carbon nanotube matrix as efficient redox mediators for enzymatic glucose biofuel cell anodes and glucose biosensors. *Biosensors and Bioelectronics*, *109*, 20–26.

Bradford, M. M. (1976). A rapid and sensitive method for the quantitation of microgram quantities of protein utilizing the principle of protein-dye binding. *Analytical Biochemistry*, *72*, 248–254.

Cosnier, S., Holzinger, M., & Le Goff, A. (2014). Recent advances in carbon nanotube-based enzymatic fuel cells. *Frontiers in Bioengineering and Biotechnology*, *2*, 1–6.

De Volder, M. F. L., Tawfick, S. H., Baughman, R. H., & Hart, A. J. (2013). Carbon nanotubes: Present and future commercial applications. *Science*, *339*, 535–539.
De Zhang, W., Shen, L., Phang, I. Y., & Liu, T. (2004). Carbon nanotubes reinforced nylon-6 composite prepared by simple melt-compounding. *Macromolecules*, *37*, 256–259.
Dresselhaus, M. S., Dresselhaus, G., & Saito, R. (1995). Physics of carbon nanotubes. *Carbon*, *33*, 883–891.
Erol, O., Uyan, I., Hatip, M., Yilmaz, C., Tekinay, A. B., & Guler, M. O. (2018). Recent advances in bioactive 1D and 2D carbon nanomaterials for biomedical applications. *Nanomedicine: Nanotechnology, Biology, and Medicine*, *14*, 2433–2454.
Feng, W., & Ji, P. (2011). Enzymes immobilized on carbon nanotubes. *Biotechnology Advances*, *29*, 889–895.
Galano, A. (2010). Carbon nanotubes: Promising agents against free radicals. *Nanoscale*, *2*, 373–380.
Ganazzoli, F., & Raffaini, G. (2019). Classical atomistic simulations of protein adsorption on carbon nanomaterials. *Current Opinion in Colloid and Interface Science*, *41*, 11–26.
Gao, Y., & Kyratzis, I. (2008). Covalent immobilization of proteins on carbon nanotubes using the cross-linker 1-ethyl-3-(3-dimethylaminopropyl)carbodiimide—A critical assessment. *Bioconjugate Chemistry*, *19*, 1945–1950.
Georgakilas, V., Bourlinos, A., Gournis, D., Tsoufis, T., Trapalis, C., Mateo-Alonso, A., et al. (2008). Multipurpose organically modified carbon nanotubes: From functionalization to nanotube composites. *Journal of the American Chemical Society*, *130*, 8733–8740.
Greenfield, N. J. (2006). Using circular dichroism spectra to estimate protein secondary structure. *Nature Protocols*, *1*, 2876–2890.
Guo, T., Nikolaev, P., Rinzler, A. G., Tomanek, D., Colbert, D. T., & Smalley, R. E. (1995). Self-assembly of tubular fullerenes. *Journal of Physical Chemistry*, *99*, 10694–10697.
Guo, T., Nikolaev, P., Thess, A., Colbert, D. T., & Smalley, R. E. (1995). Catalytic growth of single-walled manotubes by laser vaporization. *Chemical Physics Letters*, *243*, 49–54.
Guzik, U., Hupert-Kocurek, K., & Wojcieszynska, D. (2014). Immobilization as a strategy for improving enzyme properties—Application to oxidoreductases. *Molecules*, *19*, 8995–9018.
Hammes, G. G. (2015). Vibrations in macromolecules. In *Methods of biochemical analysis (Second)*. John Wiley & Sons, Inc.
He, H., Pham-Huy, L. A., Dramou, P., Xiao, D., Zuo, P., & Pham-Huy, C. (2013). Carbon nanotubes: Applications in pharmacy and medicine. *BioMed Research International*, *2013*, 12.
Iijima, S. (1991). Helical microtubules of graphitic carbon. *Nature*, *354*, 56–58.
Iijima, S. (2002). Carbon nanotubes: Past, present, and future. *Physica B: Condensed Matter*, *323*, 1–5.
Jeon, I., Chang, D. W., Kumar Ashok, N., & Baek, J.-B. (2011). Functionalization of carbon nanotubes. In S. Yellampalli (Ed.), *Carbon nanotubes—Polymer nanocomposites* (pp. 91–110). InTech.
Kaushik, B. K., & Majumder, M. K. (2015). Carbon nanotube: Properties and applications. In *SpringerBriefs in applied sciences and technology*. *Carbon nanotube based VLSI interconnects* (pp. 17–37). New Delhi: Springer.
Ke, C., Li, X., Huang, S., Xu, L., & Yan, Y. (2014). Enhancing enzyme activity and enantioselectivity of Burkholderia cepacia lipase via immobilization on modified multi-walled carbon nanotubes. *RSC Advances*, *4*, 57810–57818.
Kelly, S. M., Jess, T. J., & Price, N. C. (2005). How to study proteins by circular dichroism. *Biochimica et Biophysica Acta*, *1751*, 119–139.
Kim, J., Grate, J. W., & Wang, P. (2006). Nanostructures for enzyme stabilization. *Chemical Engineering Science*, *61*, 1017–1026.

Kim, J., Grate, J. W., & Wang, P. (2008). Nanobiocatalysis and its potential applications. *Trends in Biotechnology*, *26*, 639–646.

Kim, B. C., Lee, I., Kwon, S. J., Wee, Y., Kwon, K. Y., Jeon, C., et al. (2017). Fabrication of enzyme-based coatings on intact multi-walled carbon nanotubes as highly effective electrodes in biofuel cells. *Scientific Reports*, 7, 1–10.

Kim, J., Park, G., Lee, S., Hwang, S. W., Min, N., & Lee, K. M. (2017). Single wall carbon nanotube electrode system capable of quantitative detection of CD4 + T cells. *Biosensors and Bioelectronics*, *90*, 238–244.

Kong, J., & Yu, S. (2007). Fourier transform infrared spectroscopic analysis of protein secondary structures. *Acta Biochimica et Biophysica Sinica*, *39*, 549–559.

Kyotani, T., & Tomita, A. (1995). Formation of ultrafine carbon tubes by using an anodic aluminum oxide film as a template. *Chemistry of Materials*, 7, 1994–1995.

Mallakpour, S., & Soltanian, S. (2016). Surface functionalization of carbon nanotubes: Fabrication and applications. *RSC Advances*, *6*, 109916–109935.

Markiton, M., Boncel, S., Janas, D., & Chrobok, A. (2017). Highly active nanobiocatalyst from lipase noncovalently immobilized on multiwalled carbon nanotubes for Baeyer-Villiger synthesis of lactones. *ACS Sustainable Chemistry & Engineering*, *5*, 1685–1691.

Mehra, N. K., Mishra, V., & Jain, N. K. (2014). A review of ligand tethered surface engineered carbon nanotubes. *Biomaterials*, *35*, 1267–1283.

Meng, L., Fu, C., & Lu, Q. (2009). Advanced technology for functionalization of carbon nanotubes. *Progress in Natural Science*, *19*, 801–810.

Miller, J. M., Dunn, B., Valentine, J. S., & Zink, J. I. (1996). Synthesis conditions for encapsulating cytochrome *c* and catalase in SiO_2 sol-gel materials. *Journal of Non-Crystalline Solids*, *202*, 279–289.

Moyo, M., Okonkwo, J. O., & Agyei, N. M. (2014). An amperometric biosensor based on horseradish peroxidase immobilized onto maize tassel-multi-walled carbon nanotubes modified glassy carbon electrode for determination of heavy metal ions in aqueous solution. *Enzyme and Microbial Technology*, *56*, 28–34.

Nagaraju, K., Reddy, R., & Reddy, N. (2015). A review on protein functionalized carbon nanotubes. *Journal of Applied Biomaterials and Functional Materials*, *13*, e301–e312.

Orfanakis, G., Patila, M., Catzikonstantinou, A. V., Lyra, K.-M., Kouloumpis, A., Spyrou, K., et al. (2018). Hybrid nanomaterials of magnetic iron nanoparticles and graphene oxide as matrices for the immobilization of β-glucosidase: Synthesis, characterization, and biocatalytic properties. *Frontiers in Materials*, *5*, 1–11.

Othman, A. M., González-Domínguez, E., Sanromán, Á., Correa-Duarte, M., & Moldes, D. (2016). Immobilization of laccase on functionalized multiwalled carbon nanotube membranes and application for dye decolorization. *RSC Advances*, *6*, 114690–114697.

Patila, M., Diamanti, E. K., Bergouni, D., Polydera, A. C., Gournis, D., & Stamatis, H. (2018). Preparation and biochemical characterisation of nanoconjugates of functionalized carbon nanotubes and cytochrome *c*. *Nanomedicine Research Journal*, *3*, 10–18.

Patila, M., Kouloumpis, A., Gournis, D., Rudolf, P., & Stamatis, H. (2016). Laccase-functionalized graphene oxide assemblies as efficient nanobiocatalysts for oxidation reactions. *Sensors (Switzerland)*, *16*, 1–14.

Patila, M., Orfanakis, G., Polydera, A., Pavlidis, I., & Stamatis, H. (2017). Graphene-based nanobiocatalytic systems. In P. Grunwald (Ed.), *Biocatalysis and nanotechnology* (pp. 243–277). Pan Stanford Pte. Ltd.

Patila, M., Pavlidis, I. V., Diamanti, E. K., Katapodis, P., Gournis, D., & Stamatis, H. (2013). Enhancement of cytochrome *c* catalytic behaviour by affecting the heme environment using functionalized carbon-based nanomaterials. *Process Biochemistry*, *48*, 1010–1017.

Patila, M., Pavlidis, I. V., Kouloumpis, A., Dimos, K., Spyrou, K., Katapodis, P., et al. (2016). Graphene oxide derivatives with variable alkyl chain length and terminal functional groups as supports for stabilization of cytochrome *c*. *International Journal of Biological Macromolecules*, *84*, 227–235.

Pavlidis, I. V., Patila, M., Bornscheuer, U. T., Gournis, D., & Stamatis, H. (2014). Graphene-based nanobiocatalytic systems: Recent advances and future prospects. *Trends in Biotechnology*, *32*, 312–320.

Pavlidis, I. V., Patila, M., Polydera, A. C., Gournis, D., & Stamatis, H. (2014). Immobilization of enzymes and other biomolecules on graphene. In V. Georgakilas (Ed.), *Functionalization of graphene* (pp. 139–172). Wiley-VCH Verlag & Co.

Pavlidis, I. V., Tsoufis, T., Enotiadis, A., Gournis, D., & Stamatis, H. (2010). Functionalized multi-wall carbon nanotubes for lipase immobilization. *Advanced Engineering Materials*, *12*, 179–183.

Pavlidis, I. V., Vorhaben, T., Gournis, D., Papadopoulos, G. K., Bornscheuer, U. T., & Stamatis, H. (2012). Regulation of catalytic behaviour of hydrolases through interactions with functionalized carbon-based nanomaterials. *Journal of Nanoparticle Research*, *14*, 842–852.

Pavlidis, I. V., Vorhaben, T., Tsoufis, T., Rudolf, P., Bornscheuer, U. T., Gournis, D., et al. (2012). Development of effective nanobiocatalytic systems through the immobilization of hydrolases on functionalized carbon-based nanomaterials. *Bioresource Technology*, *115*, 164–171.

Pedrosa, V. A., Paliwal, S., Balasubramanian, S., Nepal, D., Davis, V., Wild, J., et al. (2010). Enhanced stability of enzyme organophosphate hydrolase interfaced on the carbon nanotubes. *Colloids and Surfaces B: Biointerfaces*, 77, 69–74.

Pelton, J. T., & McLean, L. R. (2000). Spectroscopic methods for analysis of protein secondary structure. *Analytical Biochemistry*, *277*, 167–176.

Poudel, Y. R., & Li, W. (2018). Synthesis, properties, and applications of carbon nanotubes filled with foreign materials: A review. *Materials Today Physics*, 7, 7–34.

Prasad, S., Maiti, N. C., Mazumdar, S., & Mitra, S. (2002). Reaction of hydrogen peroxide and peroxidase activity in carboxymethylated cytochrome *c*: Spectroscopic and kinetic studies. *Biochimica et Biophysica Acta*, *1596*, 63–75.

Prestrelski, S. J., Tedeschi, N., Arakawa, T., & Carpenter, J. F. (1993). Dehydration-induced conformational transitions in proteins and their inhibition by stabilizers. *Biophysical Journal*, *65*, 661–671.

Raghavendra, T., Basak, A., Manocha, L. M., Shah, A. R., & Madamwar, D. (2013). Robust nanobioconjugates of Candida antarctica lipase B—Multiwalled carbon nanotubes: Characterization and application for multiple usages in non-aqueous biocatalysis. *Bioresource Technology*, *140*, 103–110.

Ray, M., Chatterjee, S., Das, T., Bhattacharyya, S., Ayyub, P., & Mazumdar, S. (2011). Conjugation of cytochrome c with hydrogen titanate nanotubes: Novel conformational state with implications for apoptosis. *Nanotechnology*, *22*, 415705.

Sahoo, S., Chitturi, V. R., Agarwal, R., Jiang, J. W., & Katiyar, R. S. (2014). Thermal conductivity of freestanding single wall carbon nanotube sheet by Raman spectroscopy. *ACS Applied Materials and Interfaces*, *6*, 19958–19965.

Saifuddin, N., Raziah, A. Z., & Junizah, A. R. (2013). Carbon nanotubes: A review on structure and their interaction with proteins. *Journal of Chemistry*, *2013*, 18.

Schnorr, J. M., & Swager, T. M. (2011). Emerging applications of carbon nanotubes. *Chemistry of Materials*, *23*, 646–657.

Sen, R., Suzuki, S., Kataura, H., & Achiba, Y. (2001). Growth of single-walled carbon nanotubes from the condensed phase. *Chemical Physics Letters*, *349*, 383–388.

Serefoglou, E., Litina, K., Gournis, D., Kalogeris, E., Tzialla, A. A., Pavlidis, I. V., et al. (2008). Smectite clays as solid supports for immobilization of β-glucosidase: Synthesis, characterization, and biochemical properties. *Chemistry of Materials*, *20*, 4106–4115.

Sharma, P., Kumar Mehra, N., Jain, K., & Jain, N. K. (2016). Biomedical applications of carbon nanotubes: A critical review. *Current Drug Delivery*, *13*, 796–817.
Sireesha, M., Jagadeesh Babu, V., Kranthi Kiran, A. S., & Ramakrishna, S. (2018). A review on carbon nanotubes in biosensor devices and their applications in medicine. *Nano*, *4*, 36–57.
Smiljanic, O., Stansfield, B. L., Dodelet, J. P., Serventi, A., & Désilets, S. (2002). Gas-phase synthesis of SWNT by an atmospheric pressure plasma jet. *Chemical Physics Letters*, *356*, 189–193.
Stavyiannoudaki, V., Vamvakaki, V., & Chaniotakis, N. (2009). Comparison of protein immobilisation methods onto oxidised and native carbon nanofibres for optimum biosensor development. *Analytical and Bioanalytical Chemistry*, *395*, 429–435.
Steiner, S. A., III, Baumann, T. F., Bayer, B. C., Blume, R., Worsley, M. A., Moberlychan, W. J., et al. (2009). Nanoscale zirconia as a nonmetallic catalyst for graphitization of carbon and growth of single- and multiwall carbon nanotubes. *Journal of the American Chemical Society*, *131*, 12144–12154.
Subrizi, F., Crucianelli, M., Grossi, V., Passacantando, M., Pesci, L., & Saladino, R. (2014). Carbon nanotubes as activating tyrosinase supports for the selective synthesis of catechols. *ACS Catalysis*, *4*, 810–822.
Sun, B. M., Cao, W. H., Guo, Y. H., Wang, Y., Luo, J. T., & Jiang, P. (2013). Flame pyrolysis synthesis of self-oriented carbon nanotubes. *AIP Advances*, *3*, 092102. https://doi.org/10.1063/1.4821097.
Taqieddin, E., & Amiji, M. (2004). Enzyme immobilization in novel alginate-chitosan core-shell microcapsules. *Biomaterials*, *25*, 1937–1945.
Triantafyllidis, K. S., Karakoulia, S. A., Gournis, D., Delimitis, A., Nalbandian, L., Maccallini, E., et al. (2008). Formation of carbon nanotubes on iron/cobalt oxides supported on zeolite-Y: Effect of zeolite textural properties and particle morphology. *Microporous and Mesoporous Materials*, *110*, 128–140.
Tsoufis, T., Xidas, P., Jankovic, L., Gournis, D., Saranti, A., Bakas, T., et al. (2007). Catalytic production of carbon nanotubes over Fe-Ni bimetallic catalysts supported on MgO. *Diamond and Related Materials*, *16*, 155–160.
Tzialla, A. A., Pavlidis, I. V., Felicissimo, M. P., Rudolf, P., Gournis, D., & Stamatis, H. (2010). Lipase immobilization on smectite nanoclays: Characterization and application to the epoxidation of α-pinene. *Bioresource Technology*, *101*, 1587–1594.
Venyaminov, S. Y., & Kalnin, N. N. (1990). Quantitative IR spectrophotometry of peptide compounds in water (H2O) solutions. I. Spectral parameters of amino acid residue absorption bands. *Biopolymers*, *30*, 1243–1257.
Verma, M. L., Naebe, M., Barrow, C. J., & Puri, M. (2013). Enzyme immobilisation on amino-functionalised multi-walled carbon nanotubes: Structural and biocatalytic characterisation. *PLoS One*, *8*, 16–18.
Wang, Y., Iqbal, Z., & Malhotra, S. V. (2005). Functionalization of carbon nanotubes with amines and enzymes. *Chemical Physics Letters*, *402*, 96–101.
Whitmore, L., & Wallace, B. A. (2008). Protein secondary structure analyses from circular dichroism spectroscopy: Methods and reference databases. *Biopolymers*, *89*, 392–400.
Willner, I., Yan, Y. M., Willner, B., & Tel-Vered, R. (2009). Integrated enzyme-based biofuel cells—A review. *Fuel Cells*, *9*, 7–24.
Wolfenden, B. S., & Willson, R. L. (1982). Radical-cations as reference chromogens in kinetic studies of one-electron transfer reactions: Pulse radiolysis studies of 2,2'-azinobis-(3-ethyl benzthiazoline-6-sulphonate). *Journal of the Chemical Society, Perkin Transactions*, *11*, 805–812.
Wu, Z., Chen, Z., Du, X., Logan, J. M., Sippel, J., Nikolou, M., et al. (2004). Transparent, conductive carbon nanotube films. *Science*, *305*, 1273–1276.
Zhang, C., & Cai, X. (2019). Immobilization of horseradish peroxidase on Fe_3O_4/nanotubes composites for biocatalysis-degradation of phenol. *Composite Interfaces*, *26*, 379–396.

Zhang, D. H., Yuwen, L. X., & Peng, L. J. (2013). Parameters affecting the performance of immobilized enzyme. *Journal of Chemistry*, *2013*, 7.

Zhu, Z. (2017). An overview of carbon nanotubes and graphene for biosensing applications. *Nano-Micro Letters*, *9*, 25.

Zhu, L., Xu, L., Tan, L., Tan, H., Yang, S., & Yao, S. (2013). Direct electrochemistry of cholesterol oxidase immobilized on gold nanoparticles-decorated multiwalled carbon nanotubes and cholesterol sensing. *Talanta*, *106*, 192–199.

Zniszczoł, A., Herman, A. P., Szymańska, K., Mrowiec-Białoń, J., Walczak, K. Z., Jarzebski, A., et al. (2016). Covalently immobilized lipase on aminoalkyl-, carboxy- and hydroxy-multi-wall carbon nanotubes in the enantioselective synthesis of Solketal esters. *Enzyme and Microbial Technology*, 87–88, 61–69.

CHAPTER THIRTEEN

Biocatalytic hydrogenations on carbon supports

Lisa A. Thompson[a], Jack S. Rowbotham[a], Holly A. Reeve[a], Ceren Zor[b], Nicole Grobert[b], Kylie A. Vincent[a,*]

[a]Department of Chemistry, University of Oxford, Inorganic Chemistry Laboratory, Oxford, United Kingdom
[b]Department of Materials, University of Oxford, Oxford, United Kingdom
[*]Corresponding author: e-mail address: kylie.vincent@chem.ox.ac.uk

Contents

Abstract

We describe the use of carbon as a versatile support for H_2-driven redox biocatalysis for NADH-dependent C=X bond reductions in batch and flow reactions. In each case, carbon is providing an electronic link between enzymes for H_2 oxidation and reduction of the biological cofactor NAD^+ as well as a support for a multi-enzyme biocatalysis system. Carbon nanopowders offer high surface areas for enzyme immobilization and good dispersion in aqueous solution for heterogeneous batch reactions. Difficulties

Methods in Enzymology, Volume 630
ISSN 0076-6879
https://doi.org/10.1016/bs.mie.2019.10.017

in handling multi-wall carbon nanotubes in aqueous solution are overcome by growing them on quartz tubes to form carbon nanotube column reactors, and we show that these facilitate simple translation of H_2-driven biocatalysis into flow processes. Using this flow reactor design, high conversions (90%) and total enzyme turnover numbers up to 54,000 could be achieved. Use of an entirely heterogeneous biocatalysis system simplifies recovery and re-use of the enzymes; combined with highly atom-efficient cofactor recycling, this means that high product purity can be achieved. We demonstrate these methods as platform approaches for overcoming challenges with NADH-dependent biocatalysis.

1. Introduction

Biocatalysis is becoming an increasingly powerful tool for chemical synthesis, with importance in the production of complex fine chemicals where the exquisite selectivity of enzymes is often unmatched by traditional chemical methods (Sheldon & Woodley, 2018; Woodley, 2008). In particular, biocatalytic hydrogenation-type reactions, in which a prochiral C=X bond is enantioselectively reduced, offer significant advantages over traditional metal catalyzed routes, but are yet to out-compete them due to two-key process limitations. First, most redox biocatalysts for hydrogenation-type reactions rely on NAD(P)H as a reducing equivalent, and an appropriate enzymatic cofactor recycling system is required such that the expensive cofactor can be used in catalytic quantities. Second, difficulties with removing enzymes from solution compromise product purity and mean that the cost of these often multi-enzyme systems can be prohibitively high. One approach to overcoming these limitations is to use enzymes immobilized on solid supports, as this facilitates enzyme separation and re-use. For some classes of enzyme, immobilization on a solid support leads to higher stability and enhanced tolerance to temperature and organic media compared to handling in solution; however, the supports are often expensive or require customized modification prior to enzyme immobilization (Liese & Hilterhaus, 2013; Rodrigues, Ortiz, Berenguer-Murcia, Torres, & Fernández-Lafuente, 2013; Sheldon & van Pelt, 2013). At present, there are no clear 'go to' methods for heterogeneous biocatalysis for redox reactions where the need for multi-enzyme systems further complicates immobilization and implementation strategies. This has promoted significant interest in development of widely applicable methods for immobilization of enzymes on solid supports and for successful demonstration of platforms for implementation of heterogeneous biocatalytic hydrogenation reactions.

We introduce carbon as an enzyme support for redox biocatalysis. Traditional metal catalysts for hydrogenation reactions are often used on carbon supports where carbon offers high surface areas and is relatively inert under reaction conditions. Carbon has also been used extensively as an electrode material in a wide range of applications spanning fuel cells, biosensors and electroanalytical techniques. Here carbon is low cost, available in a wide range of forms and is fairly innocent toward competing redox chemistry (Zhang et al., 2016), making it an excellent candidate for catalyst immobilization.

Carbon has long been used as an electrode material for studying electrode-immobilized redox enzymes in the technique known as protein film electrochemistry (Blanford & Armstrong, 2006; Leger & Bertrand, 2008). Certain specialized enzymes contain electron relay centers, for example Fe-S clusters, which allow rapid transfer of electrons between the enzyme active site and the protein-electrode interface. Carbon has also been shown as a support to electronically link two redox active enzymes such that electrons released during a biocatalytic oxidation can be transferred to a second enzyme for a catalytic reduction reaction (Vincent et al., 2007).

We have demonstrated biocatalytic hydrogenations with H_2-driven cofactor recycling using hydrogenase, NAD^+ reductase and an NADH-dependent enzyme co-immobilized on a carbon support (Reeve, Lauterbach, Ash, Lenz, & Vincent, 2012; Reeve, Lauterbach, Lenz, & Vincent, 2015). The hydrogenase oxidizes H_2 to protons and electrons, which are transferred from the catalytic site to the surface of the enzyme, via a chain of Fe-S clusters. The NAD^+ reductase uses electrons and a proton from solution for reduction of NAD^+ to NADH for supply to the NADH-dependent enzyme for a C=X bond reduction. By co-immobilizing the enzymes onto an electronically conducting support, electrons from H_2 oxidation can be used for NAD^+ reduction and the entire biocatalytic system can be easily removed from solution and re-used. This biocatalytic system has a dual need for carbon as a support material and as an electron transfer method and tackles both key challenges with implementation of redox biocatalysis.

We demonstrate that a large variety of carbon types can be used for enzyme immobilization and that different carbon materials open up the possibility of different implementation strategies. For example, we show the suitability of carbon black nanoparticles for use of this biocatalytic hydrogenation system in batch reactors and then focus on multi-wall carbon nanotubes (MWCNTs) grown on a quartz substrate as the support material for translation of this H_2-driven biocatalysis approach into a flow reactor.

2. Carbon immobilized enzymes for redox biocatalysis

2.1 Carbon supports for redox enzymes

There are literature reports of enzyme immobilization onto a range of carbon allotropes including graphite, graphene, amorphous carbon, fullerenes, single-wall carbon nanotubes (SWCNTs), multi-wall carbon nanotubes (MWCNTs), carbon fibers, carbon papers, carbon nanobuds, carbon nanofoams (Grigoras, 2017; Min, Kim, Park, & Yoo, 2012). For study of specialized redox enzymes that contain electron relay centers, where carbon is acting as the enzyme support and an electron donor/acceptor, pyrolytic graphite has been most widely used. Pyrolytic graphite edge electrodes offer a balance between highly ordered graphitic domains which facilitate electron movement, with a 'rough' edge plane which has been shown to aid enzyme immobilization by simple adsorption (Quinson et al., 2014). Graphite electrodes are fairly innocent toward electrocatalytic reactions in the potential window relevant for investigating biological electrocatalysis, at least under anaerobic conditions. This is demonstrated by a cyclic voltammogram recorded at a unmodified pyrolytic graphite electrode in aqueous solution, Fig. 1A, solid gray line. This means that the electro-catalytic activity of a wide

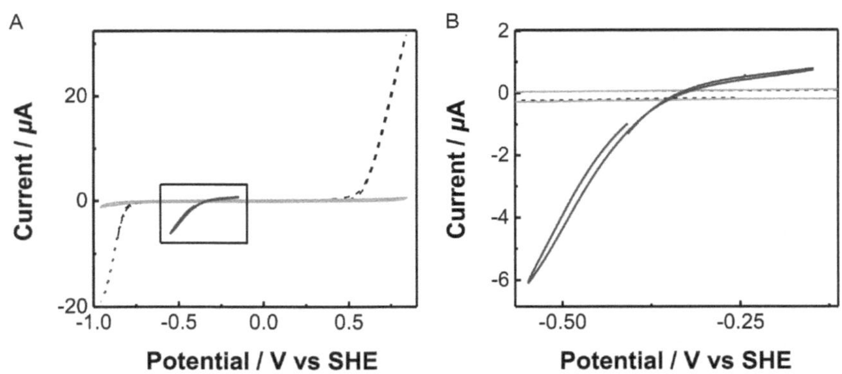

Fig. 1 Electrochemical cyclic voltammetry experiments to demonstrate the reduction of NAD^+ to NADH and oxidation of NADH to NAD^+ by an unmodified electrode (- - - -) and by a NAD^+ reductase moiety immobilized on a carbon electrode surface (——). The experiments were performed in phosphate buffer (pH 7) with $[NAD^+]=[NADH]=$ 1 mM. A cyclic voltammogram recorded at an unmodified carbon electrode in the absence of NAD^+ and NADH (——) is used as a control experiment. In all cases the electrode was rotated at 2000 rpm. Panel (B) shows the expanded region of the box in (A). Potentials are quoted vs the standard hydrogen electrode (SHE).

range of different immobilized enzymes has been observed using protein film electrochemistry, as shown in Fig. 1B for NAD^+/NADH cycling by an NAD^+ reductase. In contrast, NAD^+ reduction and NADH oxidation at carbon electrodes requires large overpotentials, Fig. 1A, dotted line. For these electroanalytical applications, the main requirements are for highly reproducible carbon electrode surfaces to aid reproducible background cyclic voltammograms and for highly efficient electron transfer between the electrode and the protein such that the enzyme active site chemistry is the rate limiting step.

Initial successful reports of electronically linked redox enzymes used carbon particles prepared by abrasion of pyrolytic graphite, since methods for achieving good enzyme adsorption were already established for this carbon material. This technique was demonstrated by Armstrong and co-workers for H_2-driven nitrate to nitrite conversion (Vincent et al., 2007), H_2-driven fumarate reduction (Vincent et al., 2007) and for the water-gas shift reaction (Lazarus et al., 2009). We went on to demonstrate H_2-driven NADH recycling by combining hydrogenase and NAD^+ reductase on pyrolytic graphite particles (Reeve et al., 2012). However, this form of carbon is expensive and due to the enzyme preferentially adsorbing at the 'edge'-like surfaces, relatively low enzyme loading is achieved. Furthermore, the particles must be freshly prepared before use, are non-uniform and are relatively large, leading to mass transport limitations in batch reactions where they do not stay well-suspended.

2.2 Carbon supports for H_2-driven NADH generation in batch reactions

Our H_2-driven cofactor recycling system combines a hydrogenase for H_2 oxidation and an NAD^+ reductase for NAD^+ to NADH conversion, co-immobilized onto carbon particles such that electrons from H_2 oxidation can be used for NAD^+ reduction (Fig. 2A). The technology relies on the carbon for electron transfer and for enzyme immobilization to create a heterogeneous biocatalyst system. A suite of carbon supports including carbon paper, carbon nanopowder, carbon black and MWCNTs were trialed (Reeve et al., 2015). The hydrogenase and NAD^+ reductase were immobilized by simple adsorption at 4 °C. The enzyme-modified supports were exposed to a solution of H_2-saturated buffered solution containing oxidized cofactor, NAD^+, and in each case high conversion of NAD^+ to NADH was observed (Fig. 2B) (Reeve et al., 2015). Control reactions show that the hydrogenase and NAD^+ reductase must be co-immobilized

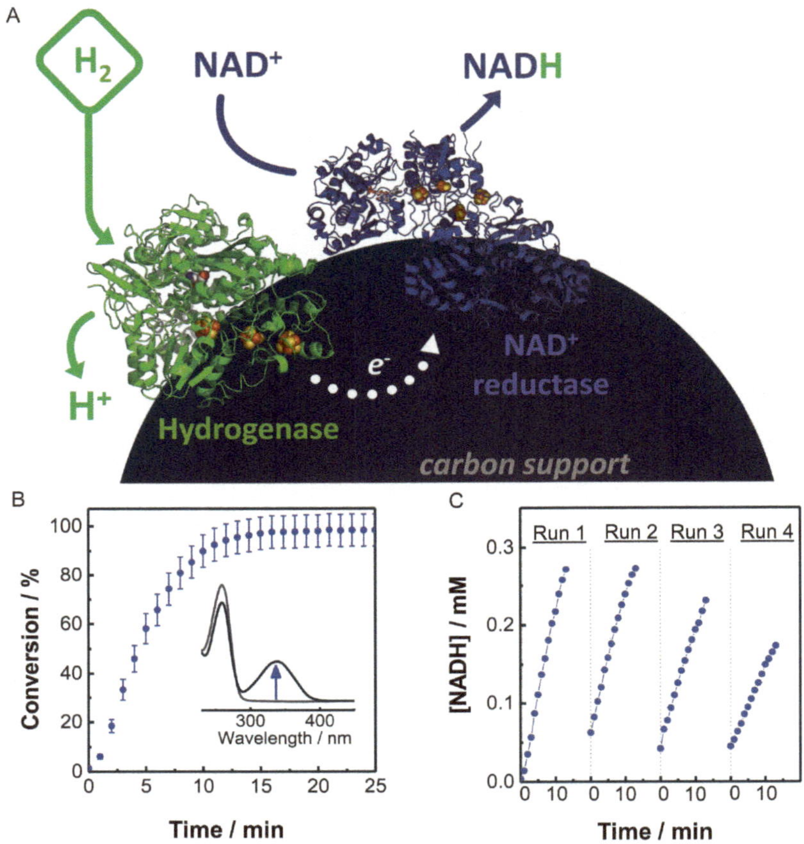

Fig. 2 Using co-immobilized hydrogenase and NAD^+ reductase on a carbon support for the H_2-driven reduction of NAD^+. (A) General scheme for the catalytic system, highlighting the importance of the carbon material for electron transfer between the two enzymes. (B) A typical reaction profile, showing the conversion of NAD^+ to NADH by the heterogeneous biocatalytic system as measured by UV-spectrophotometry (inset). (C) Activity of the enzyme particles when recovered and re-used over several runs. Reactions were carried out in Tris-HCl (50 mM, pH 8.0) at 30 °C, with an initial [NAD^+] of (B) 70 μM, and (C) 1 mM.

for H_2-driven NADH generation to take place (Reeve et al., 2015). These results therefore demonstrate that this range of carbon materials is suitable for enzyme adsorption and that the hydrogenase and NAD^+ reductase are able to exchange electrons with each carbon support and therefore sustain their electro-catalytic activity.

Carbon nanotubes (CNTs) have been shown as enzyme supports for H_2/H^+ cycling by immobilized hydrogenases (Alonso-Lomillo et al., 2007;

Feng & Ji, 2011; Kihara et al., 2011; McDonald et al., 2007) because they display good electron transfer and enzyme immobilization capabilities. However, CNTs are not suitable for study under batch conditions due to their inherent hydrophobicity, which leads to poor dispersion in aqueous solution.

An ideal carbon support for application in batch reactions would have high surface area, good electrical conductivity, be easily suspended in solution but also easily removed from solution for catalyst recovery and re-use. Carbon black agglomerated nanopowders present many advantages over other types of carbon supports and are commonly used as a support for metal catalysts in fuel cells and in catalytic hydrogenation reactions (Asedegbega-Nieto, Guerrero-Ruiz, & Rodríguez-Ramos, 2006; Li, Wang, Chen, Waje, & Yan, 2006). These powders are cheap and readily available, and are amorphous materials with no graphitic planar-like regions, meaning that they have higher surface area suitable for enzyme immobilization compared to graphitic carbon materials. The nanopowders are also easier to suspend in solution compared to CNTs.

We chose Black Pearls 2000 (BP, Cabot Corporation) for use in this work. Additionally to BP particles being easily dispersed and therefore minimizing mass transport limitations during operation of the heterogeneous catalyst, the enzyme-modified particles are also easily separated from solution by centrifugation or sedimentation. The results in Fig. 2C show that the particles are recovered from solution by centrifugation and re-used in fresh reaction solution with only a modest drop in activity.

Additionally, we have demonstrated the ability of the system to work in both directions, for the reduction of NAD^+ by H_2, or the oxidation of NADH under a N_2 atmosphere (Reeve et al., 2017). Variation of the reaction parameters has shown that the system can operate within a pH range of 6–8 and can even be used in water, without the presence of buffer, and is tolerant to low partial pressures of H_2 (<1 bar) (Reeve et al., 2015).

2.3 Spectroscopy techniques for investigation of carbon black as an enzyme support

Spectroelectrochemical study of a hydrogenase with a NiFe active site on BP particles (handled as an electrode) showed that all adsorbed hydrogenase molecules were in electron transfer contact with the carbon surface and responded to a change in potential at the carbon. This indicates that hydrogenase is well-coupled electronically to the carbon, perhaps due to the highly porous nature of BP providing a high degree of contact between enzyme and carbon material (Hidalgo, Ash, Healy, & Vincent, 2015). The pore size is in the range of 4–6 nm and the hydrogenase has a diameter of 5 nm (Quinson et al., 2014), suggesting that the hydrogenase may be able to diffuse into pores.

As discussed, the electron transfer properties of carbon allows for direct electronic communication between the two enzymes. This electron transfer through the conductive carbon was confirmed by separating the two enzymes by a sheet of carbon paper and monitoring changes in spectral signatures reporting on the redox state of the hydrogenase active site as NADH or NAD^+ were introduced in different ratios to the electronically coupled NAD^+ reductase. The hydrogenase became reduced when NADH was supplied to the NAD^+ reductase, confirming the electron flow between the two enzymes through the carbon (Reeve et al., 2017).

We have demonstrated BP as a suitable enzyme support for batch reactions due to its high surface area, excellent enzyme adsorption capabilities, ease of use as a well-dispersed catalyst suspension and the possibilities for catalyst recovery. In the following section we will discuss use of this heterogeneous biocatalyst as a cofactor recycling system to drive NADH-dependent biocatalysis.

2.4 Coupling to NADH-dependent reactions

The co-adsorbed hydrogenase and NAD^+ reductase system is extended to H_2-driven C=X bond reductions by introducing an NADH-dependent enzyme, Fig. 3. This system allows highly atom-efficient redox biocatalysis

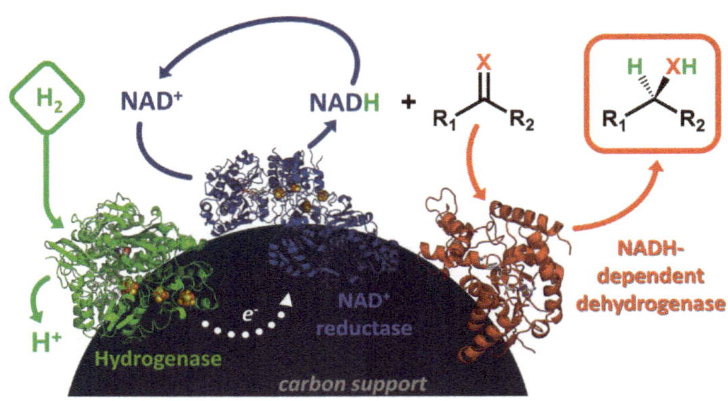

Fig. 3 A scheme detailing the heterogeneous H_2-driven cofactor recycling system applied to hydrogenation of C=X bonds (where X = O, NR, or CR_2).

with both electrons and protons from H_2 oxidation being incorporated into the final product by transfer of hydride (H^-) via the cofactor, and H^+ from solution.

This H_2-driven cofactor recycling has significant advantages over conventional strategies for implementing NADH-dependent biocatalysts which rely on use of a cofactor recycling enzyme such as glucose dehydrogenase or formate dehydrogenase and a supply of super stoichiometric carbon based reductant. The presence of this reductant and the resulting waste oxidation product can complicate downstream processing. In our system the NADH-dependent enzyme can be used in solution because it does not rely on electron transfer with the carbon support but co-adsorption has the benefit of making the entire biocatalytic system heterogeneous, simplifying catalyst recovery. Therefore, this heterogeneous biocatalytic hydrogenation system overcomes two-key limitations of redox biocatalysis.

As an illustrative example, the H_2-driven NADH recycling system was used to supply NADH to an (*S*)-selective alcohol dehydrogenase (ADH) in order to convert a prochiral ketone (acetophenone) to a chiral alcohol ((*S*)-1-phenylethanol), with very high enantiomeric excess (*ee*), Fig. 4

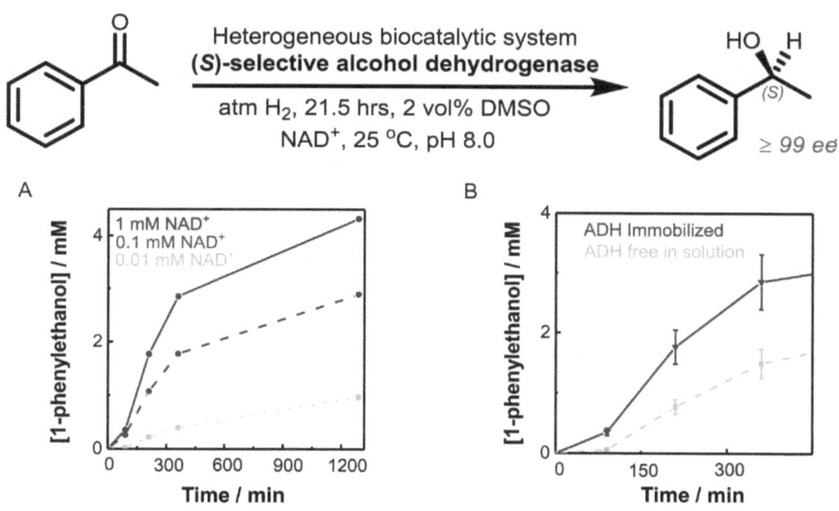

Fig. 4 Application of H_2-driven cofactor recycling for the reduction of acetophenone by an (*S*)-selective alcohol dehydrogenase (ADH). Each experiment was carried out in Tris-HCl buffered water (50 mM, pH 8.0) containing acetophenone (10 mM) and 2 vol% DMSO, with shaking at 500 rpm. (A) Influence of cofactor concentration on the activity of the biocatalytic system. (B) Comparison of the H_2-driven reduction of acetophenone when the ADH is freely suspended in solution relative to being co-immobilized on carbon ($[NAD^+]=1$ mM).

(Reeve et al., 2015). The results in Fig. 4A demonstrate that high conversion of acetophenone to phenylethanol can be achieved under mild reaction conditions (pH 8, 1 bar H_2) and that significant activity is retained as the cofactor concentration is decreased from 1 to 0.01 mM (0.1–0.001 mol% compared to acetophenone). Using this strategy, the NADH could be recycled >600 times, and the NAD^+ reductase enzyme achieved a total turnover number on the order of 10^5 (Reeve et al., 2015). We have also demonstrated that there is a significant increase in the rate of (*S*)-1-phenylethanol production when the ADH is co-immobilized, compared to when it is freely suspended in solution, as demonstrated in Fig. 4B.

Beyond alcohol dehydrogenases, the heterogeneous H_2-driven cofactor recycling strategy is, in principle, compatible with any NADH-dependent enzyme. Accordingly, this approach has been demonstrated for the conversion of pyruvate to L-lactate (by L-lactate dehydrogenase) (Reeve et al., 2012) and alanine (by L-alanine dehydrogenase) (Zor et al., 2017). Work is now ongoing to couple the H_2-driven cofactor recycling catalyst to the expanding family of enzymes used in the production of chiral amines (such as reductive aminases, amine dehydrogenases, and imine reductases) (Aleku et al., 2017), as well as to ene-reductase enzymes for the asymmetric reduction of a range of activated C=C bonds (Dominguez, Schell, Bisagni, & Kalthoff, 2016).

So far, we have found that BP particles are suitable as a support for a wide range of NADH-dependent enzymes which readily adsorb to the carbon surface without extensive pre-treatment or modification, and with surprising stability given the absence of a covalent attachment. This work therefore suggests that this approach could be a suitable platform for implementing biocatalytic hydrogenation-type reactions in batch reactors. The following section will develop these themes further and explore the possibility of extending the heterogeneous H_2-driven biocatalysis system into a flow process (Zor et al., 2017).

3. Carbon nanotube supported biocatalysis in flow

Continuous flow technology has been increasing in popularity in the pharmaceutical and fine chemical industries over the past few decades as a greener, more efficient synthesis method that allows access to a wider range of chemistries than is possible in batch (Bogdan & Dombrowski, 2019; Plutschack, Pieber, Gilmore, & Seeberger, 2017). Flow is also a valuable methodology for heterogeneous catalysis, where the catalyst is typically constrained within a column-type reactor and the reagent solution is passed through it. Exposing reagent solutions to higher catalyst loadings through the reactor often leads to improved yields in shorter reaction times, and the lack of mechanical mixing prevents degradation of the catalyst, leading to increased lifetimes. These advantages have led to interest in implementing immobilized biocatalysts in flow reactors (Britton, Majumdar, & Weiss, 2018; Thompson, Peñafiel, Cosgrove, & Turner, 2019).

Carbon supported catalysts can be implemented in flow reactors in packed bed columns or as slurries. Catalytic hydrogenation is a key area for translation into flow reactors due to safer implementation of H_2 gas (Cossar, Hizartzidis, Simone, McCluskey, & Gordon, 2015; Irfan, Glasnov, & Kappe, 2011). In the previous section, nanoscale carbon materials were highlighted as suitable supports for H_2-driven biocatalysis due to their high surface areas and simple dispersion. However, these supports do not lend themselves to application in packed bed reactors where larger scale carbons are required to prevent high back pressures.

In this study, flow reactors presented us with a way to use the favorable properties of MWCNTs as a support for biocatalysis (good enzyme adsorption and electron transfer properties), while overcoming the issues with poor dispersion of MWCNTs in solution that hinder batch operation. We have shown that the MWCNTs themselves could be pre-supported on the inner surface of a flow column, and then the enzymes directly adsorbed onto the MWCNTs within the reactor, removing the need for suspending the MWCNTs in solution. We have demonstrated this flow approach for NAD^+ reduction and NADH-dependent biocatalysis.

3.1 Flow reactor set-up

Carbon nanotube lined column (MWCNT-column) reactors were prepared by growing MWCNTs onto a piece of quartz tubing. The MWCNTs were removed from the outer edge, leaving the inside wall coated with aligned MWCNTs (Fig. 5A). Enzymes can be immobilized by adsorption onto the MWCNTs by passing enzyme-containing solution through the MWCNT-column.

The overall flow set-up consisted of a sealed reservoir vessel containing the H_2-saturated reaction solution, a pump to circulate the solution through the system, and an enzyme-modified MWCNT-column reactor (Fig. 5B). The system components were connected using Marprene or PTFE tubing and PTFE or PEEK fittings and connectors. The flow system was used in either re-circulating or single-pass mode.

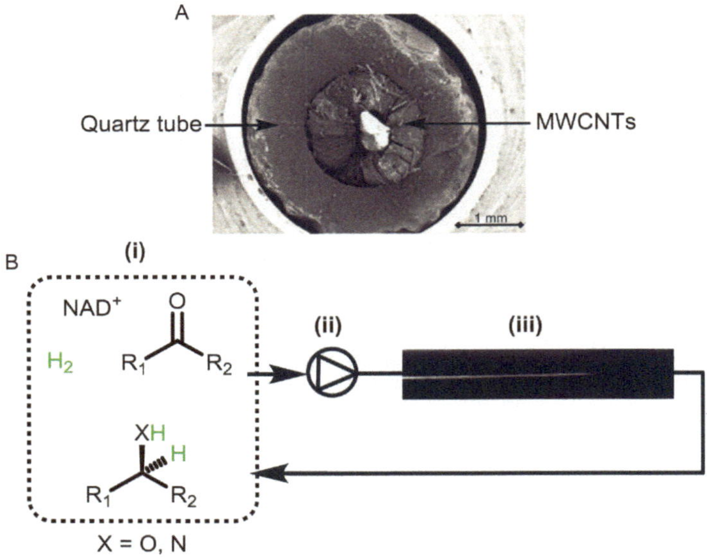

Fig. 5 (A) Scanning electron microscopy (SEM) image of MWCNT-column reactor cross-section, showing quartz tubing lined with MWCNTs. (B) Schematic representation of a re-circulating flow reactor for heterogeneous biocatalytic hydrogenation, with built-in cofactor recycling: (i) reagent/product reservoir; (ii) pump; (iii) enzyme-modified MWCNT-column reactor.

3.2 Flow reaction for NAD^+ to NADH conversion

The suitability of the MWCNT-column reactor for H_2-driven biocatalysis in flow was initially demonstrated for H_2-driven NADH generation, with a combination of hydrogenase and NAD^+ reductase co-immobilized within the reactor. Energy dispersive X-ray spectroscopic (EDX) analysis of enzyme-modified MWCNTs showed increased Fe content, consistent with the Fe-S clusters present in the hydrogenase and NAD^+ reductase enzymes, confirming that the enzymes were immobilized on the MWCNTs. A H_2-saturated solution of NAD^+ was circulated through the reactor at $25\,\mu L\,min^{-1}$, with a continuous flow of H_2 across the headspace of the reservoir vessel (Zor et al., 2017).

The results in Fig. 6 show significant conversion of NAD^+ to NADH, demonstrating that the hydrogenase and NAD^+ reductase are co-immobilized in the MWCNT-column such that they are electronically connected by the MWCNT network.

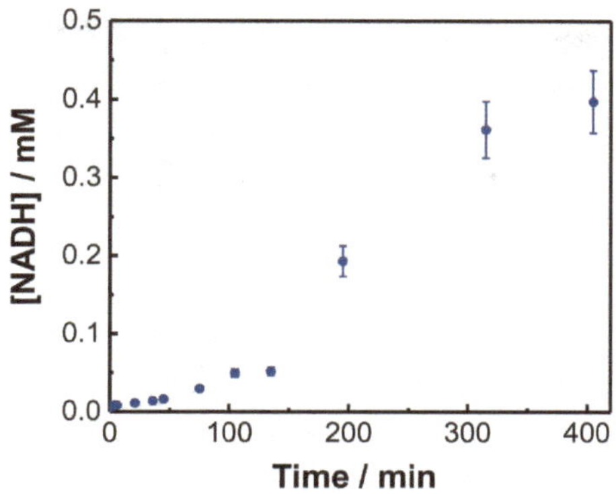

Fig. 6 Time series for H_2-driven NADH production using a MWCNT-column reactor with co-immobilized hydrogenase and NAD^+ reductase. A H_2-saturated cofactor solution was re-circulated through the column and monitored using in line UV-vis spectroscopy.

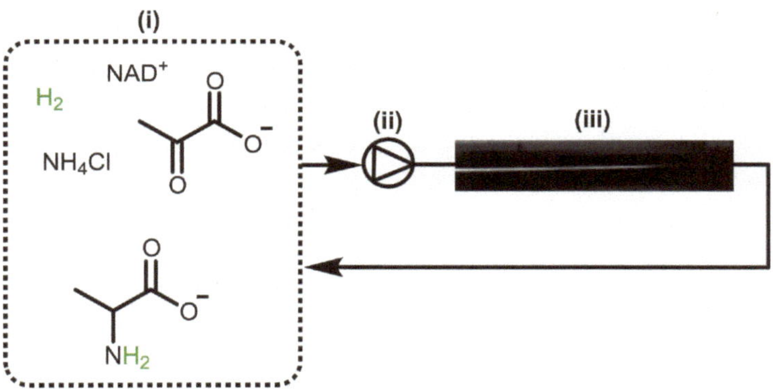

Fig. 7 Re-circulating flow reactor for heterogeneous biocatalytic reductive amination of pyruvate, with built-in H_2-driven cofactor recycling: (i) reagent/product reservoir; (ii) pump; (iii) enzyme-modified MWCNT-column reactor.

3.3 Re-circulating flow reactor for reductive amination

The biocatalytic reductive amination of pyruvate to produce alanine with H_2-driven cofactor recycling was investigated in the re-circulating flow system (Fig. 7), using a combination of hydrogenase, NAD^+ reductase and an NADH-dependent L-alanine dehydrogenase (L-AlaDH) co-immobilized in the MWCNT-column reactor. The H_2-saturated reaction solution containing cofactor and the pyruvate reagent was re-circulated, and after 26 h, 90% conversion of 2 mM pyruvate to alanine was achieved. In an additional experiment, 40% conversion of 12.5 mM pyruvate was achieved under the same reactor conditions, providing a total turnover number of 54,000 for the NAD^+ reductase (Zor et al., 2017).

The system was also demonstrated for reduction of acetophenone to 1-phenylethanol using a combination of hydrogenase, NAD^+ reductase and an NADH-dependent (*S*)-selective ADH co-immobilized in the MWCNT-column reactor. After 26 h, 30% conversion of acetophenone to (*S*)-1-phenylethanol was achieved with >99% *ee* (Zor et al., 2017).

3.4 Single-pass flow reactor

Further development of the system provided a demonstration of a single-pass flow reaction for the reduction of acetophenone (Fig. 8). In a 58 min residence time (tRes), 8% conversion of 1 mM acetophenone to (*S*)-1-phenylethanol (>99% *ee*) was achieved. Due to the low solubility of H_2 gas in aqueous solution it is likely that the H_2 availability was limiting, leading to lower conversions, even at low substrate concentrations.

Challenges with enzyme immobilization and cofactor recycling have previously hindered implementation of redox biocatalysis in flow. The results in this section demonstrate a method for using MWCNTs as a support for H_2-driven redox biocatalysis and represent a platform for implementing NADH-dependent enzymes in flow reactions. The system we describe may simplify translation of NADH-dependent biocatalysis into continuous processes.

Fig. 8 Single-pass flow reactor for heterogeneous biocatalytic ketone reduction, with built-in H_2-driven cofactor recycling.

4. Methods

4.1 General reagents and conditions

All commercial reagents were used as received, unless specified otherwise. Buffer salts were obtained from Sigma-Aldrich, nicotinamide cofactors were bought from Prozomix, and carbon black particles (Black Pearls 2000, BP) were purchased from Cabot Corporation. All solutions were prepared with MilliQ water (Millipore, 18 MΩcm). When required, solutions were deoxygenated by sparging with N_2 for 1 h prior to use. All other reagents and solvents were purchased from Sigma-Aldrich. All immobilizations and reactions were set up in a N_2-filled glovebox containing <1 ppm O_2 (Glove Box Technologies, Ltd).

4.2 Enzymes for H_2-driven biocatalysis

4.2.1 Hydrogenase

The membrane-bound hydrogenase 2 (Hyd-2) from *Escherichia coli* (100 kDa) was used throughout, and was expressed and purified according to previously published methods (Lukey et al., 2010).

4.2.2 NAD^+ reductase

The NAD^+ reductase used was an I64A variant of the soluble hydrogenase (177 kDa) from *Ralstonia eutropha*, in which the inherent hydrogenase moiety is inactivated, and was expressed and purified according to previously published methods (Lauterbach & Lenz, 2013).

4.2.3 NADH-dependent enzymes

All of the NADH-dependent oxidoreductase enzymes were from commercial suppliers and used as received, without further purification. The (*S*)-selective alcohol dehydrogenase was ADH105 from Johnson Matthey, and the L-alanine dehydrogenase was A7189 (*Bacillus subtilis*) from Sigma-Aldrich.

4.3 Analytical techniques

4.3.1 Analysis of cofactors by UV-visible spectroscopy

Cofactor concentrations were established by UV-visible spectroscopy with an Agilent Cary 60 spectrophotometer. Conversion of NAD^+ to NADH was determined using the ratio of absorbance at 260 nm (Abs_{260}) and 340 nm (Abs_{340}) and comparing to calibration curves for the Abs_{260}/Abs_{340} vs the ratio of oxidized:reduced cofactor ($[NAD^+]/[NADH]$) (Reeve et al., 2015).

4.3.2 Analysis of NADH-dependent ketone reductions

4.3.2.1 Analysis methods for conversion

At specified time points, samples were removed from the reaction and MeCN was added (to 20 vol%) prior to centrifugation (10 min, 7000 × *g*) to extract the organic product. The samples were further passed through a centrifugal protein concentrator (Amicon) to remove any remaining unadsorbed enzyme and diluted fivefold with MilliQ water. Samples were subsequently analyzed by HPLC using a Shimadzu UFLC Prominence instrument, equipped with a LC-20AD pump, SPD-20A dual wavelength detector and a Chromolith Performance 100-3 mm column (10 cm × 3 mm, 2 μm particle size) according to the methods below. Commercial analytical standards were used to confirm the identity of the peaks and to establish response factors for the various analytes in order to calculate their relative concentrations in the original reaction solution.

HPLC method [acetophenone/1-phenylethanol]: Isocratic flow: $1\,\mathrm{mL\,min^{-1}}$, H_2O/MeCN (80:20 v/v), run time: 7.5 min, 40 °C. Detection was carried out at 210 nm, $tR_{1\text{-phenylethanol}} = 2.1$ min, $tR_{\text{acetophenone}} = 3.4$ min.

HPLC method [pyruvate/alanine]: Gradient flow: $0.75\,\mathrm{mL\,min^{-1}}$, H_2O/MeCN from 5:95 to 80:20 v/v linearly over 20 min, 40 °C. Detection was carried out at 210 nm, $tR_{\text{pyruvate}} = 7.2$ min, $tR_{\text{alanine}} = 12.5$ min.

4.3.2.2 Analysis methods for enantiomeric excess (*ee*)

The enantiomeric excess (*ee*) of the 1-phenylethanol product obtained from the enzymatic reduction of acetophenone was ascertained by chiral-phase HPLC. Following removal of the catalyst by the method described above, the reaction mixture was extracted into an equal volume of heptane/^{i}PrOH (99:1 v/v) and the organic phase transferred to a glass vial for analysis on a Shimadzu UFLC Prominence HPLC instrument, equipped with a LC-20AD pump, SPD-20A dual wavelength detector and a Chiralpak IA column (15 cm × 4.6 mm, 5 μm particle size) according to the method below. The identity of the relevant peaks was confirmed by comparison with commercial analytical standards.

HPLC method [(R)-1-phenylethanol/(S)-1-phenylethanol]: Isocratic flow: $1\,\mathrm{mL\,min^{-1}}$, heptane/^{i}PrOH (99:1 v/v), run time: 15 min, 40 °C. Detection was carried out at 210 nm, $tR_{\text{acetophenone}} = 4.9$ min, $tR_{(R)\text{-1-phenylethanol}} = 11.9$ min, $tR_{(S)\text{-1-phenylethanol}} = 12.5$ min.

4.3.3 Quantification of catalyst performance

In order to quantify the rate of conversion of H_2-driven NAD^+ reduction, the ratio of [NAD^+]:[NADH] in the reaction solution at various time points was established according to the method described in Section 4.3.1. The gradient of the slope of the initial period of a plot of conversion vs time could subsequently be used to quantify the activity of the enzyme-modified carbon particles per μg of immobilized NAD^+ reductase. This calculation assumes that all of the NAD^+ reductase originally combined with the carbon support was fully adsorbed and active, and therefore represents a measure of the lower limit of the activity of the immobilized NAD^+-reducing moiety.

When an additional NADH-dependent enzyme was included (such as an ADH or L-AlaDH), the performance of the entire coupled biocatalytic system could be determined by following the conversion of reactant to product at time intervals, according to the methods outlined in Section 4.3.2.

4.4 Batch methods

4.4.1 Enzyme immobilization on different carbon supports

In a typical immobilization procedure, carbon materials were suspended in MilliQ water or a suitable buffer (such as Tris-HCl, 50 mM, pH 8.0) to produce a 20 mg mL^{-1} solution. This was then sonicated for between 45 and 90 min, with periodic manual agitation of the suspension using a pipette, to break up larger particles. The resulting fine carbon suspension was then added to a solution of the enzymes to be immobilized, usually the hydrogenase and NAD^+ reductase. The enzyme-carbon mixture was then stored at 4 °C, to allow adsorption of the enzymes. After 1 h, the particle suspension was centrifuged (7000 × *g*, 5 min) and the supernatant removed by pipette, leaving behind the enzyme-modified carbon support. The catalyst could then be used directly, washed further by re-suspension in buffer, centrifugation and removal of the washings, or have additional enzymes immobilized in a similar manner. In a typical experiment, between 2 and 20 μg of hydrogenase and NAD^+ reductase were each immobilized per 100 μg of carbon support.

4.4.2 Batch reaction protocols

In a typical batch reaction, enzyme-modified carbon supports (prepared according to the procedure in Section 4.4.1) were added to a suitable reaction vessel, containing H_2-saturated NAD^+ solution in buffer (such as Tris-HCl, 50 mM, pH 8.0). H_2-saturated solutions were prepared by bubbling H_2 into a sealed falcon tube, containing the NAD^+ solution, for at least 1 h. Additional reagents and enzymes were then added, with 2 vol% dimethylsulfoxide (DMSO) being added as a co-solvent for reactions involving organic substrates. The reaction vessel could be a sealed cuvette with bubbling H_2, a pressure vessel with 2–5 bar H_2 pressure, or a microcentrifuge tube with a supply of H_2 gas across the headspace. The reaction could be agitated by conventional mechanical stirring, shaking, or by allowing the H_2 to bubble through the reaction solution. Reactions were conducted at between 20 and 35 °C and for up to 24 h, as required. When reactions were conducted in a cuvette, the reaction was monitored directly by UV-spectrophotometry as detailed in Section 4.3.1, otherwise, samples were taken at regular intervals for analysis by the methods outlined in Section 4.3. If required for re-use, the enzyme-modified particles were separated from the solution by centrifugation (7000 × *g*, 5 min) and removal of the supernatant, before being added to fresh reaction solution.

4.5 Flow methods

4.5.1 Preparation of carbon nanotube column reactor

MWCNT-column reactors were produced using quartz tubes (20 mm length, 2 mm inner diameter, 4 mm outer diameter, Robson Scientific) as substrates in aerosol assisted chemical vapor deposition (AACVD) for the production of vertically aligned MWCNTs (Meysami, Dillon, Koós, Aslam, & Grobert, 2013; Meysami, Koós, Dillon, Dutta, & Grobert, 2015; Meysami, Koós, Dillon, & Grobert, 2013). These short quartz tubes were placed inside a larger quartz reaction tube, inside a tubular furnace. 5 wt% ferrocene in toluene was used as the precursor solution. MWCNTs were then grown over 90 min at 850 °C, using Ar as the carrier gas. At the end of the AACVD experiment, the furnace was allowed to cool to room temperature and the MWCNT-coated quartz tubes were removed. MWCNTs that had formed on the outside of the quartz tubes were mechanically removed to yield 20 mm MWCNT-column reactors (Fig. 9) (Zor et al., 2017).

4.5.2 Carbon nanotube characterization methods

Scanning electron microscopy (SEM) images of the MWCNTs were obtained using a JEOL 840F microscope, with the secondary electron detector operated at 5 kV acceleration voltage. Energy dispersive X-ray

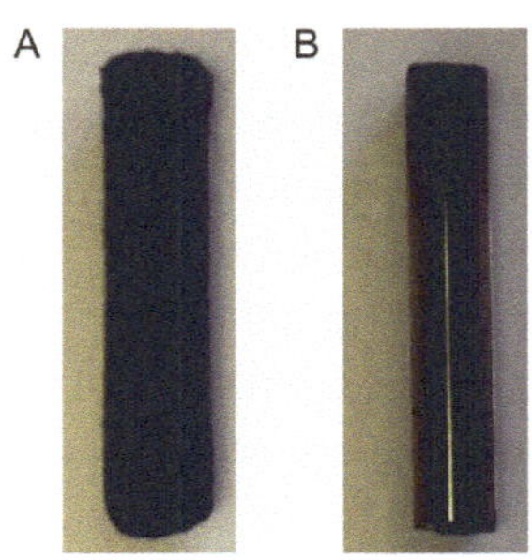

Fig. 9 (A) Photograph of MWCNT-coated quartz tube after synthesis; (B) After removal of MWCNT from the outside of the quartz tube.

spectroscopy (EDX) was also used to determine elemental changes in the MWCNTs after enzyme immobilization and use in flow experiments. EDX mapping was obtained using a Zeiss Merlin field emission analytical scanning electron microscope at 15 kV.

4.5.3 Flow reactor equipment

1. Peristaltic pump (Watson Marlow 120 U) or Syringe pump (Asia, Syrris)
2. Enzyme-modified MWCNT-column reactor
3. Marprene tubing (1/32″ inner diameter, 1/16″ outer diameter, Watson Marlow)
4. PTFE tubing (1/32″ inner diameter, 1/16″ outer diameter, Supelco)
5. Fittings and connections (PTFE or PEEK)
6. Reservoir flask/vessel

4.5.4 Enzyme immobilization in carbon nanotube column reactor

An unmodified MWCNT-column reactor was connected to the peristaltic pump, using a combination of PTFE and Marprene tubing. The required enzymes were mixed together and then pumped into the MWCNT-column reactor. The MWCNT-column reactor was then disconnected from the pump and stored at 4 °C to allow the enzymes to adsorb. After 90 min, the MWCNT-column reactor was re-connected to the peristaltic pump, and Bis-tris buffer (50 mM, pH 6.0) was pumped through the system to remove any unadsorbed enzymes from the column.

4.5.5 Flow reaction for cofactor conversion

The enzyme-modified MWCNT-column reactor was prepared according to Section 4.5.6, using a mixture of hydrogenase and NAD^+ reductase enzymes, and connected to the peristaltic pump prior to carrying out the reaction. The reaction solution comprising NAD^+ (0.5 mM) in Bis-tris buffer (50 mM, pH 6.0) was H_2-saturated, at atmospheric pressure, in the reservoir vessel for 1 h. The reaction solution was then circulated through the flow reactor (25 μL min^{-1}) for 400 min, while continuously flowing H_2 across the headspace of the reservoir vessel. NADH production was monitored at 340 nm continuously throughout the reaction by flowing into a low volume cuvette (path length 1 cm, Hellma Analytics) before re-circulating into the reactor, and conversion of NAD^+ to NADH was measured at the end-point of the reaction.

4.5.6 Flow reaction for reductive amination of pyruvate

The enzyme-modified MWCNT-column reactor was prepared according to the above procedure, using a mixture of hydrogenase (120 μg), NAD^+ reductase (48 μg) and NADH-dependent L-AlaDH (300 μg), and connected to the peristaltic pump prior to carrying out the reaction. The reaction solution comprising NAD^+ (1 mM), pyruvate (1.99 mM) and ammonium chloride (100 mM) in Bis-tris buffer (50 mM, pH 6.0) was H_2-saturated in the reservoir vessel for 1 h. The reaction solution was then circulated through the flow reactor (25 μL min^{-1}) for 26 h.

4.5.7 Flow reaction for acetophenone reduction

The enzyme-modified MWCNT-column reactor was prepared according to the above procedure, using a mixture of hydrogenase (90 μg), NAD^+ reductase (64 μg) and NADH-dependent ADH105 (1 mg), and connected to the peristaltic pump prior to carrying out the reaction. The reaction solution comprising NAD^+ (0.5 mM) and acetophenone (8.7 mM) in Bis-tris buffer (50 mM, pH 6.0) was H_2-saturated, at atmospheric pressure, in the reservoir vessel for 1 h. The reaction solution was then circulated through the flow reactor (25 μL min^{-1}) for 26 h.

4.5.8 Single-pass flow reaction for acetophenone reduction

An unmodified MWCNT-column reactor was connected to the syringe pump (Asia, Syrris), using PTFE tubing and standard HPLC fittings. A mixture of hydrogenase (70 μg), NAD^+ reductase (22 μg) and NADH-dependent ADH105 (600 μg) were pumped into the MWCNT-column reactor. The MWCNT-column reactor was then disconnected from the syringe pump and stored at 4 °C to allow the enzymes to adsorb. After 90 min, the MWCNT-column reactor was re-connected to the syringe pump, and Tris-HCl buffer (100 mM, pH 8.0) was pumped through the system to remove any unadsorbed enzyme from the column.

The reaction solution comprising NAD^+ (0.1 mM) and acetophenone (1 mM) in Tris-HCl buffer (100 mM, pH 8.0) was H_2-saturated, at atmospheric pressure, in the reservoir vessel for 1 h. The reaction solution was then pumped through the flow reactor (1.3 μL min^{-1}, tRes = 58 min), without re-circulation.

5. Conclusions

In summary, we have taken advantage of the range of carbon materials available to demonstrate H_2-driven biocatalysis in batch and flow reactions using a system of electronically linked redox enzymes. We use the high surface area and simple dispersion of carbon black for batch reactions and demonstrate that the heterogeneous catalysts can be removed from solution and re-used. For translation of H_2-driven biocatalysis into flow reactions, we take advantage of the ability to grow MWCNTs on a quartz tube substrate. This overcomes problems with handling MWCNTs in solution as well as problems with using nanoparticles in packed bed reactors. Combined with the high atom efficiency offered by using H_2 gas as a reductant for biocatalytic reductions, these platform approaches overcome the key challenges for implementing NADH-dependent reactions. We show that a range of carbon materials are suitable for immobilization of hydrogenase, NAD^+ reductase and a number of NADH-dependent enzymes, and further demonstrate that carbon is a cheap, versatile support for development of heterogeneous biocatalytic processes.

Acknowledgments

L.A.T., J.S.R., H.A.R. and K.A.V. are supported by the Engineering and Physical Sciences Research Council grant EP/N013514/1 under the Innovate UK/BBSRC/EPSRC Industrial Biotechnology Catalyst scheme (to H.A.R. and K.A.V.) and INSPIRE award EP/J015202/1 (to N.G. and K.A.V.). N.G. is supported by the Royal Society, European Research Council (ERC-2009-StG-240500-DEDIGROWTH, ERC-2015-POC-680559-CONDUCT, ERC-2011-POC-309786-DEVICE, and ERC-2016-POC-754748-OxfordNano) and the European Commission (CONTACT-FP7-PEOPLE-ITN-2008-238363, BCN-Tubes-FP6-2004-NMP-TI-4-033350). We are grateful to Dr. Oliver Lenz and Dr. Lars Lauterbach (Technical University, Berlin) for preparation of NAD^+ reductase and soluble hydrogenase, and to Prof. Fraser Armstrong and Ms. Elena Nomerotskaia (University of Oxford) for samples of *E. coli* hydrogenases, and to Dr. Beatriz Dominguez (Johnson Matthey) for providing ADH105.

References

Aleku, G. A., France, S. P., Man, H., Mangas-Sanchez, J., Montgomery, S. L., Sharma, M., et al. (2017). A reductive aminase from Aspergillus oryzae. *Nature Chemistry*, *9*(10), 961–969. https://doi.org/10.1038/nchem.2782.

Alonso-Lomillo, M. A., Rüdiger, O., Maroto-Valiente, A., Velez, M., Rodríguez-Ramos, I., Muñoz, F. J., et al. (2007). Hydrogenase-coated carbon nanotubes for efficient H_2 oxidation. *Nano Letters*, 7(6), 1603–1608. https://doi.org/10.1021/nl070519u.

Asedegbega-Nieto, E., Guerrero-Ruiz, A., & Rodríguez-Ramos, I. (2006). Modification of the stereoselectivity in the citral hydrogenation by application of carbon nanotubes as support of the Pt particles. *Carbon*, *44*(4), 804–806. https://doi.org/10.1016/j.carbon.2005.10.043.

Blanford, C. F., & Armstrong, F. A. (2006). The pyrolytic graphite surface as an enzyme substrate: Microscopic and spectroscopic studies. *Journal of Solid State Electrochemistry*, *10*(10), 826–832. https://doi.org/10.1007/s10008-006-0183-2.

Bogdan, A. R., & Dombrowski, A. W. (2019). Emerging trends in flow chemistry and applications to the pharmaceutical industry. *Journal of Medicinal Chemistry*, *62*, 6422–6468. https://doi.org/10.1021/acs.jmedchem.8b01760.

Britton, J., Majumdar, S., & Weiss, G. A. (2018). Continuous flow biocatalysis. *Chemical Society Reviews*, *47*(15), 5891–5918. https://doi.org/10.1039/C7CS00906B.

Cossar, P. J., Hizartzidis, L., Simone, M. I., McCluskey, A., & Gordon, C. P. (2015). The expanding utility of continuous flow hydrogenation. *Organic & Biomolecular Chemistry*, *13*(26), 7119–7130. https://doi.org/10.1039/c5ob01067e.

Dominguez, B. B., Schell, U., Bisagni, S., & Kalthoff, T. (2016). Reduction of activated carbon-carbon double bonds using highly active and enantioselective double bond reductases. *Johnson Matthey Technology Review*, *60*(4), 243–249. https://doi.org/10.1595/205651316X692572.

Feng, W., & Ji, P. (2011). Enzymes immobilized on carbon nanotubes. *Biotechnology Advances*, *29*(6), 889–895. https://doi.org/10.1016/j.biotechadv.2011.07.007.

Grigoras, A. G. (2017). Catalase immobilization—A review. *Biochemical Engineering Journal*, *117*, 1–20. https://doi.org/10.1016/j.bej.2016.10.021.

Hidalgo, R., Ash, P. A., Healy, A. J., & Vincent, K. A. (2015). Infrared spectroscopy during electrocatalytic turnover reveals the Ni-L active site state during H_2 oxidation by a NiFe hydrogenase. *Angewandte Chemie International Edition*, *54*(24), 7110–7113. https://doi.org/10.1002/anie.201502338.

Irfan, M., Glasnov, T. N., & Kappe, C. O. (2011). Heterogeneous catalytic hydrogenation reactions in continuous-flow reactors. *ChemSusChem*, *4*(3), 300–316. https://doi.org/10.1002/cssc.201000354.

Kihara, T., Liu, X.-Y., Nakamura, C., Park, K.-M., Han, S.-W., Qian, D.-J., et al. (2011). Direct electron transfer to hydrogenase for catalytic hydrogen production using a single-walled carbon nanotube forest. *International Journal of Hydrogen Energy*, *36*(13), 7523–7529. https://doi.org/10.1016/j.ijhydene.2011.03.135.

Lauterbach, L., & Lenz, O. (2013). Catalytic production of hydrogen peroxide and water by oxygen-tolerant [NiFe]-hydrogenase during H_2 cycling in the presence of O_2. *Journal of the American Chemical Society*, *135*(47), 17897–17905. https://doi.org/10.1021/ja408420d.

Lazarus, O., Woolerton, T. W., Parkin, A., Lukey, M. J., Reisner, E., Seravalli, J., et al. (2009). Water-Gas shift reaction catalyzed by redox enzymes on conducting graphite platelets. *Journal of the American Chemical Society*, *131*(40). 14154. https://doi.org/10.1021/ja905797w.

Leger, C., & Bertrand, P. (2008). Direct electrochemistry of redox enzymes as a tool for mechanistic studies. *Chemical Reviews*, *108*(7), 2379–2438. https://doi.org/10.1021/cr0680742.

Li, W., Wang, X., Chen, Z., Waje, M., & Yan, Y. (2006). Pt−Ru supported on double-walled carbon nanotubes as high-performance anode catalysts for direct methanol fuel cells. *The Journal of Physical Chemistry B*, *110*(31), 15353–15358. https://doi.org/10.1021/jp0623443.

Liese, A., & Hilterhaus, L. (2013). Evaluation of immobilized enzymes for industrial applications. *Chemical Society Reviews*, *42*(15), 6236–6249. https://doi.org/10.1039/C3CS35511J.

Lukey, M. J., Parkin, A., Roessler, M. M., Murphy, B. J., Harmer, J., Palmer, T., et al. (2010). How Escherichia coli is equipped to oxidize hydrogen under different redox conditions. *The Journal of Biological Chemistry*, *285*(6), 3928–3938. https://doi.org/10.1074/jbc.M109.067751.

McDonald, T. J., Svedruzic, D., Kim, Y.-H., Blackburn, J. L., Zhang, S. B., King, P. W., et al. (2007). Wiring-up hydrogenase with single-walled carbon nanotubes. *Nano Letters*, *7*(11), 3528–3534. https://doi.org/10.1021/nl072319o.

Meysami, S. S., Dillon, F., Koós, A. A., Aslam, Z., & Grobert, N. (2013). Aerosol-assisted chemical vapour deposition synthesis of multi-wall carbon nanotubes: I. Mapping the reactor. *Carbon*, *58*, 151–158. https://doi.org/10.1016/j.carbon.2013.02.044.

Meysami, S. S., Koós, A. A., Dillon, F., Dutta, M., & Grobert, N. (2015). Aerosol-assisted chemical vapour deposition synthesis of multi-wall carbon nanotubes: III. Towards upscaling. *Carbon*, *88*, 148–156. https://doi.org/10.1016/j.carbon.2015.02.045.

Meysami, S. S., Koós, A. A., Dillon, F., & Grobert, N. (2013). Aerosol-assisted chemical vapour deposition synthesis of multi-wall carbon nanotubes: II. An analytical study. *Carbon*, *58*, 159–169. https://doi.org/10.1016/j.carbon.2013.02.041.

Min, K., Kim, J., Park, K., & Yoo, Y. J. (2012). Enzyme immobilization on carbon nanomaterials: Loading density investigation and zeta potential analysis. *Journal of Molecular Catalysis B: Enzymatic*, *83*, 87–93. https://doi.org/10.1016/j.molcatb.2012.07.009.

Plutschack, M. B., Pieber, B., Gilmore, K., & Seeberger, P. H. (2017). The Hitchhiker's guide to flow chemistry. *Chemical Reviews*, *117*(18), 11796–11893. https://doi.org/10.1021/acs.chemrev.7b00183.

Quinson, J., Hidalgo, R., Ash, P. A., Dillon, F., Groberta, N., & Vincent, K. A. (2014). Comparison of carbon materials as electrodes for enzyme electrocatalysis: Hydrogenase as a case study. *Faraday Discussions*, *172*, 473–496.

Reeve, H. A., Ash, P. A., Park, H., Huang, A., Posidias, M., Tomlinson, C., et al. (2017). Enzymes as modular catalysts for redox half reactions in H_2-powered chemical synthesis: From biology to technology. *The Biochemical Journal*, *474*(2), 215–230. https://doi.org/10.1042/BCJ20160513.

Reeve, H. A., Lauterbach, L., Ash, P. A., Lenz, O., & Vincent, K. A. (2012). A modular system for regeneration of NAD cofactors using graphite particles modified with hydrogenase and diaphorase moieties. *Chemical Communications*, *48*(10), 1589–1591. https://doi.org/10.1039/C1CC14826E.

Reeve, H. A., Lauterbach, L., Lenz, O., & Vincent, K. A. (2015). Enzyme-modified particles for selective biocatalytic hydrogenation by hydrogen-driven NADH recycling. *ChemCatChem*, 7(21), 3480–3487. https://doi.org/10.1002/cctc.201500766.

Rodrigues, R. C., Ortiz, C., Berenguer-Murcia, Á., Torres, R., & Fernández-Lafuente, R. (2013). Modifying enzyme activity and selectivity by immobilization. *Chemical Society Reviews*, *42*(15), 6290–6307. https://doi.org/10.1039/C2CS35231A.

Sheldon, R. a., & van Pelt, S. (2013). Enzyme immobilisation in biocatalysis: Why, what and how. *Chemical Society Reviews*, *42*, 6223–6235. https://doi.org/10.1039/c3cs60075k.

Sheldon, R. A., & Woodley, J. M. (2018). Role of biocatalysis in sustainable chemistry. *Chemical Reviews*, *118*(2), 801–838. https://doi.org/10.1021/acs.chemrev.7b00203.

Thompson, M. P., Peñafiel, I., Cosgrove, S. C., & Turner, N. J. (2019). Biocatalysis using immobilized enzymes in continuous flow for the synthesis of fine chemicals. *Organic Process Research & Development*, *23*(1), 9–18. https://doi.org/10.1021/acs.oprd.8b00305.

Vincent, K. A., Li, X., Blanford, C. F., Belsey, N. A., Weiner, J. H., & Armstrong, F. A. (2007). Enzymatic catalysis on conducting graphite particles. *Nature Chemical Biology*, *3*(12), 761–762. https://doi.org/10.1038/nchembio.2007.47.

Woodley, J. M. (2008). New opportunities for biocatalysis: Making pharmaceutical processes greener. *Trends in Biotechnology*, *26*(6), 321–327. https://doi.org/10.1016/j.tibtech.2008.03.004.

Zhang, W., Zhu, S., Luque, R., Han, S., Hu, L., & Xu, G. (2016). Recent development of carbon electrode materials and their bioanalytical and environmental applications. *Chemical Society Reviews*, *45*, 715–752. https://doi.org/10.1039/c5cs00297d.

Zor, C., Reeve, H. A., Quinson, J., Thompson, L. A., Lonsdale, T. H., Dillon, F., et al. (2017). H_2-Driven biocatalytic hydrogenation in continuous flow using enzyme-modified carbon nanotube columns. *Chemical Communications*, *53*(71), 9839–9841. https://doi.org/10.1039/C7CC04465H.

CHAPTER FOURTEEN

Nano-immobilized cellulases for biomass processing with application in biofuel production

Reinu E. Abraham, Munish Puri*
Centre for Marine Bioproducts Development, College of Medicine and Public Health, Flinders University, Bedford Park, SA, Australia
*Corresponding author: e-mail address: munish.puri@flinders.edu.au

Contents

Abstract

Several researches have focused on the enzymatic pretreatment of lignocellulose biomass to produce fermentable sugars that can lead to ethanol production thus facilitating pathways for sustainable biofuel production. Enzymes are fundamental to the pretreatment process, however, are required in larger quantities during pretreatment process thus influencing biofuel production cost. Immobilization of enzymes to a suitable support/matrix could enhance its stability, and reusability thus containing cost.

Methods in Enzymology, Volume 630
ISSN 0076-6879
https://doi.org/10.1016/bs.mie.2019.09.006

This chapter focuses on developing an advanced technology for immobilizing enzymes to nanomaterials; variety of nanomaterials used for immobilization, nature of enzyme/protein nanomaterial interactions, methods of enzyme immobilization, and factors affecting mode of interaction for achieving hydrolysis of microcrystalline cellulose and natural cellulosic substrate. The binding of enzyme (94%) to a nanomaterial was established by spectroscopy techniques. The kinetics study, conducted at optimum pH (pH 4) and temperature (50 °C for free and 60 °C immobilized enzyme), exhibited improvement in immobilized enzyme properties. The immobilized enzyme retained up to 50% of its enzyme activity in five consecutive cycles. This chapter advocates the use of nano-immobilized enzymes in biomass hydrolysis for biofuel production.

1. Introduction

1.1 Role of enzyme(s) in bioprocessing

The increasing demand on fossil fuel consumption and depleting reserves has raised its prices globally and calls for an alternative solution. Thus, the application of renewable sources for the production of biofuels has gained increased attention in recent times. The utilization of untapped lignocellulosic waste biomass to produce lignocellulosic ethanol is a promising solution to depleting fossil fuel reserve (Dixit, Basu, Puri, Mathur, & Barrow, 2019; Puri, Abraham, & Barrow, 2012). This process comprises simple conversion of cellulosic polymer into fermentable sugar and then its fermentation to produce ethanol. However, lignocellulosic biomass is robust and tough in its structure due to intertwining with various complex compounds. The lignocellulosic biomass is composed of lignin, cellulose, hemicellulose, pectin, ash, minerals and proteins that cross-links to form rigid and strong bonds (Abraham, Verma, Barrow, & Puri, 2014; Correa et al., 2019).

The utilization of lignocellulosic biomass to produce biofuel occurs in various stage(s) that includes (i) pretreatment of lignocellulose to open its rigid structure and break the strong glycosidic bonds. (ii) Thereafter, its hydrolysis into fermentable sugars using group of hydrolytic enzymes that slowly digests the polysaccharides and produce monomers. And (iii) this is followed by the fermentation of these monomeric sugar into ethanol using a yeast/bacterium (Chandel et al., 2019; Viccaro, Cozzi, Rocchi, & Romano, 2019). Despite these sequential processing technologies are mapped there are a range of challenges that are encountered during the processing of the biomass (Li, Sakuragi, & Makino, 2019). These challenges broadens with the nature of the biomass that includes hardwood, softwood, and grasses; the synergistic action of enzymes and its high efficiency to produce fermentable sugars, various inhibitors that are formed are a result of biomass digestion which greatly interferes between the enzymatic digestion and converts the monomers into acids and furfurals (Abraham, Barrow, & Puri, 2013). In addition, process too is dictated by challenges such as the conversion efficiency of yeast that affects yield and quality of ethanol produced; and ethanol tolerance duration the course of fermentation that could result in the higher production cost (Raud, Kikas, Sippula, & Shurpali, 2019).

Various strategies have been applied to develop an economic conversion pathway for bioethanol production, which includes combination of various pretreatment approaches, such as simultaneous enzyme hydrolysis and fermentation, immobilization of enzyme(s) and using recombinant yeast to improve its efficiency (Majidian, Tabatabaei, Zeinolabedini, Naghshbandi, & Chisti, 2018). The dosage of enzyme during the pretreatment step to produce fermentable sugars is one of the major costs that contributes toward high price of enzyme. Enzymes play a crucial role in the bioprocessing of biomass, based on enzyme substrate interaction leading to enzyme substrate complex that leads to product formation, which depends on the activity of enzymes and supportive reaction conditions (Singh, Mathur, Barrow, & Puri, 2018; Singh, Mathur, Gupta, Puri, & Puri, 2018).

In order to economize enzyme consumption in the pretreatment process, enzyme may be engineered or immobilized to improve its properties. Enzymes such as cellulase (comprised of following activities β-glucosidase, exo/endoglucanase and hemicellulase), xylanase, and pectinase are major group of hydrolytic enzymes that work synergistically and used for the lignocellulose hydrolysis either as sole or in combinations for biofuel production (Bala & Singh, 2019; Singh, Mathur, Barrow, & Puri, 2018; Singh, Mathur, Gupta, et al., 2018).

Immobilization of an enzyme refers to confinement of an enzyme in a certain region of defined space with retention of its activity so that it can be used repeatedly without experiencing a loss of an activity (Puri, Verma, & Barrow, 2013). It's considered that the application of these enzymes when immobilized onto a matrix (also referred to as "support" in the literature) increases its stability, reusability and easier product separation since some matrix used for enzyme immobilization are magnetic in nature. The immobilization techniques involve binding of an enzyme to a matrix, which is usually activated by a cross-linker that attaches to suitable binding sites on the enzyme. The immobilized enzymes provide better thermostability over free enzymes that makes them superior over free enzyme, confirming long term storage stability (Amaly et al., 2018).

1.2 Type of nanomaterials used for enzyme immobilization

The matrix used for enzyme immobilization varies and includes nanofibers, nanoparticles (NPs), organic metal frames while provides similar function (Khoshnevisan et al., 2017). Even though the nanomaterials are quite small, but it provides high surface area for enzyme to attach (Wu, Zeng, Ding, Tu, & Xia, 2019). The properties of nanomaterial, which includes surface area, binding capacity, stability, non-reacting properties, toxicity with other materials and easy recovery, are highly significant, as alteration in any of these properties would hinder the enzymatic reaction and later the biomass hydrolysis (Raza, Yong, & Deng, 2019; Yuan, Yang, Xue, Feng, & Jiang, 2016). Nanomaterials used for enzyme immobilization are mentioned below in a Table 1.

The enzyme can bind to the matrix in various ways that includes entrapment, cross-linking, adsorption and encapsulation (Poorakbar et al., 2018). The attachment can be done using a cross-linker which binds matrix and enzyme together as shown in Fig. 1.

However, immobilization of enzyme on a carbon nanotube (CNT) has also been studied but not extensively and rapidly emerging area (Feng & Ji, 2011). The fabrication of CNTs take places through various methods including chemical vapor method, discharge, and laser ablation. Previous research suggests that the CNTS provide high strength rheological properties, thermal and electrical properties (Zhang & Henthorn, 2010). CNTs can also be a good candidate to develop biosensors for detecting glucose (Wang et al., 2003). Laccase, an enzyme used in biomass peroxidation, was immobilized over MWCNTs by adsorption. Biocatalytic performance of immobilized enzyme is dependant mostly on the intrinsic properties of both enzyme and support, immobilization technique and immobilization conditions employed. Upon optimization, a decrease in thermal stability of the immobilized enzyme was observed.

1.3 Enzyme immobilization and hydrolysis

Cellulases, from *Aspergillus niger,* was immobilized onto functionalized multiwalled carbon nanotubes (MWCNTs) via carbodiimide coupling. The immobilization of cellulase on MWCNTs offer unique advantages including enhanced electronics properties, a large edge to basal plane ratio,

Table 1 List of some common nano-supports used for immobilizing enzymes.

Nanosupport (matrix)	Enzyme	References
Zinc magnetite nanoparticle; magnetic nanoparticle; acrylonitrile copolymer nanofiber	Cellulase	Abraham et al. (2014), Cheng and Chang (2013), and Hegedüs, Hancsók, and Nagy (2012)
Fe_3O_4-coated chitosan material; glutaraldehyde-alginate beads; carbon nanotube	Xylanase	Liu, Dai, Guan, and Xu (2014), Pal and Khanum (2011), and Shah and Gupta (2008)
Silicon oxide	β-glucosidase	Singh, Zhang, Nguyen, Jeya, and Lee (2011)
SiO_2 microparticles with carbon nanotube-derived mesopores	Lipase from *Thermomyces lanuginosus*	Kumar et al. (2019)
Gold nanoparticles and carbon nanotube	1-(3-(Dimethylamino)-propyl)-3-ethylcarbodiimide hydrochloride (EDC) and N-hydroxysulfosuccinimide (NHS)	Zeng et al. (2019)
Carbon nanotubes	Glucose oxidase from *Aspergillus niger*	Ghica, Pauliukaite, Fatibello-Filho, and Brett (2009)
Multi-walled carbon nanotubes (MWCNTs)	Cellulase	Ahmad and Khare (2018)

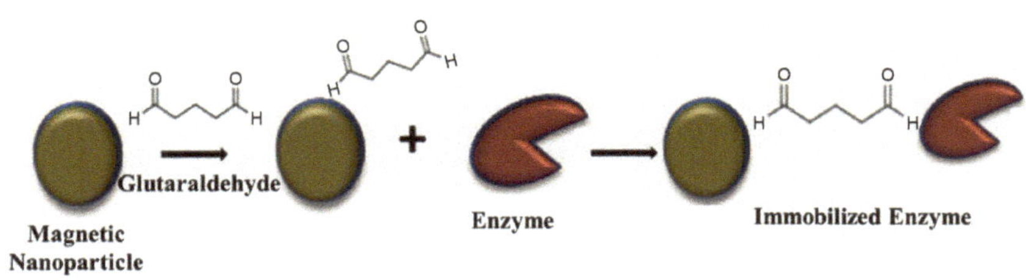

Fig. 1 Schematic representation of binding of a nanoparticle to an enzyme (NP was activated using a glutaraldehyde, which acts as a cross-linker). *This figure is reproduced from Abraham, R. E., Verma, M. L., Barrow, C. J., & Puri, M. (2014). Suitability of magnetic nanoparticle immobilized cellulases in enhancing enzymatic saccharification of pretreated hemp biomass.* Biotechnology for Biofuels, *7, 90.*

and rapid electrode kinetics. The immobilized preparation reused for 10 × without much loss in enzyme activity and resulted in efficient cellulose hydrolysis (Ahmad & Khare, 2018).

A nanoplatform (multifunctional hyperbranched polyglycerol-grafted) consisting of a magnetic nanoparticle supported by hyperbranched polyglycerol has been studied for xylanase immobilization. A variety of analytical tools (FT-IR, TGA and elemental analysis) confirmed binding an enzyme to the magnetic nanoparticle. The immobilized enzyme exhibited substantial improvement in reusability, and storage stability of xylanase making it suitable for wide range of application (Landarani-Isfahani, Taheri-Kafrani, & Razmjou, 2015).

Purified novel β-glucosidase (44 KDa) from *Streptomyces griseus* was successfully immobilized onto zinc oxide nanoparticles by adsorption. The immobilized enzyme was characterized in terms of pH (6.9), temperature (69 °C) and could be reused up to 10 × without experiencing a significant loss of activity. Nanoparticle immobilized glucosidase exhibited greater thermostability and used for alkyl glucosides hydrolysis (Kumar, Ryan, & Henehan, 2017).

2. Experimental

2.1 Chemicals and reagents

Following chemicals were used in the process

1. Iron (III) chloride hexahydrate ($FeCl_3.6H_2O$)
2. Iron (II) chloride tetrahydrate ($FeCl_2.4H_2O$)
3. Zinc chloride ($ZnCl_2$)
4. Sodium hydroxide (NaOH)
5. Carboxymethyl cellulose (CMC)
6. Lignocellulosic biomass (Hemp hurds)
7. 3,5-Dinitrosalicylic acid (DNS)
8. Cellulase enzyme (1–2 mg)
9. Sodium acetate
10. Glutaraldehyde
11. Potassium hydrogen phthalate
12. Carboxymethyl cellulose
13. Sodium hydroxide
14. Sodium citrate
15. Potassium phosphate

2.2 Equipment

1. Freeze dryer
2. Water bath
3. Orbital shaker
4. Scanning Electron Microscope (SEM)
5. Fourier-transform infrared spectroscopy (FTIR)
6. Micropipette
7. UV-spectrophotometer

2.3 Preparation of nanomaterials w.r.t. magnetic nanoparticles

The preparation of nanomaterial varies with type of material and its application. However, the fabrication of a magnetic nanomaterial requires additional step of doping into magnetite and following testing of magnetic properties to determine magnetization of nanoparticles. In a below mentioned protocol, we have determined the procedure to prepare a magnetic nanomaterial. The particle size distribution of iron oxide is shown as an example in Fig. 2 which represents the size profile of a nanoparticle.

2.3.1. Mix aqueous solution of iron (III) chloride hexahydrate ($FeCl_3.6H_2O$) and iron (II) chloride tetrahydrate ($FeCl_2.4H_2O$) and zinc chloride ($ZnCl_2$) in molar ratio Fe3+:Fe2+:Zn2+=2.0:0.6:0.4. The dopping of zinc into magnetite was conducted to increase the saturation magnetization of nanoparticles.

2.3.2. Aqueous solution of sodium hydroxide (NaOH) was added subsequently to neutralize the pH of the above mixture which form precipitates.

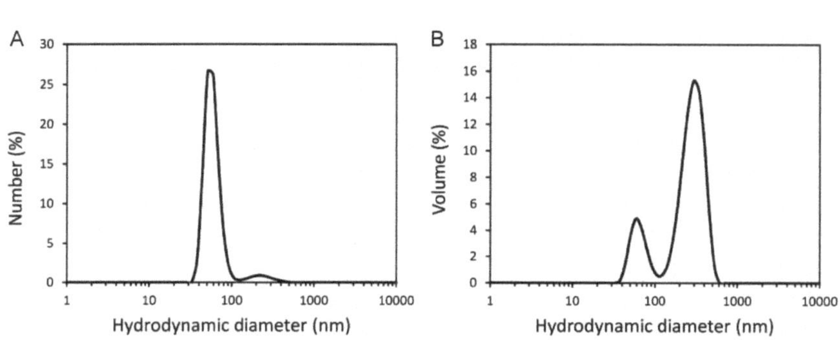

Fig. 2 The hydrodynamic particle size distribution of iron oxide nanoparticles; (A) in number-weighted and (B) volume-weighted distributions.

2.3.3. The obtained precipitates were subjected to hydrothermal treatment (water bath) for 12h at 150 °C.

2.3.4. After the hydrothermal treatment the precipitates were thoroughly rinsed with deionized water.

2.3.5. The precipitates were freeze dried at −80 °C and 0.014 mbar for 24 h and the magnetic nanoparticle was collected and stored at room temperature.

3. Immobilization of cellulases

In this study, enzyme cellulase was immobilized on a synthesized magnetic nanoparticle(s). The cellulase enzyme follows Koshland-type mechanism that acts in two dimension and classified in glycosyl hydrolase family according to Carbohydrate-Active enZYmes Database (CAZy). Cellulase enzymes are further subgrouped into Endoglucanases/Endocellulase, Exoglucanase/Exocellulase and β-glucosidase which works in synergistic action to cleave the bonds and reduce polymeric chain into monomers (Rabinovich, Melnick, & Bolobova, 2002). However, genetically modified cellulase with the synergistic enzyme can be quite efficient in hydrolysing pretreated biomass. The binding efficiency of nanomaterial can be determined by estimating the percentage of enzyme bound to the enzyme added into the system. Higher the protein bound onto the nano-support higher will be the binding efficiency.

$$\text{Binding efficiency (\%)} = \frac{\text{Total amount of protein binded}}{\text{Total amount of protein added}} * 100$$

Cellulase was immobilized onto an activated nano-support and the protocol is described below:

3.1 Preparation of immobilized cellulase

3.1.1. The magnetic nanoparticle at a concentration of 5 mg/mL was suspended in deionized water.

3.1.2. This suspension was sonicated for 1 h to obtain a homogenous solution and thereafter suspended in 1 M glutaraldehyde solution to activate the nanoparticle.

3.1.3. This suspension was incubated at 25 °C for 1 h in an orbital shaker at 250 rpm.

3.1.4. The activated nanomaterial support was thoroughly rinsed thrice and suspended in sodium acetate buffer. The washing was conducted by placing a magnet next to suspension vial, which attracts all the nanomaterial leaving behind the suspension liquid.

3.1.5. The immobilization of enzyme onto the activated nanoparticle was achieved by incubating the activated suspension with 5 mg/mL of cellulase enzyme for 2 h at 25 °C.

3.1.6. The immobilized enzyme nanomaterial was separated using a magnet and the suspension was used to estimate the unbound protein.

3.1.7. The immobilized mixture was thoroughly washed thrice with deionized water twice and the wash was collected.

3.1.8. The immobilized suspension and washing liquid was used to estimate the unbound and removed protein by Bradford method (Bradford, 1976).

3.1.9. The binding efficiency of the immobilized enzyme was calculated according to the formula and plot as shown in Fig. 3.

$$\text{Binding efficiency } (\%) = \frac{\text{Total amount of protein binded}}{\text{Total amount of protein added}} * 100$$

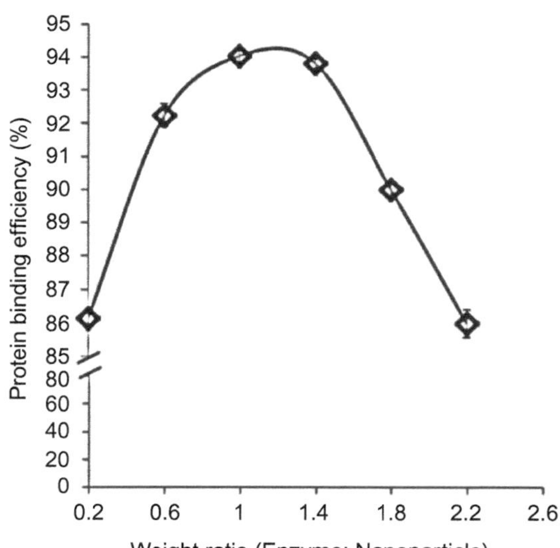

Fig. 3 Binding efficiency of cellulase onto magnetic nanoparticles with varying concentration of protein loading (Abraham et al., 2014).

3.2 Enzyme assay for β-glucosidase

Beta-glucosidase (BGL) or cellobiase is an enzyme that acts on cellobiose to further break down the structure into monomeric units of glucose. The enzyme assay as immobilized and free enzyme is given below (Verma, Chaudhary, Tsuzuki, Barrow, & Puri, 2013):

3.2.1. The reaction mixture contained 2.9 mL of 0.02 M sodium acetate buffer at pH 4.0, the substrate used for this study was 10.0 mM *p*-nitrophenyl-*b*-D-glucopyranoside (0.09 mL).

3.2.2. Enzyme solution containing 2 U of enzyme (Beta-glucosidase (EC 3.2.1.21) from *A. niger* (60 U/g) was used) and was incubated at 60 °C for 5 min before being added into the substrate. This allows enzyme to activate in the system. Once the enzyme was activated the substrate was added into the enzyme tube and incubated at 60 °C for 10 min. The reaction was stopped by adding 0.5 M sodium carbonate (0.5 mL) into the reaction mixture.

3.2.3. The released product was 4-nitrophenol which was measured using UV–vis spectrophotometer (Shimadzu, Japan) at wavelength 405 nm and its concentration calculated from a plot constructed for standard *p*-nitrophenol. The standard curve of *p*-nitrophenol was prepared before the enzyme assay.

3.2.4. One unit of enzyme activity is defined as 1 mol of *p*-nitrophenol liberated per minute at pH 4.0, and a temperature of 60 °C. Activity of immobilized BGL (5 mg) was measured with pNPG (equal amounts corresponding to free enzyme) and cellobiose at 60 °C.

3.2.5. For immobilized enzyme, the reaction was continued, and the aliquots were removed at intervals of 4 h to measure the degree of cellobiose hydrolysis. All enzyme assays were performed in triplicate reported as mean values ± SD.

3.3 Enzyme assay for xylanase

Xylanase activity was measured according to modified Bailey, Biely, and Poutanen (1992).

3.3.1. The reaction contained 900 μL of 1% (w/v) solubilized birchwood xylan solution.

3.3.2. To this solution 100 μL enzyme solution (xylanase/hemicellulose) was added in a test tube and incubated at 50 °C for 5 min in water bath.

3.3.3. The reaction was stopped by adding 1.5 mL dinitrosalicylic acid (DNS) reagent and boiled for 5 min.

3.3.4. The absorbance was measured at 540 nm using UV–vis spectrophotometer (Shimadzhu, Japan) and the reaction was terminated at zero time in the control tubes.

3.3.5. The standard curve for the enzyme assay was prepared using 0–500 μg xylose. An autozero was set using buffer solution.

3.3.6. The activity of xylanase enzyme is defined as the amount of enzyme that produced 1 μmol of reducing sugars equivalent to xylose per minute under the assay conditions described.

3.4 Kinetics properties of immobilized enzyme

3.4.1. The kinetic study of immobilized enzyme is estimated by enzyme assay such as CMC assay (carboxymethyl assay) at 50 °C.

3.4.2. The reaction was set at 50 °C using 0.5 mL immobilized and free enzyme (about 20 CMC units) and 0.5 mL of 2% (w/v) substrate (CMC) dissolved in 0.1 M sodium acetate buffer (pH 4.0) and incubated for 30 min.

3.4.3. This reaction was stopped by using 3 mL of DNS reagent and boiled 5 min in a vigorously boiling water bath.

3.4.4. The concentration of glucose released was estimated at 540 nm using a spectrophotometer and the concentration was estimated against the glucose standard curve.

3.4.5. The same experiment was repeated in a range of temperature (40–80 °C), pH (3.5–6), substrate concentration (Carboxymethyl cellulose—0.5–2.5%, w/v) and buffer (sodium citrate, sodium acetate, phosphate buffer) to estimate its kinetics and optimize best condition of immobilized enzyme as shown in Fig. 4A and B.

3.5 Characterization of nano-immobilized enzyme

The structure of immobilized enzyme onto a nano-support can be characterized using analytical methods which confirms its attachment onto the surface of nano-support. Techniques including scanning electron microscopy (SEM), Transmission electron microscopy (TEM), Fourier-transform infrared spectroscopy (FTIR), atomic force microscopy (AFM), Circular dichroism (CD) spectroscopy and mass spectroscopy are some of the common confirmatory techniques used to confirm the immobilization (Medina, Sáez Del Bosque, Frías, Sánchez De Rojas, & Medina, 2017). In this write-up, we have provided information about TEM and FT-IR spectroscopy.

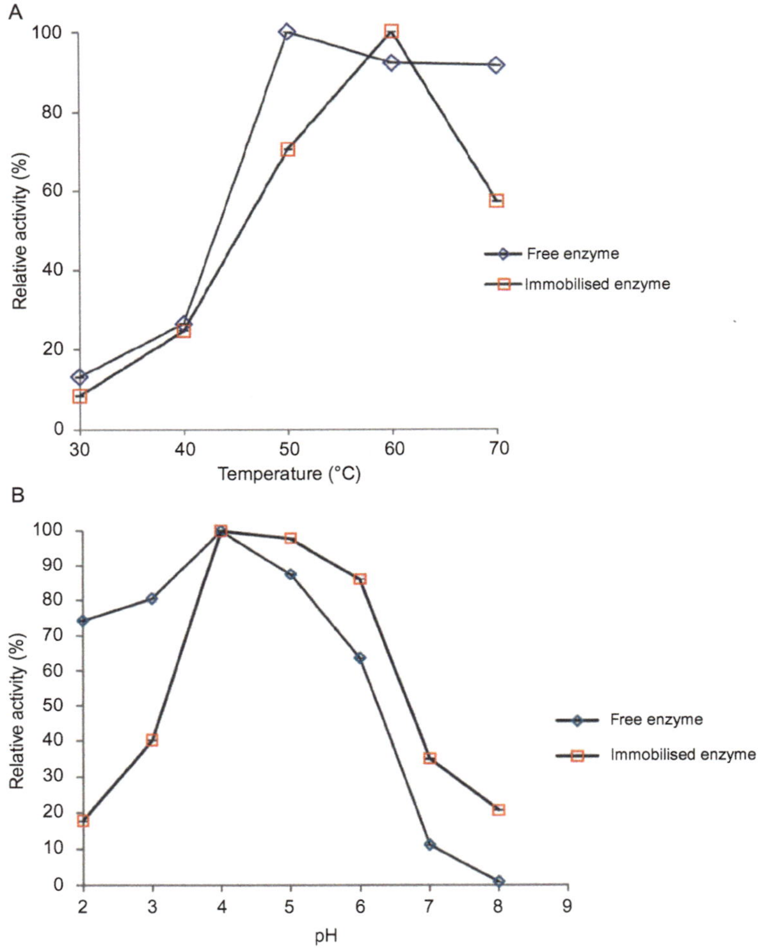

Fig. 4 (A) Optimization of temperature and (B) pH as a result of relative activity of free and immobilized enzyme (Abraham et al., 2014).

3.5.1 Transmission electron microscopy (TEM)

The confirmation study of immobilization can be achieved using Transmission electron microscopy (TEM) as shown in Fig. 5A and B using an example of magnetic iron oxide nanoparticles. The difference is clearly evidenced in Fig. 5A and B, the activated nanomaterial has separated and after immobilization a thin layer of enzyme on the surface of nanomaterial has appeared indicating the attachment of enzyme onto the surface of nanomaterial.

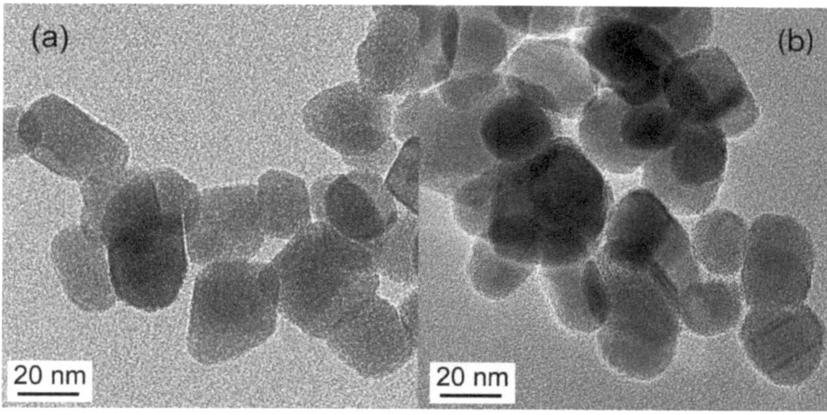

Fig. 5 Transmission electron microscopy images of free (A) and enzyme immobilized (B) iron oxide nanoparticles (Verma et al., 2013).

3.5.2 Fourier transform infrared spectroscopy (FTIR) of immobilized enzyme

FTIR is a technique, which is quick, easy and economical compared to other sophisticated characterization techniques. This technique can be used for the confirmation of immobilization which is seen as additional peaks on the FTIR spectrum (as seen in Fig. 6). The magnetic nanomaterial has relatively flat spectrum, however, after immobilization the peaks have appeared in the spectrum that are aligning with the peaks coming from cellulase enzyme. The methodology for conducting an FTIR experiment is described below:

3.5.2.1 The ATR-FTIR spectroscopy was used to determine the attachment of enzyme (cellulase) onto the activated nanoparticle.

3.5.2.2 The spectrum was recorded using an FTIR spectrometer (BrukerOptik GmbH, Ettingen, Germany). The FTIR instrument was equipped with the detector deuterated triglycine sulfate (DTGS) with a single reflection diamond ATR sampling module (Platinum ATR QuickSnap™).

3.5.2.3 The scanning range was set from 2200 to 400 cm^{-1} with a scanning resolution of 4 cm^{-1} and 64 scans per sample on the OPUS 6.0 suite (Bruker) software.

3.5.2.4 A test run was conducted before the sample run to evaluate the instrument.

3.5.2.5 Once the instrument indicates it as a successful run. Then the samples (immobilized enzyme, free enzyme and nanoparticle) can be

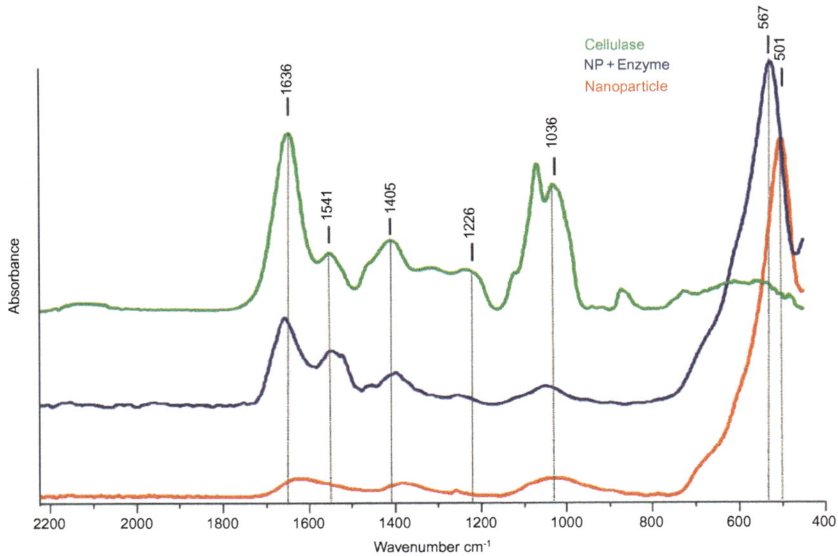

Fig. 6 Comparison of immobilized enzyme, free and nanomaterial bound enzyme using FTIR (Verma et al., 2013).

placed on top of the diamond crystal carefully. Now press the lock keeping a note the samples are right above the diamond crystal.

3.5.2.6 The peaks were overlaid on top of each other and then the major peaks were compared and marked on the graph for easy comparison on the software.

4. Reusability of immobilized enzyme

This study determines the reusability of immobilized enzyme. The objective of this study was to achieve a high level of enzyme reusability. This allows cost-efficiency of biomass bioprocessing due to less intake of enzyme. Reusability leads to less use of fresh enzyme in the bioprocessing with easier separation. The protocol of reusability study is mentioned below:

4.1 The reusability of immobilized enzyme is determined at its optimized condition (highest activity given by enzyme-pH, temperature, and substrate concentration).

4.2 The immobilized preparation was thoroughly washed with deionized water and the same CMC assay (step 4.3) was repeated until the relative activity of enzyme reaches to 50%.

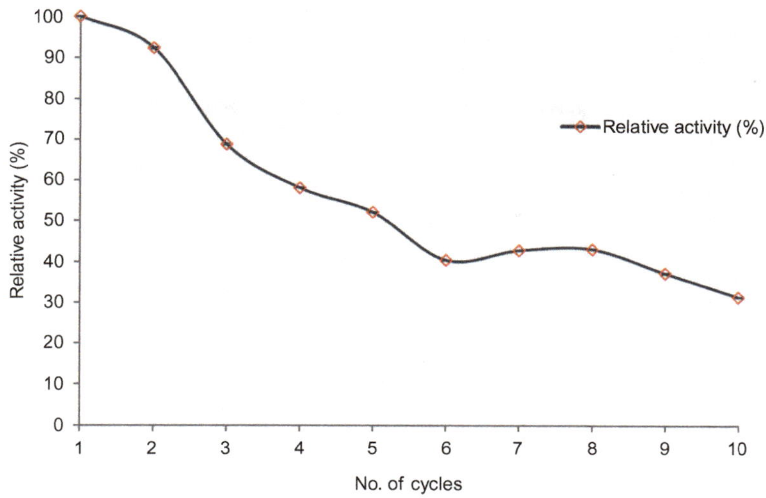

Fig. 7 Reusability of immobilized enzyme: Relative activity of the immobilized enzyme is plotted against the number of reusable cycles.

4.3 After every thorough wash immobilized enzyme was suspended in 0.5 mL of buffer and 0.5 mL of substrate and the CMC assay cycles were repeated.

4.4 The first cycle of the enzyme was taken as the 100% activity and the remaining cycles were plotted against it as shown in Fig. 7.

5. Application of nanomaterial immobilized enzyme

The application of a magnetic nonmaterial for enzyme immobilization makes it an excellent model, as it conforms easier handling and recovery at the later stage (Abraham et al., 2014). However, the nanomaterial should exhibit strong magnetic properties in order to make the bioprocessing efficient. The immobilization of enzyme and its hydrolysis includes various steps such as preparation of nanomaterial, immobilization of enzyme, kinetic study of enzyme, (as explained in Sections 2–4) and pretreatment of biomass and hydrolysis of biomass using immobilized enzyme (Verma et al., 2013).

5.1 Pretreatment of biomass and enzyme hydrolysis

Pretreatment of biomass was conducted to break glycosidic bonds and open the structure of biomass. Hemp hurds (*Cannabis sativa;* Industrial hemp variety) biomass was used in this study. Biomass was pretreated using sodium hydroxide. The pretreatment was carried inside an

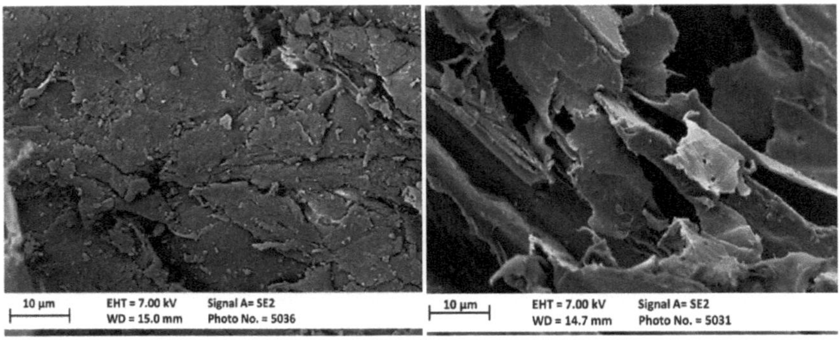

Fig. 8 SEM images of lignocellulose biomass (A) raw/untreated biomass (hemp hurds) and (B) pretreated with sodium hydroxide (Abraham et al., 2014).

autoclave (60 min at 120 °C). The structure of the hemp hurds break opens under high temperature and pressure and the presence of sodium hydroxide removes hemicellulose and lignin. The effect of pretreatment on hemp biomass is shown in Fig. 8 as an example using scanning electron microscope (SEM).

Fig. 8 clearly demonstrates the opening of the hemp structure after pretreatment compared to the control. The raw hemp hurd image shows a rigid structure and rough surface. The enzymes used for this study was cellulase from *Trichoderma reesei* procured from Sigma-Aldrich. The single enzyme was powerful enough to hydrolyse the structure of the biomass and break into glucose monomer unit. The pretreatment procedure conducted on lignocellulosic biomass is mentioned below:

5.1.1 The lignocellulosic biomass (hemp hurd was taken as an example) was first pretreated to open the structure of biomass using sodium hydroxide (1%, w/v).

5.1.2 The dried and milled hemp powder was weighing 2%, w/v was added in a schott bottle along with the sodium hydroxide (1%, w/v) and 100 mL of water. This slurry was then autoclaved at 121 °C for 60 min and taken out once completely cooled.

5.1.3 The pretreated biomass was separated from the slurry using a filter paper, and then thoroughly washed thrice. This biomass was then dried at 50 °C.

5.1.4 The dried biomass weighing (1:1, 1:2, 1:3 and 1:4 ratio of Enzyme: substrate) was added into the flask and another flask containing CMC was taken and the immobilized enzyme (5 mg) was added and incubated in an orbital shaker and allowed to shake for 48 h at 50 °C in a volume of 25 mL.

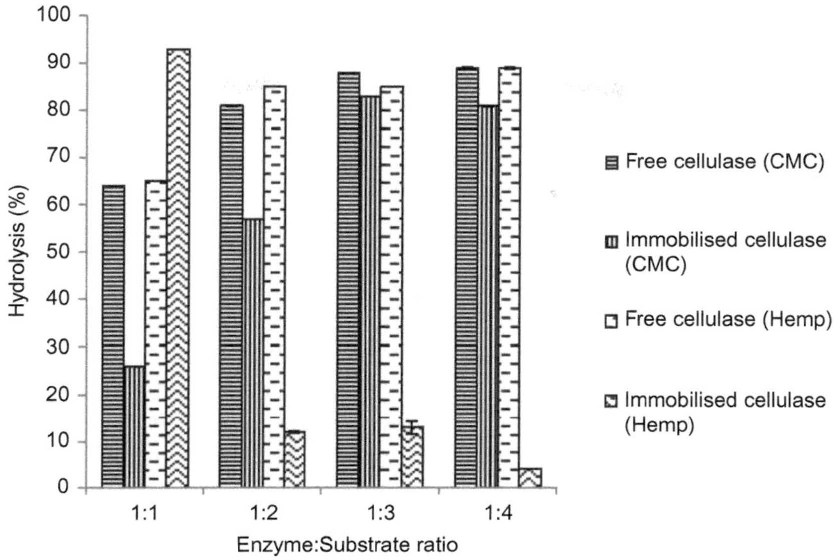

Fig. 9 Enzymatic hydrolysis of CMC and pretreated lignocellulosic biomass by free and immobilized enzyme w.r.t. percentage of cellulose hydrolysis (Abraham et al., 2014).

5.1.5 Sample were taken out after an interval of every 12h and DNS reagent was added in a volume of 3mL and glucose estimation was done as mentioned in Section 3.4.

5.1.6 The hydrolysis/digestion percentage was calculated using below formula and the graph was plotted against the percentage of hydrolysis with different enzyme: substrate ratio as shown in Fig. 9.

$$\text{Cellulose digested} = \text{Glucose concentration}{*}v\ (\text{total reaction volume})\ {*}0.9\ (\text{correction factor})$$

$$\text{Cellulose hydrolysis } (\%) = \frac{\text{Amount of cellulose digested}}{\text{Amount of cellulose added}} {*}100$$

6. Conclusion

The immobilization of cellulase onto an activated magnetic nanoparticle exhibiting better substrate hydrolysis was demonstrated. FTIR spectroscopy was used for the confirmation of enzyme binding onto the nanoparticle. The activity of free and nanomaterial immobilized enzyme was compared at optimum pH and temperature conditions. The reusability of nanoimmobilized enzyme exhibited five cycles of substrate hydrolysis and easier separation leading to improved product yields. Enzyme immobilization on a nanomaterial is a potential method in hydrolysing biomass. This technique brings cost-effectivity by improving efficient hydrolysis of the biomass for biofuel production.

Acknowledgments

Authors thank publishers of the journals Bioresource Technology (Elsevier Press) and Biotechnology and Biofuels (Springer Nature Press) for granting permission to use various figures in the given write-up.

Conflict of interest

The authors of this chapter have no conflict of interest.

References

Abraham, R. E., Barrow, C. J., & Puri, M. (2013). Relationship to reducing sugar production and scanning electron microscope structure to pretreated hemp hurd biomass (Cannabis sativa). *Biomass and Bioenergy*, *58*, 180–187.

Abraham, R. E., Verma, M. L., Barrow, C. J., & Puri, M. (2014). Suitability of magnetic nanoparticle immobilized cellulases in enhancing enzymatic saccharification of pretreated hemp biomass. *Biotechnology for Biofuels*, *7*, 90.

Ahmad, R., & Khare, S. K. (2018). Immobilization of *Aspergillus niger* cellulase on a multiwall carbon nanotube for cellulose hydrolysis. *Bioresource Technology*, *252*, 72–75.

Amaly, N., Si, Y., Chen, Y., El-Moghazy, A. Y., Zhao, C., Zhang, R., et al. (2018). Reusable anionic sulfonate functionalized nanofibrous membranes for cellulase enzyme adsorption and separation. *Colloids and Surfaces B: Biointerfaces*, *170*, 588–595.

Bailey, M. J., Biely, P., & Poutanen, K. (1992). Interlaboratory testing of methods for assay of xylanase activity. *Journal of Biotechnology*, *23*, 257–270.

Bala, A., & Singh, B. (2019). Cellulolytic and xylanolytic enzymes of thermophiles for the production of renewable biofuels. *Renewable Energy*, *136*, 1231–1244.

Bradford, M. M. (1976). A rapid and sensitive method for the quantitation of microgram quantities of protein utilizing the principle of protein-dye binding. *Analytical Biochemistry*, *72*, 248–254.

Chandel, A. K., Albarelli, J. Q., Santos, D. T., Chndawat, S. P., Puri, M., & Meireles, M. A. (2019). Comparative analysis of key technologies for cellulosic ethanol production from Brazillian sugarcane bagasse at a commercial scale. *Biofuels, Bioproducts and Biorefining*, *13*, 994–1014.

Cheng, C., & Chang, K.-C. (2013). Development of immobilized cellulase through functionalized gold nano-particles for glucose production by continuous hydrolysis of waste bamboo chopsticks. *Enzyme and Microbial Technology*, *53*(6), 444–451.

Correa, D. F., Beyer, H. L., Fargione, J. E., Hill, J. D., Possingham, H. P., Thomas-Hall, S. R., et al. (2019). Towards the implementation of sustainable biofuel production systems. *Renewable and Sustainable Energy Reviews*, *107*, 250–263.

Dixit, P., Basu, B., Puri, M., Mathur, A. S., & Barrow, C. J. (2019). A screening approach for assessing lytic polysaccharide monooxygenase activity in fingal strains. *Biotechnology for Biofuels*, *12*, 185.

Feng, W., & Ji, P. (2011). Enzymes immobilized on carbon nanotubes. *Biotechnology Advances*, *29*, 889–895.

Ghica, M. E., Pauliukaite, R., Fatibello-Filho, O., & Brett, C. M. A. (2009). Application of functionalised carbon nanotubes immobilized into chitosan films in amperometric enzyme biosensors. *Sensors and Actuators B: Chemical*, *142*, 308–315.

Hegedüs, I., Hancsók, J., & Nagy, E. (2012). Stabilization of the cellulase enzyme complex as enzyme nanoparticle. *Applied Biochemistry and Biotechnology*, *168*(6), 1372–1383.

Khoshnevisan, K., Vakhshiteh, F., Barkhi, M., Baharifar, H., Poor-Akbar, E., Zari, N., et al. (2017). Immobilization of cellulase enzyme onto magnetic nanoparticles: Applications and recent advances. *Molecular Catalysis*, *442*, 66–73.

Kumar, A., Park, G. D., Patel, S. K. S., Kondaveeti, S., Otari, S., Anwar, M. Z., et al. (2019). SiO_2 microparticles with carbon nanotube-derived mesopores as an efficient support for enzyme immobilization. *Chemical Engineering Journal*, *359*, 1252–1264.

Kumar, P., Ryan, B., & Henehan, G. T. M. (2017). β-glucosidase from *Streptomyces griseus*: Nanoparticle immobilization and application to alkyl glucoside synthesis. *Protein Expression and Purification*, *132*, 164–170.

Landarani-Isfahani, A., Taheri-Kafrani, A., & Razmjou, A. (2015). Xylanase immobilized on multifunctional hyperbranched polyglycerol-grafted magnetic nanoparticles: An efficient and robust biocatalyst. *Langmuir*, *31*, 9219–9227.

Li, P., Sakuragi, K., & Makino, H. (2019). Extraction techniques in sustainable biofuel production: A concise review. *Fuel Processing Technology*, *193*, 295–303.

Liu, M.-q., Dai, X.-j., Guan, R.-f., & Xu, X. (2014). Immobilization of *Aspergillus niger* xylanase A on Fe_3O_4-coated chitosan magnetic nanoparticles for xylooligosaccharide preparation. *Catalysis Communications*, *55*, 6–10.

Majidian, P., Tabatabaei, M., Zeinolabedini, M., Naghshbandi, M. P., & Chisti, Y. (2018). Metabolic engineering of microorganisms for biofuel production. *Renewable and Sustainable Energy Reviews*, *82*, 3863–3885.

Medina, J. M., Sáez Del Bosque, I. F., Frías, M., Sánchez De Rojas, M. I., & Medina, C. (2017). Characterization and valorisation of biomass waste as a possible addition in eco-cement design. *Materials and Structures*, *50*, 207.

Pal, A., & Khanum, F. (2011). Efficacy of xylanase purified from *Aspergillus niger* DFR-5 alone and in combination with pectinase and cellulase to improve yield and clarity of pineapple juice. *Journal of Food Science and Technology*, *48*(5), 560–568.

Poorakbar, E., Shafiee, A., Saboury, A. A., Rad, B. L., Khoshnevisan, K., Ma'mani, L., et al. (2018). Synthesis of magnetic gold mesoporous silica nanoparticles core shell for cellulase enzyme immobilization: Improvement of enzymatic activity and thermal stability. *Process Biochemistry*, *71*, 92–100.

Puri, M., Abraham, R. E., & Barrow, C. J. (2012). Biofuel production: Prospects, challenges and feedstock in Australia. *Renewable and Sustainable Energy Reviews*, *16*, 6022–6031.

Puri, M., Verma, M. L., & Barrow, C. J. (2013). Enzyme immobilization on nanomaterial for biofuel production. *Trends in Biotechnology*, *31*, 215–216.

Rabinovich, M. L., Melnick, M. S., & Bolobova, A. V. (2002). The structure and mechanism of action of cellulolytic enzymes. *Biochemistry (Moscow)*, *67*, 850–871.

Raud, M., Kikas, T., Sippula, O., & Shurpali, N. J. (2019). Potentials and challenges in lignocellulosic biofuel production technology. *Renewable and Sustainable Energy Reviews*, *111*, 44–56.

Raza, S., Yong, X., & Deng, J. (2019). Immobilizing cellulase on multi-layered magnetic hollow particles: Preparation, bio-catalysis and adsorption performances. *Microporous and Mesoporous Materials*, *285*, 112–119.

Shah, S., & Gupta, M. N. (2008). Simultaneous refolding, purification and immobilization of xylanase with multi-walled carbon nanotubes. *Biochimica et Biophysica Acta (BBA)—Proteins and Proteomics*, *1784*(2), 363–367.

Singh, N., Mathur, A. S., Barrow, C. J., & Puri, M. (2018). Enhanced cellulosic ethanol production via consolidated bioporcessing by *Clostridium thermocellum* ATCC 31924. *Bioresource Technology*, *250*, 860–867.

Singh, N., Mathur, A. S., Gupta, R. P., Puri, S. K., & Puri, M. (2018). Consolidating bioprocessing at high temperature. Chapter 20 In R. Singhania, A. Agrawal, R. K. Sukumaran, & P. Kumar (Eds.), *Waste to Wealth* (p. 2018). Springer Press.

Singh, R. K., Zhang, Y.-W., Nguyen, N.-P.-T., Jeya, M., & Lee, J.-K. (2011). Covalent immobilization of β-1,4-glucosidase from *Agaricus arvensis* onto functionalized silicon oxide nanoparticles. *Applied Microbiology and Biotechnology*, *89*(2), 337–344.

Verma, M. L., Chaudhary, R., Tsuzuki, T., Barrow, C. J., & Puri, M. (2013). Immobilization of β-glucosidase on a magnetic nanoparticle improves thermostability: Application in cellobiose hydrolysis. *Bioresource Technology*, *135*, 2–6.

Viccaro, M., Cozzi, M., Rocchi, B., & Romano, S. (2019). Conservation agriculture to promote inland biofuel production in Italy: An economic assessment of rapeseed straight vegetable oil as a self-supply agricultural biofuel. *Journal of Cleaner Production*, *217*, 153–161.

Wang, S. G., Zhang, Q., Wang, R., Yoon, S. F., Ahn, J., Yang, D. J., et al. (2003). Multi-walled carbon nanotubes for the immobilization of enzyme in glucose biosensors. *Electrochemistry Communications*, *5*, 800–803.

Wu, L. M., Zeng, Q. H., Ding, R., Tu, P. P., & Xia, M. Z. (2019). Fe_3O_4/acid activated montmorillonite/cellulase composites: Preparation, structure, and enzyme activity. *Applied Clay Science*, *179*, 105129.

Yuan, B., Yang, X.-Q., Xue, L.-W., Feng, Y.-N., & Jiang, J.-H. (2016). A novel recycling system for nano-magnetic molecular imprinting immobilized cellulases: Synergistic recovery of anthocyanin from fruit and vegetable waste. *Bioresource Technology*, *222*, 14–23.

Zeng, K., Wei, D., Zhang, Z., Meng, H., Huang, Z., & Zhang, X. (2019). Enhanced competitive immunomagnetic beads assay with gold nanoparticles and carbon nanotube-assisted multiple enzyme probes. *Sensors and Actuators B: Chemical*, *292*, 196–202.

Zhang, P., & Henthorn, D. B. (2010). Synthesis of PEGylated single wall carbon nanotubes by a photoinitiated graft from polymerization. *AICHE Journal*, *56*, 1610–1615.

CHAPTER FIFTEEN

Few biomedical applications of carbon nanotubes

Neelam Yadav[a,b], Manshi Tyagi[c], Shikha Wadhwa[d], Ashish Mathur[d], Jagriti Narang[c,*]

[a]Department of Biotechnology, Deenbandhu Chhotu Ram University of Science and Technology, Sonepat, India
[b]Centre for Biotechnology, Maharishi Dayanand University, Rohtak, India
[c]Department of Biotechnology, School of Chemical and Life Sciences, Jamia Hamdard, New Delhi, India
[d]Amity Institute of Nanotechnology, Amity University, Noida, Uttar Pradesh, India
[*]Corresponding author: e-mail address: jags_biotech@yahoo.co.in

Contents

Abstract

Nanotubes of carbon are allotropic form of carbon material that rolled to form a cylindrical structure that may be singlewalled carbon nanotubes (SWCNTs) and multiwalled carbon nanotubes (MWCNTs) depending upon the number of carbon layers. These carbon nanotubes have exhibited characteristics properties such as electrical, optical, thermal and mechanical. Carbon nanotubes can be employed for immobilization matrix for biomolecules such as an enzyme, nucleic acid, etc. Enzymes can be immobilized onto carbon nanotubes via absorption or covalent bonding. Various enzymatic based biosensors are also developed for the detection of various analytes. Present chapter mainly emphasizes characteristics of carbon nanotubes, their preparation methods, purification and exploitation of CNTs as an immobilization matrix for theranostic applications.

Methods in Enzymology, Volume 630
ISSN 0076-6879
https://doi.org/10.1016/bs.mie.2019.11.005

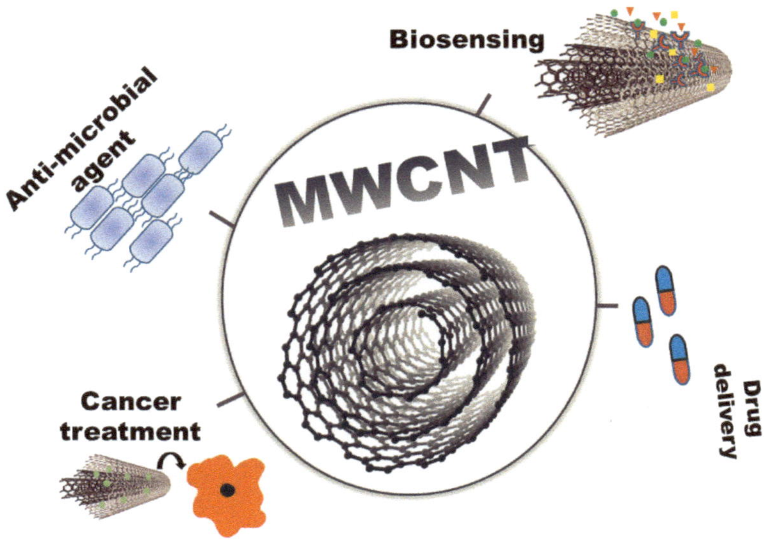

1. Introduction

Nanotechnology has been a recent and advanced technology which gets merged into material sciences. Materials with an average diameter less than 100 nm are termed as nanoparticles (Seifert, 1994). Carbon nanomaterials have stimulated immense interest in scientists because of enhanced physical and chemical features such as electrical, optical, thermal and mechanical properties (Craciun, Russo, Yamamoto, & Tarucha, 2011; Hu, Wang, & Hu, 2010; Neto, Guinea, Peres, Novoselov, & Geim, 2009). Among carbon nanomaterials, carbon nanotubes are most exploited for various applications. Carbon nanotubes are a cylindrical shaped allotropes of carbon which arises due to rolled graphene planes. Carbon nanotubes have become the main exponent of nanotechnology after their discovery in recent history by Iijima in 1991 using High-Resolution Electron Microscopy (HREM) (Iijima, 1991). Carbon nanotubes were first described as carbon shoots obtained in an arc evaporation test by Iijima (1991).

Multiwalled carbon nanotubes is consisting of many co-axial cylinders which is of sheets of graphene covering a hollow core structure. The dimensions of MWCNT are its length ranges from one to numerous micrometer while the inner diameter of cylindrical structure ranges in between from 1 to 3 nm and outer diameter of cylindrical structure ranges between 2 and 100 nm. The nano-dimensions are restricted to diameter (Rastogi et al., 2014). Multiwalled carbon nanotubes have exceptionally enhanced features in terms of both physically and chemically and properties are applied in a plethora of applications including sorbent material (Bhanjana, Dilbaghi, Kim, & Kumar, 2017), drug delivery system, solar cell electrode (Leu et al., 2016), cancer therapy, and diagnostic system. This chapter underlines some researches which proved that MWCNT has played a vital role in biomedical engineering due to its exceptional characteristic features. The chapter also summarizes the synthesis, purification and properties of MWCNTs in nutshell.

2. Synthesis and purification of MWCNTs

Carbon nanotubes preparation comprises the conversion of different precursors of carbon into rolled cylindrical structure which is known as carbon nanotubes. The external factors which greatly impact the preparation of carbon nanotubes are pressure and temperature. The aspect ratio of nanotubes depends on the surrounding experiment conditions. Arc discharge and laser ablation are two methods which are most commonly used in the preparation of carbon nanotubes.

The first most common method which is applied for the preparation of carbon nanotubes is Arc discharge method. The method involves the use of two graphitic electrodes which are distantly separated and immersed in an inert gas such as mercury. The inert gas is meant for cooling and condensation. The graphite electrodes act as a cathode and anode. When arc discharge or voltage is applied across the two electrodes then soot get deposited inside the chamber which is fullerenes while carbon nanotubes gets deposited on the cathode.

In laser ablation methods, a laser pulse is being used to vaporize the graphitic carbon in the presence of inert gas. The vaporized carbon is being collected on the cooling chamber. These methods require very high-temperature which sometimes becomes difficult to maintain in laboratory conditions and also less amount of deposits obtained. Therefore, nowadays aforementioned methods are being replaced by chemical vapor deposition techniques that require low-temperature as compared to the arc discharge method.

In chemical vapor deposition method, orientation, aspect ratio, density and purity of CNTs can be tuned as per the requirements of the experiment. In the chemical vapor deposition technique, the carbon precursor which is being used is hydrocarbon sources. Herein the hydrocarbons are converted into gaseous forms by using energy sources. Hydrocarbons are exposed to high temperature which are converted into pure form of carbon and then reaches to the heated and catalyst coated substrate which causes the change in the configuration of carbon material. Then carbon nanotubes are being formed. Nowadays this method is commonly used in the industries for MWCNTs synthesis. The method has many advantageous features which make it famous for the MWCNT synthesis such as low-temperature requirements, tunable, large amount, high purity and low energy consumption.

Further information about the preparation of CNTs can be explored in the article on carbon nanotubes synthesis and properties (Kingston & Simard, 2003; Terrones, 2004). Purity of MWCNTs also played a vital role as impurity of MWCNTs may lead to bring pseudo results in biomedical applications. Therefore, purification of MWCNTs is required as the synthesized MWCNTs may be associated with some carbonaceous or metallic impurities (Hou, Liu, & Cheng, 2008). Huang, Wang, Luo, and Wei (2003) produced ultrahigh purified multiwalled carbon nanotubes through a catalytical vapor deposition method and subsequently annealed at vacuum pressure ranging from 0.001 to 10 Pa and temperature higher than 2000 °C between 1500 and 2150 °C. This high-temperature annealing purification technique achieved highest multiwalled carbon nanotubes purification of about 99.9% (Huang et al., 2003). Purity of carbon nanotubes can be evaluated using Raman spectroscopy (Dillon, Yudasaka, & Dresselhaus, 2004) while near-infrared spectroscopy can be used for quantitative assessment (Zhao et al., 2004).

As the application of multiwalled carbon nanotubes are gradually related to some of their specific properties, a summarized overview of their properties plays an important role before exploring their potential biomedical applications. Carbon nanotubes possess exceptional features which makes them a good candidate for biomedical applications such as unique structural, optical, chemical, mechanical, thermal and enhanced electronic properties. Its enhanced conducting properties make it very suitable as a sensing interface for sensor applications. The conductivity of MWCNTs depends on its orientation and its aspect ratio. In addition, concentration of carbon nanotubes is also required very less in order to produce the high electrical conductivities (Jariwala, Sangwan, Lauhon, Marks, & Hersam, 2013; Pham et al., 2017). Carbon nanotubes showed large surface to volume ratio which made them an ideal candidate for successful immobilization and conjugation of biomolecules. Furthermore, the CNTs have functional groups which provide ample opportunity to form a strong linkage between the conjugates (Power, Gorey, Chandra, & Chapman, 2018).

3. Biomedical applications

CNTs can be exploited for many diagnostic applications as carbon nanotubes have many advantageous features which make them more promising candidate for biomedical application as it is biocompatible, fast electron transfer kinetics, light weight, chemical inertness, high tensile strength, light weight, antimicrobial properties, exposed functional groups for easy conjugation and can be mass produced (Smart, Cassady, Lu, & Martin, 2006). Some researchers stressed on toxicity of carbon nanotube although the toxicity of carbon nanotube depends on its size, amount and time of exposure (Chatterjee et al., 2014; Chen et al., 2015). In spite of this CNTs also played a significant role in fabrication of sensors, drug targeting, cancer treatment an antimicrobial activity. All these applications of MWCNTs are discussed in brief from ocean of CNTs study.

4. CNTs as immobilization matrix for biomolecules

Unique properties of multiwalled carbon nanotubes such as increased electron transfer kinetics, electrochemically accessible area, strong immobilization capability, increased surface area, good chemical inertness, and substantial mechanical strength make them extremely attractive in various enzyme immobilization and electrochemical biosensor application (Baughman, Zakhidov, & De Heer, 2002). Enzyme can be immobilized onto CNTs surface via physical adsorption or chemical adsorption. Physical adsorption method is due to the hydrophobicity which is provided by CNT. CNT exhibits hydrophobicity which acts as supporting component for the immobilization of biomolecules via physical adsorption. For instance, lysozyme conjugates with CNTs using hydrophobic interactions due to π-π stacking associations in side walls of CNTs and aromatic rings of lysozyme (Saifuddin, Raziah, & Junizah, 2013). CNTs have the capability of the addition of functional group, i.e., —COOH and form carboxylated CNTs. Carboxyl group of CNT forms a bond with the amino group of protein and forms covalent binding with the enzyme. CNTs have capabilities of crosslinking using glutaraldehyde. For example, Sarma, Vatsyayan, Goswami, and Minteer (2009) have developed an amperometric glucose sensor which consist of glucose oxidase (GOx) enzyme immobilized onto functional nano-interface, i.e., carbon nanotubes and platinum nanoparticles for the detection of glucose in sample.

Multiwalled carbon nanotubes have remarkable bio-compatible potential as it provides high surface area which causes better immobilization of enzyme and increases the stability by maintaining the integrity of enzyme which made it suitable for the construction of enzymatic based biosensors (Marega et al., 2013). The aforesaid properties of CNTs leads to the construction of many enzymatic based electrochemical biosensors. Electrochemical devices measure the change which arises after the interaction of analyte and the working electrode. The change can be observed in the form of current, impedance or voltage (Duarte, Justino, Freitas, Duarte, & Rocha-Santos, 2014). Electrochemical devices have distinct features such as fast response, high sensitivity, economical miniaturization and integration potential which make them suitable for many application such as biological, environmental and pharmaceutical (Duarte et al., 2014). Electrochemical biosensors are further classified as amperometric, potentiometric, and impedimetric (Duarte et al., 2014). Carbon nanotubes are used as electrode in various sensor as they display greater electron transfer capabilities (Yang, Denno, Pyakurel, & Venton, 2015). Electronic properties of multiwalled carbon nanotubes are significance of graphene sheet curvature. MWCNTs are electrochemically active due to the rich π-electron cloud outside the surface of tubes (Baptista, Belhout, Giordani, & Quinn, 2015; Wang & Dai, 2015). MWCNTs have unique property of facilitating the electron transfer between the electroactive species present in the solution and the electrode interface (Futra et al., 2016; Primo, Oviedo, Sánchez, Rubianes, & Rivas, 2014; Zhang, Liu, & Chai, 2012; Zhao et al., 2016).

Oliveira et al. (2013) constructed an amperometric sensor for the analysis of pirmicarb pesticide. Herein this approach MWCNTs were used for the immobilization of enzyme laccase which acts on substrate 4-aminophenol. The MWCNT paste electrode was formed by dispersing the enzyme laccase (3% w/w) and the binder paraffin (60:40%; MWCNT:Binder) was also used in this approach. The enzyme immobilized onto MWCNT showed better catalytic property, kinetic performance and with enhanced electron transfer kinetics. The biosensor modified with the laccase and MWCNT showed linear range 9.9×10^{-7} to 1.15×10^{-5} mol/L. The response was determined on the basis of inhibition activity of laccase enzyme. The stability of the biosensor was found to be 1 month as enzyme maintained its catalytic efficiency up to 1 month. The biosensor was also applied in the spiked tomato and lettuce samples and it showed good analytical recovery. The developed biosensor showed anti-interference capability as the presence of other compounds in real samples showed insignificant affect on the electroanalysis of pirimicarb.

Li, Zhao, Shi, Han, and Xiao (2017) developed an amperometric enzymic biosensor comprising 3D graphene and MWCNTs nanocomposites and acetylcholinesterase enzyme which was employed for the detection of carbamates pesticides. The biosensor was developed by homogenously mixing the graphene oxide and MWCNTs (ratio 1:3). The prepared mixture was ultra-sonicated in DI water. Afterward, the developed mixture was homogeneously deposited onto the glassy carbon electrode and kept for drying. Then the modified electrode was activated electrochemically by using 0.5 M sulfuric acid by varying the potential from −0.6 to 1.0 V for successive 17 cycles. The activated electrode showed a good response as compared to a non-activated electrode. The nanocomposites on the developed sensor proved to be a good immobilization matrix for the enzyme. The enzyme acetylcholinesterase was loaded onto the activated 3D graphene/MWCNTs/GCE and was kept for drying. The immobilized enzyme showed a high affinity toward the substrate as depicted from its Km value. The prepared device was employed for the analysis of carbofuran in the range of $0.05–1 \times 10^{-9}$ g/mL and paraoxon in the range of $1–104 \times 10^{-9}$ g/mL.

Nayak, Santhosh, and Ramaprabhu (2014) described the synthesis of the immobilization matrix using graphene oxide (100 mg), MWCNTs (5 mg) and $HAuCl_4$ (27 mg). Aforesaid materials were homogeneously mixed together and were spread on glass Petri dish and sunlight was focused using the convex mirror. The increase in temperature causes exfoliation of graphene oxide converted the metal salt to metal nanoparticles. Physical examination such as the release of a gaseous product and volume expansion also confirmed the synthesis of nanoparticles. The developed material has the capability to effectively immobilize the biomolecules and increases electrical conductivity. The fabrication of bioelectrode was done by immobilizing the glucose oxidase (5 mL; 15 mg/mL) onto the surface and was employed for the analysis of both hydrogen peroxide and glucose. The prepared device was able to detect the hydrogen peroxide in the concentration ranging from 1 to 62 mM and 50 μM to 20 mM for glucose. The immobilized enzyme also showed great affinity toward the substrate as compared to the free enzyme. The developed biosensor showed anti-interferants capability toward serum metabolites.

Malik, Chaudhary, and Pundir (2019) developed a enzymic biosensor based on carboxylated MWCNT, copper nanoparticles and polyaniline for the detection of pyruvate. The enzyme which was immobilized onto the nanoparticles modified electrode was pyruvate oxidase (POD). Herein this approach, carboxylated MWCNTs were used in order to covalently attach the enzyme on to nanoparticles modified surface. The-COOH gp of MWCNTs ensures covalent binding with the —NH_2 gp of enzyme which facilitate robust immobilization of enzyme onto the surface. The detection range of the developed device was found to be 0.1–2000 μM. The results of biosensors showed good correlation with the spectrophotometric method. The sensor has the ability to be applied in serum samples for the detection of pyruvate.

Christwardana and Frattini (2018) developed a biosensor comprising carbon nanotubes coated with polyethyleneimine (PEI) for the detection of glucose. The carbon nanotube with PEI polymer forms entrapment matrix to entrap enzyme. The entrapped enzyme was much protected from detachment and denaturation. The entrapment matrix offered much more stability and prevents thermal degradation of the developed sensor. Thermal degradation is a major problem which affects the electron mobility and detection of glucose. Therefore, the developed matrix using CNT and PEI can be viable option for construction of enzyme-based biosensors.

Tang, Li, Wu, Gong, and Zeng (2014) described the formation of a very complex nanostructure using MWCNT, polythiophene, chitosan and PTFE for the immobilization of enzyme and acting as sensing matrix. The thiophene monomer was electropolymerized onto the glassy carbon surface. The sensor was developed sequentially as the first thiophene was electropolymerized, -MWCNT, GOx and then finally chitosan-PTFE film was deposited and forms parallel multi-component reaction system (PMRS). The prepared device was able to detect in the concentration ranging from 0.04 to 2.5 mM. The authors indicated that the development of this complex nano architecture using MWCNTs provide enhanced specificity, sensitivity and stability. The MWCNTs-based nanocomposites act as best interface for supporting biomolecules and as a transducer for the development of third generation biosensor. Du et al. (2015) described the use of CNT along with the silica for the co-immobilization of two enzymes, i.e., α-amylase and glucoamylase. The developed matrix offers a synergistic effect of biocompatibility and faster diffusion of small molecules. As silica offers the best immobilization matrix for biomolecules and protects the integrity while CNTs, in addition, offers more conductivity, enabling movement of molecules from outside or inside of the developed matrix. A sol-gel layer with entrapped CNT—α-amylase was formed outside the CNT-glucoamylase onto silica microspheres. The developed matrix offers more stability and prevents the leakage of enzymes from the developed matrix.

Khan et al. (2019) described the immobilization of lipase enzyme onto the surface of CNT by adsorption. The immobilized enzyme showed better performance in its activity. The enzyme activity gets increased upon increase in concentration of enzyme. However additional amount of enzyme may not be supported by the developed matrix due to the saturation in pore binding on fictionalized MWCNTs. The immobilized enzyme onto the developed biosensor was found to have optimum pH 6.0. Garlet et al. (2014) utilizes the carbon nanotubes as an effective immobilization matrix for the non-covalent attachment of inulinase enzyme. The results depicted that there is fast absorption and the ratio of enzyme and adsorbent is 1:460 and loading capacity is 51047 U/g. The immobilized enzyme showed 100% of its activity even after storage of 35 days. This proved that carbon nanotubes proved to be an effective adsorbent which protected the integrity and stability of enzyme. Shishkova et al. (2018) described the use of synthesized SWCNTs for immobilizing the enzyme superoxide dismutase. The functional SWCNT-enzyme complex possesses the anti-oxidant properties and can be used in nanomedicine for delivering active SOD to cells and tissues. Liu et al. (2019) designed an electrochemical nanosensor in which aptamer was conjugated with the nanocomposites for the sensitive and specific detection of β-estradiol. The employed nanocomposites was gold nanoparticles and carbon nanotubes which act as a sensing interface and immobilization matrix for the biorecognition element, i.e., aptamer.

Basu, Sah, Pradhan, and Bhattacharya (2019) fabricated a nano-enabled biosensor comprising poly-L-lysine functionalized MWCNTs decorated on reduced graphene nanosheets for the specific analysis of cholesterol. The developed sensor has the capability to detect the trace levels of cholesterol in femtomolar (Basu et al., 2019). Eldhose et al. determine the capability of polylactic/c-MWCNTs nanocomposite to be developed into microneedle array (MNA's) based electrochemical sensor and the potential to determine dermal sensing. Their result expresses that micro molding synthesized solvent cast nanocomposite film can easily make MNA's. The developed MNAs showed strong strength which can easily bear stresses and is much suitable for skin insertion. While during their electrochemical characterization using ascorbic acid sample they were able to achieve limit of detection up to 180 μM.

5. Non-enzymatic sensor

Beden, Hamidi, Palecek, et al. (2015) developed an electrochemical sensor for subnanomolar detection of dopamine using electroactive adducts.

The sensor was modified with the MWCNTs and AuNPs for improved analytical parameters. The sensors showed a better response when the electrode was modified with the nano-hybrids. The sensor showed a very good and broad linear range and low detection limit (Beden et al., 2015). Palisoc, Natividad, De Jesus, and Carlos (2018) developed a highly sensitive nanosensor having anchored nanocomposites as an interface for the electron transfer. The nanocomposites used here are silver nanoparticles and MWCNT. Along with the nanocomposite, Nafion membrane was also used to develop a robust interface. The sensor was employed for the detection of heavy metals such as lead and cadmium in vegetable samples. The developed sensor has the ability to detect the trace level of heavy metals in real samples. The data presented in the paper concluded that the washed vegetables contain fewer amounts of heavy metals as compared to unwashed vegetables (Palisoc et al., 2018).

6. Enzymatic biosensors developed in our lab using MWCNTs

In our laboratory we have constructed an electrochemical device for the measurement of glucose using SWCNTs and MWCNTs (Alhans et al., 2018). The experimental setup used foe aforementioned device has been discussed in following section:

6.1 Experimental system

6.1.1 Chemicals and reagents

(i) Anhydrous β-D-(+)-glucose
(ii) Glucose oxidase (GOx 147.9 U/mg; Source: *Aspergillus niger*)
(iii) Potassium ferricyanide
(iv) Methanol
(v) SWCNT and MWCNT

6.1.2 Equipments

(a) Potentiostat/Galvanostat by Auto lab
(b) Weighing balance
(c) pH meter
(d) Pipettes (20–1000 μL)
(e) Magnetic stirrer
(f) Magnetic beads
(g) Ultracentrifuge
(h) Water bath

(i) Ultraviolet (UV)-spectrophotometer
(j) Distilled water
(k) Glass wares (flasks, beakers, glass rod)

6.1.3 Development of SWCNTs and MWCNTs-based gold printed circuit board electrodes (Au-PCB) by anchoring GOx

The drop-casting procedure was employed for the fabrication of amperometric biosensor independent of markers or mediators. For this purified SWCNT and MWCNT were allowed to dissolve into methanol. Now for uniform mixing, this preparation was ultra-sonicated for 25 min. After this, a drop of 10 μL was deposited at the surface of gold pad which was used as a working electrode. Afterwards the gold electrodes was left undisturbed for 1 h at room temperature (RT) for the evaporation of methanol from the mixture. For the development of SWCNTs and MWCNTs based gold printed circuit board electrodes (Au-PCB) by anchoring GOx, 1 mg/mL GOx was coated onto CNT/Au-PCB electrodes. The electrodes were enzymatically modified by drop-casting of GOx (5 μL) and was left undisturbed for a day at 4 °C for the effectual immobilization of GOx onto the developed matrix. Finally, a chamber was assigned to augment the effective area of the sensor. Therefore, this chamber was exposed to potassium ferricyanide that was acting as an electron transfer mediator. The aforementioned chamber was filled with 400 μL poly methyl methacrylate (PMMA). This entire working system was consisting of a two-electrode configuration system consisting of GOx anchored onto MWCNTs or SWCNTs (working electrode) and the counter electrode (Scheme 1) (Alhans et al., 2018). The schematic representation of the fabrication of Au-PCB electrode gas been depicted in Scheme 1.

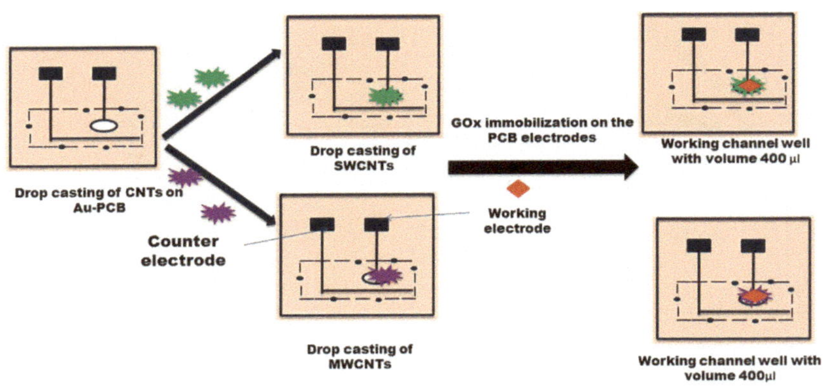

Scheme 1 The schematic representation of fabrication of Au-PCB electrode.

6.1.4 Determination of glucose by SWCNTs and MWCNTs-based gold printed circuit board electrodes (Au-PCB) by anchoring GOx

The circular region present at Au-PCB electrode on which nanotubes and enzyme was drop deposited act as an effectual surface and responsible for the redox activity. For individual electrodes, we prepared the spiked sample solutions of glucose (1–100 mM) and 5 mM ferricyanide $[Fe(CN)_6]^{3-}$ separately in 1:1 ratio. For the electrochemical investigation, 400 μL of the spiked sample was added into the well. Consequently, the GOx enzyme catalyzes the oxidation of glucose to gluconic acid and a simultaneous reduction of the $[Fe(CN)_6]^{3-}$ into $[Fe(CN)_6]^{4-}$. Chronoamperometry and cyclic voltammetry techniques were used for the quantification of ferrocyanide $[Fe(CN)_6]^{4-}$ ions.

$$\text{Glucose} + 2\text{K}\,[\text{Fe(CN)}] + \text{HO} \rightarrow \text{Gluconic acid} + 2\text{K}\,[\text{Fe(CN)}]$$

6.1.5 Characterization of SWCNTs and MWCNTs-based gold printed circuit board electrodes (Au-PCB) by anchoring GOx using cyclic voltammetry (CV) and electrochemical impedance spectroscopy (EIS)

Electrochemical characterization of SWCNTs and MWCNTs-based gold printed circuit board electrodes (Au-PCB) by anchoring GOx and bare gold printed circuit board electrodes performed by Electrochemical impedance spectroscopy (EIS) and cyclic voltammetry (CV) (Fig. 1). From Fig. 1 lower

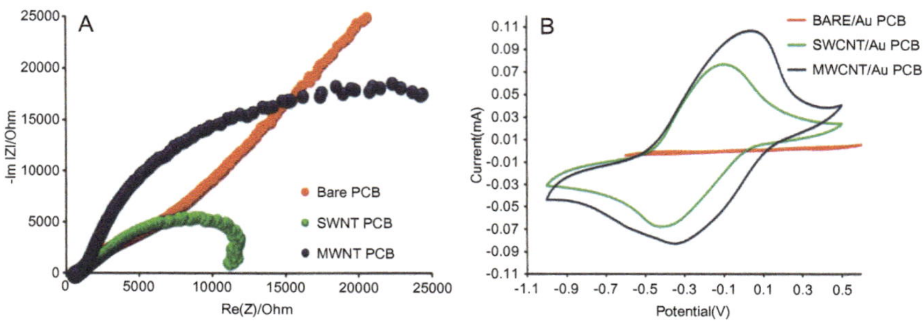

Fig. 1 (A) EIS spectra of bare PCB, SWCNT/Au-PCB and MWCNT/Au-PCB in 5 mM $[Fe(CN)_6]^{3-}$. (B) CV of bare PCB, SWCNT/Au-PCB and MWCNT/Au-PCB in 5 mM $[K_3Fe(CN)_6]$. *Source: Alhans, R., Singh, A., Singhal, C., Narang, J., Wadhwa, S., & Mathur, A. (2018). Comparative analysis of single-walled and multi-walled carbon nanotubes for electrochemical sensing of glucose on gold printed circuit boards.* Materials Science and Engineering: C 90, *273–279.*

oxidation and reduction peaks of SWCNT/Au-PCB was observed at −0.12 and −0.48 V while MWCNT/Au-PCB showed peaks 0.038 and −0.34 V, respectively.

The electrochemical behavior of SWCNTs and MWCNTs-based gold printed circuit board electrodes (Au-PCB) using CV were studied in 5 mM $[Fe(CN)_6]^{3-}$ with −1.0 to 0.5 V potential range. During EIS investigation the impedance graph of SWCNTs and MWCNTs-based gold printed circuit board electrodes (Au-PCB), SWCNTs and MWCNTs-based gold printed circuit board electrodes (Au-PCB) by anchoring GOx and bare gold printed circuit board electrodes (Au-PCB) was recorded in 5 mM potassium ferro/ferri cyanide (amplitude 0.10 mV) (Fig. 2). From Fig. 2A and B it can be concluded that there is a proportional relation between peak currents to the square root of the scan rates during electrochemical reactions of glucose. Therefore, for the detection of glucose, EIS and CV showed better results.

6.1.6 Investigation of glucose by chronoamperometry

The technique chronoamperometry has shown unique advantages viz easy to operate, reliable outcomes, sensitive and fast response time for the detection of target analyte. Therefore, in view of these advantages we detected glucose in the spiked samples using aforementioned technique. During investigation 500 mV step potential was applied to the working electrode and current was measured as a function of time. Thus, faradic steady state current was measured as an average of current values obtained in 100–120 s and chronoamperometric curves for GOx/SWCNT/Au-PCB and GOx/MWCNT/Au-PCB were plotted at different glucose concentrations. Cottrell equation was used for the evaluation of glucose amount.

7. Conclusion

Carbon nanotubes are rolling cylindrical structure. Multiwalled carbon nanotubes have shown improved electrical conductivity, large surface area, better chemical consistency and mechanical strength. As carbon nanotubes have shown unique advantages and therefore, acting as carrier for the effective delivery of biomolecules like antibiotics, proteins, DNA, RNA, immunoactive compounds and lectins. Biosensors based on carbon nanotubes have shown better reproducibility, sensitivity, reliability and economic. Employment of MWCNTs in biosensing technology makes them more sensitive, specific and conducive and can be used in diagnosis of

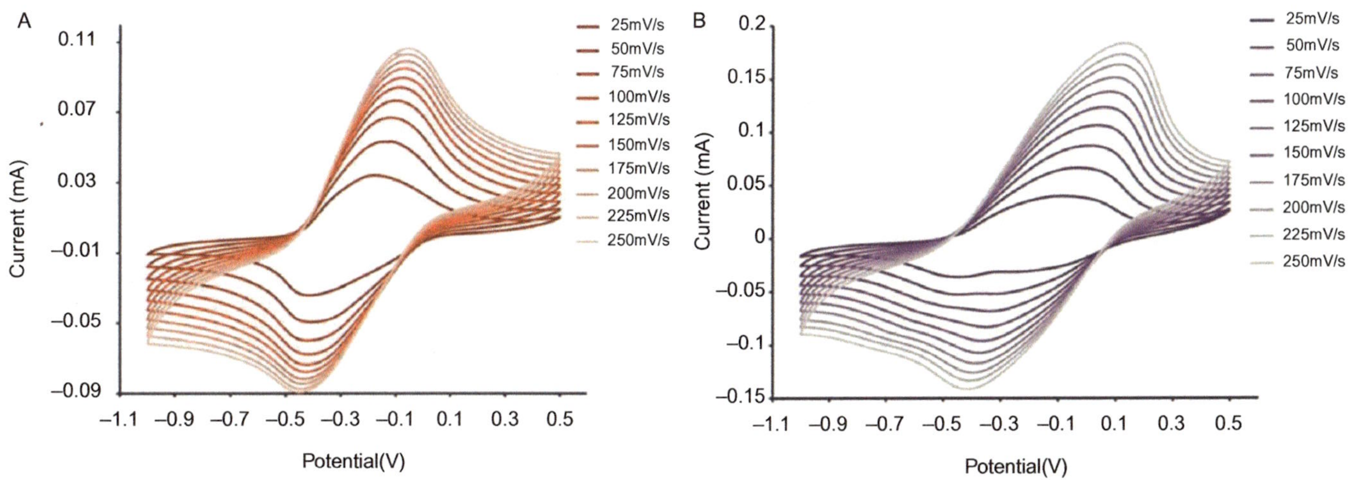

Fig. 2 (A) CV of GOx/SWCNT/Au-PCB electrode at different scan rates ranging between 25 and 250 $mV\,s^{-1}$, in the presence of glucose in 5 mM $[Fe(CN)_6]^{3-}$. (B) CV of GOx/MWCNT/Au-PCB electrodes at different scan rates ranging from 25 to 250 $mV\,s^{-1}$. *Reprinted with Alhans, R., Singh, A., Singhal, C., Narang, J., Wadhwa, S., & Mathur, A. (2018). Comparative analysis of single-walled and multi-walled carbon nanotubes for electrochemical sensing of glucose on gold printed circuit boards.* Materials Science and Engineering: C 90, *273–279.*

diseases and treatment, for example, cancer. MWCNTs based biosensors have been used to detect many analytes such as pyridoxine, dopamine ascorbic acid, uric acid and many more. Furthermore, carbon nanotubes have also exhibited antimicrobial activity. Therefore, future research should be focus on investigating more efficient CNT based devices for the betterment of human health.

References

Alhans, R., Singh, A., Singhal, C., Narang, J., Wadhwa, S., & Mathur, A. (2018). Comparative analysis of single-walled and multi-walled carbon nanotubes for electrochemical sensing of glucose on gold printed circuit boards. *Materials Science and Engineering: C*, *90*, 273–279.

Baptista, F. R., Belhout, S. A., Giordani, S., & Quinn, S. J. (2015). Recent developments in carbon nanomaterial sensors. *Chemical Society Reviews*, *44*(13), 4433–4453.

Basu, A. K., Sah, A. N., Pradhan, A., & Bhattacharya, S. (2019). Poly-L-lysine functionalised MWCNT-rGO nanosheets based 3-d hybrid structure for femtomolar level cholesterol detection using cantilever based sensing platform. *Scientific Reports*, *9*(1), 3686.

Baughman, R. H., Zakhidov, A. A., & De Heer, W. A. (2002). Carbon nanotubes—The route toward applications. *Science*, *297*(5582), 787–792.

Beden, N., Hamidi, H., Palecek, E., et al. (2015). Subnanomolar determination of dopamine by electrochemical sensor based on Au-NPs-MWCNTs and mannopentose-Os (VI) adducts. *Insights in Analytical Electrochemistry*, *1*, 2.

Bhanjana, G., Dilbaghi, N., Kim, K. H., & Kumar, S. (2017). Carbon nanotubes as sorbent material for removal of cadmium. *Journal of Molecular Liquids*, *242*, 966–970.

Chatterjee, N., Yang, J., Kim, H. M., Jo, E., Kim, P. J., Choi, K., et al. (2014). Potential toxicity of differential functionalized multiwalled carbon nanotubes (MWCNT) in human cell line (BEAS2B) and Caenorhabditis elegans. *Journal of Toxicology and Environmental Health, Part A*, 77(22–24), 1399–1408.

Chen, R., Zhang, L., Ge, C., Tseng, M. T., Bai, R., Qu, Y., et al. (2015). Subchronic toxicity and cardiovascular responses in spontaneously hypertensive rats after exposure to multiwalled carbon nanotubes by intratracheal instillation. *Chemical Research in Toxicology*, *28*(3), 440–450.

Christwardana, M., & Frattini, D. (2018). Electrochemical study of enzymatic glucose sensors biocatalyst: Thermal degradation after long-term storage. *Chemosensors*, *6*, 53.

Craciun, M. F., Russo, S., Yamamoto, M., & Tarucha, S. (2011). Tuneable electronic properties in graphene. *Nano Today*, *6*(1), 42–60.

Dillon, A. C., Yudasaka, M., & Dresselhaus, M. S. (2004). Employing Raman spectroscopy to qualitatively evaluate the purity of carbon single-wall nanotube materials. *Journal of Nanoscience and Nanotechnology*, *4*(7), 691–703.

Du, K., Sun, J., Zhou, X., Feng, W., Jiang, X., & Ji, P. (2015). A two-enzyme immobilization approach using carbon nanotubes/silica as support. *Biotechnology Progress*, *31*(1), 42–47.

Duarte, K., Justino, C. I., Freitas, A. C., Duarte, A. C., & Rocha-Santos, T. A. (2014). Direct-reading methods for analysis of volatile organic compounds and nanoparticles in workplace air. *TrAC Trends in Analytical Chemistry*, *53*, 21–32.

Futra, D., Heng, L. Y., Jaapar, M. Z., Ulianas, A., Saeedfar, K., & Ling, T. L. (2016). A novel electrochemical sensor for 17β-estradiol from molecularly imprinted polymeric microspheres and multi-walled carbon nanotubes grafted with gold nanoparticles. *Analytical Methods*, *8*(6), 1381–1389.

Garlet, T. B., Weber, C. T., Klaic, R., Foletto, E. L., Jahn, S. L., Mazutti, M. A., et al. (2014). Carbon nanotubes as supports for inulinase immobilization. *Molecules*, *19*(9), 14615–14624.

Hou, P. X., Liu, C., & Cheng, H. M. (2008). Purification of carbon nanotubes. *Carbon*, *46*(15), 2003–2025.

Hu, Y. H., Wang, H., & Hu, B. (2010). Thinnest two-dimensional nanomaterial—Graphene for solar energy. *Chemsuschem*, *3*(7), 782–796.

Huang, W., Wang, Y., Luo, G., & Wei, F. (2003). 99.9% purity multi-walled carbon nanotubes by vacuum high-temperature annealing. *Carbon*, *41*(13), 2585–2590.

Iijima, S. (1991). Helical microtubules of graphitic carbon. *Nature*, *354*(6348), 56.

Jariwala, D., Sangwan, V. K., Lauhon, L. J., Marks, T. J., & Hersam, M. C. (2013). Carbon nanomaterials for electronics, optoelectronics, photovoltaics, and sensing. *Chemical Society Reviews*, *42*(7), 2824–2860.

Khan, A. K., et al. (2019). *IOP conference series: Materials science engineering*. 495 012055.

Kingston, C. T., & Simard, B. (2003). Fabrication of carbon nanotubes. *Analytical Letters*, *36*(15), 3119–3145.

Leu, Y. A., Yeh, M. H., Lin, L. Y., Li, T. J., Chang, L. Y., Shen, S. Y., et al. (2016). Thermally stable boron-doped multiwalled carbon nanotubes as a Pt-free counter electrode for dye-sensitized solar cells. *ACS Sustainable Chemistry & Engineering*, *5*(1), 537–546.

Li, Y., Zhao, R., Shi, L., Han, G., & Xiao, Y. (2017). Acetylcholinesterase biosensor based on electrochemically inducing 3D graphene oxide network/multi-walled carbon nanotube composites for detection of pesticides. *RSC Advances*, 7, 53570–53577.

Liu, X., Deng, K., Wang, H., Li, C., Zhang, S., & Huang, H. (2019). Aptamer based ratiometric electrochemical sensing of 17β-estradiol using an electrode modified with gold nanoparticles, thionine, and multiwalled carbon nanotubes. *Microchimica Acta*, *186*, 347.

Malik, M., Chaudhary, R., & Pundir, C. S. (2019). Construction of an amperometric pyruvate biosensor based on enzyme bound to a nanocomposite and its comparison with enzyme nanoparticles bound to electrode. *International Journal of Applied Sciences and Biotechnology*, 7(2), 195–206.

Marega, R., De Leo, F., Pineux, F., Sgrignani, J., Magistrato, A., Naik, A. D., et al. (2013). Functionalized Fe-filled multiwalled carbon nanotubes as multifunctional scaffolds for magnetization of cancer cells. *Advanced Functional Materials*, *23*(25), 3173–3184.

Nayak, P., Santhosh, P. N., & Ramaprabhu, S. (2014). Synthesis of Au-MWCNT–graphene hybrid composite for the rapid detection of H_2O_2 and glucose. *RSC Advances*, *4*, 41670.

Neto, A. C., Guinea, F., Peres, N. M., Novoselov, K. S., & Geim, A. K. (2009). The electronic properties of graphene. *Reviews of Modern Physics*, *81*(1), 109.

Oliveira, T. M., Barroso, F. M., Morais, S., de Lima-Neto, P., Correia, A. N., Oliveira, M. B., et al. (2013). Biosensor based on multi-walled carbon nanotubes paste electrode modified with laccase for pirimicarb pesticide quantification. *Talanta*, *106*, 137–143.

Palisoc, S. T., Natividad, M. T., De Jesus, N., & Carlos, J. (2018). Highly sensitive AgNP/MWCNT/Nafion modified GCE-based sensor for the determination of heavy metals in organic and non-organic vegetables. *Scientific Reports*, *8*(1), 17445.

Pham, V. D., Repain, V., Chacon, C., Bellec, A., Girard, Y., Rousset, S., et al. (2017). Properties of functionalized carbon nanotubes and their interaction with a metallic substrate investigated by scanning tunneling microscopy. *The Journal of Physical Chemistry C*, *121*(43), 24264–24271.

Power, A. C., Gorey, B., Chandra, S., & Chapman, J. (2018). Carbon nanomaterials and their application to electrochemical sensors: A review. *Nanotechnology Reviews*, 7(1), 19–41.

Primo, E. N., Oviedo, M. B., Sánchez, C. G., Rubianes, M. D., & Rivas, G. A. (2014). Bioelectrochemical sensing of promethazine with bamboo-type multiwalled carbon nanotubes dispersed in calf-thymus double stranded DNA. *Bioelectrochemistry*, *99*, 8–16.

Rastogi, V., Yadav, P., Bhattacharya, S. S., Mishra, A. K., Verma, N., Verma, A., et al. (2014). Carbon nanotubes: An emerging drug carrier for targeting cancer cells. *Journal of drug delivery*, *2014*, 1–23.

Saifuddin, N., Raziah, A. Z., & Junizah, A. R. (2013). Carbon nanotubes: A review on structure and their interaction with proteins. *Journal of Chemistry*, *2013*, 18 pp.

Sarma, A. K., Vatsyayan, P., Goswami, P., & Minteer, S. D. (2009). Recent advances in material science for developing enzyme electrodes. *Biosensors and Bioelectronics*, *24*(8), 2313–2322.

Seifert, G. (1994). In G. Schmid (Ed.), *Clusters and colloids. From theory to applications.* Weinheim: VCH Verlagsgesellschaft. 546 p., 232 figs., 37 tabs., hard cover, ISBN 3-527-29043-5. Crystal research and technology, 29(7), 956–956.

Shishkova, D. K., Khodyrevskaya, Yu. I., Kutikhin, A. G., Rybakov, M. S., Mukhamadiyarov, R. A., & Shandakov, S. D. (2018). Preparation of a functional enzyme–carbon nanotube complex by the immobilization of superoxide dismutase on single-wall carbon nanotubes. *Nanotechnologies in Russia*, *13*(7–8), 349–355.

Smart, S. K., Cassady, A. I., Lu, G. Q., & Martin, D. J. (2006). The biocompatibility of carbon nanotubes. *Carbon*, *44*(6), 1034–1047.

Tang, W., Li, L., Wu, L., Gong, J., & Zeng, X. (2014). Glucose biosensor based on a glassy carbon electrode modified with polythionine and multiwalled carbon nanotubes. *PLoS One*, *9*, e95030.

Terrones, M. (2004). Carbon nanotubes: Synthesis and properties, electronic devices and other emerging applications. *International Materials Reviews*, *49*(6), 325–377.

Wang, Z., & Dai, Z. (2015). Carbon nanomaterial-based electrochemical biosensors: An overview. *Nanoscale*, 7(15), 6420–6431.

Yang, C., Denno, M. E., Pyakurel, P., & Venton, B. J. (2015). Recent trends in carbon nanomaterial-based electrochemical sensors for biomolecules: A review. *Analytica Chimica Acta*, *887*, 17–37.

Zhang, H., Liu, G., & Chai, C. (2012). A novel amperometric sensor based on screen-printed electrode modified with multi-walled carbon nanotubes and molecularly imprinted membrane for rapid determination of ractopamine in pig urine. *Sensors and Actuators B: Chemical*, *168*, 103–110.

Zhao, Y., Fan, L., Hong, B., Ren, J., Zhang, M., Que, Q., et al. (2016). Nonenzymatic detection of glucose using three-dimensional PtNi nanoclusters electrodeposited on the multiwalled carbon nanotubes. *Sensors and Actuators B: Chemical*, *231*, 800–810.

Zhao, B., Itkis, M. E., Niyogi, S., Hu, H., Perea, D. E., & Haddon, R. C. (2004). Extinction coefficients and purity of single-walled carbon nanotubes. *Journal of Nanoscience and Nanotechnology*, *4*(8), 995–1004.

CHAPTER SIXTEEN

Multiwalled carbon nanotubes bound beta-galactosidase: It's activity, stability and reusability

Maryam Khan, Qayyum Husain*
Department of Biochemistry, Faculty of Life Sciences, Aligarh Muslim University, Aligarh, UP, India
*Corresponding author: e-mail address: qayyumbiochem@gmail.com

Contents

Abstract

Carbon nanotubes (CNTs) based biosensors are recognized to be a next generation building block for ultrasensitive and fast biosensing systems. This article starting with a brief history on CNTs provides an overview on the recent expansion of research in the field of CNT-based biosensors. This is followed by the discussion on structure and properties related to CNTs. Furthermore, the basic and some newly developed synthetic methods of CNTs are summarized. In this chapter, we used polyaniline cobalt multiwalled CNTs to immobilize β-galactosidase, by adopting both noncovalent and

Methods in Enzymology, Volume 630
ISSN 0076-6879
https://doi.org/10.1016/bs.mie.2019.10.018

covalent strategies. Herein, the methodologies of both techniques have been discussed in detail. The η (effectiveness factor) values for nanocomposite bound β-galactosidase by physical adsorption and covalent method were calculated to be 0.93 and 0.97, respectively. The covalently bound β-galactosidase retained 92% activity even after its 10th successive reuse as compared to the adsorbed enzyme which exhibited only 74% of its initial activity. CNT armored enzymes demonstrated remarkably high catalytic stability at both sides of temperature and pH-optima along with easy recovery from the reaction medium which can be utilized in various biotechnological applications. Lastly, the scientific and technological challenges in the field are discussed at the end of this chapter.

Abbreviations

AC	activated carbon
AOx	alcohol oxidase
APS	ammonium persulfate
CNTs	carbon nanotubes
CVD	chemical vapor deposition
DWCNTs	double walled carbon nanotubes
EDTA	ethylene diamine tetra acetic acid
Fc	ferrocene
FTIR	Fourier transform infrared spectroscopy
GA	glutaraldehyde
GCE	glassy carbon electrode
GOx	glucose oxidase
HRP	horse radish peroxidase
MNPs	magnetic nanoparticles
MWCNTs	multiwalled carbon nanotubes
NC	nanocomposite
NM	nanomaterial
NPs	nanoparticles
ONPG	o-nitrophenyl β-D-galactopyranoside
PEI	polyethylenimine
SAED	selection area electron diffraction
SDS	sodium dodecyl sulfate
SEM	scanning electron microscopy
SWCNHs	single walled carbon nanohorns
SWCNTs	single walled carbon nanotubes
TEM	transmission electron microscopy

1. Introduction

The catalytic function of β-galactosidase, βGS (EC.3.2.1.23) is to hydrolyze lactose into its monomers that is glucose and galactose. It has wide applications in food processing industry, such as to enhance sweetness, solubility, flavor and digestibility of dairy products (Panesar, Kumari, & Panesar, 2010; Singh, Kumar, Mittal, & Mehta, 2016). However, majority of enzymes are fairly unstable and their industrial application is often restricted by a lack of long-term operational stability and the technically challenging recovery process (Sheldon & van Pelt, 2013).

In order to make enzyme utilization in biotechnological processes more favorable, different methods for cost reduction have been put into practice and, immobilization is one of them. Immobilization process enhances stability of the enzyme under both storage and operational conditions. Most importantly, it provides a facile separation of the enzyme from the product hence protein contamination of the product is minimized or avoided altogether (Saifuddin, Raziah, & Junizah, 2013). Immobilization also facilitates enzyme recovery which thereby helps to reduce the cost because the enzyme can be used repeatedly (Khan & Husain, 2018; Sheldon & van Pelt, 2013). Also, it can preclude intramolecular processes, e.g., autolysis, proteolysis and aggregation. The choice of support and the method of immobilization used are the utmost factors taken into consideration prior immobilization.

Recent advances in nanotechnology have provided a wealth of diverse nano-scaffolds that can be used as support for enzyme immobilization. Nano-biocatalysis integrating the biocatalyst and nanoscale materials is drawing great attention as innovative technology. The premise of using nanoscale structures for enzyme immobilization is to reduce diffusion limitations and maximize the functional surface area to increase enzyme loading. Among the various nanomaterials (NMs) used till date, carbon-based NMs have emerged to be one of the most promising ones as they possess honeycomb like structure, thereby providing larger surface area for efficient enzyme loading (Khan, Husain, & Naqvi, 2016).

1.1 Structure and properties of CNTs

Carbon nanotubes (CNTs), also known as buckytubes, were discovered in 1991 when the Japanese electron microscopist Sumio Iijima published his first report on multi-walled carbon nanotubes (MWCNTs), eventually followed by the discovery of single-walled CNTs, SWCNTs (Ijima, 1991). Since then, CNTs have emerged most intensively studied nanostructured materials as evidenced by thousands of papers published on it every year (Balasubramanian & Burghard, 2005). Due to the exceptional properties they possess, they have been employed in a variety of applications that attracts both academic and industrial interest. CNTs are long hollow tubular structures made up of sp^2 carbon atoms and a nanoscale diameter with one or more walls. The well ordered arrangement of carbon atoms linked via sp^2 bonds provides them the strength and stiffness.

Depending on the number of walls, CNTs can be classified into SWCNTs and MWCNTs. Generally, CNTs are comprised of several concentric rings or cylinders of increasing diameter commonly known as MWCNTs. They can be extended to a diameter of about 100 nm and the distance between the two walls is similar to the distance between the two graphene layers in graphite, 3.5 Å (Balasubramanian & Burghard, 2005). Double walled CNTs (DWCNTs) bridge the gap between SWCNTs and MWCNTs by possessing the properties of both kinds of CNTs. They are made of only two concentric cylinders. More specifically, DWCNTs resemble SWCNTs with respect to their small diameter, length and ability to form bundles, however, their mechanical stability is remarkably high compared to SWCNTs, when covalently functionalized (Yang, Chen, Ren, Zhang, & Yang, 2015). Furthermore, the outer wall of DWCNTs can be functionalized without affecting mechanical and electrochemical properties of the inner tube similar to MWCNTs (Pumera, 2007).

CNTs are well suited for a wide variety of challenging biomedical applications in the field of medicine and engineering; Fig. 1 (Chen & Chatterjee, 2013; Gomez-Gualdrón, Burgos, Yu, & Balbuena, 2011). CNTs-based

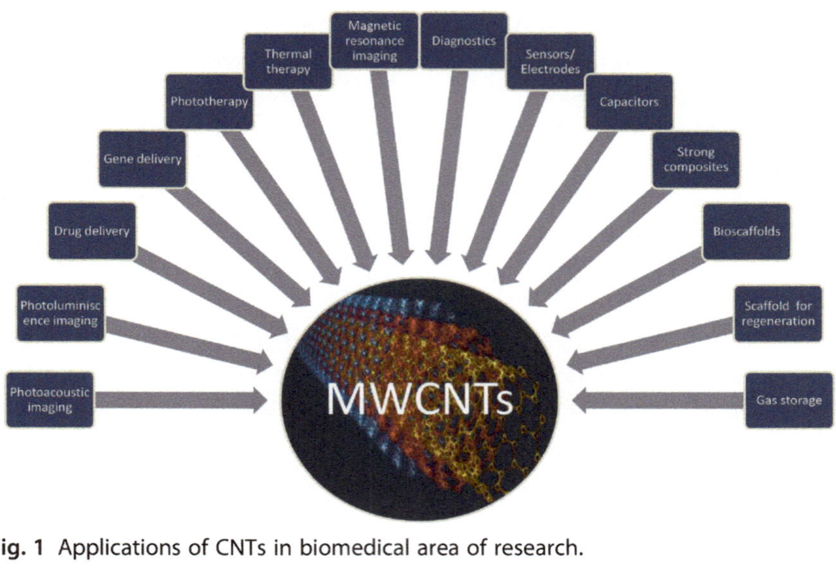

Fig. 1 Applications of CNTs in biomedical area of research.

materials can have a profound impact on electronic and optoelectronic devices, chemical sensors, nanocomposites, optical biosensor and energy storage (Arugula & Simonian, 2014; Boujtita, 2014; Pisoschi, 2013). It is well known that the material properties (e.g., electrical or optical) of CNTs are very sensitive to be affected by exposure to biomolecules and this has led to the investigation, by a number of groups of these materials as sensing elements for biosensors (Hernandez-Ibanez et al., 2016; Jasim, Wajidullah, Shi, Lin, & Yang, 2017). Additionally, their unique structure provides an exceptional electric-current-carrying capacity along their length (1000 × higher than copper wires) which leads to attractive CNT-based new electrode material (Abdalla, Al-Marzouki, Al-Ghamdi, & Abdel-Daiem, 2015). The properties possessed by CNTs, such as fast electron transportation high thermal conductivity, excellent mechanical flexibility and good biocompatibility, make them an ideal choice in the fabrication of sensitive electrochemical biosensor. New applications of CNT-based NM in electrochemical sensors and biosensors are now on the increase with the advent of nanotechnology (Andreescu, Njagi, & Ispas, 2008).

1.2 Synthesis and functionalization of CNTs

A number of different routes to synthesize CNTs have been demonstrated over the recent years. The three main synthesis techniques for CNTs production are arc discharge, laser ablation and chemical vapor deposition, CVD (Gougis, Tabet-Aoul, Ma, & Mohamedi, 2014; Park, Vosguerichian, & Bao, 2013). CNTs generally obtained by the arc method or hydrocarbon pyrolysis are multi-walled, having several graphitic sheets or layers. Single-walled carbon nanohorns (SWCNHs) are horn-shaped single-walled tubules with a conical tip. They are generally synthesized by laser ablation of pure graphite without using metal catalyst with high production rate and yield, and typically form radial aggregates (Manawi, Ihsanullah, Samara, Al-Ansari, & Atieh, 2018). Table 1 demonstrates various methodological processes employed in the synthesis of CNTs along with their outcome in detail.

One of the major drawbacks associated with CNTs to be used in biosensor and biomedical applications is their tendency to aggregate in bundles through strong interactions which are difficult to disrupt rendering these materials incapable of achieving their full potential. However, functionalization of CNTs by attaching functional groups on their surface helps to solubilize them and hence facilitate their use in a myriad of applications

Table 1 Detailed description of synthetic methods of CNTs.

Method	Process	Result	Reference
Arc discharge	Two graphite rods in water bath at different voltage	High yield CNTs of good quality	Lakshmi and Khan (2014)
	Two graphite electrodes submerged in different media	Nanocarbon structures of different dimensions	Vasu, Pramoda, Moses, Govindaraj, and Rao (2014)
	Instead of metal catalyst and vacuum devices, strong oxidizing agents HNO_3/H_2O_2 are used	Improves purity of MWCNTs	Singhal, Singh, Singla, and Dharamvir (2013)
	The discharge is maintained in a magnetic field	High quality MWCNTs	Singh, Dharamvir, and Singhal (2013)
	During synthesis physical forces are applied	Relatively straight and defect free MWCNTs	Berkmans, Ramakrishnan, Jain, and Haridoss (2013)
Laser ablation	Binary catalysts combined with Fe, Co and Ni (transition metals)	Different types of carbon nanostructures	Jedrzejewska et al. (2013)
	Irradiation of CO_2 laser in continuous wave mode onto a boron containing graphite target at room temperature	Fine crystalline structure of MWCNTs	Yuge et al. (2012)
CVD	Co used as catalyst (at 700°C) with the use of hydrogen to acetylene gas ratio 25:25 Scm	High yield MWCNTs	Kishore and Pandurangan (2012)
	Using Ni/MgO as catalyst and CH_4 in microfluidized bed photochemical deposition	High purity MWCNTs with less defect	Fu, Cui, Wei, and Ma (2014)
Solvothermal	At low-temperature (180°C)	MWCNTs bundles	Krishnamurthy and Agarwal (2013)
	At temperature 200°C for 10h	Magnetic MWCNTs with alterable structure	Xiao et al. (2012)

(O'connell et al., 2001; Tasis, Tagmatarchis, Georgakilas, & Prato, 2003). In fact, the functionalization of CNTs has become a pre-requisite in order to enable facile fabrication of nanodevices. Among all the functionalization approaches, physical, chemical or combined modifications have been exploited for their homogeneous dispersion in common solvents to improve their solubility and application.

Physical modification utilize the mechanical means such as ultrasonic, milling, crushing and friction to activate CNTs surface in order to change their surface morphology (Fujigaya & Nakashima, 2008). This method can increase CNTs' internal energy and surface activity and therefore make the tubes react with or attached to other materials, to attain the purpose of the surface modification. At present, the large shear force or ultrasonic processing is often used to disperse CNTs (Hasan, Crawford, & Ivanova, 2013; Spitalsky et al., 2009).

One possibility of CNT functionalization is filling the inner cavity with molecules, known as endohedral functionalization (Huang et al., 2013; Ye, Jo, Jeong, & Lee, 2013). Additionally, metallofullerenes were encapsulated in CNTs as well as noble metals (e.g., Au, Ag, Pt, Pd), yielding metallic nanowires (Huang et al., 2013; Meng et al., 2013). In general, chemical approach can be either covalent or non-covalent in nature. Various ways by which we can covalently functionalize CNTs have been demonstrated in Fig. 2. Non-covalent type of functionalization possesses several

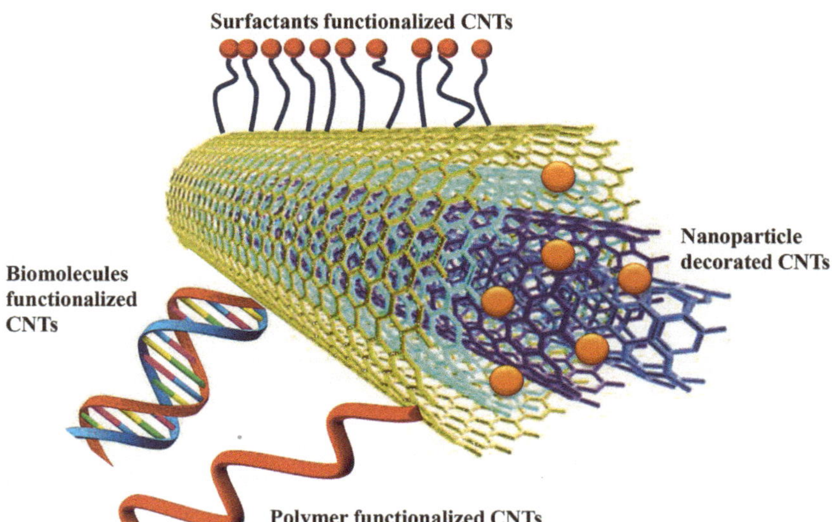

Fig. 2 Different methods of covalent functionalization of CNTs.

advantages over covalent method such as the preservation of structural homogeneity and electrical properties of CNTs. However, these procedures involve ultrasonication, centrifugation and filtration steps.

For ideal covalent or non-covalent functionalization coating on CNTs, some points have to be kept in mind (Tîlmaciu & Morris, 2015). First, the coating used should be non-toxic and biocompatible in nature. Second, it should be stable enough to avoid detachment from the nanotube's surface in biological solutions, and lastly it should bear active functional groups so as to provide interaction of CNTs with other molecules in order to prepare subsequent CNT bioconjugate. To summarize, it has been proven through various studies that surface functionalized CNTs can behave in a biologically different and safer manner, compared to their pristine counterparts (Mananghaya, Santos, & Yu, 2017; Marchesan, Kostarelos, Bianco, & Prato, 2015).

1.3 Application in biosensing

With the outburst of nanotechnology which includes NM with distinguished physicochemical properties, a new class of biosensors referred to as nanobiosensors have been developed. These types of biosensors have unprecedented combination of NM exhibiting properties notably their small size and large surface/volume ratio along with the functionality of "macro"- biosensors in order to facilitate sensitivity and lower detection limits (Holzinger, LeGoff, & Cosnier, 2014). While designing biosensors, the immobilization of enzymes on the transducer surface is an important and decisive step. Many studies have been performed on the immobilization of enzymes due to the known benefits of the enhancement in the property of the enzyme which is convenient in the handling, ease of separation, reduced product cost and conceivable increase in thermal and pH stability (Bucko et al., 2012; Khan, Husain, & Ahmad, 2018). Among various industrially important enzymes, βGS has been extensively utilized as a prototype enzyme due to its adaptability as a biosensor for lactose detection and to provide lactose-free milk to lactose intolerance patients worldwide.

The selection of appropriate immobilization technique affects the biosensor performance like sensitivity, selectivity and stability, as it affects the enzyme orientation, loading, mobility and structure. Nanoparticles-based enzyme immobilization and their potential applications in biosensors have been described in a review paper (Ansari & Husain, 2012). Furthermore, different kinds of techniques applied for the immobilization of βGS and its significances in food industry has been explained in detail in a review article (Panesar et al., 2010).

To obtain the best immobilized enzyme, it is highly recommended that the characteristics of support and enzyme should be known. 3D conformation of an enzyme is liable to alterations because it is balanced by the optimization of counter forces like, hydrophilic interactions, Columbic force, hydrogen bonds, and van der Waals interactions. Essentially, enzymes can pivot inside out while using a hydrophobic support, losing its activity because proteins in general are hydrophobic in their interior and hydrophilic in the exterior (Netto, Toma, & Andrade, 2013). These immobilization techniques have been mainly employed for βGS to advance electrochemical, optical, or gravimetric enzymatic biosensors.

In recent years, CNTs have received great attention as promising carbon-based nanoelectronic devices. Fig. 3A indicates the data based on the articles published related to CNT-biosensors in the past 15 years and according to the list of publications, China occupies a prime position in developing such biosensors based on CNTs (Fig. 3B). The large surface area and good electrical conductivity of CNTs allow them to act as "electron wire" between the redox center of an enzyme or protein and an electrode's surface, which make them an excellent material for the design of electrochemical biosensors (Husain, 2017, 2018a; Mundra, Wu, Sauer, Dordick, & Kane, 2014). Table 2 shows the recent developments in biosensors based on SWCNT/MWCNT, while Table 3 lists different enzyme biosensors based particularly on MWCNTs, their respective analyte, linearity range and limit of detection.

Because of the unprecedented properties possessed by CNTs, as mentioned in previous section, a large number of CNT-conjugates have been designed for detection of DNA biomarkers, cell surface sugars, protein receptors, hormones, antibodies, enzymes and various metabolites (Tîlmaciu & Morris, 2015). Depending on their mechanism of target recognition and transduction, these biosensors are broadly subdivided into electronic transducers, electrochemical CNT-biosensors, immunosensors, and optical CNT-based biosensors. Functionalized CNTs have been used as electrochemical sensors and many review papers concerning CNTs-electrochemical sensors have been highlighted (Dai, 2007; Daniel et al., 2007; Gao, Guo, Liua, & Huang, 2012; Husain, 2018a; Jacobs, Peairs, & Venton, 2010; Yang et al., 2015).

Das and Goswami (2013) immobilized an octameric alcohol oxidase (AOx) (Mr 675 kDa) obtained from *Pichia pastoris* on MWCNT-nafion (MWCNT-Nf) matrix. The complex encapsulated with polyethylenimine (PEI) on gold electrode, showed a redox peak at 0.21 V (vs Ag/AgCl electrode at pH 7.5) for oxidation of alcohol. The electron transfer rate constant

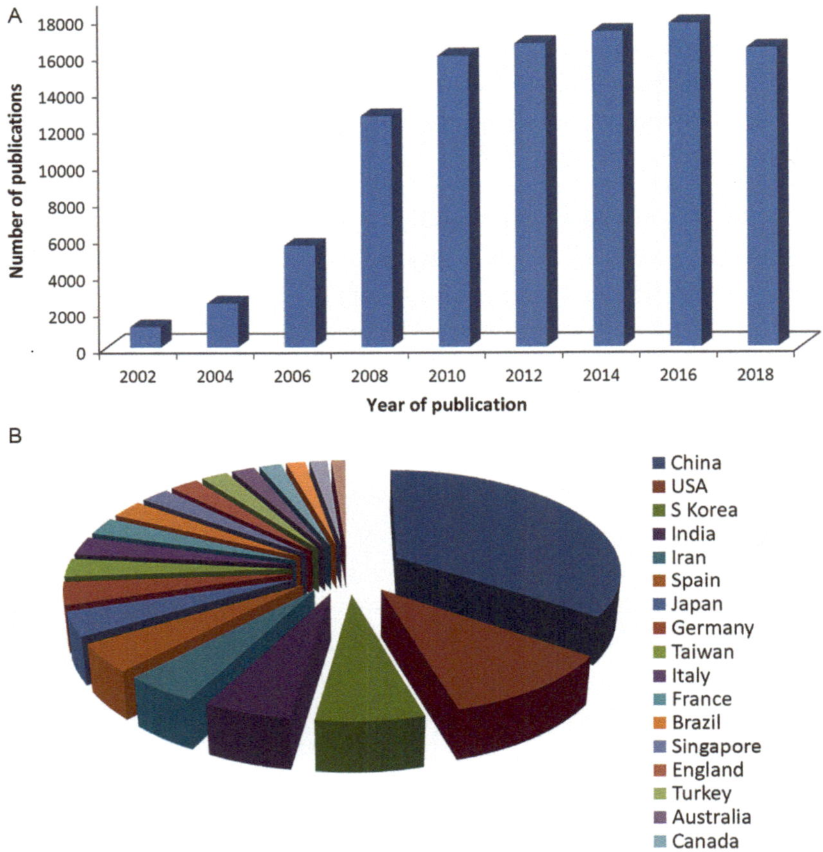

Fig. 3 (A) Graphical representation of number of published papers of CNT-based biosensors in last 15 years. This data was searched using Google scholar and includes regular articles, review articles and book chapters. (B) Top 20 countries of research community working on CNTs-based biosensors.

and surface coverage of the immobilized enzyme were $1.69 \pm 0.15\,s^{-1}$ and $2.43 \times 10^{-12}\,mol\,cm^{-2}$, respectively. Au-MWCNT-Nf-AOx-PEI bioelectrodes for alcohol exhibited a linear response in the range of 8–42 μM and detection limit of 5 μM. Moreover, these bioelectrodes retained ~90% of the original response even after 4 weeks of storage in potassium phosphate buffer, pH 7.5 at 4 °C.

Chinnadayyala et al. (2014) developed a third generation biosensor for improved detection of alcohol in liquid samples. They reported an alcohol oxidase, AOx with a two-fold increase in efficiency (K_{cat}/K_m) by physical entrapment of the activator ferrocene (Fc) in the protein matrix through a

Table 2 Recent developments in biosensors based on SWCNT/MWCNT.

CNT biosensor	Type of biosensor	Analyte	Reference
CNTs	Electrochemical	Ions, metabolites and protein bio-markers	Wang and Dai (2015)
MnO_2-CNT/Pt	Enzymatic	Catechin	Vilian et al. (2015)
MWCNTs	Cholesterol	Cholesterol in blood	Li, Liao, Hu, Ma, and Wu (2005)
Chitosan-CNTs	Electrochemical	Nitric oxide	Santos, Rodrigues, Laranjinha, and Barbosa (2013)
Functionalized-MWCNTs	Electrochemical	Epinephrine	Prasad, Prasad, Tiwari, and Madhuri (2013)
Functionalized-MWCNTs	Electronic	Rat striatum	Kress et al. (2014)
Chitosan-SWCNTs	Electronic	Cellular nitric oxide	Li, Liu, et al. (2005)
SWCNTs	Electronic	Epidermal carcinoma cells	Jin et al. (2010)
Pt/CNTs	Amperometric biosensor	Cysteine	Fei et al. (2005)
CNT arrays	Amperometric biosensor	Biomarker	Yuichi et al. (2009)
MWCNTs-biopolar electrode	Amperometric	Prostate-specific antigen	Feng, Pan, Zhang, Zhu, and Chen (2004)
MWCNTs-P450	Amperometric	Breast cancer	Baj-Rossi, De Micheli, and Carrara (2012)
Tricosane/f-SWCNTs	Electrochemical	Lung cancer	Liu et al. (2011)
D-(+) galactose-SWCNTs	Electrochemical	Cancer marker galactin-3	Park et al. (2011)
	Electrochemical	HeLa and HL60	Zheng, Zhang, Zou, and Jun-Jie (2012)
SWCNTs/MWCNTs	Amperometric	Prostate cancer, PSA detection in blood samples	Shobha and Muniraj (2015)
Au-Ag/MWCNTs	Amperometric	MGC-803 gastric cancer cells	Zhang et al. (2014)

Table 3 Various MWCNTs—enzyme based biosensors, their analytes, linearity range of measurement and detection limit.

Sensor	Analyte	Limit of detection	Linearity range	Sensitivity	Reference
Glucose oxidase/NH_2-MWCNT	Glucose	9 μM	17–650 μM		Khodadadei, Ghourchian, Soltanieh, Hosseinalipour, and Mortazavi (2014)
MWCNT-nafion HRP-Chitosan ferrocene entrapped polyethylenimine	Ethanol	2 μM	5–3000 μM	1 μM	Chinnadayyala, Kakoti, Santhosh, and Goswami (2014)
MWCNT-nafion HRP-Chitosan alcohol oxidase polyethylenimine	Ethanol	5 μM	8–42 μM	0.264 μM	Das and Goswami (2013)
Au NP-MWCNTs	Uric acid	–	0.01–0.8 mM	10 and 20 mg L^{-1}	Chauhan and Pundir (2011)
MWCNT-acetyl choline esterase	Carbofuran	–	10^{-10}–10^{-3} g L^{-1}	10^{-12} g L^{-1}	Zhang, Li, Ma, Xiong, and Qu (2013)
Carboxylated MWCNT sorghum leaf oxalate oxidase	Oxalate	–	8.4–272 μM	0.0113 μA μM^{-1}	Yadav, Devi, Kumari, Yadav, and Pundir (2011)
MWCNT—Lysozyme	Lysozyme	–	2.5 mM $(Fe CN)_6^{3-/4-}$	42.09 μg mL^{-1}	Silva et al. (2014)
MWCNT-nafion cysteamine modified tyrosinase	Dopamine	–	0.05–100 μM	1 μM	Canbay and Akyilmaz (2014)
Glucose oxidase-ZnO NPs-MWCNT	Glucose	–	0.1–6 μM	12.5 mA m^{-1} cm^{-2}	Wang et al. (2009)
Glucose oxidase-silicon dioxide Fe_3O_4@SiO_2/MWCNT	Glucose	–	1 μM–30 mM	800 nM	Baby and Ramaprabhu (2010)
Au NPSs Graphene MWCNT	Glucose	–	4.8 μM	29.72 mA mM^{-1} cm^{-2}	Vanegas et al. (2014)
Ni-MWCNT	Glucose	0.021 μM	0.05–12 mM	70 mA mM^{-1} cm^{-2}	Başkaya et al. (2017)
Palladium Ni-MWCNT	Glucose	0.014 μM	0.01–1.00 mM	90 mA mM^{-1} cm^{-2}	Koskun, Şavk, Şen, and Şen (2018)

simple microwave-based partial unfolding technique. The sensor was fabricated by immobilizing Fe entrapped alcohol oxidase, FcAOx and sol–gel chitosan film coated horseradish peroxidase (HRP) on MWCNT modified GCE through layer-by-layer technique. The amperometric biosensor exhibited a linear response to alcohol in the range of 5.0×10^{-6}–$30 \times 10^{-4}\,\text{mol}\,\text{L}^{-1}$ with a detection limit of $2.3 \times 10^{-6}\,\text{mol}\,\text{L}^{-1}$, and a sensitivity of $150\,\mu\text{A}\,\text{mM}^{-1}\,\text{cm}^{-2}$. The biosensor response was steady for 28 successive measurements completed in a period of 5 h and retained ~90% of the original response even after 4 weeks when stored at 4 °C.

In a study MWCNTs were amine-functionalized by a process of dielectric barrier discharge plasma treatment. The amine-functionalized MWCNTs were subsequently fixed on glassy carbon electrode (GCE) and glucose oxidase (GOX) was immobilized as a model enzyme on the modified GCE. The obtained enzyme electrode was used for glucose biosensing with a linear range of 17–646 μM and sensitivity of $12.3\,\mu\text{A}/\text{mM}\,\text{cm}^2$. Based on the signal to noise ratio of 3, the detection limit was estimated to be 9 μM. The Michaelis–Menten constant for immobilized GOX was found to be 480 μM (Khodadadei et al., 2014).

Functionalized MWCNT supported with highly monodisperse Ni NP modified on GCE (Ni@f-MWCNT/GCE) were synthesized through microwave assisted method and examined for non-enzymatic glucose sensing in ionic liquids by cyclic voltammetry and chronoamperometry. The results demonstrated a good linear span of 0.05–12.0 mM and a detection boundary of 0.021 μM. Furthermore, in the amperometric signal of the electrodes after 200th cycles, no significant change was observed (Başkaya et al., 2017).

In a more recent study, a novel glucose sensor was prepared by in situ reduction technique. A highly sensitive activated carbon (AC) decorated on monodisperse nickel and palladium alloy NC were fixed on a GCE (Ni-Pd@AC/GCE NCs) and GOX was immobilized on the modified carbon electrode. In addition, the Ni-Pd@AC/GCE NC showed a very low detection limit of 0.014 μM, a wide linear range of $0.01\,\text{mM}^{-1}\,\text{mA}$ and a very high sensitivity of $90\,\text{mA}\,\text{mM}^{-1}\,\text{cm}^{-2}$ (Koskun et al., 2018).

The combination of polymers with carbon NM for the preparation of chemical sensors has opened new avenues of research due to their biocompatibility, excellent sensitivity and selectivity. Among the various conducting polymers, polyaniline (PANI) is one of the most considerable, due to its high electrical conductivity, unsophisticated synthesis, and environment friendly nature (Das, Prusty, Adhikary, & Kalra, 2015). In recent past, it has gained much recognition in biosensor applications, since it acts as an efficient mediator for electron transfer in the redox reaction. Cobalt (Co), because of its electrical and thermal properties is also extensively employed for the development of nanoparticle (NP) based biosensors. Therefore, we synthesized polyaniline doped cobalt MWCNTs NC (PANI/Co/MWCNT-NC) in order to achieve combinatorial effect for successful immobilization of βGS isolated from *Aspergillus oryzae*.

2. Methodology part

2.1 Synthesis of PANI/Co/MWCNT-NC

General safety: Safety glasses, latex, or nitrile gloves and lab coat.

Chemicals and reagents

1. Multiwalled carbon nanotubes (purity ≥95% wt%, 10 nm diameter)—Central Drug House Private Limited, New Delhi, India
2. Sodium dodecyl sulfate (SDS)—Central Drug House Private Limited, New Delhi, India
3. Ammonium persulfate (APS)—Central Drug House Private Limited, New Delhi, India

Equipment

1. Beaker (50–500 mL)
2. Stirrer/hot plate
3. Analytical balances
4. pH meter
5. Sonicator
6. Magnetic stirrer with bar
7. Hot air oven

Methodology

Method adopted by Giri, Das, and Kalra (2012) after some minor modification was used for the synthesis of PANI/Co/MWCNT-NC.

1. In a beaker, mix 150 mL of 1 M HCl and 50 mL of 0.2 M sodium dodecyl sulfate and subsequently add 60 mg of MWCNT into it.
2. Sonicate the mixture for 30 min at room temperature.
3. In another beaker, take 1 mL of aniline monomer in 50 mL of 1 M HCl and add calculated amount of $CoCl_2$ (3%, w/v) by stirring continuously to achieve a homogenous suspension.
4. Mix both the solutions prepared in the previous steps and sonicate for another 10 min.

5. Add 50 mL of 1 M HCl containing 2.04 g of APS drop by drop in stirring condition to the solution for the complete polymerization of the monomer.
6. Incubate the mixture at 0–5 °C for 24 h.
7. Finally, filter the resulting precipitate and wash with distilled water and ethanol.
8. Dry the precipitate well in a vacuum oven at 60 °C for 48 h for further studies.

2.2 Characterization of PANI/Co/MWCNT-NC

General safety: Safety glasses, latex, or nitrile gloves and lab coat.

Chemicals and reagents

1. Potassium bromide (AR Grade)—Sisco Research Laboratories (SRL), India
2. Sodium acetate (AR Grade) buffer (pH 4.5, 0.1 M)—SRL, India
3. PANI/Co/MWCNT-NC (diluted solutions)

Equipment

1. Carbon coated copper grids-300 mesh
2. JEOL-2100 transmission electron microscope, Japan
3. Sputter coater-JEOL JFC 1600, Japan
4. JSM-6510 LV scanning electron microscope, Japan
5. FTIR-Perkin Elmer instrument, USA (Spectrum-II, wave no. 4000-400).

Methodology

The free and enzyme bound NC preparations were characterized by TEM (JEOL-2100, 200 κV), SEM (JSM-6510 LV, 15 κV) and FT-IR (Perkin Elmer instrument, USA, Spectrum- II, wave no. 4000-400).

1. For TEM studies, prepare the samples by drop coating diluted solutions on carbon coated copper grids at normal atmospheric conditions.
2. Prepare small samples from PANI/Co/MWCNT-NC and air dried immobilized βGS for SEM studies. Mount these samples on carbon tape on a copper stub, followed by 60 nm of gold deposition by means of a sputter coater.
3. For FT-IR, crush the samples and thoroughly mix with KBr to form a transparent pellet.
4. Finally, UV–vis spectrophotometer (Shimadzu, UV-1800) was acquired to measure the enzyme concentration.

The immobilization procedure as to how the enzyme was attached on/to the NC has been described in the subsequent section.

2.3 Immobilization of βGS on PANI/Co/MWCNT-NC

General safety: Safety glasses, latex, or nitrile gloves, lab coat.

Chemicals

1. Sodium acetate (AR Grade) buffer (pH 4.5, 0.1 M)—SRL, India
2. PANI/Co/MWCNT-NC (600 mg)
3. βGS (*Aspergillus oryzae*)—Sigma-Aldrich Chemicals Co. (USA)

Materials and equipment

1. Falcons (15 mL)—Tarson, India
2. Microtubes (1.5 mL)—Eppendorf, India
3. Shaking incubator—Biogen. India
4. Micro-pipettes (20.0–100.0 μL, 0.1–1.0 mL)
5. Ultrasonicator—Life Care
6. Magnet (6000G)
7. UV–vis spectrophotometer (Shimadzu, UV-1800)
8. Centrifuge-REMI Cooling Centrifuge

Methodology

1. First, vary the concentration of PANI/Co/MWCNT-NC from 2 to 18 mg and then independently adsorb free βGS on varying concentrations by continuously shaking the mixture in sodium acetate buffer, pH 4.5 at 25 °C.
2. Second, differ the enzyme concentration from 1 $mg mL^{-1}$ to 5 $mg mL^{-1}$, keeping the NC concentration (10 mg) fixed.
3. In second set of experiments, where βGS is covalently attached to the support, vary NC concentration from 2 to 10 mg, while keeping the glutaraldehyde (GA) concentration fixed at 2%.
4. Furthermore, vary the concentration of GA from 0.5% to 3.0% (v/v) by maintaining the concentration of NC at 6 mg.
5. Calculate immobilization yield in all cases and take the highest yielding parameters to progress further.
6. For simple adsorption method, suspend and sonicate PANI/Co/MWCNT-NC (250 mg) in 0.1 M sodium acetate buffer, pH 4.5 for 20 min.
7. For covalent method, mix NC (250 mg) with 2%, (v/v) GA in continuous stirring condition for 2 h.
8. Centrifuge the activated support at 3000 × *g* for 10 min at 4 °C and wash thrice with assay buffer of 0.1 M sodium acetate, pH 4.5 in order to remove traces of GA.
9. Load βGS (24.67 U) individually in both types of colloidal suspensions.

10. Stir the mixture slowly for 5 h (simple adsorption) and 2 h (covalent) and then centrifuge at 3000 × g for 10 min at 4 °C.
11. Remove the unbound enzyme by washing thrice with assay buffer and preserve the precipitate at 4 °C for further studies.
12. Also, store the supernatant and the washes at 4 °C for the calculation of immobilization yield (Husain, Ansari, Alam, & Azam, 2011).
13. Determination of immobilization yield.
    The immobilization efficiency was determined in terms of immobilization yield as follows:

$$\text{Immobilization yield } (\%) = \text{A} - \text{B}/\text{A} \times 100\%$$

where A is the initially added activity of βGS in the immobilization system and B is the residual enzyme in the supernatant and washing solution after the immobilization process.

2.4 Stability studies of immobilized βGS

Chemicals and reagents

1. βGS—Sigma-Aldrich Chemicals Co. (USA)
2. Glycine-HCl (pH 2.0 and 3.0)—SRL, India
3. Sodium acetate buffers, pH (4.0, 4.5 and 5.0)—SRL, India
4. Sodium phosphate buffers, pH (6.0 and 7.0)—SRL, India
5. Tris HCl buffers, pH (8.0 and 9.0)—SRL, India
6. Crushed ice
7. Galactose (1.0–5.0%, w/v)—SRL, India
8. *o*-Nitrophenyl β-D galactopyranoside, ONPG (0.1–1.0 mM)—SRL, India

Equipment

1. Water bath (30–110 °C)—Scientech Instruments, Delhi
2. Ice maker and crusher—Labman
3. Ice box
4. Microtubes (1.5 mL)—Eppendorf, India
5. UV–vis spectrophotometer (Shimadzu, UV-1800)

Methodology

1. *pH-activity profiles* (Ansari & Husain, 2010)
   (a) Prepare different buffers of varying pH including glycine–HCl (pH 2.0 and 3.0), sodium acetate (pH 4.0, 4.5 and 5.0), sodium phosphate (pH 6.0 and 7.0), and Tris–HCl (pH 8.0 and 9.0).
   (b) Set the molarity of each buffer as 0.1 M.

(c) Incubate the free and both types of immobilized enzyme preparations (0.04 U) in the buffers of different pH for 30 min and then perform activity assay by monitoring their activity in a spectrophotometer at 405 nm.

(d) Assume the activity at pH 4.5 as control (100%) for the calculation of residual percent activity.

2. *Temperature-activity profiles* (Ansari & Husain, 2010)

(a) Assay the activity of free and immobilized βGS (0.04 U) at different temperatures starting from 20 °C to 80 °C in 0.1 M sodium acetate buffer, pH 4.5 for 15 min.

(b) Consider the enzyme activity at 50 °C as control (100%) for the calculation of residual percent activity for free and both types of immobilized enzyme at other temperatures.

3. *Thermal denaturation* (Ansari & Husain, 2010)

(a) Incubate free and immobilized βGS (0.04 U) at 60 °C in 0.1 M sodium acetate buffer, pH 4.5 for varying times.

(b) Take out aliquots of each preparation (20 μL) in triplicates at gaps of 15 min and chill them quickly in crushed ice for 5 min.

(c) Bring the enzyme to room temperature and measure their activity.

(d) Define the activity of enzyme without incubation at 60 °C as control (100%) for the calculation of residual percent activity at each of the respective reactions.

4. *Effect of galactose* (Ansari & Husain, 2010)

(a) Incubate free and immobilized enzyme preparations (0.04 U) in the presence of various concentrations of galactose (1.0–5.0%, w/v).

(b) Evaluate the activity of enzymes in 0.1 M sodium acetate buffer, pH 4.5 at 37 °C by performing routine enzyme assay.

(c) Consider the activity of enzyme without added galactose as control (100%) for the calculation of remaining percent activity.

5. *Reusability of immobilized enzyme* (Husain et al., 2011)

(a) Take immobilized enzyme preparations (0.04 U) in triplicates for assaying the activity of the enzyme after repeated use.

(b) Perform the reaction in 1.5 mL microtubes for 10 min at 37 °C

(c) Post-incubation, centrifuge the mixture at 3000 × *g* for 10 min at 4 °C.

(d) Decant the supernatant and promptly add 0.1 M sodium carbonate into it, in order to terminate the reaction.

(e) After each assay, wash the immobilized enzyme with 0.1 M sodium acetate buffer, pH 4.5, by centrifugation at 3000 × *g* for 10 min.

(f) Store the obtained pellet in assay buffer at 4 °C and repeat this process for 10 successive days.
(g) Consider the activity determined after the first cycle as control (100%) for the calculation of residual percent activity after repeated use.

6. *Determination of kinetic parameters*

   Evaluate the kinetic parameters K_m and V_{max} of free and immobilized βGS from Michaelis–Menten plot by measuring their initial rates by varying the concentrations of the substrate, ONPG (0.1 − 1.0 mM) in 0.1 M sodium acetate buffer, pH 4.5 (Khan et al., 2016).

2.5 Genotoxicity assessment of PANI/Co/MWCNT-NC

Reagents and chemicals

1. pBR322 DNA—Thermo Fischer scientific (USA)
2. Ethylene diamine tetra acetic acid (EDTA)—SRL, India
3. Tris—SRL, India
4. Bromophenol Blue—Merck & Co., USA
5. Glycerol—Merck & Co., USA
6. Ethidium bromide—Sigma-Aldrich Chemicals Co. (USA)

Equipment

1. Beakers (500–1000 mL)
2. Incubator
3. Microtubes (1.5 mL) with stand-Eppendorf, India
4. Electrophoretic gel apparatus
5. UV-transilluminator

Methodology

Plasmid nicking assay (Khan, Husain, & Ansari, 2013).

(a) Prepare a reaction mixture (30 μL) containing 10 mM Tris–HCl buffer, pH 7.5, pBR322 DNA (0.5 μg) with free enzyme (0.04 U), PANI/Co/MWCNT-NC (50 μg), and immobilized enzyme (10 μL). Incubate this for 2 h at 37 °C.
(b) Add 10 μL of solution containing 40 mM EDTA, 0.05% (w/v) bromophenol blue (tracking dye), and 50% (v/v) glycerol after incubation.
(c) Perform electrophoresis in submarine 1.0% (w/v) agarose gel.
(d) View and photograph ethidium bromide stained gel under a UV-transilluminator.

2.6 Assay of βGS

The ONPG hydrolyzing activity of free and bound βGS can be assayed by measuring the release of *o*-nitrophenol from ONPG at 405 nm (Ansari & Husain, 2012).

(a) Perform the reaction by continuous shaking in a total volume of 2.0 mL containing 1.60 mL, 1.65 mL and 1.68 mL of 0.1 M sodium acetate buffer for free, simply adsorbed and covalently immobilized enzymes, respectively, pH 4.5, 0.04 U βGS (Free—100 μL, adsorbed—50 μL, cross-linked—20 μL) and 0.3 mL of 20 mM ONPG at 37 °C for 15 min.

(b) Terminate the reaction by the addition of 2.0 mL of 1.0 M sodium carbonate solution and read product formation spectrophotometrically at 405 nm.

(c) The one unit (1.0 U) of βGS activity is defined as the amount of enzyme that catalyzes the release of 1.0 μmol of *o*-nitrophenol ($\varepsilon_m = 4500\ L^{-1}\ mol^{-1}\ cm^{-1}\ min^{-1}$) under standard assay conditions.

2.7 Estimation of protein

(a) Calculate protein concentration according to the method as described by Lowry, Rosenbrough, Farr, and Randall (1951).

(b) Use Bovine serum albumin as a standard.

2.8 Statistical analysis

(a) Each value represents the mean for three independent experiments.

(b) Perform statistical analysis for comparison between free and both types of immobilized βGS by one way ANOVA test.

(c) Plot the data expressed in various studies using origin 6.1 and express with standard deviation of error (±).

(d) Lastly, consider P-values ≤ 0.05 as statistically significant.

3. Results and discussion

3.1 Characterization studies of free and βGS bound PANI/Co/MWCNT-NC

Native and βGS bound PANI/Co/MWCNT-NC was characterized by transmission electron microscopy (TEM), scanning electron microscopy (SEM) and Fourier transform infrared (FT-IR). The TEM images of the free and the enzyme bound NC were illustrated in Fig. 4A–C. The coating of PANI on the CNT was confirmed by the presence of thin border all along

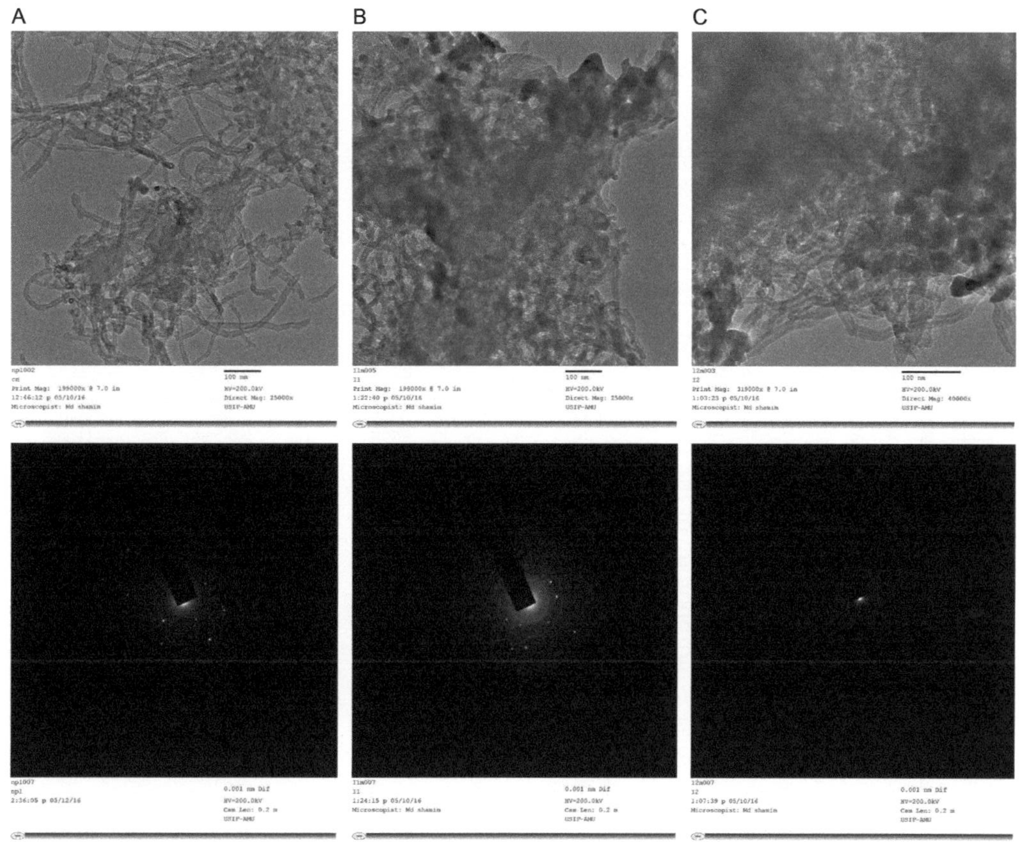

Fig. 4 TEM images and SAED patterns for free and enzyme bound NC. The figure shows (A) free PANI/Co/MWCNT-NC, (B) NC adsorbed βGS, and (C) NC covalently bound βGS. *Adapted from Khan, M., Husain, Q., & Bushra, R. (2017). Immobilization of β-galactosidase on surface modified cobalt/multiwalled carbon nanotube nanocomposite improves enzyme stability and resistance to inhibitor.* International Journal of Biological Macromolecules, 105, *693–701.*

the tubular structures (Zhao et al., 2012). The particle size of the NC was found in the range of 18–23 nm. The incorporation of Co can be visualized in TEM images by the presence of small black spots between the PANI coated CNTs. The binding of the enzyme on the surface of NC via adsorption and covalent binding methods has been compared in Fig. 4B and C. The covalent coupling method allowed binding of more enzymes compared to the adsorbed βGS. The selection area electron diffraction (SAED) pattern indicates that the NC is crystalline in nature. However, the immobilization process altered SAED pattern. Post-immobilization, the adsorption procedure made the NC highly crystalline in nature with concentric rings clearly visible in the SAED pattern (Fig. 4B), whereas in the case of GA treated NC, its crystallinity was decreased. Adsorption method is simple and does not involve any use of chemicals, while the covalent binding necessitates the use of GA and covalent bonds are formed in between the enzyme and the matrix. The surface morphology of the bound and free NC was studied by SEM. According to the SEM images, the surface characteristic of the NC was significantly changed upon immobilization of the enzyme by both the methods (Fig. 5). Rigid binding was observed in case of cross-linked enzyme.

The FT-IR spectra were recorded for free PANI/Co/MWCNT-NC, NC adsorbed enzyme and βGS covalently attached to the NC in the range of 500–4000 cm^{-1} (Fig. 6). The characteristic peaks at 1493 and 1570 cm^{-1} for benzenoic and quinonic type rings signify the presence of PANI in the NC, respectively (Zhao et al., 2012). Peak at 1300 cm^{-1} corresponds to

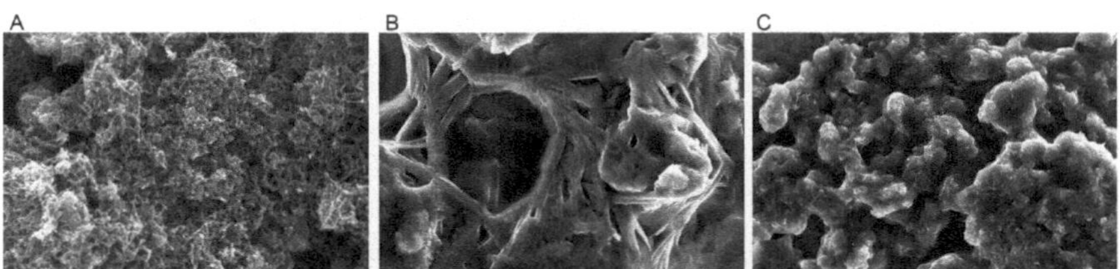

Fig. 5 SEM images for free and enzyme bound NC. The figure illustrates (A) free PANI/Co/MWCNT-NC, (B) NC adsorbed βGS, and (C) NC covalently bound βGS. *Adapted from Khan, M., Husain, Q., & Bushra, R. (2017). Immobilization of β-galactosidase on surface modified cobalt/multiwalled carbon nanotube nanocomposite improves enzyme stability and resistance to inhibitor.* International Journal of Biological Macromolecules, 105, *693–701.*

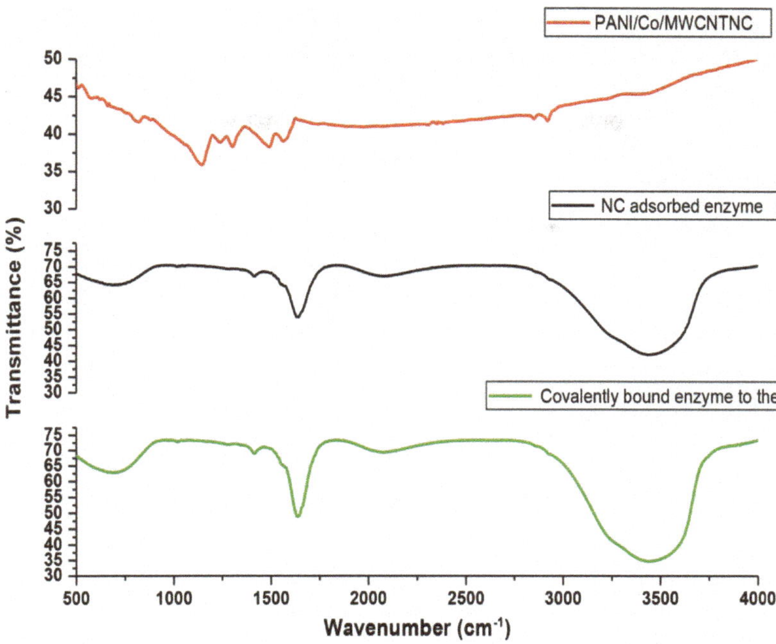

Fig. 6 FT-IR spectra for free and enzyme bound NC. The figure depicts spectra of free PANI/Co/MWCNT-NC, NC adsorbed and NC covalently bound βGS. *Adapted from Khan, M., Husain, Q., & Bushra, R. (2017). Immobilization of β-galactosidase on surface modified cobalt/multiwalled carbon nanotube nanocomposite improves enzyme stability and resistance to inhibitor.* International Journal of Biological Macromolecules, 105, *693–701.*

C—N stretching bond. The band which appeared at 1238 cm^{-1} in the spectrum obtained for NC indicated that Co interacts with the nitrogen atoms in the PANI chain (Giri et al., 2012). The characteristic peak for MWCNT was observed at 1148 cm^{-1} which corresponds to C=C. The binding of βGS on PANI/Co/MWCNT-NC was confirmed by the absorption bands obtained for adsorbed and covalently attached βGS. The stretching of carbonyl group of βGS was observed in both types of immobilized enzyme preparations as viewed by broadening of peaks from 3131 to 3653 cm^{-1}. Similar findings have been reported earlier, where broadening of peaks was observed from 3220 to 3447 cm^{-1} (Khan et al., 2016). The peak at 1647 cm^{-1} further signifies the presence of amide linkage. However, the peak intensity decreased in case of covalently immobilized enzyme ensuring a strong interaction between protein and support. Identical results have been observed when the same enzyme was immobilized on functionalized magnetic nano graphene (Shafi, Khan, Khan, & Husain, 2019).

3.2 Immobilization studies of adsorbed and covalently bound βGS

The adsorption of protein onto CNT involves both hydrophobic and electrostatic interactions. Hydrophobic interactions involve the binding of side chains of hydrophobic amino acids of proteins with sidewall of CNT. In electrostatic interactions, the π-electrons on the surface of CNT interact with π-electrons of the aromatic ring of amino acids (e.g., phenylalanine and tryptophan). Furthermore, the negative charges present on the surface of support interact with positive charges of the enzyme. This type of immobilization is also referred to as immobilization by ionic exchange. Apart from electrostatic and hydrophobic interactions, hydrogen bonds and non-specific adsorption can also play a significant role in the adsorption of enzyme onto CNT surfaces (Husain, 2016, 2018b; Saifuddin et al., 2013).

GA is one of the most widely used cross-linking agents for surface modification of enzyme and the support. It interacts with primary amino groups of protein by intermolecular cross-linking the enzyme molecules immobilized in/on any type of support. GA can be used in two ways to covalently bind the enzyme with the support. In the first method, initially enzyme is adsorbed on the aminated support which is followed by covalent coupling via GA (Ali, Husain, Alam, & Ahmad, 2017). The advantage of this method is to provide rigidity and multi-point covalent attachment while the drawback lies in the loss of enzyme activity and stability, because the surface amino groups of the enzyme are modified. The second method involves the treatment of pre-activated aminated support by GA. The functionalized NC is then allowed to react with the free amino group of the enzyme (Ashraf & Husain, 2011a, 2011b; Barbosa, Torres, Ortiz, & Fernandez-Lafuente, 2012; Siddiqui & Husain, 2019).

In this work, second method has been employed in which GA was used to create reactive aldehyde groups on the support. These free aldehyde groups present on the activated support subsequently reacts with exposed amino groups of the enzyme. In the covalent method, two parameters were analyzed. The maximum immobilization yield of 97% was achieved when 6 mg of dry weight of support was treated with 2% (v/v) GA solution. GA has been used to in two ways to covalently bind the enzyme with NC. In first method, initially βGS was adsorbed on the aminated support which is

Table 4 Immobilization of βGS on PANI/Co/MWCNT-NC.

Method of immobilization	Enzyme activity loaded, X, (U)	Enzyme activity in washes, Y, (U)	Enzyme activity bound on PANI/Co/MWCNT-NC		Immobilization yield (%), B/A × 100
			Theoretical (X − Y) = A	Actual/ immobilized = B	
Adsorption	24.67	0.087	24.58	22.84	93 ± 0.73
Covalent binding	24.67	0.002	24.66	24.00	97 ± 0.54

Each value represents the mean for three independent experiments performed in duplicates, with average standard deviations, ≤5%.
Adapted from Khan, M., Husain, Q., & Bushra, R. (2017). Immobilization of β-galactosidase on surface modified cobalt/multiwalled carbon nanotube nanocomposite improves enzyme stability and resistance to inhibitor. *International Journal of Biological Macromolecules, 105*, 693–701.

followed by covalent coupling with GA. At higher concentration of GA, there was a minor loss in immobilization yield. It might be due to higher number of aldehyde groups or protein molecules cross-linking with each other. These conformational changes in the protein molecules might decrease its activity (Ali et al., 2017; Zhu et al., 2012).

Table 4 demonstrates the immobilization yield of βGS obtained via both methods. Schematic representation of immobilization of βGS on functionalized-CNTs by both methods has been shown in Fig. 7. Several earlier workers have also employed GA as a cross-linking agent in the process of enzyme immobilization (Musthapha, Akhtar, Khan, & Husain, 2004; Zhu et al., 2012). The use of GA, as cross linker has prevented leaching of enzyme from the support. There was a marginal increase in the immobilization yield in case of covalently bound enzyme. The improvement in immobilization yield of the enzymes by GA coupling has also been described by earlier workers (Ashraf & Husain, 2010, Gibriel, Amin, Yassien, El Banna, & Khaled, 2014).

3.3 Stability, reusability and kinetic studies of free and PANI/Co/MWCNT-NC bound βGS

3.3.1 Effect of pH and temperature

pH-activity profile for soluble and immobilized βGS has been illustrated in Fig. 8A. The soluble enzyme exhibited a sharp pH-optimum at 4.5, while immobilized enzyme showed a broadening in pH-optima. The covalently attached βGS demonstrated broader pH-optima compared to adsorbed enzyme. At pH 2, the free enzyme retained only 19% of the original activity,

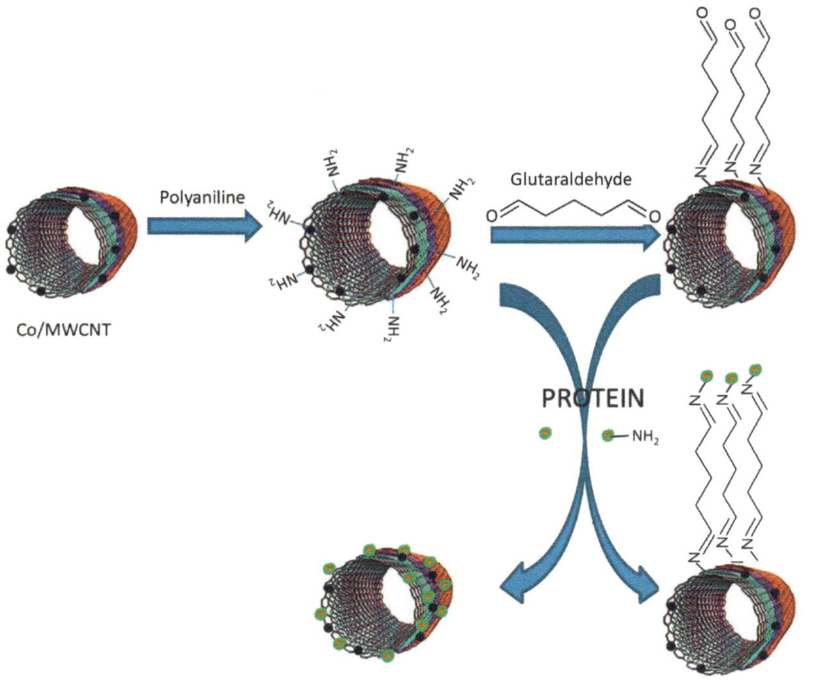

Fig. 7 Over all reaction steps involved in the functionalization of PANI/Co/MWCNT-NC and subsequent immobilization of βGS by physical adsorption and covalent attachment. *Adapted from Khan, M., Husain, Q., & Bushra, R. (2017). Immobilization of β-galactosidase on surface modified cobalt/multiwalled carbon nanotube nanocomposite improves enzyme stability and resistance to inhibitor.* International Journal of Biological Macromolecules, 105, *693–701.*

while the adsorbed enzyme could maintain 51% activity and covalently bound enzyme retained about 68% activity under similar experimental conditions. A shift in pH-optima after immobilization of enzymes has also been earlier reported by various researchers (Bayraktar, Serilmez, Karakas, Bicak-Celem, & Onal, 2011; Shafi et al., 2019). Depending on the nature and surface charges on the support as well as the enzyme, the pH in the vicinity of enzyme molecule may differ and hence alter the pH-optima of the immobilized enzyme.

Effect of temperature on the activity of free, NC adsorbed and covalently attached βGS was investigated in the range of 20–80 °C (Fig. 8B). Temperature-optima for soluble and NC adsorbed enzymes was found to be same at 50 °C. However, the covalently bound enzyme exhibited

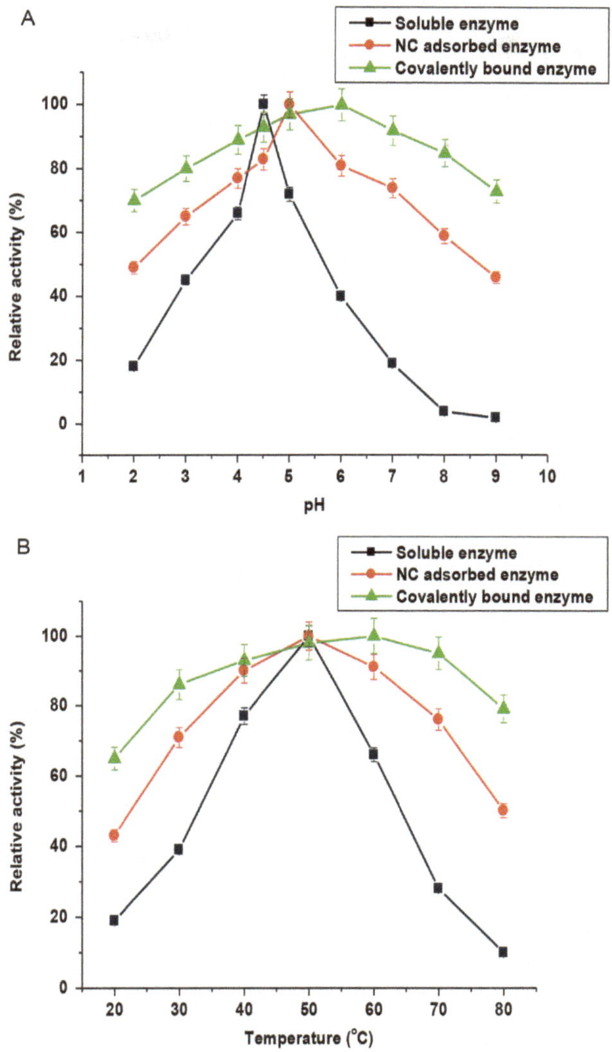

Fig. 8 Stability studies of free and PANI-Co-MWCNT-NC bound βGS. The figure shows (A) pH-activity and (B) temperature-activity profiles. Data represents mean ± standard deviation of three individual experiments performed in duplicates, *P*-value ≤ 0.05. *Adapted from Khan, M., Husain, Q., & Bushra, R. (2017). Immobilization of β-galactosidase on surface modified cobalt/multiwalled carbon nanotube nanocomposite improves enzyme stability and resistance to inhibitor.* International Journal of Biological Macromolecules, 105, *693–701.*

maximum activity at 60 °C. The soluble enzyme lost 90% activity when incubated at 80 °C for 15 min, whereas the NC adsorbed enzyme retained 50% activity and the βGS covalently immobilized to NC retained about 80% of the maximum activity under identical incubation conditions. Ansari and Husain (2010) have earlier shown that there was a shift in temperature-optima from 50 °C to 60 °C in case of cross-linked Con-A-cellulose adsorbed βGS. The stability and activity retention exhibited by the immobilized enzyme in a broad range of temperature can be explained by the phenomenon of multi-point attachment which provides rigidity to the enzyme, thereby preventing enzyme conformational changes. The increased enzymatic activity under those conditions, however, is not caused by the generation of a more active enzyme form, but by avoiding distortion of the enzyme structure (Ansari & Husain, 2011). Guidini, Fischer, Santana, Cardoso, and Ribeiro (2010) immobilized *Aspergillus oryzae* βGS on the ionic exchange resin Duolite A568 by ionic-binding and covalent coupling method. They found that the covalently bound enzyme was significantly more stable against the extreme conditions of pH and temperature compared to simply adsorbed enzyme.

Fig. 9A illustrates the thermal denaturation plots for soluble and immobilized enzyme preparations. The soluble enzyme lost almost 95% of the initial activity after 2 h incubation at 60 °C. However, the adsorbed enzyme retained 46% activity, while the covalently immobilized βGS appeared more stable and maintained 78% of its activity at the end of 2 h incubation. The immobilization matrix imparts high rigidity to the enzyme, thus providing enzyme a greater resistance against heat inactivation.

3.3.2 Effect of galactose

For the construction of biosensors, greater importance is given on the activity of the immobilized enzyme in the presence of higher concentration of galactose which is considered to be a competitive inhibitor for βGS catalyzed reaction (Haider & Husain, 2009; Park & Oh, 2010). Fig. 9B explains the effect of increasing concentrations of galactose on the activity of free and immobilized βGS. The βGS covalently bound to NC was found to be most resistant against galactose inhibition and thus it retained 70% of its initial activity in the presence of 5% galactose. However NC adsorbed βGS nearly lost 66% of its activity. The interaction between the matrix and enzyme plays a vital role in the protection of enzyme inhibition. Rigidity of the enzyme due to multi-point covalent attachment may be the possible reason for

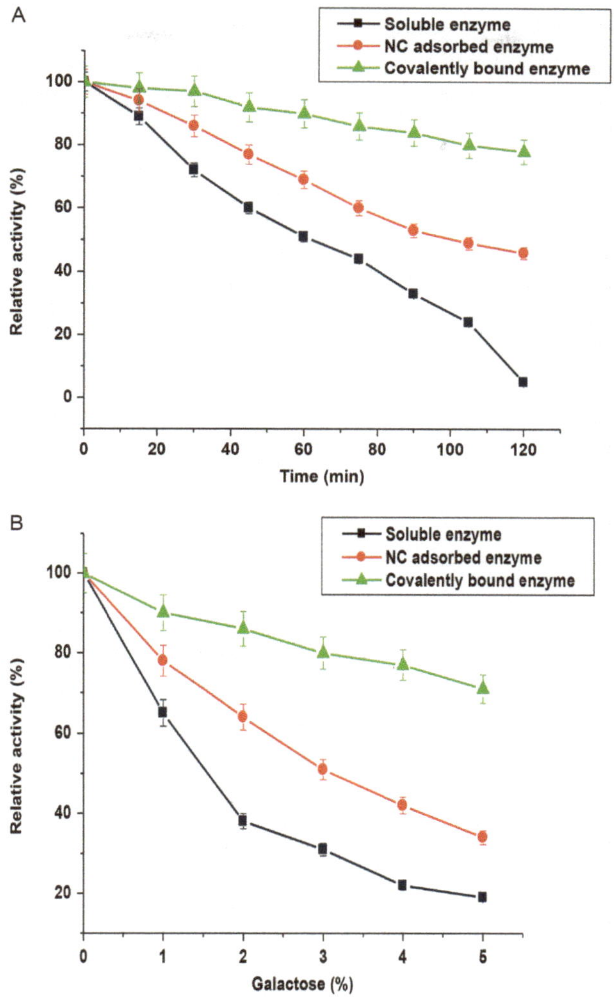

Fig. 9 Stress inactivation studies of free and PANI-Co-MWCNT-NC bound enzyme. The figure depicts (A) effect of temperature and (B) effect of galactose. Data represents mean ± standard deviation of three individual experiments performed in duplicates, *P*-value ≤ 0.05. *Adapted from Khan, M., Husain, Q., & Bushra, R. (2017). Immobilization of β-galactosidase on surface modified cobalt/multiwalled carbon nanotube nanocomposite improves enzyme stability and resistance to inhibitor.* International Journal of Biological Macromolecules, 105, *693–701.*

providing resistance against inhibitor in βGS covalently immobilized on NC due to more protection in its native conformation (Khan, Husain, & Bushra, 2017; Rodrigues, Ortiz, Berenguer-Murcia, Torres, & Fernandez-Lafuente, 2013).

3.3.3 Determination of kinetic parameters

The K_i, K_m and V_{max} values obtained for soluble and immobilized enzyme preparations are listed in Table 5. The K_{iapp} values were found to significantly increase in βGS covalently bound to NC. These values infer that the covalently immobilized enzyme was comparatively more stable as compared to the adsorbed or the soluble enzyme and less affected by galactose inhibition. The K_m values were found to decrease from 0.5 to 0.4 mM for adsorbed enzyme and 0.3 mM for covalently attached enzyme, indicating higher substrate-enzyme affinity (Mateo et al., 2004). A slight decrease in V_{max} values was also observed in case of immobilized enzyme suggesting a decrease in activity retention by the enzyme during binding. Ye, Jiang, and Xu (2007) have immobilized lipase on chitosan-modified poly(acrylonitrile-co-maleic acid) membrane surface and they observed a similar pattern in kinetic parameters by showing a decrease in K_m and a very slight decrease in V_{max} in the bound enzyme compared to the native form.

3.3.4 Reusability of immobilized enzyme

Reusability of enzyme provides cost effectiveness for its use in an economically viable enzyme catalyzed process (Haider & Husain, 2007). The PANI/Co/MWCNT-NC adsorbed enzyme retained about 74% activity while the covalently bound βGS maintained 92% of its initial activity even after 10th successive reuse (Fig. 10A). The covalently immobilized βGS was found to be more rigid than the adsorbed enzyme because of the covalent linkages formed between the NC and enzyme which subsequently reduces the chances of leaching of the enzyme from the support or stabilizing native conformation. The activity of the βGS immobilized on GA treated κ-carrageenan gel beads retained 60% of its initial activity even after 20 reuses (Elnashar et al., 2014).

Table 5 Kinetic parameters of free and PANI/Co/MWCNT-NC bound βGS.

Enzyme	K_i (mM)	K_m (mM)	V_{max} (μmol min^{-1} mL^{-1})
Free βGS	0.5 ± 0.14	0.5 ± 0.12	0.04 ± 0.005
Adsorbed βGS	1.1 ± 0.32	0.4 ± 0.08	0.04 ± 0.004
Cross-linked βGS	11.1 ± 1.3	0.3 ± 0.05	0.03 ± 0.002

Each value represents the mean of three independent experiments performed in duplicates with average standard deviation of ≤5%.

Adapted from Khan, M., Husain, Q., & Bushra, R. (2017). Immobilization of β-galactosidase on surface modified cobalt/multiwalled carbon nanotube nanocomposite improves enzyme stability and resistance to inhibitor. *International Journal of Biological Macromolecules, 105*, 693–701.

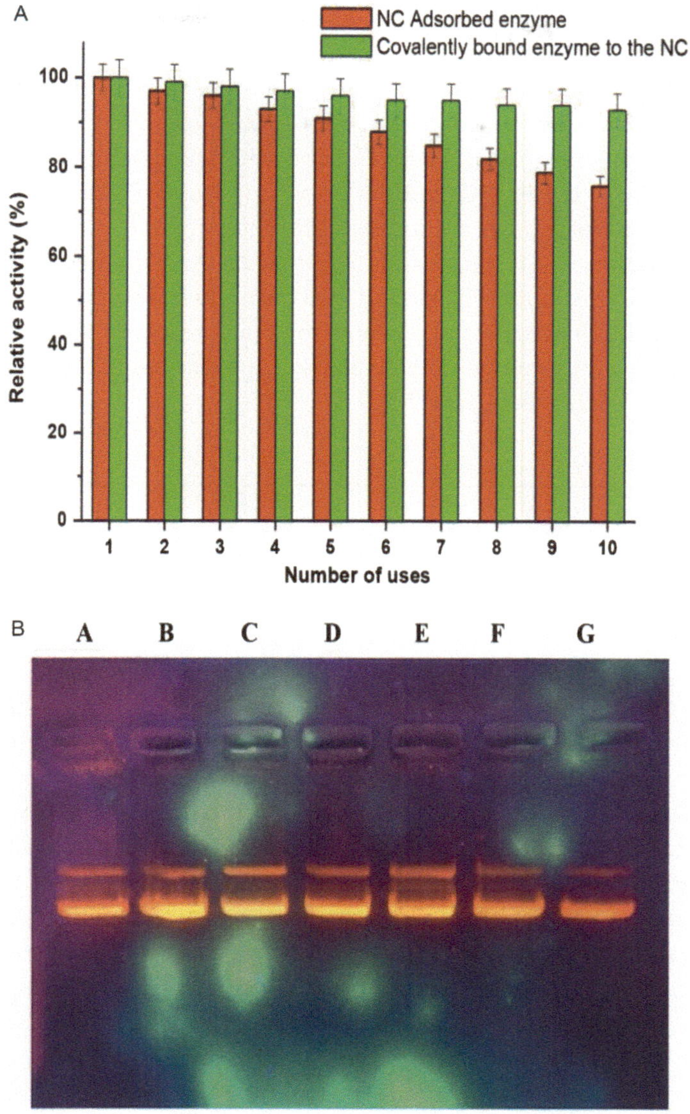

Fig. 10 Reusability and genotoxicity assessment of free and enzyme bound NC. The figure illustrates (A) reusability of PANI-Co-MWCNT-NC adsorbed and covalently attached βGS, *P*-value $\leq$ 0.05 was considered statistically significant. (B) Plasmid nicking toxicity assay: Lane A denotes pBR322 DNA alone, Lane B & C signify pBR322 DNA with PANI-Co-MWCNT-NC (50 μg), Lane D & E denotes pBR322 DNA incubated with adsorbed enzyme, while Lane F & G contain pBR322 DNA with covalently bound βGS. *Adapted from Khan, M., Husain, Q., & Bushra, R. (2017). Immobilization of β-galactosidase on surface modified cobalt/multiwalled carbon nanotube nanocomposite improves enzyme stability and resistance to inhibitor.* International Journal of Biological Macromolecules, 105, *693–701.*

3.4 Genotoxicity assessment of free and βGS bound PANI/Co/MWCNT-NC

To assess the genotoxicity of the support, plasmid nicking assay was performed. It is one of the most sensitive assays to detect the effect of various toxicants on the integrity of the DNA. The pBR322 plasmid was incubated in the presence of free NC and both types of immobilized enzyme preparations. Fig. 10B shows the agarose gel with similar band pattern in all the cases, which was insignificantly different to that of control. The NC matrix used for the immobilization of βGS did not induce any kind of breakage in pBR322 DNA revealing its non-toxic nature hence it is compatible and safe to be employed in biotechnological applications (Khan, Husain, & Ahmad, 2018; Khan et al., 2017).

4. Conclusion

With the growing demands of our society, in particular healthcare issues associated with aging of the population, next generation medical diagnostics require implementation of rapid, sensitive and cost effective alternatives to the more traditional immunological assays currently used. Similarly, the detection of pathogens and toxic compounds in our environment constitute major threats and sensitive detection procedures are required to address concerns for global security. An ever growing number of biosensing devices and strategies have been developed since the Clark oxygen electrode came to existence in the 1950s. However, NM has clearly provided new and attractive platforms to engineer biosensors with enhanced sensitivity and performance. In particular, the structural, electronic, and optical properties combined by CNTs offer a wealth of opportunities to develop new generation nano-tools for biosensing applications, especially for probing disease markers. The electronic properties of CNTs are extremely well suited for electrochemical, amperometric, simpedometric, or field-effect transduction of signals, but their particular optical properties and their ability to penetrate readily through biological membranes also make them suitable candidates for the development of photoacoustic imaging sensors. However, when proposing a biosensor, it is always recommended to take into consideration that nanoscale materials, although small can be sensed by the body as intruders and consequently attacked. Therefore, issues related to their length, diameter, lifetime, stability, durability, mechanical properties, body adjustment, and toxicity must be evaluated by a scientific and systematic method. Indeed, when used in their pristine state, directly after synthesis, CNTs contain impurities and can have harmful effects. Nonetheless, when purified and surface-functionalized, their toxicity is drastically decreased and they represent optimal platforms for all kinds of applications. CNTs have comparable dimensions to redox proteins and can be used as effective electrical wiring/connectors with redox enzymes. However, better control of the chemical and physical properties of the CNT-biosensors is still needed. For example, the separation process for different types of CNTs, the miniaturization of the sensors and the in vivo stability need to be addressed to meet feature requirements.

Eventually, a clear understanding of the structure–function relationships will be convenient to support and guide experimental efforts toward the most promising carbon nanostructures with improved biosensing abilities. Thus, fabrication of new CNTs-based optical biosensors and synthesis of new CNT fibers is still a focus of future works in biosensing field. Clearly, the development of CNT-based biosensors is a multi-faceted challenging work that needs a well cooperation between materials scientists, who are developing new materials, and engineers, who are committed to fabricate the new micro/nanodevices such as bionanosensors. Therefore, the greatest and the last challenge is how to build a joint platform that allows the efficient discussion, learning and cooperating between physicists, chemists, and electrical/mechanical engineers.

Although recent works have made many amazing advances in this field, there are still some challenges and further works may mainly include the following five aspects. The first one is cost effectiveness. The high price of CNT materials and CNT-based enzymatic biosensors has become the main problem which seriously restricts their development and widespread applications (Yan, Chan-Park, & Zhang, 2007). Reducing these costs is a complex challenge and will almost certainly require the integration of materials research and nanotechnology. Non-enzymatic sensor based on CNTs is a potential choice which has been attracted by some research groups (Zhu et al., 2012). Novel nanotechnologies such as nanoimprint lithography and soft lithography are also helpful. Both chemical and physical properties of functionalized CNTs strongly depend on the ambient conditions, such as temperature and pH. Thus, the second challenge is how to prolong the thermal stability and lifetime of CNT-based biosensors in application of CNTs. Using NC materials instead of expensive and fragile enzymes to detect biomolecule has shown improved thermal tolerance, long-term stability, and cost effectiveness. Except that, some studies have indicated that the sensitivity and the detection limits is significantly enhanced using NC materials combined with CNTs and metal (Au, Pt, etc.) for biosensor applications. With the sustainable development of new NM combined with CNTs, characterization of these new materials at the molecular level is essential and a key scientific challenge. In parallel with experimental studies, molecular modeling must be further developed as a tool to predict the performance of new materials for a given biomolecule. Such computational methods will enable a quick evaluation of these novel materials.

References

Abdalla, S., Al-Marzouki, F., Al-Ghamdi, A. A., & Abdel-Daiem, A. (2015). Different technical applications of carbon nanotubes. *Nanoscale Research Letters*, *10*, 358–369.

Ali, M., Husain, Q., Alam, N., & Ahmad, M. (2017). Enhanced catalytic activity and stability of ginger peroxidise immobilized on amino-functionalized silica coated titanium dioxide nanocomposites: A cost effective tool for bioremediation. *Water, Air, & Soil Pollution*, *228*, 22–34.

Andreescu, S., Njagi, J., & Ispas, C. (2008). Chapter 7—Nanostructured materials for enzyme immobilization and biosensors. In V. Erokhin, M. K. Ram, & O. Yavuz (Eds.), *The new frontiers of organic and composite nanotechnology* (pp. 355–394). Amsterdam: Elsevier.

Ansari, S. A., & Husain, Q. (2010). Lactose hydrolysis by β galactosidase immobilized on concanavalin A-cellulose in batch and continuous mode. *Journal of Molecular Catalysis B: Enzymatic*, *63*, 68–74.

Ansari, S. A., & Husain, Q. (2011). Immobilization of *Aspergillus oryzae* β-galactosidase on concanavalin A-layered calcium alginate-cellulose beads and its application in lactose hydrolysis in continuous spiral bed reactors. *Polish Journal of Chemical Technology*, *13*, 15–20.

Ansari, S. A., & Husain, Q. (2012). Potential applications of enzymes immobilized on/in nano materials: A review. *Biotechnology Advances*, *30*, 512–523.

Arugula, M. A., & Simonian, A. (2014). Novel trends in affinity biosensors: Current challenges and perspectives. *Measurement Science and Technology*, *25*, 032001.

Ashraf, H., & Husain, Q. (2010). Studies on bitter gourd peroxidase catalyzed removal of *p*-bromophenol from waste water. *Desalination*, *262*, 267–272.

Ashraf, H., & Husain, Q. (2011a). Application of immobilized peroxidase for the removal of p-bromophenol from polluted water in batch and continuous processes. *Journal of Water Reuse and Desalination*, *1*, 52–60.

Ashraf, H., & Husain, Q. (2011b). Use of DEAE cellulose adsorbed and cross linked white radish (*Raphanus sativus*) peroxidase for the removal of α-napthol in batch and continuous process. *International Biodeterioration & Biodegradation*, *64*, 27–31.

Baby, T. T., & Ramaprabhu, S. (2010). SiO_2 coated Fe_3O_4 magnetic nanoparticle dispersed multiwalled carbon nanotubes based amperometric glucose biosensor. *Talanta*, *80*, 2016–2022.

Baj-Rossi, C., De Micheli, G., & Carrara, S. (2012). Electrochemical detection of anti-breast-cancer agents in human serum by cytochrome P450-coated carbon nanotubes. *Sensors (Basel, Switzerland)*, *12*, 6520–6537.

Balasubramanian, K., & Burghard, M. (2005). Chemically functionalized carbon nanotubes. *Small*, *1*, 180–192.

Barbosa, O., Torres, R., Ortiz, C., & Fernandez-Lafuente, R. (2012). Versatility of glutaraldehyde to immobilize lipases: Effect of the immobilization protocol on the properties of lipase B from *Candida Antarctica*. *Process Biochemistry*, *47*, 1220–1227.

Başkaya, G., Yildiz, Y., Savk, A., Okyay, T. O., Eriş, S., Sert, H., et al. (2017). Rapid, sensitive, and reusable detection of glucose by highly monodisperse nickel nanoparticles decorated functionalized multi-walled carbon nanotubes. *Biosensors & Bioelectronics*, *91*, 728–733.

Bayraktar, H., Serilmez, M., Karakas, T., Bicak-Celem, E., & Onal, S. (2011). Immobilization and α-galactosidase on sepabead EC-EA and EC-HA. *International Journal of Biological Macromolecules*, *49*, 855–860.

Berkmans, J., Ramakrishnan, S., Jain, G., & Haridoss, P. (2013). Aligning carbon nanotubes, synthesized using the arc discharge technique, during and after synthesis. *Carbon*, *55*, 185–195.

Boujtita, M. (2014). Chemical and biological sensing with carbon nanotubes (CNTs). In K. C. Honeychurch (Ed.), *Nanosensors for chemical and biological applications: Sensing with nanotubes, nanowires and nanoparticles* (pp. 3–27). Elsevier.

Bucko, M., Mislovicova, D., Nahalka, J., Vikartovska, A., Sefcovicova, J., Katrlik, J., et al. (2012). Immobilization in biotechnology and biorecognition: From macro-to nanoscale systems. *Chemical Papers*, *66*, 983–998.

Canbay, E., & Akyilmaz, E. (2014). Design of a multiwalled carbon nanotube–nafion–cysteamine modified tyrosinase biosensor and its adaptation of dopamine determination. *Analytical Biochemistry*, *444*, 8–15.

Chauhan, N., & Pundir, C. S. (2011). An amperometric uric acid biosensor based on multiwalled carbon nanotube–gold nanoparticle composite. *Analytical Biochemistry*, *413*, 97–103.

Chen, A., & Chatterjee, S. (2013). Nanomaterials based electrochemical sensors for biomedical applications. *Chemical Society Reviews*, *42*, 5425–5438.

Chinnadayyala, S. R., Kakoti, A., Santhosh, M., & Goswami, P. (2014). A novel amperometric alcohol biosensor developed in a 3rd generation bioelectrode platform using peroxidase coupled ferrocene activated alcohol oxidase as biorecognition system. *Biosensors & Bioelectronics*, *55*, 120–126.

Dai, L. (2007). Electrochemical sensors based on architectural diversity of the π-conjugated structure: Recent advancements from conducting polymers and carbon nanotubes. *Australian Journal of Chemistry*, *60*, 472–483.

Daniel, S., Rao, T. P., Rao, K. S., Rani, S. U., Naidu, G. R. K., Lee, H. Y., et al. (2007). A review of DNA functionalized/grafted carbon nanotubes and their characterization. *Sensors and Actuators B: Chemical*, *122*, 672–682.

Das, M., & Goswami, P. (2013). Direct electrochemistry of alcohol oxidase using multiwalled carbon nanotube as electroactive matrix for biosensor application. *Bioelectrochemistry*, *89*, 19–25.

Das, C. K., Prusty, B., Adhikary, M. C., & Kalra, S. (2015). Chromium doped polyaniline/mwcnts nanocomposite for supercapacitor electrode materials. *International Journal of Engineering Science Research Technology*, *4*, 4–16.

Elnashar, M. M., Awad, G. E., Hassan, M. E., Eldin, M. S. M., Haroun, B. M., & Ek-Diwany, A. I. (2014). Optimal immobilization of *β*-galactosidase onto *κ*-carrageenan gel beads using response surface methodology and its applications. *The Scientific World Journal*, *2014*, 571682. https://doi.org/10.1155/2014/571682.

Fei, S., Chen, J., Yao, S., Deng, G., He, D., & Kuang, Y. (2005). Electrochemical behaviour of L-cysteine and its detection at carbon nanotube electrode modified with platinum. *Analytical Biochemistry*, *339*, 29–35.

Feng, Q. M., Pan, J. B., Zhang, H. R., Zhu, J. J., & Chen, H. Y. (2004). Disposable paper-based bipolar electrode for sensitive electrochemiluminescence detection of a cancer biomarker. *Chemical Communications (Cambridge, England)*, *50*, 10949–10951.

Fu, X., Cui, X., Wei, X., & Ma, J. (2014). Investigation of low and mild temperature for synthesis of high quality carbon nanotubes by chemical vapor deposition. *Applied Surface Science*, *292*, 645–649.

Fujigaya, T., & Nakashima, N. (2008). Methodology for homogeneous dispersion of single-walled carbon nanotubes by physical modification. *Polymer Journal, 40*, 577–589.
Gao, C., Guo, Z., Liua, J., & Huang, X. (2012). The new age of carbon nanotubes: An updated review of functionalized carbon nanotubes in electrochemical sensors. *Nanoscale, 4*, 1948–1963.
Gibriel, Y., Amin, A. A., Yassien, N. M. N., El Banna, H. L., & Khaled, F. M. (2014). Immobilization techniques for *P. aculeatum* dextranase. *International Journal of Current Microbiology and Applied Sciences, 3*, 1114–1134.
Giri, S., Das, C. K., & Kalra, S. S. (2012). Preparation and characterization of the cobalt doped polyaniline/MWCNT nanocomposites for supercapacitor application. *Journal of Materials Science Research, 1*, 3–9.
Gomez-Gualdrón, D. A., Burgos, J. C., Yu, J., & Balbuena, P. B. (2011). Chapter 5—Carbon nanotubes: Engineering biomedical applications. In V. Antonio (Ed.), *Progress in molecular biology and translational science* (pp. 175–245). Academic Press.
Gougis, M., Tabet-Aoul, A., Ma, D., & Mohamedi, M. (2014). Laser synthesis and tailor-design of nanosized gold onto carbon nanotubes for non-enzymatic electrochemical glucose sensor. *Sensors and Actuators B: Chemical, 193*, 363–369.
Guidini, Z., Fischer, J., Santana, L. N. S., Cardoso, V. L., & Ribeiro, E. J. (2010). Immobilization of *Aspergillus oryzae* β-galactosidase in ion exchange resins by combined ionic-binding method and cross-linking. *Biochemical Engineering Journal, 52*, 137–143.
Haider, T., & Husain, Q. (2007). Preparation of lactose-free milk by using salt-fractionated almond (*Amygdalus communis*) β-galactosidase. *Journal of Science and Food Agriculture, 87*, 1278–1283.
Haider, T., & Husain, Q. (2009). Immobilization of β-galactosidase from *Aspergillus oryzae* via immunoaffinity support. *Biochemical Engineering Journal, 43*, 307–314.
Hasan, J., Crawford, R. J., & Ivanova, E. P. (2013). Antibacterial surfaces: The quest for a new generation of biomaterials. *Trends in Biotechnology, 31*, 295–304.
Hernandez-Ibanez, N., Garcia-Cruz, L., Monteil, V., Foster, C. W., Banks, C. E., & Iniesta, J. (2016). Electrochemical lactate biosensor based upon chitosan/carbon nanotubes modified screen-printed graphite electrodes for the determination of lactate in embryonic cell cultures. *Biosensors & Bioelectronics*, 77, 1168–1174.
Holzinger, M., LeGoff, A., & Cosnier, S. (2014). Nanomaterials for biosensing applications: A review. *Frontiers in Chemistry, 2*, 63–72.
Huang, J., Dong, Z., Li, Y., Li, J., Wang, J., Yang, H., et al. (2013). High performance non-enzymatic glucose biosensor based on copper nanowires–carbon nanotubes hybrid for intracellular glucose study. *Sensors and Actuators B: Chemical, 182*, 618–624.
Husain, Q. (2016). Magnetic nanoparticles as a tool for the immobilization/stabilization of hydrolases and their applications. An overview. *Biointerface Research in Applied Chemistry, 6*, 1585–1606.
Husain, Q. (2017). Nanosupport bound lipases their stability and application. *Biointerface Research in Applied Chemistry*, 7, 2194–2216.
Husain, Q. (2018a). Carbon nanotubes mediated immobilized glucose oxidase biosensors as an effective and sensitive analytical tool. *Biointerface Research in Applied Chemistry, 8*, 3060–3074.
Husain, Q. (2018b). Nanocarriers immobilized proteases and their industrial applications. An overview. *Journal of Nanoscience and Nanotechnology, 18*, 486–499.
Husain, Q., Ansari, S. A., Alam, F., & Azam, A. (2011). Immobilization of *Aspergillus oryzae* β- galactosidase on zinc oxide nanoparticles via simple adsorption mechanism. *International Journal of Biological Macromolecules, 49*, 37–43.
Ijima, S. (1991). Helical microtubules of graphitic carbon. *Nature, 354*, 56–58.
Jacobs, C. B., Peairs, M. J., & Venton, B. J. (2010). Review: Carbon nanotube based electrochemical sensors for biomolecules. *Analytica Chimica Acta, 662*, 105–127.

Jasim, A., Wajidullah, M., Shi, Z., Lin, X., & Yang, G. (2017). Fabrication of bacterial cellulose/polyaniline/single-walled carbon nanotubes membrane for potential application as biosensor. *Carbohydrate Polymers*, *163*, 62–69.

Jedrzejewska, A., Bachmatiuk, A., Ibrahim, I., Srekova, H., Nganou, C., Schuchner, F., et al. (2013). A systematic and comparative study of binary metal catalysts for carbon nanotube fabrication using CVD and laser evaporation. *Fullerenes, Nanotubes, and Carbon Nanostructures*, *21*, 273–285.

Jin, H., Heller, D. A., Kalbacova, M., Kim, J. H., Zhang, J., Boghossian, A. A., et al. (2010). Detection of single-molecule H_2O_2 signalling from epidermal growth factor receptor using fluorescent singlewalled carbon nanotubes. *Nature Nanotechnology*, *5*, 302–309.

Khan, M., & Husain, Q. (2018). Effect of metal ions present in milk on the functional and structural integrity of polyaniline chitosan nanocomposites bound β-galactosidases: A multi-spectroscopic approach. *Food Chemistry*, *279*, 312–320.

Khan, M., Husain, Q., & Ahmad, N. (2018). Elucidating the binding efficacy of β-galactosidase on polyaniline chitosan nanocomposite and polyaniline chitosan silver nanocomposite: Activity and molecular docking insights. *Journal of Chemical Technology and Biotechnology*, *94*, 837–849. https://doi.org/10.1002/jctb.5831.

Khan, M. J., Husain, Q., & Ansari, S. A. (2013). Polyaniline-assisted silver nanoparticles: A novel support for the immobilization of α-amylase. *Applied Microbiology and Biotechnology*, *97*, 1513–1522.

Khan, M., Husain, Q., & Bushra, R. (2017). Immobilization of β-galactosidase on surface modified cobalt/multiwalled carbon nanotube nanocomposite improves enzyme stability and resistance to inhibitor. *International Journal of Biological Macromolecules*, *105*, 693–701.

Khan, M., Husain, Q., & Naqvi, A. H. (2016). Graphene based magnetic nanocomposites as versatile carriers for high yield immobilization and stabilization of β-galactosidase. *RSC Advances*, *6*, 53493–53503.

Khodadadei, F., Ghourchian, H., Soltanieh, M., Hosseinalipour, M., & Mortazavi, Y. (2014). Rapid and clean amine functionalization of carbon nanotubes in a dielectric barrier discharge reactor for biosensor development. *Electrochimica Acta*, *115*, 378–385.

Kishore, C. S., & Pandurangan, A. (2012). Electrophoretic deposition of cobalt catalyst layer over stainless steel for the high yield synthesis of carbon nanotubes. *Applied Surface Science*, *258*(20), 7936–7942.

Koskun, Y., Şavk, A., Şen, B., & Şen, F. (2018). Highly sensitive glucose sensor based on monodisperse palladium nickel/activated carbon nanocomposites. *Analytica Chimica Acta*, *1010*, 37–43.

Kress, G. J., Shu, H. J., Yu, A., Yu, A., Benz, A., Harmon, S., et al. (2014). Fast phasic release properties of dopamine studied with a channel biosensor. *The Journal of Neuroscience*, *34*, 11792–11802.

Krishnamurthy, G., & Agarwal, S. (2013). Optimization of reaction conditions for high yield synthesis of carbon nanotube bundles by low-temperature solvothermal process and study of their H_2 storage capacity. *Bulletin of the Korean Chemical Society*, *34*, 3046–3054.

Lakshmi, G., & Khan, S. A. (2014). Synthesis of CNTs by arc discharge method in water bath. In R. K. Sharma (Ed.), *Physics of semiconductor devices* (pp. 601–602). Springer.

Li, G., Liao, J. M., Hu, G. Q., Ma, N. Z., & Wu, P. J. (2005). Study of carbon nanotube modified biosensor for monitoring total cholesterol in blood. *Biosensors & Bioelectronics*, *20*, 2140–2144.

Li, J., Liu, Q., Liu, Y., Liu, S., & Yao, S. (2005). DNA biosensor based on chitosan film doped with carbon nanotubes. *Analytical Biochemistry*, *346*, 107–114.

Liu, F. L., Xiao, P., Fang, H. L., Dai, H. F., Qiao, L., & Zhang, Y. H. (2011). Singlewalled carbon nanotube-based biosensors for the detection of volatile organic compounds of lung cancer. *Physica E: Low-dimensional Systems and Nanostructures*, *44*, 367–372.

Lowry, O. H., Rosenbrough, N. J., Farr, A. L., & Randall, R. J. (1951). Protein measurement with folin phenol reagent. *The Journal of Biological Chemistry*, *193*, 265–275.

Mananghaya, M. R., Santos, G. N., & Yu, D. N. (2017). Solubility of amide functionalized single wall carbon nanotubes: A quantum mechanical study. *Journal of Molecular Liquids*, *242*, 1208–1214.

Manawi, Y. M., Ihsanullah, Samara, A., Al-Ansari, T., & Atieh, M. A. (2018). A review of carbon nanomaterials synthesis via chemical vapour deposition (CVD) method. *Materials (Basel, Switzerland)*, *11*, 822–851.

Marchesan, S., Kostarelos, K., Bianco, A., & Prato, M. (2015). The winding road for carbon nanotubes in nanomedicine. *Materials Today*, *18*, 12–19.

Mateo, C., Monti, R., Pessela, B. C., Feuntes, M., Torres, R., Guisan, J. M., et al. (2004). Immobilization of lactase from *Kluyveromyces lactis* greatly reduces the inhibition promoted by glucose Full hydrolysis of lactose in milk. *Biotechnology Progress*, *20*, 1259–1262.

Meng, F., Shi, W., Sun, Y., Zhu, X., Wu, G., Ruan, C., et al. (2013). Nonenzymatic biosensor based on Cu_2O nanoparticles deposited on polypyrrole nanowires for improving detection range. *Biosensors & Bioelectronics*, *42*, 141–147.

Mundra, R. V., Wu, X., Sauer, J., Dordick, J. S., & Kane, R. S. (2014). Nanotubes in biological applications. *Current Opinion in Biotechnology*, *28*, 25–32.

Musthapha, S., Akhtar, S., Khan, A. A., & Husain, Q. (2004). An economical, simple and high yield procedure for the immobilization/stabilization of peroxidises using turnip roots. *Journal of Scientific and Industrial Research*, *63*, 540–547.

Netto, C. G., Toma, H. E., & Andrade, L. H. (2013). Superparamagnetic nanoparticles as versatile carriers and supporting materials for enzymes. *Journal of Molecular Catalysis B: Enzymatic*, *85*, 71–92.

O'connell, M. J., Boul, P., Ericson, L. M., Huffman, C., Wang, Y., Haroz, E., et al. (2001). Reversible water-solubilization of single-walled carbon nanotubes by polymer wrapping. *Chemical Physics Letters*, *342*, 265–271.

Panesar, P. S., Kumari, S., & Panesar, R. (2010). Potential applications of immobilized β-galactosidase in food processing industries. *Enzyme Research*, *2010*, 1–16.

Park, Y. K., Bold, B., Lee, W. K., Jeon, M. H., An, K. H., & Jeong, S. Y. (2011). D-(β)-galactose conjugated single-walled carbon nanotubes as new chemical probes for electrochemical biosensors for the cancer marker galectin-3. *International Journal of Molecular Sciences*, *12*, 2946–2957.

Park, A. R., & Oh, D. K. (2010). Effects of galactose and glucose on the hydrolysis reaction of a thermostable β-galactosidase from *Caldicellulosiruptor saccharolyticus*. *Applied Microbiology and Biotechnology*, *85*, 1427–1435.

Park, S., Vosguerichian, M., & Bao, Z. (2013). A review of fabrication and applications of carbon nanotube film-based flexible electronics. *Nanoscale*, *5*, 1727–1752.

Pisoschi, A. M. (2013). Biosensors as bio-based materials in chemical analysis: A review. *Journal of Biobased Materials and Bioenergy*, *7*, 19–38.

Prasad, B. B., Prasad, A., Tiwari, M. P., & Madhuri, R. (2013). Multiwalled carbon nanotubes bearing 'terminal monomeric unit' for the fabrication of epinephrine imprinted polymer-based electrochemical sensor. *Biosensors & Bioelectronics*, *45*, 114–122.

Pumera, M. (2007). Electrochemical properties of double wall carbon nanotube electrodes. *Nanoscale Research Letters*, *2*, 87–93.

Rodrigues, R. C., Ortiz, C., Berenguer-Murcia, A., Torres, R., & Fernandez-Lafuente, R. (2013). Modifying enzyme activity and selectivity by immobilization. *Chemical Society Reviews*, *42*, 6290–6307.

Saifuddin, N., Raziah, A. Z., & Junizah, A. R. (2013). Carbon nanotubes: A review on structure and their interaction with proteins. *Journal of Chemistry*, *2013*, 676–815.

Santos, R. M., Rodrigues, M. S., Laranjinha, J., & Barbosa, R. M. (2013). Biomimetic sensor based on hemin/carbon nanotubes/chitosan modified microelectrode for nitric oxide measurement in the brain. *Biosensors & Bioelectronics*, *44*, 152–159.
Shafi, A., Khan, M., Khan, M. Z., & Husain, Q. (2019). Ameliorating the activity and stability of β-galactosidase by tailoring potential nanobiocatalyst on functionalized nanographene: Headway to lactose hydrolysis. *LWT-Food Science and Technology*, *112*, 108260.
Sheldon, R. A., & van Pelt, S. (2013). Enzyme immobilisation in biocatalysis: Why, what and how? *Chemical Society Reviews*, *42*, 6223–6235.
Shobha, B. N., & Muniraj, N. J. R. (2015). Design, modeling and performance analysis of carbon nanotube with DNS strands as biosensors for prostate cancer. *Microsystem Technologies*, *21*, 791–800.
Siddiqui, I., & Husain, Q. (2019). Stabilization of polydopamine modified silver nanoparticles bound trypsin insights on protein hydrolysis. *Colloids and Surfaces. B, Biointerfaces*, *173*, 733–741.
Silva, M., Dias, A. C., Silva, B. V., Gomes-Filho, S. L., Kubota, L. T., Goulart, M. O., et al. (2014). Electrochemical detection of dengue virus NS1 protein with a poly (allylamine)/ carbon nanotube layered immunoelectrode. *Journal of Chemical Technology and Biotechnology*, *717*, 17–23.
Singh, C., Dharamvir, K., & Singhal, S. (2013). Production of carbon nanotubes by arc discharge method. In *Proceeding of international conference on recent trends in applied physics and material science: RAM 2013* (pp. 287–288), AIP Publishing.
Singh, R., Kumar, M., Mittal, A., & Mehta, P. K. (2016). Microbial enzymes: Industrial progress in 21st century. *3 Biotech*, *6*, 1–15.
Singhal, S., Singh, C., Singla, P., & Dharamvir, K. (2013). Effect of magnetic field on the growth of aligned carbon nanotubes using a metal free arc discharge method and their purification. *Solid State Phenomena*, *201*, 197–209.
Spitalsky, Z., Matejka, L., Slouf, M., Konyushenko, E. N., Kovarova, J., Zemek, J., et al. (2009). Modification of carbon nanotubes and its effect on properties of carbon nanotube/epoxy nanocomposites. *Polymer Composites*, *30*, 1378–1387.
Tasis, D., Tagmatarchis, N., Georgakilas, V., & Prato, M. (2003). Soluble carbon nanotubes. *Chemistry - A European Journal*, *9*, 4000–4008.
Tîlmaciu, C. M., & Morris, M. C. (2015). Carbon nanotube biosensors. *Frontiers in Chemistry*, *3*, 59–81.
Vanegas, D. C., Taguchi, M., Chaturvedi, P., Burrs, S., Tan, M., Yamaguchi, H., et al. (2014). A comparative study of carbon-platinum hybrid nanostructure architecture for amperometric biosensing. *Analyst*, *139*, 660–667.
Vasu, K., Pramoda, K., Moses, K., Govindaraj, A., & Rao, C. (2014). Single-walled nanohorns and other nanocarbons generated by submerged arc discharge between carbon electrodes in liquid argon and other media. *Materials Research Express*, *1*, 015001.
Vilian, A. T. E., Madhu, R., Chen, S. M., Veeramani, V., Sivakumar, M., Huh, Y. S., et al. (2015). Facile synthesis of MnO2/carbon nanotubes decorated with a nanocomposite of Pt nanoparticles as a new platform for the electrochemical detection of catechin in red wine and green tea samples. *Journal of Materials Chemistry B*, *3*, 6285–6292. https://doi.org/10.1039/c5tb00508f.
Wang, Z., & Dai, Z. (2015). Carbon nanomaterial-based electrochemical biosensors: An overview. *Nanoscale*, *7*, 6420–6431.
Wang, Y. T., Yu, L., Zhu, Z. Q., Zhang, J., Zhu, J. Z., & Fan, C. H. (2009). Improved enzyme immobilization for enhanced bioelectrocatalytic activity of glucose sensor. *Sensors and Actuators B: Chemical*, *136*, 332–337.

Xiao, D., Dramou, P., He, H., Pham-Huy, L. A., Li, H., Yao, Y., et al. (2012). Magnetic carbon nanotubes: Synthesis by a simple solvothermal process and application in magnetic targeted drug delivery system. *Journal of Nanoparticle Research, 14*, 1–12.

Yadav, S., Devi, R., Kumari, S., Yadav, S., & Pundir, C. (2011). An amperometric oxalate biosensor based on sorghum oxalate oxidase bound carboxylated multiwalled carbon nanotubes–polyaniline composite film. *Journal of Biotechnology, 151*, 212–217.

Yan, Y., Chan-Park, M. B., & Zhang, Q. (2007). Advances in carbon-nanotube assembly. *Small, 3*, 24–42.

Yang, N., Chen, X., Ren, T., Zhang, P., & Yang, D. (2015). Carbon nanotube based biosensors. *Sensors and Actuators B: Chemical, 207*, 690–715.

Ye, P., Jiang, J., & Xu, Z. K. (2007). Adsorption and activity of lipase from *Candida rugosa* on the chitosan-modified poly (acrylonitrile-co-maleic acid) membrane surface. *Colloids and Surfaces. B, Biointerfaces, 60*, 62–67.

Ye, Y., Jo, C., Jeong, I., & Lee, J. (2013). Functional mesoporous materials for energy applications: Solar cells, fuel cells, and batteries. *Nanoscale, 5*, 4584–4605.

Yuge, R., Toyama, K., Ichihashi, T., Ohkawa, T., Aoki, Y., & Manako, T. (2012). Characterization and field emission properties of multi-walled carbon nanotubes with fine crystallinity prepared by CO_2 laser ablation. *Applied Surface Science, 258*, 6958–6962.

Yuichi, T., Kenzo, M., Kazuhiko, M., Chikae, M., Takamura, Y., & Tamiya, E. (2009). Microfluidic and label-free multi-immunosensors based on carbon nanotube microelectrodes. *Japanese Journal of Applied Physics, 48*, 06FJ02.

Zhang, Y., Gao, G., Liu, H., Fu, H., Fan, J., Wang, K., et al. (2014). Identification of volatile biomarkers of gastric cancer cells and ultrasensitive electrochemical detection based on sensing interface of Au-Ag alloy coated MWCNTs. *Theranostics, 4*, 154–162.

Zhang, S., Li, S., Ma, J., Xiong, F., & Qu, S. (2013). Detection of carbofuran with immobilized acetylcholinesterase based on carbon nanotubes–chitosan modified electrode. *Journal of Nanomaterials, 2013*, 8–19.

Zhao, Y., Wu, W., Chen, J., Zou, H., Hu, L., & Chu, G. (2012). Preparation of polyaniline/multiwalled carbon nanotubes nanocomposites by high gravity chemical oxidative polymerization. *Industrial and Engineering Chemistry Research, 51*, 3811–3818.

Zheng, T. T., Zhang, R., Zou, L., & Jun-Jie, Z. (2012). A label-free cytosensor for the enhanced electrochemical detection of cancer cells using polydopamine-coated carbon nanotubes. *The Analyst, 137*, 1316–1318.

Zhu, Z., Garcia-Gancedo, L., Flewitt, A. J., Xie, H., Moussy, F., & Milne, W. I. (2012). A critical review of glucose biosensors based on carbon nanomaterials: Carbon nanotubes and graphene. *Sensors, 12*, 5996–6022.

Further reading

Barsan, M. M., Ghica, M. E., & Brett, C. M. A. (2015). Electrochemical sensors and biosensors based on redox polymer/carbon nanotube modified electrodes: A review. *Analytica Chimica Acta, 881*, 1–23.

Chen, Z., Tabakman, S. M., Goodwin, A. P., Kattah, M. G., Daranciang, D., Wang, X., et al. (2008). Protein microarrays with carbon nanotubes as multicolor Raman labels. *Nature Biotechnology, 26*, 1285–1292.

Khan, M., Husain, Q., & Asmat, S. (2018). Synthesis, characterization and applications of nano graphene armored enzymes. In C. V. Kumar (Ed.), *Methods in enzymology Vol. 609* (pp. 83–143). Academic Press.

Li, Y., Huang, F., Luo, Z., Xu, B., Wang, X., Wang, F. L., et al. (2012). A new hydrogen peroxide biosensor based on synergy of Au@$Au_2S_2O_3$ core–shell nanomaterials and multi-walled carbon nanotubes towards hemoglobin. *Electrochimica Acta, 74*, 280–286.

Vilian, A. T. E., Madhu, R., Chen, S. M., Veeramani, V., Sivakumar, M., Huh, Y. S., et al. (2015). Facile synthesis of MnO2/carbon nanotubes decorated with a nanocomposite of Pt nanoparticles as a new platform for the electrochemical detection of catechin in red wine and green tea samples. *Journal of Materials Chemistry B*, *3*, 6285–6292.
Zhang, W. D., Chen, J., Jiang, L. C., Yu, Y. X., & Zhang, J. Q. (2010). A highly sensitive nonenzymatic glucose sensor based on NiO-modified multi-walled carbon nanotubes. *Microchimica Acta*, *168*, 259–265.

CHAPTER SEVENTEEN

Exfoliated and water dispersible biocarbon nanotubes for enzymology applications

Ankarao Kalluri[a], Megan K. Puglia[a], Mansi Malhotra[a], Challa V. Kumar[a,b,*]
[a]Department of Chemistry, University of Connecticut, Storrs, CT, United States
[b]Department of Molecular and Cell Biology, University of Connecticut, Storrs, CT, United States
*Corresponding author: e-mail address: challa.kumar@uconn.edu

Contents

Abstract

In this chapter, we report a simple and facile method to armor enzymes with carbon nanotubes (CNTs) which are exfoliated, and debundled using bovine serum albumin (BSA). The fabricated CNT/BSA dispersions are biofriendly, biocompatible, defect-free,

Methods in Enzymology, Volume 630
ISSN 0076-6879
https://doi.org/10.1016/bs.mie.2019.11.010

and highly stable solutions. BSA gives maximum exfoliation efficiency, exceeding the 4 mg/mL of CNT concentration compared to any previous reports. Further, the produced bioCNT dispersions were characterized by UV-visible, Raman, circular dichroism spectroscopy, and scanning electron microscopy (SEM). Exfoliation and debundling of the bioCNT dispersions is possible due to the π–π interaction, hydrogen bonding, hydrophobic interaction, and electrostatic attractive forces driving the adsorption of BSA on CNTs surface. Protein adsorption then makes a highly stable suspension in water that can be stored for a prolonged period. CNT dispersions are stable over a wide range of pH from 3 to 10 and at 4 °C or 25 °C for more than 2 months. Here, we also report the facile, inexpensive and green-chemistry method to fabricate a buckypaper (CNT paper), composed of the high packing density, self-assembled and randomly oriented bioCNTs, and these assemblies could be used in many emerging applications like air and water purification, nanocomposites, energy storage, and biosensing. Moreover, the CNT dispersions stabilized by BSA were successfully used in enzyme binding and kinetic studies and bound enzyme retained substantial catalytic activity. The current approach may facilitate bulk production of water dispersed CNTs in both academic and industrial laboratories. This is done by a simple method of stirring, which provides new opportunities for a wider range of CNT applications.

1. Introduction

Carbon nanotubes (Iijima & Lchihashi, 1993; Tang, Han, Han, & Xu, 1991) are unique one-dimensional macromolecules consisting of a hollow cylinder of carbon structures with a diameter of 0.2–50 nm. CNTs have generated attention in the field of nanotechnology due to their remarkable electronic (Wei, Vajtai, & Ajayan, 2001), thermal (Berber, Kwon, & Tománek, 2000), physical, chemical, and mechanical (Yu et al., 2000) properties. These CNTs are incredibly stiff and have a Young's modulus close to 1 Tpa (Berber et al., 2000). Carbon nanotubes are predicted to have a high thermal conductivity of up to 6000 W mK^{-1} (Berber et al., 2000), and these are one of the world's most robust materials having strength between 50 and 100 GPa. Additionally, single-walled CNT are able to demonstrate very high mobility $10^5\ cm^2\ V^{-1}\ s^{-1}$ (Dürkop, Kim, & Fuhrer, 2004) and possible to achieve substantial current densities up to 100 $MAcm^{-2}$ (Wei et al., 2001). Due to these fascinating properties CNTs are exciting candidates for a wide range of applications like nanoelectronics (De Volder, Tawfick, Baughman, & John Hart, 2013), nanolithography (Dai et al., 1996; Gopal et al., 2010), nanocomposites (Chiou, Byun, Kim, & Huang, 2018), energy storage systems (Kokal et al., 2017), biosensing (Saifuddin, Raziah, & Junizah, 2013), therapeutics (Biju, 2014), biocatalysis

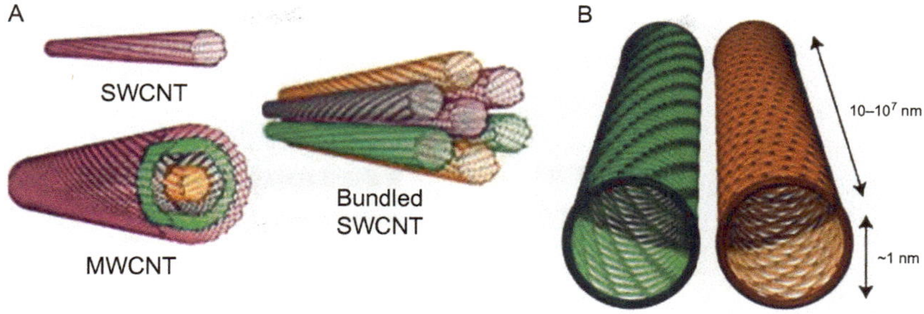

Fig. 1 (A) Single-walled carbon nanotube (SWNT), and multiwalled carbon nanotube (MWNT structures). (B) Rolled graphene represented as a carbon nanotube (Hersam, 2008). *Panels (A and B) reproduced with permission: copyright 2008, Springer Nature.*

(Azevedo et al., 2015; Zore, Kasi, & Kumar, 2017; Zore et al., 2015), biofuel (Prasad, Chen, & Chen, 2014), cancer-targeting treatment (Kavosi et al., 2018) and drug delivery systems (Bhirde et al., 2009).

Nevertheless, all these fascinating properties apply to isolated, debundled, and high-quality pristine SWCNTs, which are highly expensive and difficult to synthesize under normal lab conditions (Fig. 1). Furthermore, the main intrinsic drawback of CNTs is that they tend to aggregate into large bundles from poor dispersibility and incompatibility in the majority of solvents (including water) due to their π–π stacking and hydrophobic interactions, which make it crucial to develop methods to separate and disperse the bundles in order to utilize them for applications, such as aqueous-based biomedical devices (Oliveira et al., 2015).

The integration of CNTs with biological materials or proteins enhance the utilization potential and the properties of the biomolecules, especially in biological applications. Various methods have been proposed to solve the solubility problem of CNTs in water. For example, the integration of CNT with biological molecules achieved by surface functionalization or physical adsorption of biopolymers includes proteins (Karajanagi et al., 2006; Zhao et al., 2010), peptides (Edri & Regev, 2009; Oliveira et al., 2015), oligosaccharides (De Leo, Magistrato, & Bonifazi, 2015) or DNA (Huang, Mclean, & Zheng, 2005). The surface functionalization was improved by covalent (Zhao, Park, Han, & Lu, 2004) or non-covalent attachment and chemical oxidation with strong acids, as shown in Fig. 2A–E. The functionalization plays a pivotal role in dispersing the CNTs effectively in a suitable solvent as well as provide an appropriate interface to biological molecules or biomaterials. Furthermore, these functionalized CNTs have greater biocompatibility, bioavailability, and stability compared to their graphitized CNTs (Porter et al., 2009).

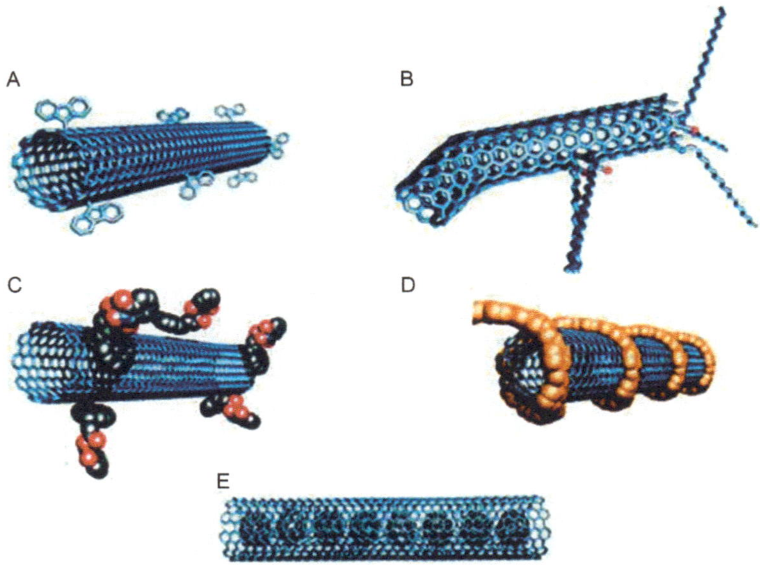

Fig. 2 Surface functionalization of CNTs via covalent binding to outer edge (A), covalent chemistry at open ends of the CNTs or at defect sites (B), non-covalent binding of wrapped surfactants (C), non-covalent binding of wrapped polymer molecules (D), molecule insertion into the CNTs (Hersam, 2008). *Panels (A–E) reproduced with permission. Copyright 2008, Springer Nature.*

Proteins, DNA, Peptide conformations play an essential role to disperse the CNTs in the aqueous solution and armoring them against denaturants. Edri et al. reported and describe the strong correlation between the conformation of the protein (BSA) and CNTs dispersion efficiency (Edri & Regev, 2008). BSA containing hydrophobic and hydrophilic pockets, pH-sensitive tertiary structure with acidic and basic residues, has high solubility in water. Due to the characteristics of the BSA protein, the conformation changes, the electrical charges of the residues depend on the pH. The relationship between BSA conformation and the recoveries of both CNT and BSA as a function of pH are shown in Fig. 3A. Kim et al. reported the facile dispersion of the CNTs prepared by mixing with the glucose oxidase solution. Further, the obtained CNT dispersions were used in facile enzyme immobilization techniques and biofuel cells (Kim et al., 2017). The authors describe the excellent dispersion of the CNTs attained by the amphiphilic nature of the enzyme's surface, the hydrophobic residues interact with the hydrophobic CNT surface, and the hydrophilic residues interact with the aqueous solution, thus preventing the CNT aggregation and leading to good CNT dispersions (Fig. 3B).

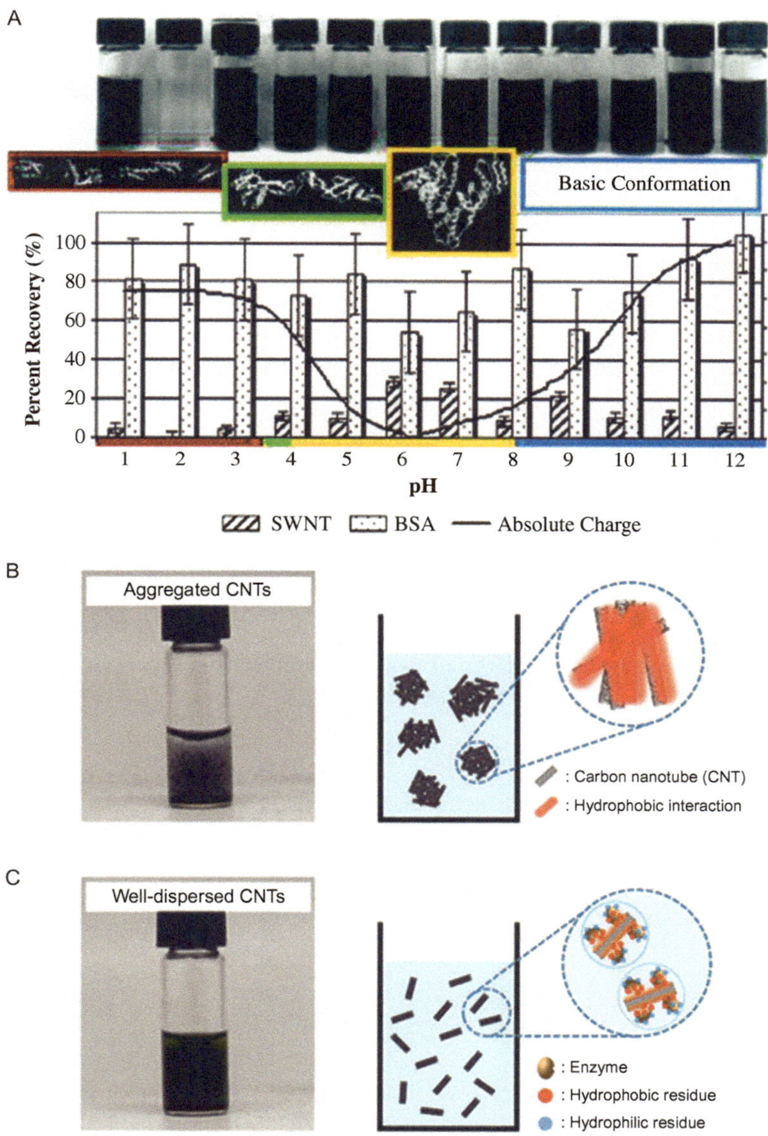

Fig. 3 (A) Effect of pH on the structure of the BSA (dotted bars) and the CNTs (striped bars). Also shown is the absolute charge on the BSA (continuous black line) at each pH. Dispersions of the BSA-CNT systems at each pH are shown above the graph (Zhao et al., 2010). (B) Image and theorized cartoon of bundled CNTs in water. (C) Image and theorized cartoon of well-dispersed CNTs in enzyme solution in water (10 mg/mL glucose oxidase) (Kim et al., 2017). *Panel (A): reproduced with permission, Copyright 2008, American Chemical Society. Panel (B): reproduced with permission, Copyright 2017, Springer Nature.*

Making exfoliated, unbundled and biocompatible water-soluble CNT dispersions of high concentrations remains an arduous and challenging task. In this chapter, we are reporting the ecofriendly, biocompatible, highly stable with extreme concentrations and water dispersible biocarbon nanotubes (bioCNT). These CNT dispersions are exfoliated and debundled through BSA by simple stirring and heating (65 °C). Here, we also reported a facile filtration method to fabricate buckypaper (Oh et al., 2015) with randomly oriented and densely packed self-assembled bioCNTs. These high-quality washed bioCNT dispersions were used for simple enzyme immobilization applications (Patole et al., 2018). This preparation of protein-based dispersion is a desirable method, compared to the sonication in surfactants or organic solvents for the CNT exfoliation and dispersion, further enhancing the tailoring of the bioCNT preparations for biomedical applications.

2. Results and discussion

2.1 Exfoliation and unbundling of bioCNT dispersion

The proposed schematic diagram (Scheme 1A) explains the highly concentrated CNT dispersions synthesized by simple mixing of the MWCNT (100 mg) in 10 mL of BSA (10 mg/mL, pH 7.0) solution at 65 °C with

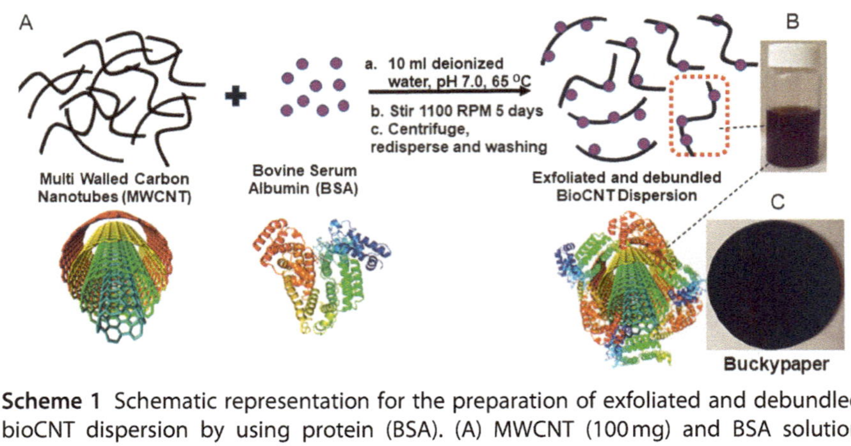

Scheme 1 Schematic representation for the preparation of exfoliated and debundled bioCNT dispersion by using protein (BSA). (A) MWCNT (100 mg) and BSA solution (10 mg/mL) mixed at pH 7.0 and stirred at 1100 RPM for 5 days at 65 °C followed by repeating the steps of a centrifuge, redisperse and washing to make bioCNT dispersion. (B) Exfoliated and unbundled bioCNT dispersion in a vial (image). (C) Circular buckypaper (bioCNT paper) prepared by vacuum filtration of bioCNT dispersion (1 mg/mL, 10 mL) through Whatman filter paper. MWCNT cartoon structure, courtesy of A.V. Krasheninnikov. Bovine serum albumin (PDB ID: 3V03).

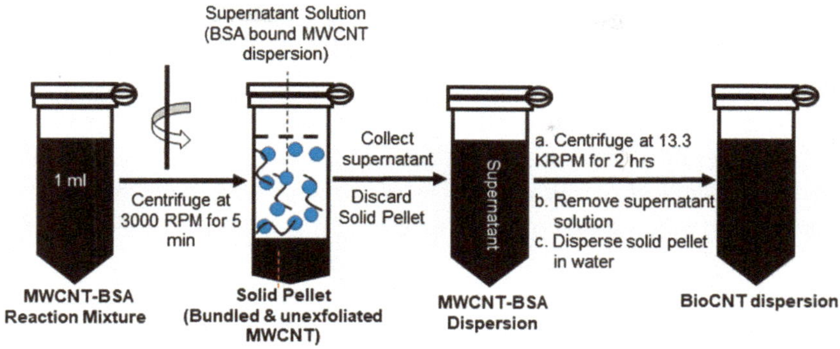

Scheme 2 Process of making bioCNT dispersion and removal of unbound (free) BSA. The reaction mixture of MWCNT with BSA centrifuged for 5 min at 3.0 KRPM and solid pellet of unexfoliated MWCNT removed. The obtained supernatant (BSA bound MWCNT) solution again centrifuged at 13.3 KRPM to get the clear and transparent solution (free BSA) and redisperse the solid pellet in deionized water to make bioCNT dispersion. Bound BSA of washed bioCNT solution confirmed through free BSA concentration of the supernatant solution. Final bioCNT contains less than 35% of the initial BSA solution after repeated washings.

stirring speed of 1100 RPM for 5 days. The CNT-BSA reaction mixture centrifuged at 3000 RPM for 5 min to separate the unreacted or aggregated CNTs. After the centrifugation step, the supernatant dispersion contains debundled and exfoliated MWCNT dispersion and excess BSA (Scheme 2). Further, the sample was centrifuged at 13.3 KRPM for 2 h to get a clear and transparent solution (free BSA) and the solid was redispersed the solid in deionized water to make bioCNT dispersion (Scheme 1B).

The percentage of the bound and unbound protein in bioCNT dispersion measured through absorbance spectroscopy. Based on the absorbance calculation, approximately 35% of the protein (BSA) attached to the CNT surface, and 65% remain in the free BSA solution phase. In a washed bioCNT solution containing 1 mg of CNT with bound 0.35 mg of protein. The obtained final dispersions have been characterized using UV-visible, Raman, circular dichroism spectroscopy techniques, scanning electron microscopy, and zeta potential measurements.

2.2 Mechanism of the exfoliation

The process involved to prepare the exfoliated and unbundled bioCNT dispersion by water-soluble protein typically includes the following steps:

stirring with heating, exfoliation, centrifugation, and accelerated phase separation followed by careful supernatant extraction. Bovine serum albumin (BSA) is a highly water-soluble protein, that is both surfactant-like and amphiphilic in nature. Peng et al. reported that hydrophobic pocket of BSA adsorbs onto the hydrophobic sidewall of the CNTs by π–π stacking interactions or Van der Waals' forces of attraction between the aromatic amino acid side chain of the protein and aromatic CNT surface (Du, Zhao, Mashayekhi, & Xing, 2014). The hydrophilic residues of the BSA are then left exposed to the water and the steric repulsive forces prevent the aggregation of the CNTs. Another hypothesis is that the stable bioCNT dispersion is obtained due to the H-bonding and polar interaction between amino acids with the CNT surface (Piao, Liu, & Li, 2012). Moreover, the electrostatic repulsive forces were confirmed through the zeta potential values of BSA adsorbed CNT dispersions which were in range of −20 to −40 mv at pH 6–10 (Fig. 5C). So, the electrostatic repulsion forces among the BSA bound CNT molecules help to keep the dispersion highly stable for extended periods (Du et al., 2014).

2.3 BioCNT characterization techniques

To avoid bundling and aggregation of MWCNT with BSA, well-dispersed bioCNT solutions are prepared with various concentrations and measured at a wavelength of 660 nm, where the absorbance of bundled CNT dispersion does not greatly contribute to the absorbance (Fig. 4A). The stock bioCNT dispersion 6 mg/mL (4 mg/mL MWCNT, 2 mg/mL BSA) obtained by oven drying of the three samples of 1 mL at 70–80 °C to get the dry mass of the samples (approximately 2 days). A linear correlation fit ($y = 0.3486x + 0.033$) with good predictability ($R^2 = 0.9985$) was obtained by the calibration curve (absorbance vs concentration) and the concentration of samples ranged from 0.1 to 2 mg/mL for the bioCNT solutions (Fig. 4B). According to Beer-Lambert's (B-L) law, the extinction coefficient of bioCNT dispersion measured at 660 nm is 0.35 mL mg^{-1} cm^{-1}, and the dispersion limit of the enzyme exfoliated CNT colloidal solutions at different pH values with % degree of debundling are compared in Table 1.

In a Raman spectrum, the peak position, shape and intensity of the peak depends on the size and number of tubes or layers, and defects present in the CNT sample. G-band is the characteristic peak of the graphitic layers, which corresponds to the tangential vibration of CNT samples. The defective graphitic structure is characterized by the D-band of the spectrum, and the

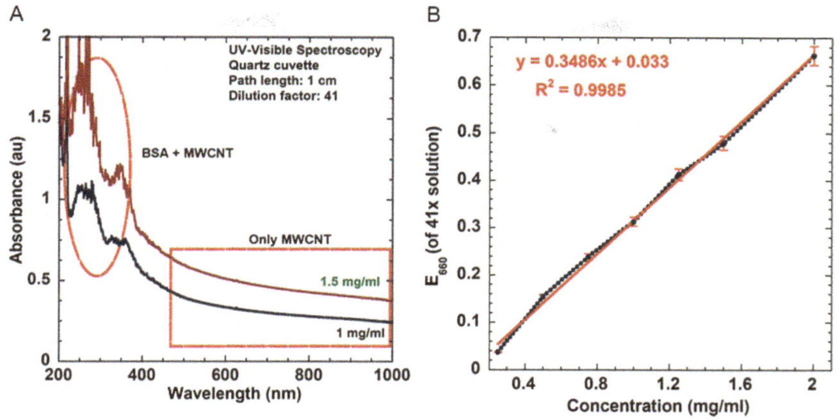

Fig. 4 (A) Absorption spectrum of the bioCNT dispersion. The graph shows the BSA and MWCNT combined absorbance at 200–400 nm and only MWCNT absorbance at 500–1000 nm range. (B) Calibration curve used to calculate the extinction coefficient of the bioCNT dispersion at pH 7.0 in deionized water. The stock bioCNT dispersion (5 mg/mL) prepared by calculating the weight of bioCNT by drying the dispersed aliquot sample in an oven at 70 °C. Error bars are the standard deviations of the three measurements.

Table 1 Comparison of the dispersion limit (D_L) values of SWCNTs in water containing different proteins at different pH values with % degree of debundling (Geckeler, 2007).

Protein	Molar mass (g/mol)	pH	Dispersion limit ($\mu g\ mL^{-1}$)	Degree of debundling (%)
Bovine serum albumin (BSA)	66,450	11.2	132.71	65
Glucose oxidase (GOD)	63,273	11.3	1.39	0
Hemoglobin (HBA)	61,933	1.7	180	55
Histone (HST)	15,316	4.2	196.78	100
Lysozyme (LSZ)	14,313	4.7	255.3	88
Myoglobin (MGB)	16,953	11.5	119.42	56
Ovalbumin (OVB)	42,861	11.2	153.40	68
Trypsin (TPS)	23,783	11.6	42.78	19

graphitic and defective band intensity (I_D/I_G) ratio is useful to measure the quality of CNT samples and radial breathing mode (RBM) which is susceptible to the smaller diameters of the carbon nanotubes (SWCNT, DWCNT). The ratio between D and G bands and the RBM mode

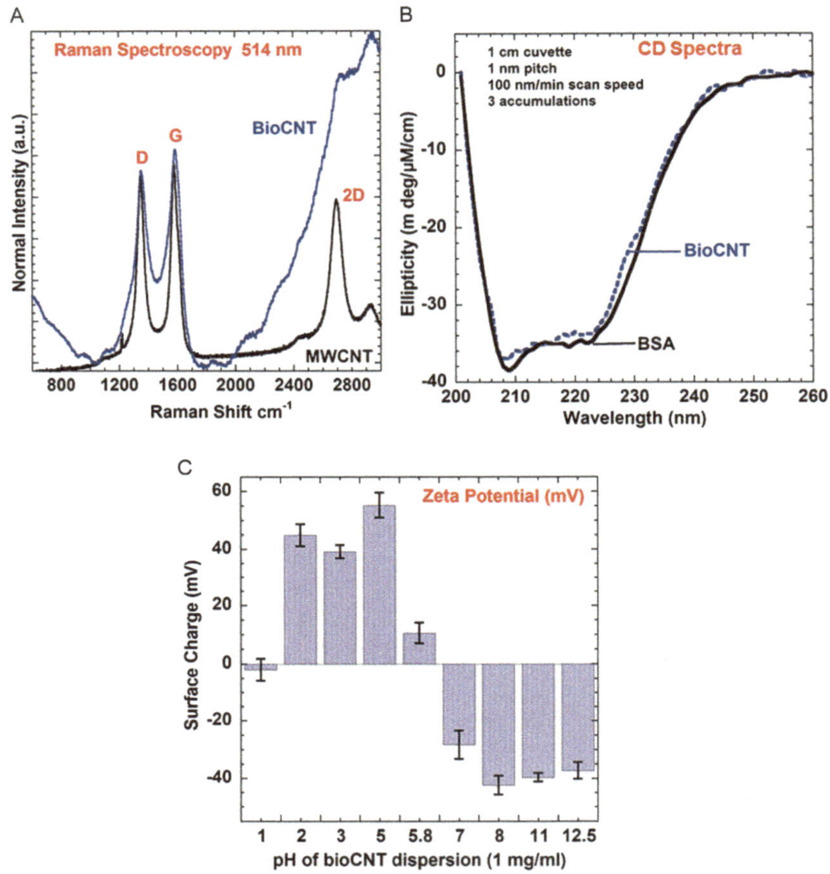

Fig. 5 (A) Raman spectroscopy (B) Circular dichroism spectra of BSA functionalized CNT surface after exfoliation compared with native BSA protein showed significant changes in secondary structure (α-helix, 208 nm). (C) Zeta potential of biofunctionalized CNT after incubation at different pHs at room temperature (after 3 h).

(Costa et al., 2008) and its relation to the diameter distribution are critical factors allowing one to distinguish between SWCNT, few walled MWCNT, and MWCNT samples. The bioCNT dispersion spectrum was compared with the MWCNT spectrum showing significant changes of the 2D peak due to the exfoliation of the CNT tubes. There is also a minor change in I_D/I_G ratio as well as Raman shift toward the left (Fig. 5A). However, further analysis of CNT dispersed in different BSA solutions are required to confirm MWCNT exfoliation into few-walled CNT samples.

The secondary structure of the free BSA and bioCNT colloidal solution was analyzed through a circular dichroism (CD) spectrophotometer. Based on the CD spectra (Fig. 5B), the α-helical content wavelength vs ellipticity at 208 nm of the bioCNT is decreased compared to the normal BSA. The decrease in α-helix conformation in bioCNT indicated that CNTs structure directly interferes with the hydrogen bond of the N—H group with the C=O group on the α-helix residue from the backbones and side chains in order to make BSA protein more readily adsorb on the CNT surface. The obtained results indicate a greater loss of α-helical structure on the hydrophobic surface compared to the hydrophilic surface (Roach, Farrar, & Perry, 2006).

The surface charge of the bioCNT dispersion is governed by the number of BSA molecules attached to the CNT surface, the protein conformations and the pH of the dispersed solution. The dispersion efficiency of the CNT depends on the pH-induced structural changes and the electrical charge of the attached BSA molecules. The BSA adsorbed CNT dispersion showed a zeta potential range from −20 to −40 mv at pH 6–12.5 (Fig. 5C). Due to these electrostatic repulsion forces the CNT colloidal dispersion remains highly stable for long periods. However, by changing the pH of the solution to highly acidic and basic conditions, there is the possibility of aggregation of the CNTs due to the degradation of BSA protein molecules, indicating the BSA plays an essential role in stabilizing and dispersing the CNTs.

Scanning electron microscopy (SEM) images revealed the surface morphology of the unbundled and exfoliated CNTs in the presence of protein. The aggregated and bundled CNT samples show poor dispersion before exfoliation (Fig. 6A), while the same CNT samples are debundled and separated as individual nanotubes after exfoliation with BSA protein (Fig. 6B). Furthermore the supernatant containing the free BSA and the exfoliated CNT dispersions appears to show the fibers completely wrapped by BSA (Fig. 6C).

3. Preparation of bioCNT-HRP complex and enzyme immobilization studies

Enzymes are a naturally occurring class of proteins, which have special properties like specificity, selectivity, and high catalytic activity. The enzymes are biofriendly and produce fewer side-products than normal chemical reactions. Due to these fascinating properties, enzymes are used in the food industry, biofuel cell production and wastewater treatment and

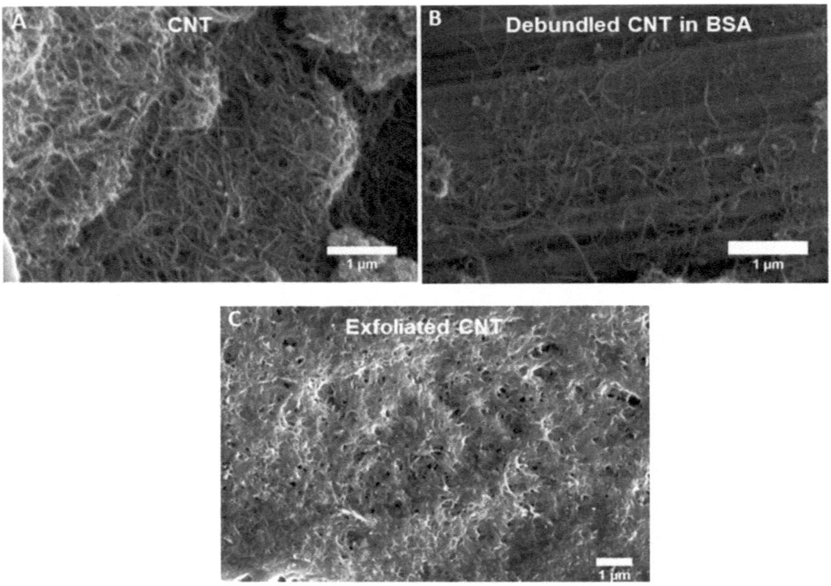

Fig. 6 SEM images showing the surface morphology of (A) aggregated native MWCNT sample before exfoliation, (B) debundled and washed bioCNT dispersion, after repeated washings and centrifugation steps of solid pellet, (C) exfoliated bioCNT dispersion (supernatant) with the BSA solution. The white scale bar in each image represents 1 μm.

biocatalysis reactions. Nanostructures and carbon-based nanocomposites are widely used as special supports for stabilization and enzyme immobilization studies. High aspect ratio and large surface area of CNTs leads to high enzyme loading capacity. Therefore, these enzyme immobilized robust nano scaffolds show increased enzyme stability and high volumetric enzyme activity. The surface functionalization process can further improve the efficiency of nanomaterials. Horseradish peroxidase (HRP) enzyme was immobilized onto the bioCNT surface by physical binding or non-covalent interactions.

3.1 Preparation of bioCNT-HRP complex

The stabilized bioCNT dispersions (2 mg/mL) are mixed with HRP enzyme (6.5 μM) and incubated for 1 day followed by centrifugation of the sample mixture, and then the bioCNT-HRP pellet at bottom of the tube is collected. The immobilized HRP enzyme concentration on the CNT surface was calculated by using the free HRP enzyme in the supernatant solution (Scheme 3). According to the calculation, approximately 30%

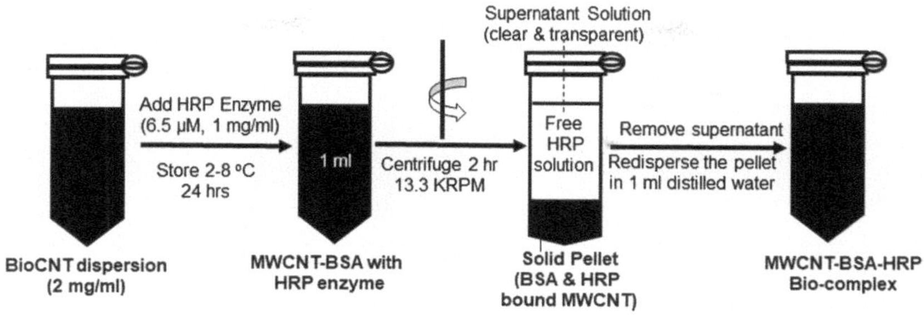

Scheme 3 Process and synthesis of bioCNT-HRP complex preparation by HRP enzyme immobilization.

of the HRP enzyme was immobilized on the bioCNT surface. Compared to the conventional enzyme immobilization techniques, the enzyme integration with nanomembrane or nanocomposites system could offer more advantages, such as high enzyme loading for improved specific activity and stability of immobilized enzymes.

3.2 Activity studies of bioCNT-HRP complex

Oxidation of guaiacol is the rate-determining step for peroxidase activity. The specific activity of the control HRP enzyme was compared with the bioCNT-HRP complex, showing that the calculated specific activity of free HRP was ~1.46 μM/s (100%) and that of bioCNT-HRP was ~1.25 μM/s, which is approximately equal to 85% relative activity of the free enzyme concentration (Fig. 7A). The turnover number (K_{cat}) is the rate constant for conversion of the enzyme-substrate complex to the product, which is determined by using the υ_{max}. The robust bioCNT-HRP enzyme dispersions remarkably functioned as an active catalyst under physiological solutions. The physical mixture of the K_{cat} (TON) of bioCNT-HRP dispersion shows a $5.0 \times 10^{-2}\,s^{-1}$, while the free HRP enzyme shows $5.6 \times 10^{-2}\,s^{-1}$ under similar conditions (Fig. 7B).

4. Methods

4.1 Preparation of MWCNT-BSA dispersion

4.1.1 Materials

1. Multiwalled carbon nanotubes, MWCNT (length 10–30 μm, Outer diameter <8 nm, purity >95% purchased from Cheap Tubes, USA).
2. Bovine Serum Albumin, BSA (Equitech-Bio, Inc., TX, BAH64–1366).
3. Distilled water (Millipore Sigma, 18 MΩ•cm resistivity).

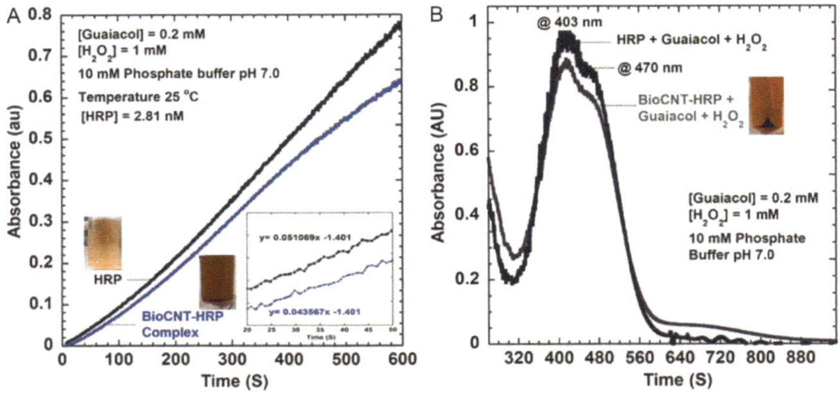

Fig. 7 Enzyme kinetic studies and Peroxidase activity of HRP bound to bioCNT using compared to free HRP enzyme (A) enzyme activity studies (inset image shows the initial rate of reaction from 20 to 50 s) (B) turn over number after 1 h.

4.1.2 Preparation method

1. Weigh 100 mg of MWCNT and transfer to 50 mL round bottom flask.
2. BSA (100 mg/mL) was dissolved in distilled water (9–10 mL) and the pH was adjusted to 7.0 by using 0.1 M HCl or 0.1 M NaOH solution and make up the final volume to 10 mL.
3. Transfer BSA solution into above 50 mL round bottom flask.
4. Stir (1000 RPM) and heat the MWCNT and BSA solution at 65–70 °C in round bottom flask up to 5 days.
5. Use a reflux condenser to prevent the evaporation of water in the round bottom flask.
6. Unexfoliated MWCNT is separated from the reaction mixture by centrifugation. MWCNT-BSA reaction mixture is transferred into a centrifuge tube and centrifuged at 3000 RPM for 5 min. The supernatant is carefully separated from the unexfoliated solid MWCNT pellet.
7. The final concentration of MWCNT in the dispersion (supernatant) can be estimated by using a UV-visible spectrophotometer.

Tip 1: MWCNT-BSA samples should be stored in 2–8 °C and benzyl alcohol (1%, 20 μL) can be added to prevent the growth of bacteria.

Tip 2: To monitor the progress of exfoliation, aliquots of 50 μL samples can be taken out at specific intervals of time (every 24 h) and analyzed using UV-Visible spectroscopy.

Tip 3: Alternatively, to avoid time-consuming centrifugation use Whatman filter paper to filter the dispersed CNTs by vacuum filtration.

4.2 Removing free protein from bioCNT solution

4.2.1 Equipment

1. Safety: Goggles, coat, nitrile gloves, ventilation hood.
2. Centrifuge (Fisher Scientific, USA).
3. Volumetric micropipettes (20, 100, 200, and 1000 μL).
4. UV-visible spectrophotometer.

4.2.2 Materials

1. Deionized water
2. Filter paper (Whatman 1 cellulose, typical 180 μm thickness, 11 μm pore size, 4.25 cm diameter)

4.2.3 Washing method

1. Begin by placing the above CNT with BSA dispersion (after centrifugation at 3000 RPM for 5 min) into centrifuge tubes.
2. Centrifuge CNT with BSA dispersion at 13.3 KRPM for 100 min or more until the supernatant is clear.
3. Collect the supernatant solution, leaving only the BSA bound CNT pellet at the bottom.
4. Save the removed supernatant (free BSA) solution and later confirm the concentration of BSA.
5. Resuspend the CNT with BSA pellet in distilled water and vortex well to dissolve the solid pellet.
6. Centrifuge this new dispersion at 13.3 KRPM for 60 min until the supernatant is clear.
7. Remove clear supernatant from the centrifuge tube, leaving only the CNT pellet at the bottom. Save the removed protein solution to later confirm the free BSA concentration.
8. Resuspend the CNT pellet in distilled water to get bioCNT dispersion
9. This obtained bioCNT dispersion will be further used in MWCNT and BSA concentration measurements.

Tip: If the MWCNT pellet is not dissolving well in water after centrifugation then sonicate the solution for 10 min to break up the pellets and redisperse the pellet into solution.

By continuous washing of CNT and BSA dispersions, more than 65% of free BSA is removed from the dispersion and 35% of BSA is bound to the CNT in bioCNT dispersion. Due to this residual BSA protein on bioCNT dispersion, the CNTs are easily soluble and resuspended in water due to the

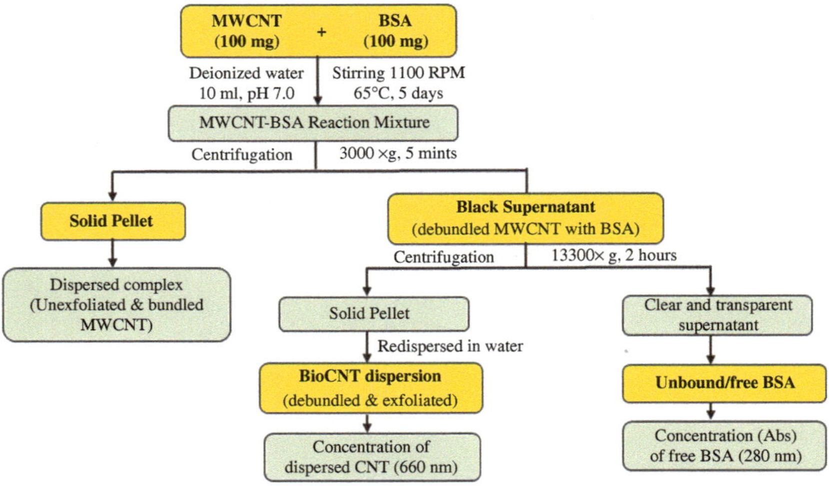

Flow chart 1 Flow chart for preparation of bioCNT dispersion through the step-by-step approach.

hydrophilic nature of bound BSA. In a washed bioCNT solution there is 1 mg of CNT with bound 0.35 mg of protein (Flow Chart 1).

4.3 UV-visible optical spectroscopy measurements

4.3.1 Equipment

1. UV-visible spectrophotometer (Agilent 8453, Santa Clara, CA)
2. Quartz cuvette having 1 cm path length

4.3.2 Method

1. Before doing the experiments, triple rinse all the glassware and quartz cuvettes with purified water and allow it to dry.
2. Aliquots of the dispersed bioCNT solution diluted with deionized water are then transfered into quartz cuvettes and the absorbance measured by using a UV-visible spectrophotometer.
3. Determine the unknown bioCNT concentration by using the extinction coefficient of 0.3486 mg^{-1} $mL\,cm^{-1}$.

Tip: Make sure the absorbance of the bioCNT dispersion or any other samples is less than one, in accordance with Beer's Law.

4.4 Raman spectroscopy

4.4.1 Equipment

1. Raman Spectrometer, 514 nm Laser with 1 μm special resolution and 50 mW power (Renishaw System 2000)

2. Laser wavelength: 514 nm
3. Laser power: 25 mW (50%)
4. Laser beam size (diameter): 0.41 μm
5. Grating: 1800 g/nm
6. Objective lens: 50 X
7. Accumulations: 10, exposure time: 10 s
8. Microscope Slides (Globe Scientific Inc., NJ)

4.4.2 Method

1. Drop-cast 50–100 μL of bioCNT (3 mg/mL) on a clean microscope slide and let the sample dry for 24 h.
2. Calibrate the Raman instrument by using a silicon wafer (521 cm^{-1}) prior to each acquisition.
3. Raman spectra collected from 600 to 3200 cm^{-1}, using a 514 nm laser Raman spectrometer.
4. A control spectrum also collected for the pristine unexfoliated MWCNT powder used in the synthesis of bioCNT dispersion.
5. Each Raman band should be normalized to the G peak and subtract the baseline by using Raman spectra software (WIRE 3.4) before doing analysis.

   Tip 1: Tap the pristine MWCNT powder onto the microscope slide and press the mound down with a second glass slide to make imaging easier.

   Tip 2: The protein (BSA) in the bioCNT can become damaged by the 514 nm laser (100% power source). To avoid this problem, reduce the laser power to half (50%, 25 mW), which will decrease the signal to noise ratio of the CNT with protein dispersion samples.

4.5 Circular dichroism (CD)

CD spectroscopy is commonly used to study the secondary and tertiary structures, and conformational stability (folding properties) of the protein solution under harsh conditions like heat, mutations, and chemical denaturing reagents. CD also measures the physical or binding interaction of proteins with nanomaterials. Secondary structure of the protein consists of α-helices and β-sheets, showing a characteristic spectrum in far-UV range from 195 to 260 nm. In CD spectra, two negative double-humped peaks at 208, 220 nm and a positive band at 194 nm are characteristic for the helical structure of a protein (Du et al., 2014; Greenfield, 2007).

4.5.1 Equipment

1. A circular dichroism (CD) spectropolarimeter (JASCO model J715)
2. 1 cm path length quartz cuvette
3. 1 nm pitch and 100 nm/min scan speed
4. 10 Accumulations

4.5.2 Materials

1. 10 mM Sodium Phosphate buffer solutions pH 7.0

4.5.3 Procedure

1. Before doing the experiments, triple rinse all the glassware and quartz cuvettes with purified water and allow it to dry.
2. The secondary structure of the protein was measured over a range of 200–260 nm by with a 1 cm path length quartz cuvette.
3. The solutions or dispersions are prepared by diluting the BSA and bioCNT with the required amount of phosphate buffer solution at pH 7.0
4. Follow the CD instructions manual and software to operate the instrument.
5. The diluted solutions were added to a 1 cm quartz cuvette to run the instrument.

4.6 Zeta potential measurements

4.6.1 Equipment

1. Zeta potential analyzer (Brookhaven zeta plus, Holtsville, NY)

4.6.2 Materials

1. Glass or polystyrene cuvette (Fisher Scientific)
2. pH adjusting reagents: Dilute HCl (0.1 M, 1 M) and NaOH (0.1 M, 1 M) solutions (Fisher Scientific)
3. 100 mM KCl solution to adjust the conductivity of a dispersed solution

4.6.3 Procedure

1. The pH of bioCNT dispersion (2 mL, ~1 mg/mL) was adjusted by using the dilute HCl and NaOH solution.
2. Adjusted samples of bioCNT dispersion (1.6 mL) were added to a polystyrene cuvette and the zeta potential analyzer electrode was submerged in the dispersion.

3. The sample's zeta potential calculated by using laser Doppler velocimetry and Smoluchowski fit of the instrument.
4. The bioCNT dispersions showed a surface potential of −28 mv at pH 7.0 and the observed isoelectric point (pI) was 5.5–6.5 (Fig. 5C) and the experimental value pI for BSA is ~5.5.

4.7 Scanning electron microscopy (SEM)

4.7.1 Equipment

1. FEI Teneo LVSEM (Thermo Fisher)

4.7.2 Method

1. Dilute dispersed solution (0.2 mg/mL) are prepared by drop cast method on copper tape affixed to SEM aluminum stubs.
2. Dry the sample solution overnight.
3. 10 kV accelerating voltage used for imaging.
4. Scale bars on images were redrawn to identical size using Image J software to improve visibility.

4.8 Storage and stability

The stability of bioCNT dispersion is one of the important parameters to be analyzed for biological applications. BioCNT dispersion can be successfully stored at 2–8 °C for more than 3 months and 25 °C for up to 1 month. Poor stability and aggregation of the CNT dispersion is indicated by the settling down of the precipitate at the bottom of the container. If the sample is stored more than 3 months without any sterilization, there is a possibility of the growth of bacteria in bioCNT dispersion due to BSA protein. The bacteria growth is prevented by adding 0.1% of the benzyl alcohol to sample dispersions.

4.9 Preparation of bioCNT-HRP complex

4.9.1 Equipment

1. UV-visible spectrometer (Agilent Technologies, Santa Clara, CA)

4.9.2 Materials

1. Horseradish peroxidase, HRP (Calzyme laboratories, Inc., USA, activity 230 U/mg, lot 65-3-100)
2. Sodium phosphate buffer 10 mM, pH 7.2
3. 1.5 mL volume Eppendorf tubes (Dot Scientific Inc.)

4.9.3 Procedure

1. Prepare HRP (~5 mg/mL) stock solutions in 10 mM phosphate buffer at pH 7 and determine the molar concentration (Soret absorbance at 403 nm) by using absorption spectroscopy.
2. Prepare a series of bioCNT-HRP complex dispersions by increasing concentrations of the HRP enzyme from 0.5 to 1.0 mg/mL with fixed bioCNT (1 mg/mL) dispersion.
3. The dispersed bioCNT-HRP complex prepared by mixing 300 μL of HRP stock solution (5 mg/mL, pH 7) and 500 μL (2 mg/mL) of bioCNT dispersion diluting with 200 μL of the phosphate buffer solution in 2 mL of Eppendorf tubes. This complex mixture mixed well and incubated for a day.
4. The final sample was centrifuged at 13.3 KRPM for 2 h to separate the bioCNT-HRP complex pellet and the unbound HRP (supernatant) was quantified by using Soret absorbance at 403 nm, this value further used to measure the immobilized (bound) HRP in the bioCNT-HRP complex. Store the samples in 2–8 °C after incubation and centrifugation steps.
5. The immobilized HRP enzyme in the bioCNT-HRP complex is approximately 20–30% (w/w) compared to the initial free HRP enzyme (1 mg/mL).

Tip 1: A shaker table can be used to mix the solution and decrease the enzyme loading time.

Tip 2: Both CNT and proteins absorb at a wavelength of 280 nm, so estimating the protein concentration using 280 nm gives large errors. Hence, we used the Soret absorbance at 403 nm to measure the absorbance of HRP in the bioCNT-HRP complex, which is more reliable and accurate compared to absorbance at 280 nm.

4.10 Activity studies of bioCNT-HRP complex

The peroxidase activity of the HRP enzyme bound to bioCNT was determined by conventional HRP assay. In this activity assay, H_2O_2 is used as a substrate and guaiacol as a chromogenic reactant. HRP activity was determined by measuring the increase in absorbance (concentration) at 470 nm corresponding to time as the colored 3,3′-dimethoxy-4,4′-biphenoquinone complex is formed. The initial rate and specific activity of the free HRP enzyme and bioCNT-HRP complex determined by the slope of the concentration at 470 nm vs time. The relative activity of HRP was calculated from the ratio of the residual activity bound HRP to the initial activity of the free HRP sample.

4.10.1 Equipment

1. UV-vis spectrophotometer

4.10.2 Materials

1. Guaiacol (Sigma-Aldrich, MKBC6310)
2. Hydrogen peroxide (Fisher Scientific, ACS 30%, 134442)

4.10.3 Procedure

1. Prepare stock solutions of guaiacol (75 mM) and H_2O_2 (25 mM) in phosphate buffer solution at pH 7.0.
2. Prepare the bioCNT-HRP complex sample as described above with the initial concentration of HRP as 6.5 μM (1 mg/mL), and the unbound HRP from the complex is removed by centrifugation (4.5 μM) up to 2 h.
3. Remove the supernatant and resuspend the bioCNT-HRP pellet in 1 mL of 10 mM phosphate buffer at pH 7.0.
4. Mix the bioCNT-HRP (0.25 μm enzyme), 0.2 mM guaiacol, and buffer (to make up to a total volume of 2 mL) in a 1 cm quartz cuvette.
5. Set the enzyme kinetics in a spectrophotometer with a steady-state stirring inside the cuvette to monitor absorbance at 470 nm for 10 min (60 s), blank the solution and then begin the data collection as 1 mM H_2O_2 is added.
6. The initial rate of the reaction is calculated by using absorbance vs time slope, which is further used to calculate the specific activity of the enzyme and bioCNT-HRP complex (typically first 30 s).
7. The calculated specific activity of free HRP was ~1.46 μM/s (100%) and that of bioCNT-HRP was ~1.25 μM/s, which is approximately equal to 85% relative activity of the free enzyme. See Flow chart 2 for the preparation of bioCNT-HRP complex and enzyme activity studies.

Tip: To prepare the guaiacol stock solution, sonicated the guaiacol in water up to 10–15 min will improve the solubility of the solution.

5. Conclusion

Here, we report a facile method for synthesizing, exfoliated, debundled water-soluble and high-quality bioCNT dispersions that are suitable biological and biomedical applications. The protein stabilized CNT dispersions are highly stable against wide range of pH, temperature and storage conditions. The CNT dispersion is stabilized by the adsorption of BSA on the CNT surface. BSA interactions with the CNT surface may be due to the π–π stacking

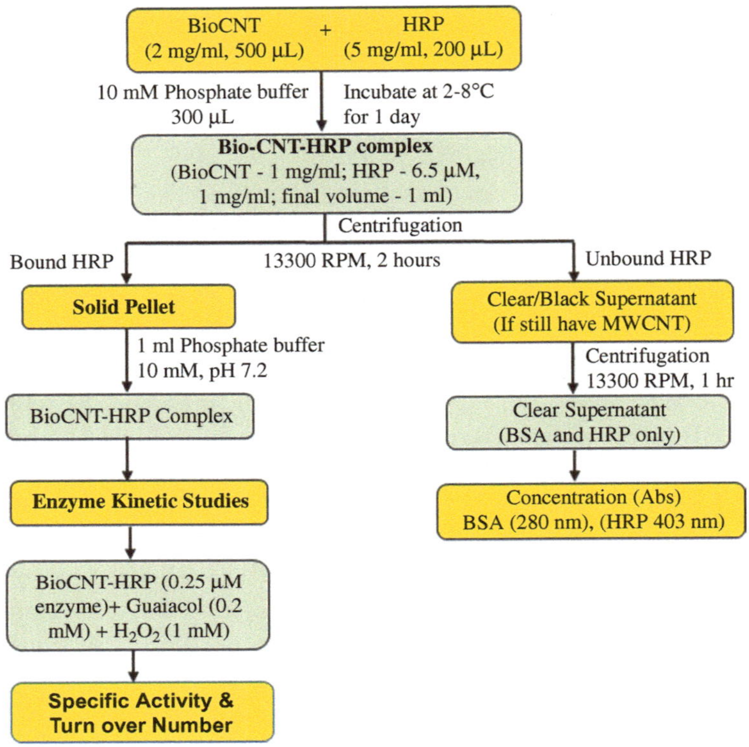

Flow chart 2 Flow chart for the preparation of the bioCNT-HRP complex and step by step approach of enzyme activity studies.

of residues, hydrophobic interactions, and electrostatic attractive forces between the CNTs and BSA. Here, we also synthesize inexpensive buckypaper (CNT paper) by a green-chemistry approach composed of self-assembled and randomly oriented bioCNTs. The fabricated buckypaper is highly flexible and conductive without any chemical bonding and the self-assembled, random oriented CNTs are held together by the protein, which acts as a binder for strong CNT entanglement. These films can be used in bio-supercapacitors and lithium ion battery (Patole et al., 2018), water filtration and bio membrane, nanocomposites, and biosensing applications.

Moreover, the CNT dispersion stabilized by the protein was successfully used in enzyme binding and kinetic studies. The CNT dispersions are very stable over a wide range of pH and temperature for more than 2 months. Storage at 4 °C improves the storage time and inhibits the bacterial growth of the bioCNT samples and storage stability can be improved by adding

0.1% of an anti-bacterial agent. The current approach may facilitate the bulk production of water dispersed CNTs in laboratories, which provides a new opportunity for biomedical and enzymatic applications.

Acknowledgments

We thank the NSF and the University of Connecticut for funding this project.

References

Azevedo, R. M., et al. (2015). A strategy for improving peroxidase stability via immobilization on surface modified multi-walled carbon nanotubes. *Journal of Chemical Technology and Biotechnology*, *90*(9), 1570–1578.

Berber, S., Kwon, Y.-k., & Tománek, D. (2000). Unusually high thermal conductivity of carbon nanotubes. *Physical Review Letters*, *84*(20), 4613–4616.

Bhirde, A. A., et al. (2009). Targeted killing of cancer cells in vivo and in vitro with EGF-directed carbon nanotube-based drug delivery. *ACS Nano*, *3*(2), 307–316.

Biju, V. (2014). Chemical modifications and bioconjugate reactions of nanomaterials for sensing, imaging, Drug Delivery and Therapy. *Chemical Society Reviews*, *43*(3), 744–764.

Chiou, K., Byun, S., Kim, J., & Huang, J. (2018). Additive-free carbon nanotube dispersions, pastes, gels, and doughs in cresols. *Proceedings of the National Academy of Sciences of the United States of America*, *115*(22), 5703–5708.

Costa, S., et al. (2008). Characterization of carbon nanotubes by Raman spectroscopy. *Materials Science-Poland*, *26*(2), 433–441.

Dai, H., et al. (1996). Nanotubes as nanoprobes in scanning probe microscopy. *Nature*, *384*(14), 147–150.

De Leo, F., Magistrato, A., & Bonifazi, D. (2015). Interfacing proteins with graphitic nanomaterials: From spontaneous attraction to tailored assemblies. *Chemical Society Reviews*, *44*(19), 6916–6953.

De Volder, M. F. L., Tawfick, S. H., Baughman, R. H., & John Hart, A. (2013). Carbon nanotubes: Present and future commercial applications. *Science*, *339*, 535–539.

Du, P., Zhao, J., Mashayekhi, H., & Xing, B. (2014). Adsorption of bovine serum albumin and lysozyme on functionalized carbon nanotubes. *Journal of Physical Chemistry C*, *118*(38), 22249–22257.

Dürkop, T., Kim, B. M., & Fuhrer, M. S. (2004). Properties and applications of high-mobility semiconducting nanotubes properties and applications of high-mobility semiconducting nanotubes. *Journal of Physics: Condensed Matter*, *16*, 553–580.

Edri, E., & Regev, O. (2008). PH effects on BSA-dispersed carbon nanotubes studied by spectroscopy-enhanced composition evaluation techniques. *Analytical Chemistry*, *80*(11), 4049–4054.

Edri, E., & Regev, O. (2009). 'Shaken, not stable': Dispersion mechanism and dynamics of protein-dispersed nanotubes studied via spectroscopy. *Langmuir*, *25*(18), 10459–10465.

Geckeler, K. E. (2007). Proteins and carbon nanotubes: Close encounter in water. *Small*, *3*(7), 1259–1265.

Gopal, N., et al. (2010). Progress in polymer science polymer nanocomposites based on functionalized carbon nanotubes. *Progress in Polymer Science*, *35*(7), 837–867.

Greenfield, N. J. (2007). Using circular dichroism spectra to estimate protein secondary structure. *Nature Protocols*, *1*(6), 2876–2890.

Hersam, M. C. (2008). Progress towards monodisperse single-walled carbon nanotubes. *Nature*, *3*, 387–394.

Huang, X., Mclean, R. S., & Zheng, M. (2005). High-resolution length sorting and purification of DNA-wrapped carbon nanotubes by size-exclusion chromatography. *Analytical Chemistry*, 77(19), 6225–6228.

Iijima, S., & Lchihashi, T. (1993). Single-shell carbon nanotubes of 1-nm diameter. *Nature*, *363*(3), 603–605.

Karajanagi, S. S., et al. (2006). Protein-assisted solubilization of single-walled carbon nanotubes. *Langmuir*, *22*(4), 1392–1395.

Kavosi, A., et al. (2018). The toxicity and therapeutic effects of single-and multi-wall carbon nanotubes on mice breast cancer. *Scientific Reports*, *8*(1), 1–12. https://doi.org/10.1038/s41598-018-26790-x.

Kim, B. C., et al. (2017). Fabrication of enzyme-based coatings on intact multi-walled carbon nanotubes as highly effective electrodes in biofuel cells. *Scientific Reports*, 7(40202), 1–10.

Kokal, R. K., et al. (2017). Solar cells with PbS quantum dot sensitized TiO_2-multiwalled carbon nanotube composites, sulfide-titania gel and tin sulfide coated C-fabric. *Physical Chemistry Chemical Physics*, *19*(38), 26330–26345.

Oh, J. Y., et al. (2015). Easy preparation of self-assembled high-density buckypaper with enhanced mechanical properties. *Nano Letters*, *15*(1), 190–197.

Oliveira, S. F., et al. (2015). Protein functionalized carbon nanomaterials for biomedical applications. *Carbon*, *95*, 767–779. https://doi.org/10.1016/j.carbon.2015.08.076.

Patole, S. P., et al. (2018). A wet-filtration-zipping approach for fabricating highly electroconductive and auxetic graphene/carbon nanotube hybrid buckypaper. *Scientific Reports*, *8*(1), 1–12. https://doi.org/10.1038/s41598-018-30009-4.

Piao, L., Liu, Q., & Li, Y. (2012). Interaction of amino acids and single-walled carbon nanotubes. *Journal of Physical Chemistry C*, *116*(2), 1724–1731.

Porter, A. E., et al. (2009). Uptake of noncytotoxic acid-treated single-walled carbon nanotubes into the cytoplasm of human macrophage cells. *ACS Nano*, *3*(6), 1485–1492.

Prasad, K. P., Chen, Y., & Chen, P. (2014). Three-dimensional graphene-carbon nanotube hybrid for high-performance enzymatic biofuel cells. *ACS Applied Materials and Interfaces*, *6*(5), 3387–3393.

Roach, P., Farrar, D., & Perry, C. C. (2006). Surface tailoring for controlled protein adsorption: Effect of topography at the nanometer scale and chemistry. *Journal of the American Chemical Society*, *128*(12), 3939–3945.

Saifuddin, N., Raziah, A. Z., & Junizah, A. R. (2013). Carbon nanotubes: A review on structure and their interaction with proteins. *Journal of Chemistry*, *2013*, 676815.

Tang, Z. R., Han, B., Han, C., & Xu, Y. J. (1991). Helical microtubules of graphitic carbon. *Nature*, *354*, 56–58.

Wei, B. Q., Vajtai, R., & Ajayan, P. M. (2001). Reliability and current carrying capacity of carbon nanotubes. *Applied Physics Letters*, *79*(20), 1172–1174.

Yu, M.-f., et al. (2000). Strength and breaking mechanism of multiwalled carbon nanotubes under tensile load. *287*, 637–641.

Zhao, J., Park, H., Han, J., & Lu, J. P. (2004). Electronic properties of carbon nanotubes with covalent sidewall functionalization. *Journal of Physical Chemistry B*, *108*(14), 4227–4230.

Zhao, X., et al. (2010). New insights into the behavior of bovine serum albumin adsorbed onto carbon nanotubes: Comprehensive spectroscopic studies. *Journal of Physical Chemistry B*, *114*(16), 5625–5631.

Zore, O. V., Kasi, R. M., & Kumar, C. V. (2017). Armored enzyme–nanohybrids and their catalytic function under challenging conditions. *Methods in Enzymology*, *590*, 1st ed. Elsevier Inc. 169–192. https://doi.org/10.1016/bs.mie.2017.02.007.

Zore, O. V., et al. (2015). Bienzyme-polymer-graphene oxide quaternary hybrid biocatalysts: Efficient substrate channeling under chemically and thermally denaturing conditions. *ACS Catalysis*, *5*(9), 4979–4988.

CHAPTER EIGHTEEN

Stabilization of phytase on multi-walled carbon nanotubes via covalent immobilization

Mohammad Pooya Naghshbandi, Hamid Moghimi*
Department of Microbial Biotechnology, School of Biology, College of Science, University of Tehran, Tehran, Iran
*Corresponding author: e-mail address: hmoghimi@ut.ac.ir

Contents

Abstract

This chapter provides one of the covalent methods to immobilize phytase, the most important feed enzyme, on Multi-walled carbon nanotube in order to improve enzyme characteristics. Various techniques such as scanning electron microscopy, FTIR and CD spectroscopy, along with zeta potential measurement were performed to confirm immobilization. Catalytic activity measurement of phytase was explained in details and enzyme loading efficiency was measured. The results indicated a 38% decrease in enzyme activity after immobilization which could because of structural changes or wrong enzyme orientation after immobilization. Also, optimum temperature and pH of free and immobilized enzyme were performed. The remaining activity of free and immobilized enzymes was also investigated after treating with higher temperatures using in the pelleting process. Immobilized phytase showed 10-fold resistance against 90 °C temperature in comparison with free phytase. Finally, the effect of NaCl on the catalytic activity of free and immobilized phytase was examined and showed interesting results. We found that the catalytic activity of immobilized could increase by about 80% in the presence of 0.2 M NaCl.

Methods in Enzymology, Volume 630
ISSN 0076-6879
https://doi.org/10.1016/bs.mie.2019.10.013

Abbreviations

3,4-POD protocatechuate 3,4-dioxygenase
AFM atomic force microscopy
AMG amyloglucosidase
BPP β-propeller phosphatase
BSA bovine serum albumin
CD circular dichroism
C-MWNT carboxylated multi-walled carbon nanotube
CNT carbon nanotube
FTIR Fourier-transform infrared
HAP histidine acid phytases
MWNT multi-walled carbon nanotube
OPH organophosphorus hydrolase
PAP purple acid phosphatase
PTP protein tyrosine phosphatase
SEM scanning electron microscopy
SWNT single-walled carbon nanotube
TEM transmission electron microscopy

1. Introduction

1.1 Phytase: Most important feed enzyme

Myo-inositol hexakisphosphate is known as phytic acid (phytate), is the storage form of phosphorus in plants (70–80% in seeds). Myo-inositol ring carrying six phosphates group and represent high negative charge in the pH of the digestive tract of monogastric animals (Kumar et al., 2015). The non-ruminants are incapable of degrading phytic acid, therefore in addition to reduce phosphorous bioavailability, phytate could chelate essential cations such as Fe^{2+}, Mg^{2+}, Ca^{2+}, Cu^{2+} and Zn^{2+} which makes the phytate an anti-nutrient component. Additionally, undigested phytate could enrich waters and lead to algal growth and eutrophication (Kumar et al., 2019).

In order to overcome the problems as mentioned earlier, Phytase enzymes are added to the animal diets. Phytase enzymes are capable of removing the phosphates from phytate. This enzyme derived from plants, animals and microorganisms. Among them, most scientific and industrial work has been conducted on microbial phytases, which showed 40% bio-efficiency more than plant-derived phytases (Bhavsar & Khire, 2014).

Phytase can be classified based on:

- The number of first dephosphorylated carbon in myo-inositol ring including carbon 3, 5 and 6 leading to subgroups of 3-phytases, 5-phytases and 6-phytases, respectively.
- Catalytic mechanisms, including four subgroups of Histidine acid phytases (HAP), purple acid phosphatase (PAP), β-propeller phosphatase (BPP), and protein tyrosine phosphatase (PTP).
- Optimal pH of activity which divided above subgroups into acid phytases (HAP, PAP and PTP) and alkaline phytases (BPP) (Joshi, 2014).

With the annual market of $500–700 million (2015–2017), Phytase dominates the market of feed enzymes. Thus, characteristics improvement of commercial phytases such as high affinity to phytate, optimum pH range of activity, and thermostability along with cheap production cost are the main desires for the final product (Sanchez & Demain, 2017; Ushasree, Jaiswal, Shyam, & Pandey, 2019). For instance, phytases should maintain their activity at 80–90 °C for 3–5 min in the pelleting process which is a standard feed treatment for poultry diets leading to easier consumption (Naghshbandi, Moghimi, & Latif, 2018). There are some development strategies in order to reach ideal Phytase, including protein engineering, immobilization, chemical modification, and pegylation. Among them, immobilization, a technical process to fixed on or within solid supports, is one of the most used strategies (Silva, Martins, Jing, Fu, & Cavaco-Paulo, 2018).

1.2 Application of nanobiotechnology in enzyme immobilization

According to Majeed et al. "Nanobiotechnology exclusively is referred to the application of nanomaterials in biotechnology" (Majeed, Bhatti, Nawaz, & Kashif, 2019). Among all applications of nanomaterials in biotechnology, protein immobilization is one of the main interests due to extensive use in a variety of industries and pharmaceutical science (Elnashar, 2010). For instance, immobilized enzymes are widely using in biosensor industries in order to generate medical diagnostic kits and detection of heavy metals in water or toxins in food. Also, drug delivery is one of the main application of immobilized protein in medicine and pharmaceutical science (Sirisha, Jain, & Jain, 2016).

Despite all extraordinary features, such as work on mild temperature, pH, and pressure condition, substrate specificity, ease of production, and biocompatibility, enzymes have some disadvantages which should be considered (Cipolatti et al., 2014). Harsh environmental conditions, e.g., high temperatures, acidic or basic pHs and presence of organic solvents could cause enzymes inactivation. Furthermore, low operational stability and shelf life along with the problems with their recovery and reusability are other disadvantages of enzymes which may make it difficult to use them (Singh, Negi, Dubey, Kumar, & Verma, 2018).

Immobilization techniques are one of the main candidates to countering enzyme disadvantages via localizing them physically in a particular region of space leading to movement limitation and specific structural alterations. Enzyme fixation may allow the properties changes of enzymes to become more stable and resistant to various environmental conditions with an extended lifetime (Naqash, Masoodi, & Rather, 2019). Different classifications of immobilization techniques are provided. Covalent Binding, adsorption, entrapment, cross-linking, encapsulation, ionic Bonding and affinity bonding are conventional enzyme immobilization methods. In general, these enzyme immobilization methods could be classified into chemical and physical techniques (Singh et al., 2018). These methods along with the characteristics of the supporting materials, e.g., hydrophilicity, low production cost, biocompatibility and resistance to harsh physiochemical conditions are most features affecting the immobilized enzyme performance (Ahmad & Sardar, 2015).

1.3 Carbon nanotubes: Extraordinary support for immobilization

As mentioned in the above section, the characteristics of the supporting materials are essential for enzyme immobilization and depend on the desired properties of the final product. Among various nanomaterials (e.g., nanoparticles, nanosheets, nanocomposites, and nonporous materials), carbon nanotubes (CNTs) are one of the promising candidates for immobilization of enzymes due to their biocompatibility as well as remarkable thermal, optical and mechanical characteristics (Naghshbandi et al., 2018). Furthermore, large surface-to-volume area allows to higher enzyme loading yield on carbon nanotubes (Yang, Denno, Pyakurel, & Venton, 2015). CNTs are rolled up graphitic sheets in a cylindrical shape with lengths in the micrometers, and diameters up to 100 nm. There are two types of CNTs, including single-walled (SWNT) and multi-walled (MWNT) carbon nanotubes (Yang et al., 2015). Both SWNT and MWNT have been used for enzyme immobilization. Although, SWNTs (central tubule without the graphitic layer) have larger surface area leading to higher enzyme loading yield, nevertheless higher dispersibility and lower production cost makes MWNTs more desirable for enzyme immobilization (Feng & Ji, 2011).

There are several methods to immobilize proteins on CNTs which divided into non-covalent and covalent attachments. Non-covalent attachments could be performed by direct physical adsorption of protein on CNTs or by the assistance of biomolecules, polymers and surfactants. In another hand, covalent attachments could be achieved by direct covalent linking of protein onto CNTs or by the help of linker molecules (Saifuddin, Raziah, & Junizah, 2012). By non-covalent immobilization, molecular structure and protein activity maintain after immobilization while in most cases, protein activity, especially the catalytic activity of enzymes were decreased after immobilization due to conformational changes in enzyme molecule. An also higher amount of loaded enzyme on CNTs is another advantage of non-covalent immobilization method. In another hand, non-covalently immobilized enzymes release and separates from CNTs in an aquatic environment. Maintaining the conjugation between enzyme and CNTs are essential to achieve nano-biocatalysts features (Nagaraju, Reddy, & Reddy, 2015; Naghshbandi et al., 2018). Therefore, in order to maintain the conjugation, covalent bonding is preferred. Also, the stability of enzyme molecular structure increases due to strong attachment between enzyme and CNTs, leading to a decrease in the natural enzyme denaturation during the time (Feng & Ji, 2011).

In this manner, several enzyme immobilization studies were performed on both SWNTs and MWNTs for various purposes. For instance, Protocatechuate 3,4-dioxygenase (Das, Hamid, & Annuar, 2016), cellulase (Mubarak et al., 2014), phytase (Naghshbandi et al., 2018), and laccase (Othman, González-Domínguez, Sanromán, Correa-Duarte, & Moldes, 2016) were covalently immobilized on MWNTs to improve their physiochemical and catalytic characteristics. Also, alkaline lipase was covalently immobilized on different modified MWNTs (MWNT-OH, MWNT-COOH and MWNT-NH_2) by Zniszczoł et al. (2016). They found different features for three versions of immobilized lipase and stated that storage stability was maintained for all MWNT-based nano-biocatalysts or even "extra-enhanced" for MWNT-OH (Zniszczoł et al., 2016). Another lipase enzyme was immobilized on amino-functionalized MWNTs and showed thermal stability improvement (Verma, Naebe, Barrow, & Puri, 2013). In 2010, Pedrosa et al. compared covalently immobilized organophosphorus hydrolase (OPH) on SWNTs and MWNTs. They argued that SWNT-OPH activity was higher than MWNT-OPH (Pedrosa et al., 2010). Likewise, amyloglucosidase (AMG) was immobilized (covalently and physically) on SWNTs and MWNTs (Cang-Rong & Pastorin, 2009).

Despite extraordinary features, there is a concern about the toxicity of CNTs in biological systems. There are several in vitro and in vivo studies on CNTs which claimed that various factors such as concentration, exposure duration and method of exposure could affect the toxicity. CNTs (even functionalized) showed the ability to cross the cell membrane and concentrate in the cytoplasm and even reach the nucleus. However, the dangerous range of concentration is varied from a study to another. These variations seem to arise mostly due to differences in the experimental protocol; most aspects of CNTs toxicity remain unclear (Francis & Devasena, 2018). Moreover, the optimum amount of phytase using in poultry feed is lower than dangerous ranges.

This chapter provides one of the covalent method procedure of phytase immobilization on multi-walled carbon nanotubes to achieve desired characteristics such as improvement in higher temperatures resistant in order to use in the pelleting process. Moreover, other enzymatic characteristics of free and immobilized phytase enzyme were examined and compared to each other. Finally, all process were provided in details.

2. Equipment, materials and procedures

General and safety equipment

1. Volumetric micropipettes (20, 200, and 1000 μL)
2. pH meter
3. Deionized water
4. UV–visible spectrophotometer
5. Stir/hot plate
6. Bench-top centrifuge
7. Goggles
8. Lab coat
9. Latex gloves
10. Ventilation hood

2.1 Modification of carboxylated-MWNT and phytase immobilization

Covalent immobilization of phytase on carboxylated-MWNT was carried out by the following manipulation: formation of an amine-reactive *O*-acylisourea intermediate via reacting MWNT–COOH first with EDC which subsequently reacts with an amine group on the surface of the protein to produce a stable amine bond (Gao & Kyratzis, 2008). Since the formed intermediate is unstable, increasing the coupling efficiency is carried out by the addition of NHS leading stabilizes the intermediate by converting it to a semistable amine-reactive NHS ester (Fig. 1).

Equipment
- Test tube 20 cm
- Ultrasonic bath
- Vortex
- Hydrophilized PTFE 0.45 μm
- Shaker

Materials
- MWNT 2 mg
- Deionized water
- MES buffer 50 and 500 mM at pH 6.1
- *N*-hydroxysuccinimide (NHS) 50 mg/mL
- 1-Ethyl-3-[3-dimethylaminopropyl] carbodiimide hydrochloride (EDC) 10 mg/mL
- Phytase solution

Procedure
1. Weight 2 mg carboxylated carbon nanotubes.
2. Add 5 mL deionized water to the MWNTs and leave at an ultrasonic bath for 1 min.
3. Add 1 mL MES 500 mM and 2.3 mL NHS 50 mg/mL then leave at an ultrasonic bath for 10 min.
4. Under fast stirring, add 1.2 mL of 10 mg/mL fresh EDC aqueous solution and continually stir the mixture (400 rpm) at room temperature for 30 min.
5. Use 0.45 mm hydrophilized PTFE membrane to filter the above mixture and remove excess EDC and NHS.
6. Wash trapped nanotubes twice with a 50 mM MES buffer solution at pH 6.1.
7. Collect trapped nanotubes in a test tube and add 9.0 mL of a 50 mM MES buffer (pH 6.1) solution then leave at an ultrasonic bath for 45 s to re-disperse activated MWNTs.
8. Quickly add 1.0 mL of different concentrations of phytase (i.e., 57–1800 mg/L) in MES buffer solution to above suspension to achieve the optimal loading rate.
9. Vortex the suspension for 1 h at room temperature.

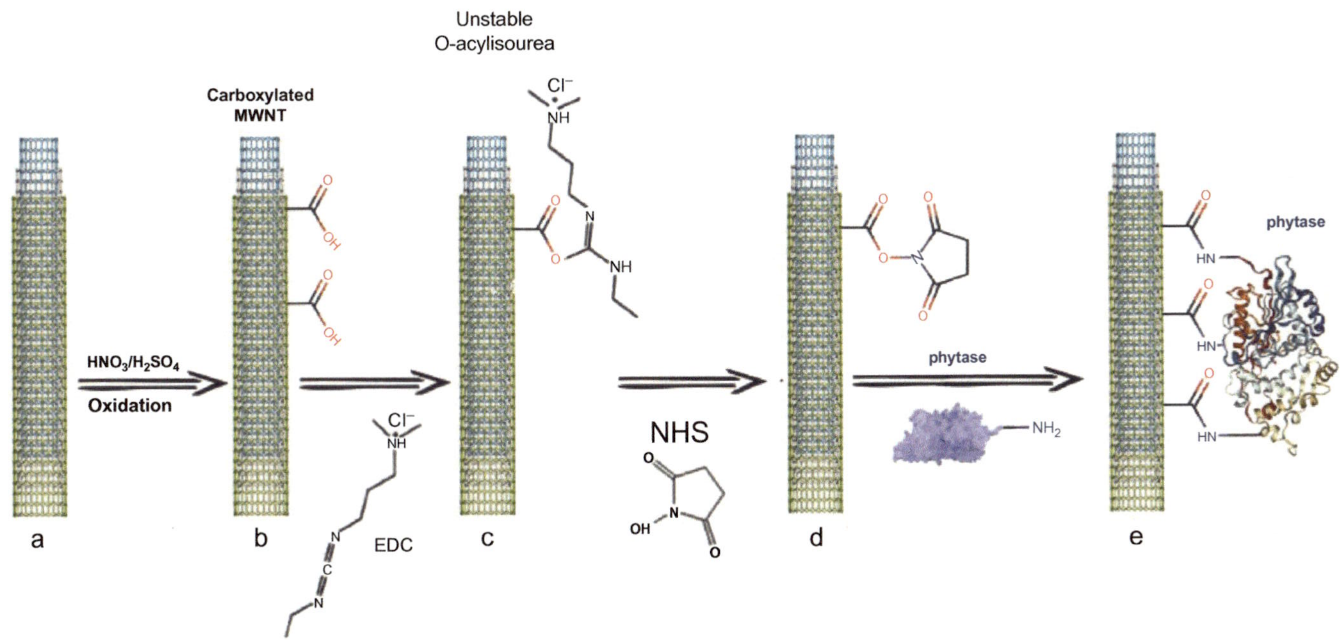

Fig. 1 Schematic view of MWNT modification and phytase covalent attachment.

10. Centrifuge and wash immobilized enzymes with 50 mM MES buffer solution (pH 6.1) to remove unbound phytase molecules (three times).
11. Disperse the immobilized phytases in 1 mL of a 50 mM MES buffer solution (pH 6.1) for analytical measurements.

It worth to mention that solubility of modified and unmodified MWNTs in water (or MES buffer) is extremely low and nanotubes aggregate even after sonication. However, after phytase immobilization, solubility improvement was observed as in Fig. 2.

2.2 Free and immobilized phytase activity assay

Activity determination of phytase is often based on colorimetric measurement of released phosphate from phytic acid (the ammonium molybdate method described by Heinohen and Lathi). Therefore, in order to determination of free and immobilized phytase, 50 μL enzymes were added to 450 μL phytic acid. Then a color reagent solution was added to react with released phosphates in the solution. Finally, one phytase unit (FTU) is defined as the amount of enzyme that liberates one μmol of inorganic phosphorus per min from 5 mM sodium phytate under standard

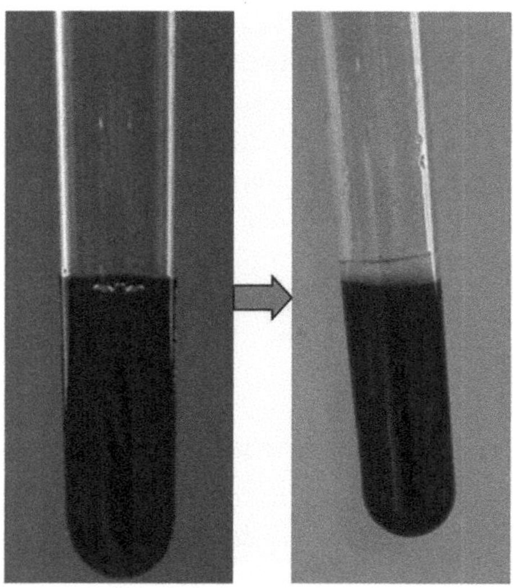

Fig. 2 Dispersibility of modified MWNT (left) and immobilized phytase (right).

conditions at 37 °C for 30 min. Comparing the catalytic activity of free and immobilized phytase provides a perspective into the performance of immobilized phytase.

Equipment

- Test tubes (10 cm)
- Water bath
- Pipet
- Spectrophotometer (UV–Vis)

Materials

- Phytic acid 5 mM
- KH_2PO_4 100 mM
- Sodium acetate acetic acid buffer (pH 5.5)
- Ammonium hepta-molybdate (5%, w/v)
- Acetone
- Sulfuric acid 5 Normal

Procedure

1. Prepare 5 mM sodium phytate in Sodium acetate acetic acid buffer (pH 5.5).
2. Prepare 2 test tubes including 450 μL of the above solution and leave them in a water bath (37 °C) for 5 min.
3. Add 50 μL distilled water to test tube one as blank and 50 μL of free/immobilized phytase to test tube 2.
4. Leave the reaction in a water bath for 30 min.
5. Immediately add 4 mL CRS (pure acetone: 5 N H_2SO_4: 10 mM ammonium molybdate with the ratio of 2:1:1, v/v/v) to each tube.
6. Measure the generated yellow color via spectrophotometer at 400 nm.
7. Compare the achieved results with a standard curve prepared with a range of 0–100 μmol potassium phosphate (KH_2PO_4) solution.

For providing reliable results, all catalytic activities were provided as specific activity (U/g). The primary results showed 62% remaining in phytase activity after immobilization.

2.3 Characterization of the immobilized enzyme

The biochemical and structural characterization of the resulting nanobiocatalysts is an essential step for the development of enzyme immobilization on carbon nanotubes. For instance, different microscopic methods such as transmission electron microscopy (TEM), scanning electron microscopy (SEM), and Atomic force microscopy (AFM), could be used to confirm the successful attachment of the enzyme on the surface of carbon nanotubes.

Also, changes in the protein secondary structures of phytase during the immobilization procedure could be determined via circular dichroism (CD) spectroscopy. Moreover, Fourier-transform infrared (FTIR) spectroscopy can be applied for detection of functional groups on the surface of C-MWNTs and immobilized phytase. Comparison of achieved peaks from their spectra could provide successful immobilization proof. In another hand, the closeness of surface charge (zeta potential) of free and immobilized enzyme compared to C-MWNT could be another proof of successful immobilization. Finally, one of the most critical characteristics of an immobilized enzyme is immobilization yield and efficiency, which could be carried out by protein quantification assay, i.e., Bradford assay.

2.3.1 Determination of nanobiocatalyst via scanning electron microscopy

Fig. 3 shows the images of the modified MWNTs and immobilized phytase captured via a scanning electron microscope (SEM). The bright spots on the surface of MWNTs are presents the immobilized phytase molecules on MWNTs.

Equipment

– Scanning electron microscope
– Test tubes (10 cm)
– Centrifuge

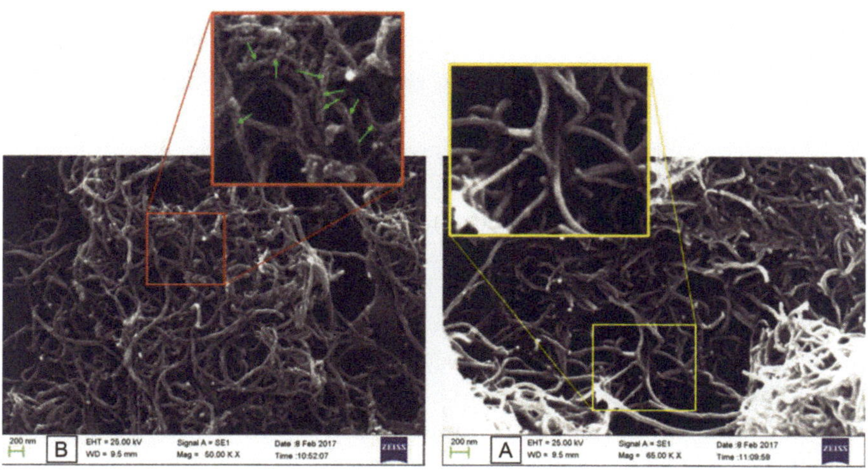

Fig. 3 Scanning electron microscopy of functionalized MWNTs (A) and phytase-MWNTs composites (B). Green arrows on part (B) are mentioning the bright spots (phytase molecules) on CNTs.

Materials
- Glutaraldehyde
- Phosphate buffer
- Osmium tetroxide
- Ethanol

Procedure
1. Fixation: Immerse immobilized phytase sample in 2% glutaraldehyde buffered with 0.1 M phosphate buffer (pH 7.2) at room temperature for 2–24 h.
2. Washing: Centrifuge and wash (three times for 10 min) the samples with 0.1 M phosphate buffer (pH 7.2).
3. Post-fixation: Immerse sample in 1–2% osmium tetroxide buffered with 0.1 M phosphate buffer (pH 7.2) at room temperature for 1–2 h.
4. Washing: Centrifuge and wash (three times for 10 min) the samples with distilled water.
5. Dehydration: prepare various concentration of ethanol to dehydrate of the samples in the following steps:
   - 25% Ethanol: 1 × 15 min
   - 50% Ethanol: 1 × 15 min
   - 75% Ethanol: 1 × 15 min
   - 90% Ethanol: 2 × 15 min
   - 100% Ethanol: 3 × 20 min
6. Leave the samples to expose to air until the alcohol evaporate.

2.3.2 Analyzing of the immobilized enzyme on C-MWNT via FTIR spectroscopy

FTIR spectra were conducted using a Bruker, vector 22 (Ettlingen, Germany) at room temperature with a wavenumber resolution of 1 cm^{-1} using KBr pellets in the frequency range of 4000–400 cm^{-1}. Also, to obtain a good signal-to-noise ratio, 32 scans were run and averaged. Fig. 4 shows the FTIR spectra of modified MWNT and immobilized phytase.

The important peaks which should be considered to confirm the successful immobilization are the following peaks at Table 1 and Fig. 4. For example, Stretching vibration at 1737 is detectable on both spectra which is one of the successful immobilization confirmation. Also, the presence of aliphatic amide bond (C—N) peaks at 1045 and 1186 of immobilized phytase spectra is another reason for successful immobilization.

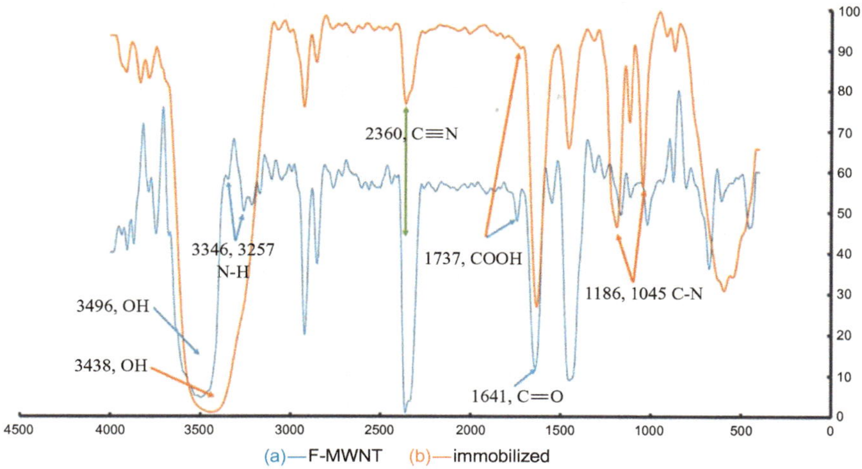

Fig. 4 FTIR spectra of functionalized MWNTs (B) and immobilized phytase (A).

Table 1 Significant peaks in spectra of modified MWNTs and immobilized phytase.

Peak (nm)	Functional group	Task
Presence of 3438, 1641 and 1737	(–OH), (–C=O) and (–COOH)	Complete oxidation of MWNTs
Stretching vibration at 1737	(–COOH)	Attachment of the enzyme molecules on the nanotubes
Presence of 3346 and 3257	(N—H)	Presence of phytase on the surface of modified MWNTs
Presence of 1045 and 1186	aliphatic amide bond (C—N)	Amidination reaction shows immobilization of phytase successfully

2.3.3 Characterization of the free and immobilized enzyme via CD spectroscopy

The UV–CD (Circular dichroism) spectrometer model-215 Aviv (New Jersey, USA) with a 10 mm cell length and bandwidth of 1 nm were used to determination of secondary structural changes of phytase after immobilization. Test solution containing 0.2 mg/mL of either free or immobilized phytase in MES buffer (50 mM, pH 6.1) was prepared and recorded over the range of 190–260 nm. MES buffer and C-MWNTs dispersed in MES buffer were used as control solutions. CDNN software (Magdeburg, Germany) was used for data analysis. Table 2 shows the difference between the percentages of free and immobilized secondary structures which indicates significant

Table 2 The percentages of secondary structure contents of free and immobilized phytase on MWNT.

	Free phytase (%)	Immobilized phytase (%)
α-Helix	35.9	20.9
β-Sheet antiparallel	7.3	11.7
β-Sheet parallel	8.3	14.1
β-Turn and random coil	13.7	14.2

Table 3 Zeta potential of free and immobilized enzymes on the support material.

Sample case 1	Zeta potential	Sample	Zeta potential	Sample case 2
Free phytase	−13.6	Free enzyme	−8.2	Candida antarctica A-type lipase (CALA)
Immobilized phytase	−12.2	Immobilized enzyme	−9.5	Immobilized CALA
Carboxylate MWNT	−7.6	Nanomaterial	−11.4	nonporous magnetic microparticles

changes in phytase secondary structure. These changes might be one of the reasons for a 38% decrease in catalytic activity after immobilization.

2.3.4 Zeta potential characterization of free and immobilized enzyme

Zetasizer Nano ZS Malvern instrument (Malvern, UK) was used to measure the free and immobilized phytase as well as C-MWNT charges. Since there is a clear correlation between zeta potential and loading efficiency of the enzyme through the immobilization process, thus zeta potential could be used as a diagnostic tool in enzyme immobilization (Schultz, Metreveli, Franzreb, Frimmel, & Syldatk, 2008). Table 3 shows two case studies of different enzyme immobilization to indicate the closeness of the free and immobilized zeta potentials in compare to nano-supports. Zeta potential closeness of free and immobilized enzymes and their difference with a zeta potential of nano-support confirms the successful immobilization.

2.3.5 Determination of enzyme loading yield and efficiency

Loading efficiency and immobilization yield are two different definitions which sometimes considered to be the same. According to Eq. (1), loading efficiency (%) is the ratio of protein content after and before immobilization.

Loading efficiency shows how much protein could be immobilized on CNTs. The rate of loading efficiency increases as the initial protein content increases until the CNTs have no longer available places for proteins. It should be mention that increasing the loading efficiency might cause a decrease in enzyme activity due to a limitation in the mobility of enzymes. In another hand, according to Eq. (2), immobilization yield (%) is the ratio of specific catalytic activity after and before immobilization.

$$\text{Loading efficiency}(\%) = \frac{\text{Immobilized enzyme protein content}\left(\frac{\mu g}{ml}\right)}{\text{Free enzyme protein content}\left(\frac{\mu g}{ml}\right)} \times 100 \quad (1)$$

$$\text{Immobilization yield}(\%) = \frac{\text{Specific activity of Immobilized Enzyme}\left(\frac{U}{g}\right)}{\text{Specific activity of Free Enzyme}\left(\frac{U}{g}\right)} \times 100 \quad (2)$$

Procedure

For determination of protein content, 100 μL of free/immobilized phytase in MES buffer was mixed in 900 μL of deionized water and then 3 mL of Bradford reagent was added to each sample. A sample containing 1 mL of deionized water was also used as a reference. The samples optical absorption were measured at 595 nm. Finally, the achieved results replaced in standard curve prepared with a range of 0–100 μg/mL bovine serum albumin (BSA). After determining the concentration of the initial phytase solution, in order to repeat the experiment and to obtain enzyme immobilization yield at different concentrations, certain concentrations were determined and the solutions were diluted by 50 mM MES buffer at pH 6.1. All diluted phytases (57–1800 μg/mL on 2 mg MWNTs) were used to measure loading efficiency and immobilization yield.

Fig. 5A shows 110 μg per mg of modified MWNTs as the highest loading efficiency recorded. Same as phytase, soybean peroxidase (SBP) and Subtilisin Carlsberg (SC), showed approximately 100–200 μg per mg of MWNTs as highest loading efficiency (Asuri et al., 2006). Also, this number for amyloglucosidase was 274 μg per mg functionalized MWNTs (Cang-Rong & Pastorin, 2009). On the contrary, some studies have reported significantly higher loading efficiencies. Fig. 5B confirmed the state as mentioned above that the decrease in enzyme mobility along with the wrong orientation of the immobilized enzyme molecules in high concentrations of the immobilized enzyme could decrease immobilization yield.

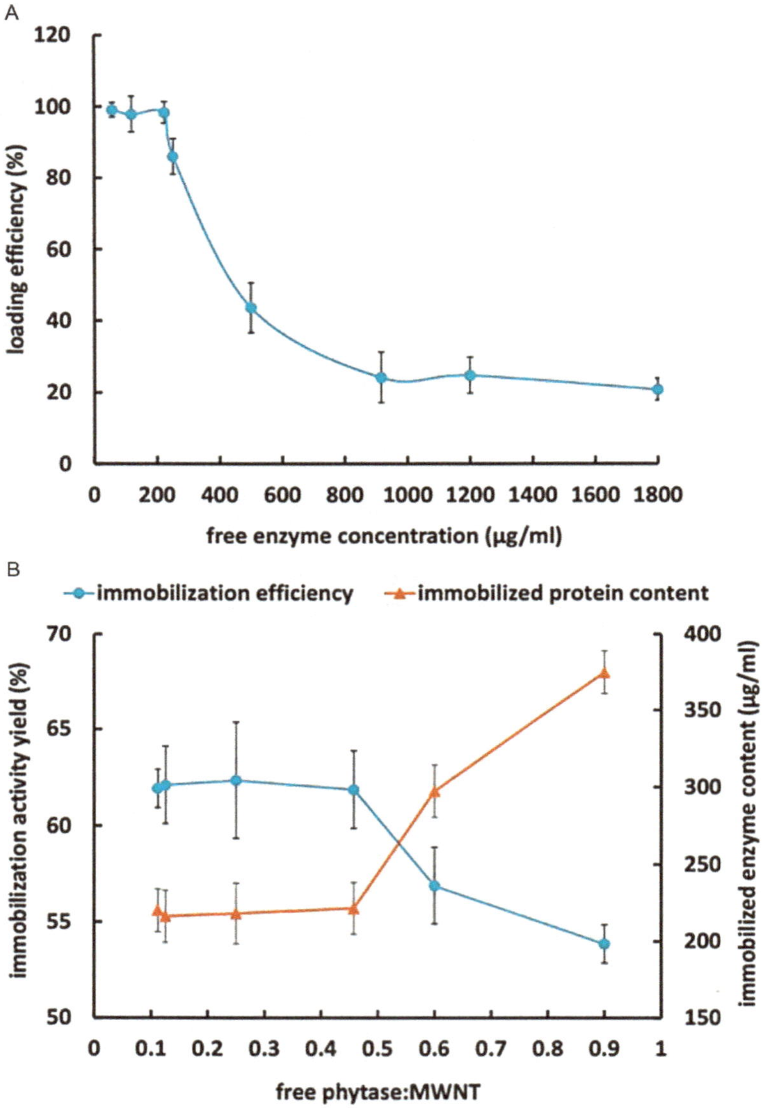

Fig. 5 Loading efficiency of immobilized phytase on F-MWNT (A) and immobilization activity yield (B).

2.4 Physiochemical and stability properties of free and immobilized phytase

There are several studies on different enzymes immobilized on various nanomaterial. Depends on enzyme and support characteristics, physicochemical properties of the immobilized enzyme might be changed.

For instance, immobilized protocatechuate 3,4-dioxygenase (3,4-POD) on MWNTs showed 60 °C as the optimum temperature in comparison with 55 °C for the free enzyme (Das et al., 2016). In another hand, although there are some changes in temperatures around optimum temperature, however, phytase and lipase immobilized on MWNTs showed the same optimum temperature with their free forms (Naghshbandi et al., 2018; Verma et al., 2013).

Different results were obtained for optimum pH of mentioned enzymes, immobilized 3,4-POD on MWNT was increased from 9 to 10 compared to free 3,4-POD and lipase showed the same number for optimum pH of free and immobilized enzyme. Impressive results found when optimum pH of phytase was shifted from 5.5 to 6.5 and 8 to 10 after immobilization on MWNT and magnesium nanoparticle respectively. However, immobilized phytase on spores of Bacillus polyfermenticus showed the same pH for free and immobilized phytase (Cho, Kim, & Pan, 2011). These findings confirm that the immobilized enzyme shows different characteristics of the different support material.

Despite different results on the optimum temperature, all different enzymes, i.e., phytase, lipase, 3,4-POD and laccase showed thermal stability against high temperatures after immobilization on carbon nanotubes.

It should be mentioned that NaCl could affect catalytic enzyme activity. For example, at specific concentrations, NaCl exhibited a positive effect on the activity of dihydrofolate reductases, HIV-1 protease, and menadione reductase as salt in the compound by inhibition of incorrect refolding. Free halophilic esterase LipC and immobilized phytase on MWNT were showed an extraordinary increase in their catalytic activity at specific concentrations of NaCl (Mozhaev, 1993; Rao et al., 2009; Zhao, 2005).

2.4.1 Determination of optimum temperature

Procedure

The optimum temperature of immobilized and free phytase was determined by carrying out the standard enzyme assay, but at different temperatures, i.e., 30, 40, 50, 55, 60, 65, and 70 °C. The substrate and enzyme were incubated at the mentioned temperatures for 30 min. Color reagent solution was then added and absorbance was measured at 400 nm. Relative enzyme activities at a selected range of temperature were measured under standard pH conditions (pH 5.5). Thus the higher activity considered 100% for relative activity and others were measured against it.

Both free and immobilized phytases showed the optimum temperature at 55 °C. However, the relative activity of immobilized phytase was still high until 65 °C, while the relative activity of the free enzyme showed a sharp downward trend after 55 °C.

2.4.2 Determination of optimum pH

Procedure

For determining optimum pH of free and immobilized phytases the enzymes along with sodium phytate incubated in different buffers, i.e., glycine-HCl (pH 2.0–3.5), sodium acetate acetic acid (pH 4.0–6.0), Tris-HCl (pH 6.5–8.0) and glycine-NaOH (pH 8.6–10.6). Relative enzyme activities at a selected range of pH were measured under standard conditions (37 °C for 30 min).

The results showed an alkali shift along with a broader range of optimum pH from 4.5 to 6.5 for immobilized phytase while optimum pH for free phytase was investigated at 5.5.

2.4.3 Investigation of high-temperature stability

Procedure

The free and immobilized phytases were separately prepared at equal amounts and concentrate in PCR vials and then incubated in thermo-cycler at 60, 70, 80 and 90 °C for 1–5 min to investigate their resistant to higher temperatures. The samples were then rested on ice for 10 min in order to bring down their temperatures before further analyses. For measuring the relative activity of phytases, the activity of non-treated enzymes was considered as the highest activity (100%). Residual enzyme activity of all treated and non-treated enzymes was measured under standard conditions (37 °C and pH 5.5 for 30 min).

A significant improvement in temperature resistant was achieved, especially at 80 and 90 °C where after 2 and 3 min treatment at 90 and 80 °C (respectively), the relative activity of free phytase reached to almost zero while immobilized phytase remains 40% activity.

2.4.4 Investigation of enzyme activity in the presence of NaCl

Procedure

Different concentrations of NaCl (from 0.05 to 3 M) were prepared in assay tubes incubated at standard conditions (37 °C and pH 5.5 for 30 min). Then 50 μL of phytase enzyme either in free or immobilized form was added to test tubes and their activity were measured. Also, phytase activity without the presence of NaCl was determined and stated as a reference activity.

The results showed that the presence of 1 M NaCl could increase free phytase activity up to 30%, while the presence of 0.2 and 0.5 M NaCl improved immobilized phytase up to 80% and 60%, respectively.

3. Conclusion

Herein, we present the use of carboxylated multi-walled carbon nanotube as nano-support for immobilization of phytase enzyme. Due to strong attachments, covalent binding via diimide-activated amidation is proposed. The detailed protocol and procedures presented to modify the MWNTs and phytase immobilization. The enzyme loading efficiency and immobilization yield were indicated by examining the protein content and catalytic activity of phytase before and after immobilization. Also, optimum temperature and pH of free and immobilized phytase along with their resistant to high temperatures were presented. Eventually, the procedure for detecting the NaCl effect on free and immobilize phytase activity was explained. The developed enzyme and nano-sûpport conjugates indicate a significant enhancement of the enzymes physiochemical characteristics in comparison with the free enzyme form. Generating efficiently immobilized phytase can contribute substantially to use in the combined feed production process, especially pelleting in high temperatures.

References

Ahmad, R., & Sardar, M. (2015). Enzyme immobilization: An overview on nanoparticles as immobilization matrix. *Biochemistry and Analytical Biochemistry*, *4*(2), 1.

Asuri, P., Karajanagi, S. S., Sellitto, E., Kim, D. Y., Kane, R. S., & Dordick, J. S. (2006). Water-soluble carbon nanotube-enzyme conjugates as functional biocatalytic formulations. *Biotechnology and Bioengineering*, *95*(5), 804–811.

Bhavsar, K., & Khire, J. (2014). Current research and future perspectives of phytase bioprocessing. *RSC Advances*, *4*(51), 26677–26691.

Cang-Rong, J. T., & Pastorin, G. (2009). The influence of carbon nanotubes on enzyme activity and structure: Investigation of different immobilization procedures through enzyme kinetics and circular dichroism studies. *Nanotechnology*, *20*(25) 255102.

Cho, E.-A., Kim, E.-J., & Pan, J.-G. (2011). Adsorption immobilization of Escherichia coli phytase on probiotic *Bacillus polyfermenticus* spores. *Enzyme and Microbial Technology*, *49*(1), 66–71.

Cipolatti, E. P., Silva, M. J. A., Klein, M., Feddern, V., Feltes, M. M. C., Oliveira, J. V., et al. (2014). Current status and trends in enzymatic nanoimmobilization. *Journal of Molecular Catalysis B: Enzymatic*, *99*, 56–67.

Das, R., Hamid, S. B. A., & Annuar, M. S. M. (2016). Highly efficient and stable novel nanobiohybrid catalyst to avert 3, 4-dihydroxybenzoic acid pollutant in water. *Scientific Reports*, *6*, 33572.
Elnashar, M. M. (2010). Immobilized molecules using biomaterials and nanobiotechnology. *Journal of Biomaterials and Nanobiotechnology*, *1*(1), 61–77.
Feng, W., & Ji, P. (2011). Enzymes immobilized on carbon nanotubes. *Biotechnology Advances*, *29*(6), 889–895.
Francis, A. P., & Devasena, T. (2018). Toxicity of carbon nanotubes: A review. *Toxicology and Industrial Health*, *34*(3), 200–210.
Gao, Y., & Kyratzis, I. (2008). Covalent immobilization of proteins on carbon nanotubes using the cross-linker 1-ethyl-3-(3-dimethylaminopropyl) carbodiimide—A critical assessment. *Bioconjugate Chemistry*, *19*(10), 1945–1950.
Joshi, J. (2014). Phytase—A key to unlock phytate complex. *International Journal of Pure and Applied Bioscience*, *2*, 304–313.
Kumar, V., Singh, D., Sangwan, P., & Gill, P. K. (2015). Management of environmental phosphorus pollution using phytases: Current challenges and future prospects. In *Applied environmental biotechnology: Present scenario and future trends* (pp. 97–114). Springer.
Kumar, V., Sinha, A. K., & Kajbaf, K. (2019). Phytic acid and phytase enzyme. In *Whole grains and their bioactives: Composition and health* (pp. 467–483): Wiley Online Library.
Majeed, M. I., Bhatti, H. N., Nawaz, H., & Kashif, M. (2019). Nanobiotechnology: Applications of nanomaterials in biological research. In *Integrating green chemistry and sustainable engineering* (pp. 581–615): Wiley Online Library.
Mozhaev, V. V. (1993). Mechanism-based strategies for protein thermostabilization. *Trends in Biotechnology*, *11*(3), 88–95.
Mubarak, N., Wong, J., Tan, K., Sahu, J., Abdullah, E., Jayakumar, N., et al. (2014). Immobilization of cellulase enzyme on functionalized multiwall carbon nanotubes. *Journal of Molecular Catalysis B: Enzymatic*, *107*, 124–131.
Nagaraju, K., Reddy, R., & Reddy, N. (2015). A review on protein functionalized carbon nanotubes. *Journal of Applied Biomaterials & Functional Materials*, *13*(4), 301–312.
Naghshbandi, M. P., Moghimi, H., & Latif, B. (2018). Covalent immobilization of phytase on the multi-walled carbon nanotubes via diimide-activated amidation: Structural and stability study. *Artificial cells, Nanomedicine, and Biotechnology*, *46*(sup1), 763–772.
Naqash, F., Masoodi, F., & Rather, S. A. (2019). Food enzymes and nanotechnology. In *Enzymes in food biotechnology* (pp. 769–784): Elsevier.
Othman, A. M., González-Domínguez, E., Sanromán, Á., Correa-Duarte, M., & Moldes, D. (2016). Immobilization of laccase on functionalized multiwalled carbon nanotube membranes and application for dye decolorization. *RSC Advances*, *6*(115), 114690–114697.
Pedrosa, V. A., Paliwal, S., Balasubramanian, S., Nepal, D., Davis, V., Wild, J., et al. (2010). Enhanced stability of enzyme organophosphate hydrolase interfaced on the carbon nanotubes. *Colloids and Surfaces B: Biointerfaces*, 77(1), 69–74.
Rao, L., Zhao, X., Pan, F., Li, Y., Xue, Y., Ma, Y., et al. (2009). Solution behavior and activity of a halophilic esterase under high salt concentration. *PLoS One*, *4*(9), e6980.
Saifuddin, N., Raziah, A., & Junizah, A. (2012). Carbon nanotubes: A review on structure and their interaction with proteins. *Journal of Chemistry*, *2013*, 676815.
Sanchez, S., & Demain, A. L. (2017). Useful microbial enzymes—An introduction. In *Biotechnology of microbial enzymes* (pp. 1–11): Elsevier.
Schultz, N., Metreveli, G., Franzreb, M., Frimmel, F. H., & Syldatk, C. (2008). Zeta potential measurement as a diagnostic tool in enzyme immobilisation. *Colloids and Surfaces B: Biointerfaces*, *66*(1), 39–44.
Silva, C., Martins, M., Jing, S., Fu, J., & Cavaco-Paulo, A. (2018). Practical insights on enzyme stabilization. *Critical Reviews in Biotechnology*, *38*(3), 335–350.

Singh, A., Negi, M. S., Dubey, A., Kumar, V., & Verma, A. (2018). Methods of enzyme immobilization and its applications in food industry. In *Enzymes in food technology* (pp. 103–124): Springer.

Sirisha, V., Jain, A., & Jain, A. (2016). Enzyme immobilization: An overview on methods, support material, and applications of immobilized enzymes. *Advances in Food and Nutrition Research*, *79*, 179–211. Elsevier.

Ushasree, M. V., Jaiswal, A. K., Shyam, K., & Pandey, A. (2019). Thermostable phytase in feed and fuel industries. *Bioresource Technology*, *278*, 400–407.

Verma, M. L., Naebe, M., Barrow, C. J., & Puri, M. (2013). Enzyme immobilisation on amino-functionalised multi-walled carbon nanotubes: Structural and biocatalytic characterisation. *PLoS One*, *8*(9), e73642.

Yang, C., Denno, M. E., Pyakurel, P., & Venton, B. J. (2015). Recent trends in carbon nanomaterial-based electrochemical sensors for biomolecules: A review. *Analytica Chimica Acta*, *887*, 17–37.

Zhao, H. (2005). Effect of ions and other compatible solutes on enzyme activity, and its implication for biocatalysis using ionic liquids. *Journal of Molecular Catalysis B: Enzymatic*, *37*(1–6), 16–25.

Zniszczoł, A., Herman, A. P., Szymańska, K., Mrowiec-Białoń, J., Walczak, K. Z., Jarzębski, A., et al. (2016). Covalently immobilized lipase on aminoalkyl-, carboxy- and hydroxy-multi-wall carbon nanotubes in the enantioselective synthesis of Solketal esters. *Enzyme and Microbial Technology*, *87*, 61–69.

CHAPTER NINETEEN

A simple magnetic nanoparticle-poly-enzyme nanobead sandwich assay for direct, ultrasensitive DNA detection

Lorico D.S. Lapitan Jr.[a], Dejian Zhou[b,*]
[a]Department of Chemical Engineering and Advanced Materials Research Laboratory, Research Center for the Natural and Applied Sciences, University of Santo Tomas, Manila, Philippines
[b]School of Chemistry and Astbury Centre for Structural Molecular Biology, University of Leeds, Leeds, United Kingdom
*Corresponding author: e-mail address: d.zhou@leeds.ac.uk

Contents

Abstract

A simple magnetic nanoparticle (MNP)-poly-enzyme nanobead sandwich assay for direct detection of ultralow levels of unlabeled target-DNA is developed. This approach

Methods in Enzymology, Volume 630
ISSN 0076-6879
https://doi.org/10.1016/bs.mie.2019.10.026

uses a capture-DNA covalently linked to a dense PEGylated polymer encapsulated MNP and a biotinylated signal-DNA to sandwich the target-DNA. A DNA ligation is then followed to offer high discrimination between the perfect-match and single-base mismatch target-DNAs. Only the presence of a perfect-match target can covalently link the biotinylated signal-DNA onto the MNP surface for subsequent binding to a polymer nanobead tagged with thousands of copies of high-activity neutravidin-horseradish peroxidase (NAV-HRP) for great enzymatic signal amplification. Combining the advantages of the dense MNP surface PEGylation to reduce non-specific adsorption (assay background) and the powerful signal amplification of poly-enzyme nanobead, this assay can directly quantify the target-DNA down to single digit attomolar with a large linear dynamic range of 5 orders of magnitude (from 10^{-18} to 10^{-13}M).

1. Introduction

Magnetic nanoparticles (MNPs) are one of the most promising nanomaterials in a wide range of biomedical applications. MNPs are widely synthesized by thermal decomposition in high boiling point organic solvents to give hydrophobic particles or by co-precipitation in an aqueous media to give water-dispersible ones (Laurent et al., 2008). The advantages of the former method is that it can offer excellent control over the particle size, shape and uniformity, although the resulting hydrophobic particles must be surface modified to make them water-dispersible prior to any biomedical applications. On the other hand, MNPs synthesized by the later method typically show broader size distributions, but the resulting MNPs are often water-dispersible and compatible to most biomedical applications. Among the many different types of MNPs, iron oxide based nanoparticles, such as magnetite (Fe_3O_4) and maghemite (γ-Fe_2O_3), have attracted significant interests in the broad biomedical and diagnostic areas because of their low-/non-toxicity and excellent biocompatibility (Akbarzadeh, Samiei, & Davaran, 2012; Laurent et al., 2008; Lee & Hyeon, 2012).

There are several highly attractive chemical physical properties which make MNPs particularly well-suited for biological and biomedical research. First, the size of MNPs can be easily tuned and optimized from a few to several hundred nanometers to match the requirement of study interest. Second, their superparamagnetic property makes them easily manipulated by an external magnetic force, allowing for easy sample retrieval from the reaction media and for further re-use. Indeed, the ability of "action at a distance" provides a tremendous advantage in a plethora of applications, such as bacterial detection (Gu, Ho, Tsang, Wang, & Xu, 2003), protein purification (Xie et al., 2010), removal of toxins (Leroux, 2007) and development of reusable biocatalysts (Garcia et al., 2011; Laurent et al., 2008). Third, the magnetic moment of MNPs can enhance the magnetic relaxation signal of nearby protons, allowing MNPs to act as a very useful MRI contrast agent (Na, Song, & Hyeon, 2009). Fourth, the MNPs can transform alternating high-frequency electromagnetic field into heat, providing a promising application in magnetic hyperthermia treatment of cancer and other diseases (Lee et al., 2011). Similar to many other nanomaterials, the nanometer size of MNPs imparts a large surface to volume ratio and a high surface energy, making them to prone to aggregation and/or oxidization in the absence of proper surface coating or stabilizing ligand. This can adversely affect their intrinsic magnetic properties and biomedical applications. In this regard, MNPs must be capped with a protective coating to improve their stability and dispersibility in water. Moreover, the coating can introduce reactive functional groups to facilitate further conjugation to biomolecules and/or targeting ligands. Several different types of MNP surface coating have been reported in the literature. The widely used ones include coating with silica, encapsulation with amphiphilic/block co-polymers, and coating with functional small, polymeric and biological ligands (Gao, Gu, & Xu, 2009; Garcia et al., 2011; Laurent et al., 2008; Singamaneni, Bliznyuk, Binek, & Tsymbal, 2011; Yang, Wu, & Hou, 2011; Zhou, Yuan, & Wei, 2011). Importantly, the large surface area of MNPs make them excellent nanoscale scaffolds for immobilizing functional biomolecules (i.e., enzymes, oligonucleotides, sugars, antibodies) for use as biomedical probes.

Enzymes are excellent biocatalysts that catalyze numerous chemical and biochemical reactions under mild conditions in an aqueous environment. Moreover, enzymes can efficient convert a specific substrate to the desired product with excellent chemo-, regio- and stereo-selectively (Bilal et al., 2017; Bilal, Rasheed, Zhao, Iqbal, & Cui, 2018). Immobilization of enzymes on solid supports such as MNPs opens a wealth of applications in both the scientific (e.g., reaction mechanisms) and commercial (e.g., healthcare and pharmaceutical products) points of view (Bornscheuer, 2003; Bornscheuer et al., 2012). They can offer high convenience in handling, rapid separation, recycle and reuse of the enzyme catalysts. In addition, their high surface area, large surface-to-volume ratio and rapid solution diffusion are some of the desirable characteristics for use as catalyst supports, allowing for high enzyme loading and minimizing diffusion limitations (Xie et al., 2009). There are three important considerations for enzyme immobilization on the MNP surfaces: (i) preserving (or even enhancing) enzyme activity after immobilization, (ii) imparting long term stability, and (iii) reducing non-specific adsorption. It is highly important to preserve the enzyme's native structure and activity after immobilization and they have sufficient stability during storage. This can be done by designing the MNP surfaces to be compatible with the enzymatic reaction and ensuring an environment that will not compromise the enzyme activity (Mitchell et al., 2002).

Enzymes can be immobilized on MNP surfaces via a number of different methods, such as physical adsorption, electrostatic interaction, covalent coupling and bio-specific interactions etc. Among which covalent linking is attractive because it can offer high stability for the enzyme-MNP conjugate. The conjugation efficiency depends strongly on the reactivity between the functional groups (e.g., –SH, $-NH_2$) on the enzyme and the MNP surface. This can be controlled using a buffer with a desired pH for the chemical linking step. Ideally, the enzyme should be chemically linked to the MNP surface with no undesirable side reactions for robust catalytic applications. Here covalent immobilization is preferred because it make highly stable enzyme-MNP conjugates that can withstand the repeated cycles of separation and washing steps without enzyme desorption. Moreover, control MNP surface enzyme loading is another important factor. Too high an enzyme loading (e.g., formation of multilayers) could restrict substrate diffusion to the inner layer enzymes, leading to reduced average turn-over rate per enzyme. This can cause a significant problem for multilayer enzymes immobilized on flat surfaces (Rauf, Zhou, Abell, Klenerman, & Kang, 2006) or large MNPs with small surface curvatures, although it can be mostly resolved by using small MNPs with high surface curvature to improve substrate accessibility (Garcia et al., 2011). Thus, it is important to determine the optimal enzyme loading that can provide high catalytic activity (Zhang, Feng, Guo, & Zhou, 2013; Zhang, Guo, Quirke, & Zhou, 2013; Zhang, Pilapong, et al., 2013). Non-specific adsorption of enzymes on MNPs is very common and must be minimized in order to reduce assay background and hence to improve sensitivity. In this regard, a careful design of the surface chemistry to make it non-sticky to enzymes and other biomolecules is critically important. Some widely used strategies include coating the surface with PEGylated (Prime & Whitesides, 1991; Zhou, Bruckbauer, Ying, Abell, & Klenerman, 2003) or zwitterionic ligands (Holmlin, Chen, Chapman, Takayama, & Whitesides, 2001; Schlenoff, 2014) and/or further blocking with small molecules or bovine serum albumin.

1.1 Enzyme immobilization strategies on magnetic nanoparticles

Enzyme immobilization on MNP surface provides an indispensable tool for separating the catalyst from the sample matrix by use of an external magnetic field. A variety of techniques have been used to prepare enzyme-MNP conjugates, including covalent bonding, physical adsorption, and biospecific interactions, etc. (Ansari & Husain, 2012; Bilal et al., 2018; Zhang & Zhou, 2012). Each method has its own advantages and also drawbacks. For example, physical adsorption is simple to do, but the resulting enzyme-MNP conjugates often lack of long term stability. On the other hand, covalent immobilization gives more robust conjugate, but it is often difficult to control the enzyme orientation to maximize active site access and to maximize the activity of immobilized enzymes (Bornscheuer, 2003). A common method used in covalent immobilization is using a heterobifunctional linker having amine and thiol dual reactive function groups, such as the commercial *N*-hydroxysuccinamide-poly(ethylene glycol)-maleimide (NHS-PEG_{12}-maleimide) shown in Scheme 1 (Kuhn, Finch, Hallahan, & Giorgio, 2007).

The activated NHS ester can readily react with nucleophiles to release the NHS group and form a stable amide bond as shown in Scheme 1. Unfortunately, the NHS esters are not very stable in aqueous solutions

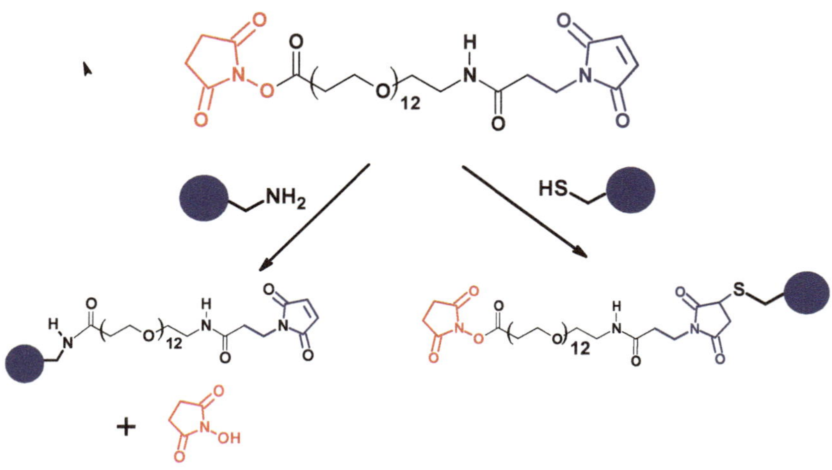

Scheme 1 The chemical structure of heterobifunctional linker, NHS-PEG_{12}-Maleimide. The group drawn in red is the amine reactive *N*-hydroxysuccinamide ester and that shown in blue is the thiol reactive maleimide. The two functional groups are linked by a flexible, hydrophilic linker containing 12 repeat units of ethylene glycol.

and have a half-live on the order of hours under physiological pH (pH 7.0–7.5) and the rate of hydrolysis increases with the increasing pH (Kalkhof & Sinz, 2008). The maleimide moiety can readily react with the free sulfhydryl groups present in the enzyme via Michael addition (Belbekhouche, Guerrouache, & Carbonnier, 2016; Ravasco, Faustino, Trindade, & Gois, 2019). The advantage of using the NHS-maleimide cross linker is that all conjugation steps are carried in aqueous media under mild physiological conditions, reducing the chances of enzyme denaturation and activity reduction. The success of this cross-linking conjugation is highly dependent on controlling the solution pH. The coupling of NHS with amines and Maleimide conjugation with free thiols are best carried out at buffered pH 7.5–8.3 and 6.5–7.3, respectively (Kalkhof & Sinz, 2008).

An important application of using heterobifunctional linkers in bioanalytical chemistry is to conjugate biomolecules on the surface of nanomaterials. An example here is a simple MNP-based DNA sandwich biosensor developed in the Zhou group (Zhang, Pilapong, et al., 2013). It uses two pairs of MNP-linked capture-DNA and enzyme-linked signal-DNA probes whose sequence are complementary to each half of its target-DNAs. Here two different enzymes, horseradish peroxidase (HRP) and alkaline phosphatase (ALP), are used to detect the two target-DNAs, respectively. The thiolated single-stranded capture-DNA was conjugated onto an aminated silica coated MNP using the NHS-PEG$_{12}$-maleimide coupling chemistry. The capture-DNA and signal-DNA-enzyme probes sandwich hybridize with a specific target-DNA, linking either an HRP or ALP to the MNP surface for enzymatic signal amplification. Using two pairs of unique capture-DNA and signal-DNA-enzyme probes, this MNP-enzyme sandwich assay can simultaneously quantitate two different target-DNAs with a detection limit down to 50 fM for each target-DNA (Zhang, Pilapong, et al., 2013).

On the other hand, the direct immobilization of enzymes on the MNP surfaces using the NHS-PEG-maleimide hetero-cross linker can be challenging. It often requires a reduction step to introduce a free thiol on the enzyme and followed by purification, several rounds of washing to remove any unbound cross-linkers, and change buffer pH to favor the specific reactions. Another issue is the limit stability of maleimide in a buffer solution: it can be hydrolyzed at pH $>$ 5.5 (Baldwin & Kiick, 2011; Kirchhof et al., 2015). This side reaction can lead to low conjugation efficiency and batch-to-batch variations. To avoid such complications, the use of strong bio-specific interactions such as the extremely strong biotin-avidin (streptavidin/neutravidin, NAV) interaction (with an equilibrium dissociation constant, $K_d \sim 10^{-15}$ M) is an attractive alternative strategy. Among the three avidin variants, NAV's isoelectric point (PI) is the close to the physiological pH and gives the lowest non-specific interactions. This property is highly beneficial for biosensing, allowing for significantly reduced background and hence high sensitivity. Notably, popular enzymes, such as horseradish peroxidase (HRP) and alkaline phosphatase (ALP), are also commercially available as their NAV-conjugates and are widely used in enzymatic assays. An additional advantage of using enzyme-NAV conjugate in biosensing is that the enzyme part is not directly involved in immobilization (hence its active site will not be blocked by steric hindrance), allowing enzyme to retain high activity and benefit sensitivity. Moreover, as a tetrameric protein, each (neutr-)avidin contains four biotin-binding sites, two on each side of the protein, this makes it very useful in assembling multilayer protein structures for bionanotechnology applications (Kim, Sohn, Zhou, Duke, & Kang, 2010; Rauf et al., 2006; Zhou et al., 2003).

1.2 Rationales of designing amphiphilic polymer for MNP encapsulation

To develop highly sensitive MNP-based assays, it is essential that the MNP has a robust and highly resistive surface coating to prevent non-specific adsorption of biomolecules, particularly enzymes, to reduce assay background. In this regard, PEGylation is one of the most useful strategies because it can provide high surface hydrophilicity and excellent resistance against non-specific adsorption to reduce background (Prime & Whitesides, 1991). Here a strategy based on encapsulation of hydrophobic MNPs using a PEGylated amphiphilic polymer is employed. The polymer encapsulation generates steric repulsions that can weaken the magnetic and van der Waals interaction among different MNPs, preventing them from forming large aggregates so as to enhance stability and water-dispersibility. Another advantage of this approach is that the PEGylated polymer can be designed to include specific functional groups for conjugation with biomolecules and increase PEG density to reduce non-specific adsorption. Here we employ a cheap and commercially available poly(maleic anhydride *alt*-1-octadecene) (PMAO, see Scheme 2) as the precursor polymer to prepare the desired PEGylated amphiphilic polymer. The PMAO has numerous long alkyl (C-18) chains which are highly hydrophobic. Meanwhile its anhydride rings are susceptible to attack by nucleophiles to give ring opening reactions. Here the PMAO is reacted

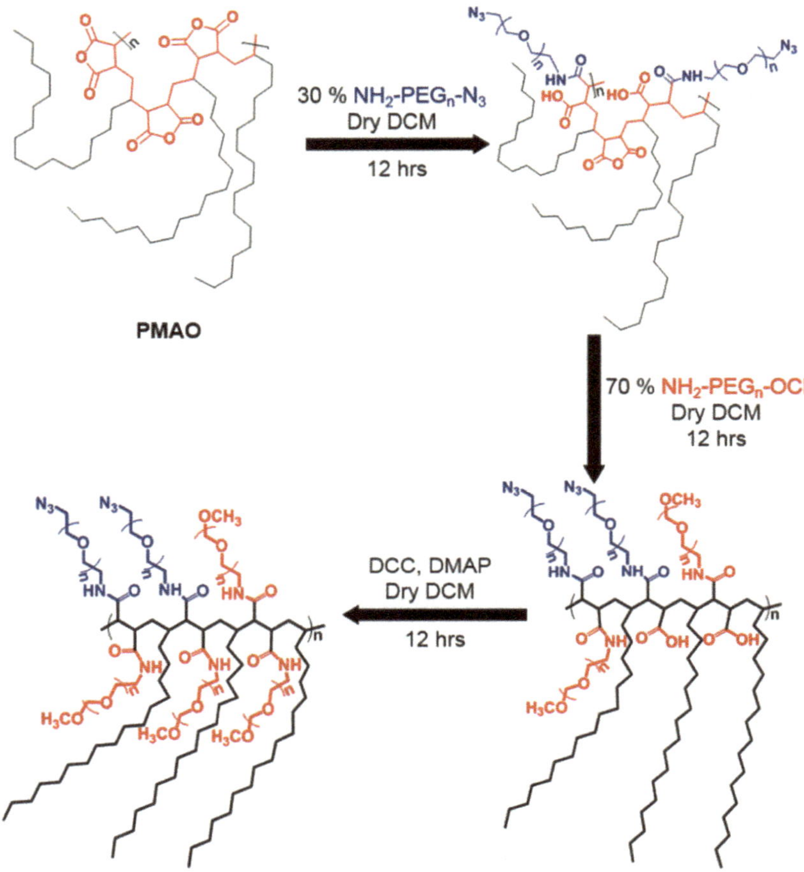

Scheme 2 Synthetic route to functional PEG-PMAO amphiphilic polymer. PMAO is first reacted with H_2N-PEG-N_3 (30% of PMAO monomer) and then H_2N-PEG-OMe followed by a DCC/DMAP mediated coupling to increase yield and PEG density. *DCC*, N,N′-dicyclohexylcarbodiimide; *DMAP*, 4-dimethylaminopyridine.

with a mixture of methoxy polyethylene glycol amine (NH_2-PEG-OCH_3) and azido-polyethylene glycol-amine (NH_2-PEG-N_3) to give an amphiphilic PEG-PMAO polymer as shown in Scheme 2.

Both of the PEGylation reagents contain an amino group which can react with the PMAO anhydride ring to form an amide linkage and a free carboxylic acid. The PEG grafting is done sequentially where NH_2-PEG-N_3 is reacted with PMAO first followed by NH_2-PEG-OCH_3 to ensure a high coupling efficiency of the PEG-N_3 for further DNA conjugation. The carboxylic acid groups formed from anhydride ring opening are further reacted with NH_2-PEG-OCH_3 via a DCC/DMAP mediated amide coupling to increase the PEG graft density. The high numbers of PEG chains in the polymer can provide a brush like structure to improve water dispersion and impart resistance against non-specific adsorption. The success of the PEG grafting onto the PMAO is confirmed from the corresponding Fourier Transform Infrared Spectroscopy (FT-IR) and Proton Nuclear Magnetic Resonance (^1H NMR) spectra. The resulting PEGylated PMAO polymer is used to encapsulate commercial hydrophobic $MnFe_2O_4$ MNPs (see next section, Lapitan, Xu, Guo, & Zhou, 2019).

1.3 Polymer encapsulation of magnetic nanoparticles

Encapsulation of MNPs with amphiphilic polymers such as the PEG-PMAO polymer here is a widely used approach to transfer and disperse hydrophobic MNPs into aqueous phase. Amphiphilic polymers contain both hydrophobic, typically long alkyl chains, and hydrophilic groups (e.g., PEG, carboxylic acid, or other charged groups) which can form hydrogen-bonding or electrostatic interactions with water molecules to impart water compatibility and dispersity. The hydrophobic groups can interact with non-polar molecules on the surface of hydrophobic MNPs via van der Waals interactions, leaving the hydrophilic groups exposed to the aqueous media to promote water solubility and stable dispersion. The flexible, amphiphilic polymer can wrap around the MNP surface with its multiple non-polar alkyl chains intercalating with the non-polar ligands on the MNP surface to form multiple van der Waals/hydrophobic interactions (Fig. 1). These multiple hydrophobic interactions are cooperative,

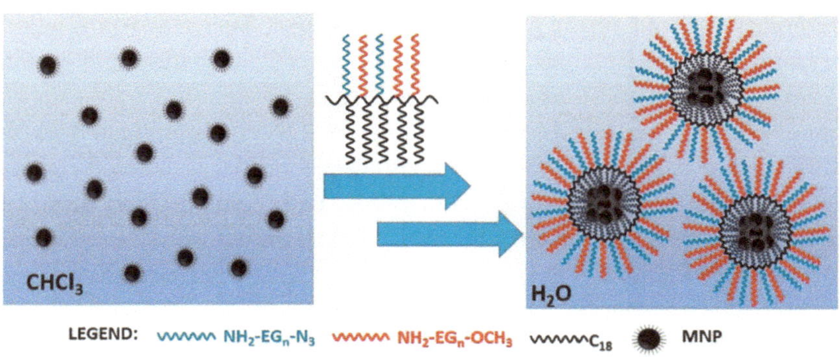

Fig. 1 Schematic polymer coating of hydrophobic magnetic nanoparticles and transfer to aqueous phase. *Figure reprinted with permission from Royal Society of Chemistry.*

resulting in strong interactions between the polymer and the MNP. Here the $MnFe_2O_4$ based MNP encapsulation is achieved via a modified two stage solvent exchange process (Tong, Hou, Ren, & Bao, 2011). The PEG-PMAO polymer and hydrophobic $MnFe_2O_4$ MNPs are first dispersed in chloroform, into which ethanol is then added to give an ethanol: $CHCl_3$ volume ratio of 1:1. After rotary evaporation to selectively remove $CHCl_3$, water is added to the MNP dispersion. After further washing with water in centrifugal filter tubes (10 k molecular weight cut off), highly stable and well-dispersed PEG-PMAO encapsulated MNPs are obtained. A transmission electron microscopy (TEM) examination of the polymer coated MNPs reveals that they appear as clustered particles rather than individual ones as shown schematically in Fig. 1 (Lapitan et al., 2019). Some degrees of MNP clustering is actually advantageous for biosensing, allowing for easier magnetic pull down and separation from the assay media (too small the MNPs cannot be easily separated magnetically). The method described here is general and can be readily extended to any hydrophobic nanoparticles regardless of the surface coating and composition of the nanoparticle core because of the non-specific nature of hydrophobic interactions.

1.4 MNP-capture-DNA conjugation

The PEG-PMAO polymer encapsulated MNPs contain numerous free azide groups on their surfaces that can be used for conjugation with a dibenzocylcooctyne (DBCO)-modified DNA via the efficient copper-free "click" chemistry (Baskin et al., 2007; Zhang, Feng, et al., 2013). This strategy is convenient because the reaction readily occurs under ambient conditions without heating or additional catalyst. This reaction relies on the use of a strained cyclooctyne which has a lower activation energy than a terminal alkyne, and thus does not require an exogenous catalyst such as copper (I) ions. A DBCO-modified capture-DNA is conjugated to the MNP surface by simple incubation at ambient conditions via the formation of a stable triazole ring highlighted in yellow in Scheme 3. The amount of the DBCO-DNA loaded onto the MNP is determined from the difference between that added and that remains in the clear supernatant after conjugation by UV–vis spectra. This conjugation method is more robust than the NHS-PEG-maleimide based cross-linking chemistry, allowing us to make the MNP-DNA conjugates reproducibly with high conjugation efficiency (Lapitan et al., 2019).

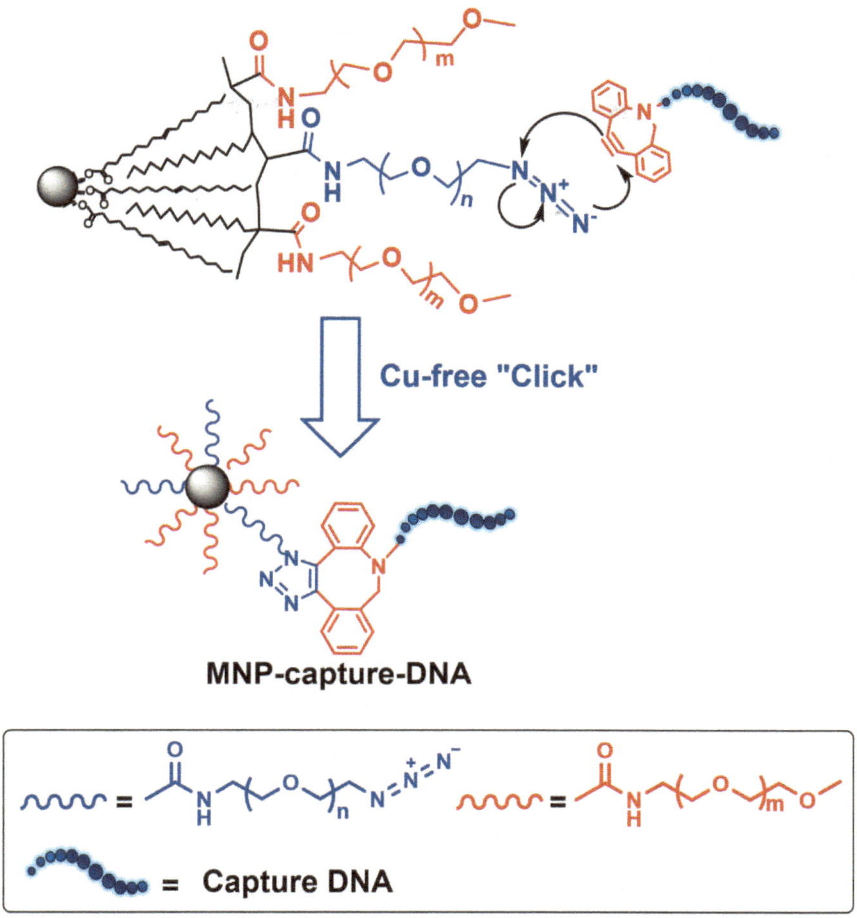

Scheme 3 Reaction mechanism of the copper-free "click" chemistry approach for the preparation of MNP-DNA conjugate using DBCO-modified capture-DNA and azide functionalized MNP.

1.5 Rationale for using neutravidin-horseradish peroxidase signal amplification

The incorporation of a redox active enzyme Horseradish peroxidase (HRP) with MNPs has led to the development of several colorimetric (Zhang, Pilapong, et al., 2013) and fluorometric biosensors (Zhang, Guo, et al., 2013). HRP is obtained from a plant source. It consists of a heme center containing an Fe(III)/Fe(II) redox couple surrounded by alpha helical protein backbones (Veitch, 2004). HRP is used extensively in biochemistry and bioassays because of its high substrate turnover rate (catalytic activity) and chemical robustness. It catalyzes the oxidation of organic substrates using hydrogen peroxides as oxygen donors. Here, the HRP catalyzed oxidation of a colorless substrate, amplex red, into a strongly colored and fluorescent product, resorufin, in the presence of H_2O_2 (see Scheme 4) is employed for fluorometric readout. The produced resorufin product can be detected through electrochemical (Lyon & Stevenson, 2006), colorimetric and fluorescence based readout strategies (Zhang, Guo, et al., 2013; Zhang, Pilapong, et al., 2013), making HRP catalyzed assays highly versatile in terms of readout methods.

However, direct immobilization of HRP and other enzymes onto solid surfaces (e.g., electrode) can cause structural changes to the enzymes and/or reduced active site accessibility due to steric hindrance (Hernández-Cancel et al., 2015; Zhang, Chi, Dong, & Wang, 1996) which can significantly reduce their catalytic activity and hence assay sensitivity. This issue is addressed here by using PEG-PMAO polymer coated MNPs, where a dense hydrophilic PEG coating on the MNP surface can prevent the enzyme to directly contact the MNP surface, thus preserving high enzyme activity. Moreover, the enzyme is further extended away from the MNP surface via a double-stranded DNA structure which greatly reduces the possibility

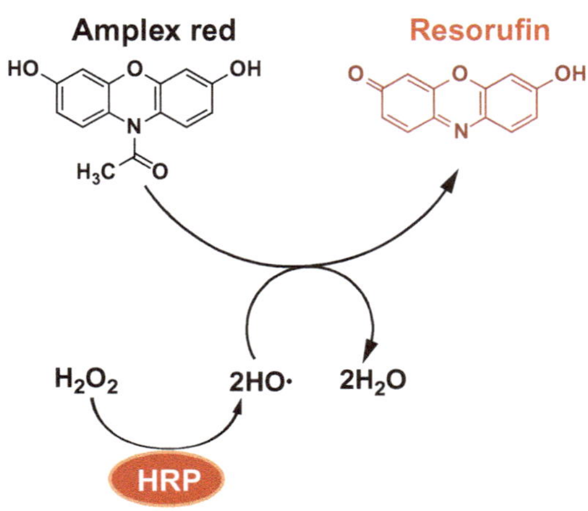

Scheme 4 Chemical transformation of Amplex red to resorufin through reaction with peroxy (OH·) radicals produced from the decomposition of H_2O_2 which catalyzed by horseradish peroxidase.

of steric blocking the active site. Furthermore, the HRP is used as NAV-conjugate here. The enzyme is linked to the MNP surface via NAV which is well-separated from the HRP active site, hence this is little disturbance to the enzyme structure after conjugation. All these allow the MNP conjugated HRPs to retain high activity. As stated in Section 1.2, the biotin-NAV interaction is very strong and robust. It can provide a facile and reliable way to immobilize NAV-conjugated enzymes onto the MNP-DNA-biotin ensemble, and allow the enzyme to retain high activity and hence high assay sensitivity (Nguyen, Sly, & Conboy, 2011).

1.6 MNP-poly-enzyme nanobead sandwich assay for DNA detection

Owing to its great signal amplification power (e.g., each HRP can turnover hundreds of copies of non-fluorescent substrates into strongly fluorescent products per second; Garcia et al., 2011), enzyme based signal amplification strategy can offer higher sensitivity than those using dye-labeled affinity reagents for detection. Our previous work has shown that by converting each captured target-DNA into an active HRP enzyme with a simple MNP-enzyme sandwich assay, it is possible to specifically detect the target-DNA down to the 50fM level using overnight enzymatic signal amplification (Zhang, Pilapong, et al., 2013). To further increase the signal amplification power, here we have designed a new poly-enzyme nanobead based signal amplifier. We first treat the aminated polystyrene nanobeads (~180nm diameter) with NHS-PEG_{12}-biotin in a $NaHCO_3$ buffer (pH 8.3) to introduce functional biotin groups (for binding to NAV-HRPs) on the nanobead surface, and then cap unreacted surface amines using NHS-PEG_{750} to reduce non-specific adsorption. The biotinylated nanobead is then treated with the NAV-HRP conjugate, tagging each nanobead with several thousand copies of NAV-HRPs via the strong biotin-NAV interaction, to form the poly-enzyme nanobead signal amplifier. As the biotin group is extended away from the nanobead surface via a long, hydrophilic PEG_{12} linker and the HRP is not directly involved in binding to nanobead surface, these ensure the nanobead tagged enzymes to retain high activity. Moreover, since each NAV contains four biotin-binding sites, two on each opposite side of the protein, the nanobead bound NAV-HRPs can still bind to other biotin-modified surfaces via its upper two free biotin-binding sites for great enzymatic signal amplification.

Here we have designed a simple MNP-based sandwich assay to convert each target-DNA into a highly active HRP enzyme or a poly-enzyme

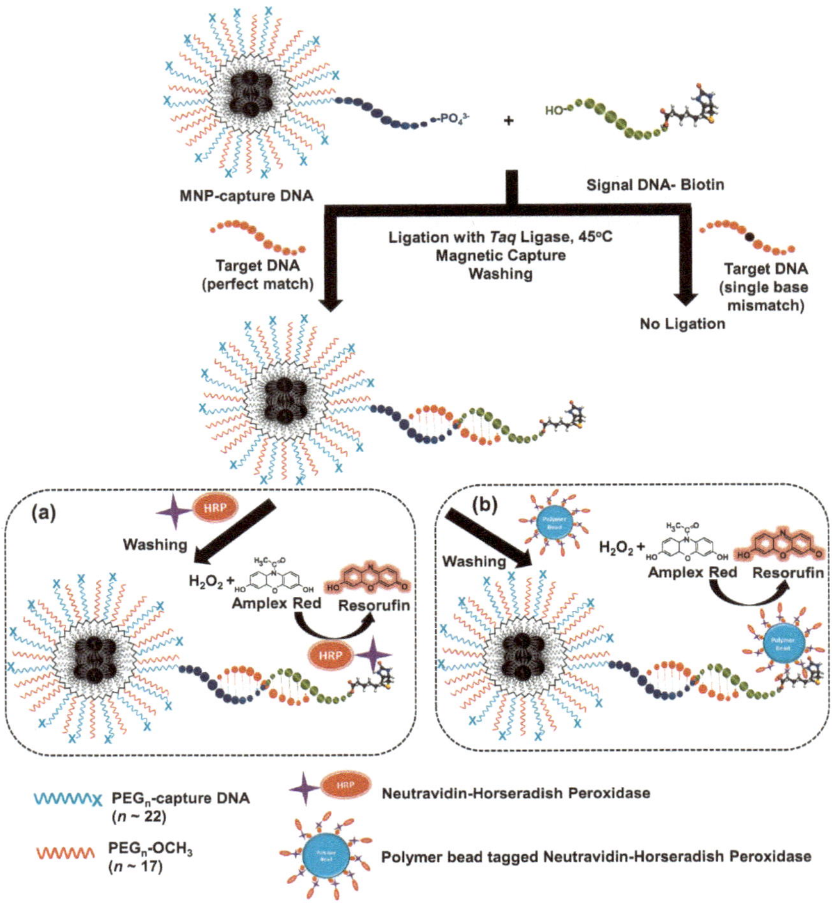

Fig. 2 Schematic illustration of our DNA sensing strategy. A MNP-linked capture-DNA and a biotinylated signal-DNA first sandwich hybridize with a target-DNA to form a nicked double-stranded structure. The signal-DNA is then covalently linked to the capture-DNA, giving a MNP-linked DNA-biotin, by Taq DNA ligase if the sequence is fully complementary, but not if there is a single-base mismatch, ensuring high single-nucleotide polymorphism (SNP) discrimination. After several rounds of magnetic separation and washing to remove any unbound signal-DNAs, then NAV-HRP (Route a) or poly-NAV-HRP-nanobead (Route b) is introduced to bind to the MNP-linked DNA-biotin for sensitive DNA detection via HRP-catalyzed enzymatic amplification using Amplex red-H_2O_2 substrates. *Figure reprinted with permission from Royal Society of Chemistry.*

nanobead for specific target-DNA detection (see Fig. 2). Our approach uses a pair of specific DNA probes, one (i.e., capture-DNA, 5′ phosphorinated) is covalently conjugated to the PEGylated PMAO polymer encapsulated MNP for capturing the target-DNA, and the other (i.e., signal-DNA) is biotinylated for specific binding to a NAV-conjugated enzyme for enzymatic signal amplification. The capture-DNA and signal-DNA sequences are complementary to each half of the target-DNA, allowing them to sandwich hybridize to the target-DNA specifically. After hybridization, a ligation step using *Taq* DNA ligase is used to covalently link the MNP-linked capture-DNA to the signal-DNA via the formation of a phosphate ester bond, ensuring that the signal-DNA (with its biotin label) will not detach from the MNP surface during the subsequent washing steps (without ligation, the nicked double-stranded DNA sandwich structure can dissociate during extensive washing steps). More importantly, the strict requirement of perfect match between the target- and probe-DNA sequences to produce covalent ligation by the *Taq* DNA ligase also greatly improves the DNA detection specificity. Only the full-match target-DNA, but not those contain a single-mismatch at the nicked double-stranded side, will give a successful ligation, allowing this sensing strategy to offer excellent discrimination between the perfect-match and single-base mismatch targets (i.e., single-nucleotide polymorphism, SNP). Since SNPs are closely associated with many human diseases such as cancer, vascular, and neurodegenerative diseases, this highly selective DNA sensor may have good potential for rapid detection and diagnosis of such diseases.

After extensive washing to remove any unbound signal-DNAs, the resulting MNP-DNA-biotin conjugates are treated with excess NAV-HRPs (Route a) or poly-NAV-HRP nanobeads (Route b) to covert each successfully ligated target-DNA (via a covalently ligated signal-DNA-biotin) into a high-activity HRP or a poly-enzyme nanobead tagging thousands of copies of HRPs for enzymatic detection. Particularly, owing to the great signal amplification power of the poly-enzyme nanobead, this simple sandwich assay has enabled the direct, label- and amplification- free detection of the target-DNA down to the low attomolar region, making it one of the most sensitive DNA sensing approaches reported in the literature (Lapitan et al., 2019). Moreover, this strategy is general, and can be readily applied to detect any target-DNA of interest by simply designing a specific pair of capture- and single-DNA probes to match the target-DNA sequence. The excellent sensitivity and specificity of this simple DNA sensor makes it highly attractive in broad biosensing and diagnostic applications.

2. Materials and methods

2.1 General equipment

Safety: Proper personal protection equipment (PPE); goggles, lab coat, nitrile gloves

1. Rotary Evaporator
2. Varian Cary 50 Bio® UV–vis Spectrophotometer, Agilent
3. Nanodrop UV–vis Spectrophotometer, Thermo Scientific
4. Dynamic Light Scattering, Malvern zetasizer-nano with BI-APD detector using a He-Ne laser at 633 nm (scattering angle: 90°)
5. Thermo Scientific Heraeus Fresco 21 microcentrifuge
6. Microcentrifuge tubes
7. Volumetric micropipettes (20, 100, 200, and 1000 μL)
8. Magnetic stirring hot plate
9. Analytical Balance
10. Magnetic separation rack
11. Bath sonicator
12. Bruker DPX300 Nuclear Magnetic Resonance (500 MHz for ^{1}H, 125 MHz for ^{13}C).
13. Bruker Daltonics MicroTOF mass spectrometer
14. PerkinElmer FT-IR spectrometer
15. General laboratory glass ware and plastic wares
16. Amicon Centrifugal Filters (10,000 MW cut off)
17. Syringe and syringe filter (0.45 μm pore size)
18. Envision® 2015 fluorescence plate reader system (Perkin Elmer)

2.2 Chemicals and reagents

1. Capture-, signal- and target-DNA probes (custom synthesized and HPLC purified), IBA GmbH.
2. Poly(maleic anhydride *alt* octa-1-decene) (PMAO, Mn = 30–50 k), Sigma-Aldrich.
3. Methoxy-polyethylene glycol amine (NH_2-PEG_n-OCH_3, MW ~ 1000, n = ~23), Sigma-Aldrich.
4. Azido polyethylene glycol amine (N_3-PEG_n-NH_2, MW ~ 1000, n = ~23), previously synthesized in house (Lapitan et al., 2019).
5. Dicyclohexylcarbodiimide (DCC, >99%), Sigma-Aldrich.
6. Dimethylaminopyridine (DMAP, >99%), Sigma-Aldrich.
7. Dry chloroform.
8. Hydrophobic $MnFe_2O_4$ MNP dry powders (nominal diameter: 50 nm), Sigma-Aldrich.
9. Amine modified polystyrene nanobead aqueous dispersion (~180 nm diameter, 2.5% solid weight), Bangs Laboratories, Inc.

10. *N*-hydroxysuccinamide-dodeca(polyethylenegylcol)-Biotin (NHS-PEG_{12}-Biotin), Thermo Scientific.
11. N-hydroxysuccinamide-polyethylenegylcol-methyl ether (NHS-$PEG_{\sim 17}$-OCH_3, MW ~750), Sigma-Aldrich.
12. Phosphate buffered saline, PBS (20 mM phosphate, 150 mM NaCl, pH 7.4).
13. PBS-tween: PBS containing 0.1% of tween 20.
14. 2-(4′Hydroxyazobenzene)-Benzoic Acid (HABA)-Biotin Assay kit, Cambridge Bioscience.
15. High-activity Neutravidin-Horseradish Peroxidase (NAV-HRP), Thermo Scientific.
16. Bovine serum albumin, Sigma-Aldrich.
17. Taq DNA ligase (40,000 U mL^{-1}), New England Biolabs.
18. 10 × DNA ligase buffer (200 mM Tris–HCl, 250 mM potassium acetate, 100 mM magnesium acetate, 10 mM NAD 1, 100 mM DTT, 1% Triton X100, pH 7.6), New England Biolabs.
19. Amplex red, Invitrogen Life technologies.

2.3 Synthesis of the PEG-PMAO polymer

1. Transfer 0.250 g (0.66 mmol monomer unit) PMAO into a clean and dry 2-neck round bottom flask (250 mL) containing a magnetic stirrer. Add 10 mL of dry $CHCl_3$ to the flask to dissolve the polymer. Seal the flask with suba seals and purge the system with nitrogen gas.
2. Add NH_2-PEG_n-N_3 (0.396 g, 0.396 mmol) dissolved in 5 mL dry $CHCl_3$ via a plastic syringe needle to the polymer solution and stir the reaction mixture for 12 h.
3. Add NH_2-PEG-OCH_3 (0.923 g, 0.923 mmol) dissolved in 5 mL dry $CHCl_3$ via a syringe needle and allow reaction mixture to stir for another 24 h.
4. Remove all of the $CHCl_3$ solvent by rotary evaporation under reduced pressure. Seal the reaction flask with a rubber septum, purge it with nitrogen gas and then add 5 mL dry CHCl3 via a syringe needle and stir to dissolve the crude polymer.
5. Cool the reaction flask in an ice bath to 0 °C. Add DCC (0.164 g, 0.792 mmol) and DMAP (0.014 g, 0.11 mmol) dissolved in 2 mL of dry $CHCl_3$ to the reaction flask via a syringe needle.
6. Allow the reaction mixture to gradually warm up to room temperature (~25 °C) and stir the reaction mixture for another 72 h.

7. Filter off the white solids (dicyclohexylurea byproduct formed in the reaction) on a Buchner funnel set-up. Transfer the clear filtrate to a round bottom flask and evaporate the solution to dryness, giving the desired PEG-PMAO polymer as a yellow oil. A portion of the crude polymer is purified by silica gel column chromatography for product characterization. N ^{1}H NMR (500 MHz, δ ppm); 0.80 (t), 1.25, 1.55–1.70 (br), 3.38–3.42, 3.40, 3.50, 3.55–3.70, 3.80. FT-IR (cm^{-1}): 2880 ($\nu_{C\text{–}H}$), 2100 (ν_{N3}), 1655 ($\nu_{C=O}$, amide), 1060 ($\nu_{C\text{–}O}$).
8. Weigh the PEG-PMAO polymer product and dissolve it in $CHCl_3$ to make a stock concentration of 100 mg mL^{-1} and keep in a refrigerator until use. The PEG-PMAO polymer is directly used for MNP encapsulation without further purification, taking the full advantage of MNP's ease of magnetic separation which allow for efficient removal of any unbound species and impurities (see Section 2.4).

2.4 PEG-PMAO encapsulation of magnetic nanoparticles

1. Weigh three batches of 5.0 mg of the hydrophobic $MnFe_2O_4$ MNP powers and transfer them each into a dry scintillation vial. Add 1.0 mL $CHCl_3$ to each vial and sonicate the samples for 10 min to improve dispersion. Do this in a fume hood cupboard.
2. Transfer 0.25, 0.50 or 0.75 mL of the PEG-PMAO polymer stock solution (100 mg mL^{-1}) prepared above into each MNP dispersion to give a final MNP:polymer weight ratio of 1:5, 1:10 or 1:15, respectively. Sonicate the MNP/polymer mixtures for another 10 min and then stir overnight at room temperature (~25 °C).
3. Add 1.25, 1.50 or 1.75 mL of ethanol to each of the above scintillation vials to make a final $CHCl_3$: ethanol volume ratio of 1:1.
4. Transfer each of the MNP dispersions to a 25 mL small round bottom flask separately. Carefully remove the $CHCl_3$ solvent by rotary evaporation. Note, the MNP dispersion must not be evaporated to dry. Only remove $CHCl_3$, but not ethanol.
5. Add ultrapure water to each of the MNP dispersion in ethanol to obtain a H_2O: ethanol volume ratio of 1:1. Transfer each of the resulting MNP dispersions to an Amicon centrifugal filter (10 k MW cut off). Remove the ethanol-water solvent by centrifugation at 4000 rpm for 10 min.
6. Wash the polymer coated MNPs with ultrapure water (0.5 mL) and followed by low speed centrifugation (4000 rpm, 5 min) and discard

the filter through water. Repeat the washing/centrifugation process 5 times to completely remove any small water-soluble impurities.

7. Add sterilized ultrapure water (0.50 mL) to disperse the remaining MNP residues and collect it as an opaque black solution.
8. Filter the above black MNP dispersion by passing through a 0.45 μm syringe filter to any remove large aggregated particles. Transfer the resulting MNP water dispersion to a 1.5 mL Eppendorf tube.
9. Centrifuge the MNP dispersion at 14,000 rpm for 15 min to obtain a black MNP pellet, discard the clear supernatant and wash the MNP pellet with ultrapure water (1 mL). Repeat the centrifugation/washing process 3 times to remove unbound polymers and other contaminants.
10. Disperse the final polymer coated MNPs in sterilized ultrapure water to obtain a black solution. Comparing results obtained at different MNP: polymer weight ratios reveals that 1:10 is the optimal ratio for MNP encapsulation. It gives high yields in successfully encapsulated MNPs and produces highly stable, well-dispersed MNPs which shows no signs of aggregation or precipitation over a period of 6 months.
11. Repeat the PEG-PMAO encapsulation process under the optimal MNP:polymer weight ratio 3 times. Combine all three batches of the polymer encapsulated MNPs to give a $\sim$10 mg mL^{-1} MNP stock which are stored in a fridge till use.

2.5 Characterization and stability of the PEG-PMAO encapsulated MNPs

The colloidal stability of the PEG-PMAO polymer coated MNPs and their hydrodynamic diameters (D_hs) at different temperatures (25–75 °C), salt contents (e.g., 0, 100, 300 mM) and solution pH (2–12) are evaluated by dynamic light scattering (DLS). Automatic instrument calibration and refractive values for the material and solvent properties are set-up for each measurement.

1. Dilute the PEG-PMAO encapsulated MNP stock to 10 μg mL^{-1} using pure water or salt solutions in standard plastic cuvettes for D_h measurement.
2. Allow DLS instrument to scan D_h repeatedly until a stable D_h size is achieved. Then take a consecutive of 10 scans for each sample. Average the data of the 10 scans to obtain the D_h (volume size) distribution histogram. Fit the data by Gaussian distribution function using Origin software to obtain the average D_h size and distribution function.

3. The PEG-PMAO polymer encapsulated MNPs show a D_h of ~230 nm in water. Its D_h size and distribution function are not affected by the presence of 300 mM NaCl. Moreover, its D_h size shows no measurable changes over a pH range of 2–12. These results demonstrate excellent colloidal stability of the PEG-PMAO encapsulated MNPs in water. The results are consist to that expected for MNPs stabilized by steric repulsion from their surface dense PEG coating, and not by electrostatic repulsion.
4. For D_h measurement at different temperature, pre-set the measuring cell compartment to the selected temperature (e.g., 25–75 °C). Transfer the MNP water-dispersion that has been pre-incubated at this temperature for at least 30 min to the cell and start the measurement as above.
5. The D_h of the PEG-PMAO encapsulated MNPs shows no measurable difference over a temperature range of 25–75 °C, confirming such PEG-PMAO encapsulated MNPs have excellent thermal stability.

2.6 DNA conjugation to PEG-PMAO coated MNPs

1. Incubate the PEG-PMAO coated MNP (~10 mg, 0.50 mL) with the DBCO capture-DNA (100 μL, 100 μM) in an Eppendorf tube at room temperature for 48 h at a constant vortex rate of 600 rpm to conjugate the capture-DNA onto the MNP via copper-free "click" reaction between the MNP surface azide and the capture-DNA terminal DBCO group.
2. Centrifuge the MNPs at 4000 rpm for 5 min to obtain a pellet. Place the Eppendorf tube in a magnetic separating rack. Carefully withdraw the clear supernatant using a micropipette and transfer it to a clean Eppendorf tube (keep it for DNA conjugation determination).
3. Wash the DNA modified MNPs with sterilized ultrapure water (0.5 mL), centrifuge at 4000 rpm for 5 min and then remove the clear supernatant. Repeat this process 3 times to completely remove any unbound capture-DNA.
4. Determine the amount of unbound capture-DNA in the supernatant by measuring its absorbance at $\lambda = 260$ nm on a nanodrop UV–vis spectrometer. The amount of capture-DNA loaded on the MNP surface is determine from the concentration difference between the capture-DNA before and after conjugation.
5. The percent capture DNA conjugation efficiency is calculated from Eq. (1):

$$\text{Conjugation Efficieny} = \frac{C_i - C_x}{C_i} \times 100 \tag{1}$$

where C_i is the initial concentration of the capture-DNA used in conjugation and C_x is the remaining capture-DNA concentration of the supernatant after conjugation. The DNA concentration is calculated from its absorbance at $\lambda = 260\,nm$ and a molar extinction coefficient of $3.03 \times 10^5\,M^{-1}\,cm^{-1}$. The dilution factors are taken into account in calculating the DNA concentration. A general high conjugation efficiency of ~70% is obtained. By varying the capture-DNA to MNP ratio, the DNA loading on the MNP can be tuned from 0.04 to $0.035\,nmol\,mg^{-1}$ (MNP).

2.7 Preparation of polymer nanobead-PEG$_{12}$-biotin

Scheme 5 shows the steps involved in preparing the poly-enzyme nanobead signal amplifier, each tagging several thousand copies of high-activity NAV-HRPs on its surface.

1. Centrifuge 100 μL amino-functionalized polystyrene nanobeads (2.54% solid by weight) at 15,000 *g* for 1 min and carefully remove and discard the clear supernatant via a micropipette.
2. Wash the nanobeads with 0.5 mL sodium bicarbonate buffer (20 mM $NaHCO_3$, pH 8.3), centrifuge at 15,000 *g* for 1 min, and then discard

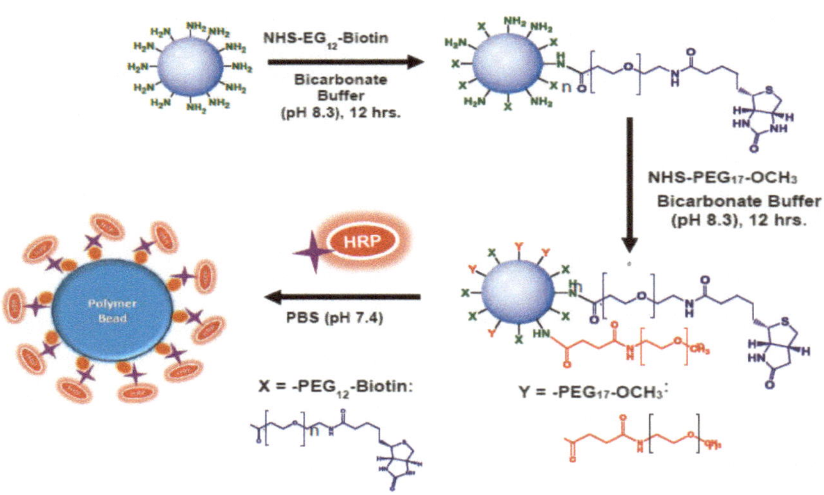

Scheme 5 Schematic process used for the preparation of the poly-enzyme nanobead signal amplifier. The nanobeads are first reacted with NHS-PEG$_{12}$-biotin to introduce functional PEG$_{12}$-biotin groups, which are then blocked with NHS-PEG$_{\sim 17}$-OMe to cap remaining free amines to reduce non-specific adsorption, and finally incubated with NAV-HRP to tag each nanobead with several thousand copies of HRPs via the strong NAV-biotin bio-specific interaction to form the poly-enzyme nanobead.

the clear supernatant. Repeat this process twice and finally disperse the nanobeads in 1.0 mL of sodium bicarbonate buffer (pH 8.3).

3. Add NHS-PEG$_{12}$-Biotin (100 μL, 250 mM in dry DMSO) to the nanobead dispersion and incubate for 12 h at a vortex rate of 600 rpm at room temperature (~25 °C) to covalently link the PEG$_{12}$-biotin groups onto the nanobead surface via the formation of stable amide bonds by reacting with its surface free amine groups.
4. Add NHS-PEG$_{\sim 17}$-OCH$_3$ (100 μL, 100 mM in dry DMSO) to the above nanobead dispersion and incubate for another 12 h at constant 600 rpm vortex at room temperature. This step is used to cap any unreacted free amines on the nanobead surface to reduce non-specific adsorption.
5. Centrifuge the nanobeads dispersion at 20,000*g* for 15 min, carefully remove the clear supernatant. Wash the nanobeads twice with 0.50 mL phosphate buffered saline (20 mM phosphate, 150 mM NaCl, pH 7.4), each is followed by centrifugation at 20,000*g* for 15 min and discarding clear supernatant. Finally disperse the biotin-functionalized nanobeads in 5.0 mL PBS to make the stock and store in a fridge until use.

2.8 Nanobead biotin valency determination and NAV–HRP tagging

1. Prepare assay solution by adding avidin to HABA solution to make 1 mL following the kit instruction and mix for 5 min.
2. Transfer 900 μL of the HABA/avidin solution into a cuvette and measure its absorbance at λ = 500 nm.
3. Add 100 μL of the biotin-modified polymer nanobead dispersion obtained above to the cuvette and thoroughly mix the solution by inversion 5 times.
4. Remove the polymer nanobeads from the incubation solution by centrifugation at 20,000*g* for 10 min. This step removes the nanobead's light scattering which can interfere the absorption measurement.
5. Read the absorbance of the clear supernatant at λ = 500 nm.
6. Determine the biotin loading on the polymer nanobeads. This is calculated from the change in absorbance at λ = 500 nm for the HABA/Avidin complex upon addition of the nanobead sample using Eq. (2):

$$A_{500} = 0.9\,(Abs_x) - \left(Abs_y\right) \tag{2}$$

where x is the absorbance of HABA/Avidin complex, and y is the absorbance of HABA/Avidin + nanobead sample. The 0.9 is the dilution factor of HABA/Avidin upon addition of sample. The amount of biotin is calculated by Eq. (3):

$$\frac{\mu mole\ biotin}{mL} = \frac{\Delta A_{500}}{34}(10) \tag{3}$$

The change in absorbance at 500 nm is divided with 34 mM which is the extinction coefficient of HABA/Avidin at 500 nm. The original biotinylated nanobead sample is diluted 10-fold in the reaction mixture. Therefore, a multiplier of 10 is used to convert the biotin concentration to its original value. Finally, the biotin valency on each nanobead is obtained from Eq. (4):

$$\frac{molarity\ of\ biotin}{molarity\ of\ polymer\ bead} \tag{4}$$

Using this method, the biotin valency per nanobead is determined to be in the range of 6000–10,000 with some degrees of batch-to-batch variation.

7. Based on its biotin valency and nanobead concentration, an equimolar amount of NAV–HRP is added to the biotin-nanobead dispersion and incubated for 30 min to make the poly-enzyme nanobead signal amplifier via the strong biotin-NAV interaction discussed in Section 1.1.
8. The poly-enzyme nanobead is diluted with PBS to make a stock concentration of 0.50 μM (based on the NAV–HRP concentration, C_{HRP}) and is stored at 4 °C in a fridge till use.

2.9 MNP-poly-enzyme nanobead sandwich assay for DNA detection

1. Rinse all sample tubes with PBS containing 1.0 mg mL^{-1} of BSA to reduce non-specific adsorption. This is important to reduce assay background. All biosensing experiments are carried out using the MNP-capture-DNA with a capture-DNA loading of ~0.35 nmol mg^{-1} (MNP) which are found to be optimal for high sensitivity and robustness.
2. Disperse 30 μg of the MNP–capture-DNA (10 μL) to Eppendorf tubes each containing 15 μL ultrapure water, 5 μL ligase buffer (10 ×), 10 μL signal-DNA (biotinylated, 500 pM), 1 μL Taq DNA ligase (10 U μL^{-1}) and 10 μL target-DNA of different concentration (e.g., 0, 10^{-18}, 10^{-17}, 10^{-16}, 10^{-15}, 10^{-14}, 10^{-13}, and 10^{-12} mol L^{-1}). For each different target-DNA concentration, three samples are prepared in parallel to improve assay robustness and reproducibility.

3. Transfer the sample tubes to a dry heating metal bath pre-heated at 45 °C and incubate for 60 min and then let the samples cool down to room temperature naturally. During this step, a pair of the MNP-capture-DNA and signal-DNA probes sandwich hybridize with a target-DNA to form a nicked duplex DNA structure. They are covalently ligated together only if their sequence are fully complementary, but not if there is a mismatch at the nicked duplex DNA site, thereby increasing signal discrimination between the full-match and single-base mismatch target-DNA. A successful ligation will produce a MNP-covalently linked biotin for binding to a NAV-HRP (Route a, Fig. 2) or a poly-NAV-HRP nanobead signal amplifier (Route b, Fig. 2), ensuring it will not detach from the MNP during the washing steps.
4. Centrifuge the MNP dispersion at 3000 rpm for 1 min and pull MNPs to the tube wall via a permanent magnet. Carefully discard the clear supernatant using a micropipette and wash the MNPs twice with PBS-Tween (200 μL), each followed by centrifugation and discarding of clear supernatant.
5. For Route a, add NAV–HRP (50 μL, 100 pM) to the MNP dispersion obtained above and incubate for 60 min. Centrifuge the samples at 5000 rpm for 1 min and pull MNPs to the tube wall via a permanent magnet and remove the clear supernatant using a micropipette. Wash the MNP residue 4 times with PBS-Tween (200 μL), once with PBS (200 μL) each is followed by centrifugation and discarding clear supernatant and finally disperse the obtained MNP-enzyme conjugates in PBS (80 μL).
6. For Route b, disperse the above MNP residues in 100 μL of the poly-enzyme nanobead signal amplifier in PBS (final HRP concentration = 1.0 pM) and incubate for 60 min at room temperature to let the poly-enzyme nanobead bind to the MNP surface biotin. Centrifuge to samples at 3000 rpm for 1 min, pull the MNPs to the tube wall with a permanent magnet, and carefully remove the clear supernatant. Wash the MNP residue with PBS-Tween (200 μL) and once with PBS (200 μL). Each washing steps contains centrifugation, magnetic pull down and careful removal of clear supernatant, and finally disperse the residue in PBS (80 μL).
7. Transfer each of the above MNP dispersions to a well of a 96 well microplate (for both Routes). Add amplex red (300 μM, 10 μL) and H_2O_2 (300 μM, 10 μL) to each well to initiate enzymatic signal amplification. Monitor the assay process by recording their fluorescence intensity-time

traces on an Envision® fluorescence plate reader system using BODIPY TMR FP 531 filter for excitation and Cy3 595 filter for recording emission intensity.

8. Fit the fluorescence intensity-time traces (typically at 0–50 min) by linear function at each target-DNA concentration to determine the fluorescence production rate. Plot the fluorescence rate as function of target-DNA concentration to give the DNA sensing calibration curve.
9. Using this method, by converting each ligated signal DNA into a NAV-HRP enzyme via Route a, this MNP-enzyme based assay can detect 10^{-13} mol L^{-1} of the unlabeled target-DNA. More importantly, by converting each ligated signal DNA into a poly-enzyme nanobead tagging several thousand copies of high-activity NAV-HRPs via Route b, this simple MNP-poly-enzyme nanobead sandwich assay can directly detect the unlabeled target-DNA down to $\sim 2 \times 10^{-18}$ mol L^{-1} with a large linear dynamic range of 5 orders of magnitude (from 10^{-18} to 10^{-13} mol L^{-1}) (Lapitan et al., 2019).

3. Concluding remarks

A simple method has been developed to convert a cheap commercial PMAO polymer into a multifunctional amphiphilic PEG-PMAO polymer by reacting PMAO with NH_2-PEG$_{\sim 23}$-OCH_3 and NH_2-PEG$_{\sim 23}$-N_3. The resulting PEG-PMAO polymer is successfully used to encapsulate commercial hydrophobic MNPs to make biocompatible and well-dispersed MNPs with excellent colloidal stability over a wide range of pH, temperature and salt content. The PEG-PMAO encapsulation allows for efficient conjugation of DBCO-modified capture-DNA onto the MNP surface via copper-free click chemistry to develop a solid platform for DNA detection and quantification via DNA sandwich strategy. A new signal amplification strategy using poly-enzyme nanobeads each tagging several thousand copies of high-activity HRPs for greatly enhanced signal amplification power is developed. Combining rapid target capture, ligation, and extremely powerful poly-enzyme nanobead signal amplification, our simple MNP-poly-enzyme nanobead sandwich assay can directly quantitate unlabeled target-DNA down to 2×10^{-18} mol L^{-1} with a large linear dynamic range of 5 orders of magnitude, placing it among the most sensitive DNA sensing methods ever reported in literature (Lapitan et al., 2019). The preparation of the poly-enzyme nanobead signal amplifier is easy, involving two simple steps:

(i) biotinylation of aminated polymer nanobeads by NHS-PEG_{12}-biotin and (ii) incubation the nanobeads with NAV-HRP. Importantly, this poly-enzyme nanobead is compatible with other biosensing and bioassay formats, including the clinical enzyme-linked immunosorbent assays (ELISA) to greatly improve the signal amplification and assay sensitivity, and hence it can impact a wide range of biosensing, bioanalytical, and diagnostic research areas.

Acknowledgments

We thank the University of Leeds for funding this project. L.D.S.L. thanks the University of Leeds for providing a Leeds International Research Scholarship (LIRS) to allow him study at University of Leeds.

References

Akbarzadeh, A., Samiei, M., & Davaran, S. (2012). Magnetic nanoparticles: Preparation, physical properties, and applications in biomedicine. *Nanoscale Research Letters*, 7(1), 144.

Ansari, S. A., & Husain, Q. (2012). Potential applications of enzymes immobilized on/in nano materials: A review. *Biotechnology Advances*, *30*(3), 512–523.

Baldwin, A. D., & Kiick, K. L. (2011). Tunable degradation of maleimide–thiol adducts in reducing environments. *Bioconjugate Chemistry*, *22*(10), 1946–1953.

Baskin, J. M., Prescher, J. A., Laughlin, S. T., Agard, N. J., Chang, P. V., Miller, I. A., et al. (2007). Copper-free click chemistry for dynamic in vivo imaging. *Proceedings of the National Academy of Sciences of the United States of America*, *104*(43), 16793–16797.

Belbekhouche, S., Guerrouache, M., & Carbonnier, B. (2016). Thiol–maleimide Michael addition click reaction: A new route to surface modification of porous polymeric monolith. *Macromolecular Chemistry and Physics*, *217*(8), 997–1006.

Bilal, M., Asgher, M., Parra-Saldivar, R., Hu, H., Wang, W., Zhang, X., et al. (2017). Immobilized ligninolytic enzymes: An innovative and environmental responsive technology to tackle dye-based industrial pollutants—A review. *The Science of the Total Environment*, *576*, 646–659.

Bilal, M., Rasheed, T., Zhao, Y., Iqbal, H. M., & Cui, J. (2018). "Smart" chemistry and its application in peroxidase immobilization using different support materials. *International Journal of Biological Macromolecules*, *119*, 278–290.

Bornscheuer, U. T. (2003). Immobilizing enzymes: How to create more suitable biocatalysts. *Angewandte Chemie International Edition*, *42*, 3336–3337.

Bornscheuer, U. T., Huisman, G. W., Kazlauskas, R. J., Lutz, S., Moore, J. C., & Robins, K. (2012). Engineering the third wave of biocatalysis. *Nature*, *485*, 185–194.

Gao, J., Gu, H., & Xu, B. (2009). Multifunctional magnetic nanoparticles: Design, synthesis, and biomedical applications. *Accounts of Chemical Research*, *42*(8), 1097–1107.

Garcia, J., Zhang, Y., Taylor, H., Cespedes, O., Webb, M. E., & Zhou, D. (2011). Multilayer enzyme-coupled magnetic nanoparticles as efficient, reusable biocatalysts and biosensors. *Nanoscale*, *3*, 3721–3730.

Gu, H., Ho, P. L., Tsang, K. W., Wang, L., & Xu, B. (2003). Using biofunctional magnetic nanoparticles to capture vancomycin-resistant enterococci and other gram-positive bacteria at ultralow concentration. *Journal of the American Chemical Society*, *125*(51), 15702–15703.

Hernández-Cancel, G., Suazo-Dávila, D., Medina-Guzmán, J., Rosado-González, M., Díaz-Vázquez, L. M., & Griebenow, K. (2015). Chemically glycosylation improves the stability of an amperometric horseradish peroxidase biosensor. *Analytica Chimica Acta*, *854*, 129–139.

Holmlin, R. E., Chen, X., Chapman, R. G., Takayama, S., & Whitesides, G. M. (2001). Zwitterionic SAMs that resist nonspecific adsorption of protein from aqueous buffer. *Langmuir*, *17*(9), 2841–2850.

Kalkhof, S., & Sinz, A. (2008). Chances and pitfalls of chemical cross-linking with amine-reactive N-hydroxysuccinimide esters. *Analytical and Bioanalytical Chemistry*, *392*(1-2), 305–312.

Kim, D. C., Sohn, J. I., Zhou, D., Duke, T. A. J., & Kang, D. J. (2010). Controlled assembly for well-defined 3D bioarchitecture using two active enzymes. *ACS Nano*, *4*(3), 1580–1586.

Kirchhof, S., Strasser, A., Wittmann, H. J., Messmann, V., Hammer, N., Goepferich, A. M., et al. (2015). New insights into the cross-linking and degradation mechanism of Diels–Alder hydrogels. *Journal of Materials Chemistry B*, *3*(3), 449–457.

Kuhn, S. J., Finch, S. K., Hallahan, D. E., & Giorgio, T. D. (2007). Facile production of multivalent enzyme-nanoparticle conjugates. *Journal of Magnetism and Magnetic Materials*, *311*(1), 68–72.

Lapitan, L. D., Xu, Y., Guo, Y., & Zhou, D. (2019). Combining magnetic nanoparticle capture and poly-enzyme nanobead amplification for ultrasensitive detection and discrimination of DNA single nucleotide polymorphisms. *Nanoscale*, *11*(3), 1195–1204.

Laurent, S., Forge, D., Port, M., Roch, A., Robic, C., Elst, L. V., et al. (2008). Magnetic iron oxide nanoparticles: Synthesis, stabilization, vectorization, physicochemical characterizations, and biological applications. *Chemical Reviews*, *108*, 2064–2110.

Lee, N., & Hyeon, T. (2012). Designed synthesis of uniformly sized iron oxide nanoparticles for efficient magnetic resonance imaging contrast agents. *Chemical Society Reviews*, *41*(7), 2575–2589.

Lee, J. H., Jang, J. T., Choi, J. S., Moon, S. H., Noh, S. H., Kim, J. W., et al. (2011). Exchange-coupled magnetic nanoparticles for efficient heat induction. *Nature Nanotechnology*, *6*(7), 418–422.

Leroux, J. C. (2007). Injectable nanocarriers for biodetoxification. *Nature Nanotechnology*, *2*(11), 679–684.

Lyon, J. L., & Stevenson, K. J. (2006). Picomolar peroxide detection using a chemically activated redox mediator and square wave voltammetry. *Analytical Chemistry*, *78*(24), 8518–8525.

Mitchell, D. T., Lee, S. B., Trofin, L., Li, N., Nevanen, T. K., Söderlund, H., et al. (2002). Smart nanotubes for bioseparations and biocatalysis. *Journal of the American Chemical Society*, *124*(40), 11864–11865.

Na, H. B., Song, I. C., & Hyeon, T. (2009). Inorganic nanoparticles for MRI contrast agents. *Advanced Materials*, *21*(21), 2133–2148.

Nguyen, T. T., Sly, K. L., & Conboy, J. C. (2011). Comparison of the energetics of avidin, streptavidin, neutrAvidin, and anti-biotin antibody binding to biotinylated lipid bilayer examined by second-harmonic generation. *Analytical Chemistry*, *84*(1), 201–208.

Prime, K. L., & Whitesides, G. M. (1991). Self-assembled organic monolayers: Model systems for studying adsorption of proteins at surfaces. *Science*, *252*, 1164–1167.

Rauf, S., Zhou, D., Abell, C., Klenerman, D., & Kang, D. J. (2006). Building three-dimensional nanostructures with active enzymes by surface templated layer-by-layer assembly. *Chemical Communications*, 1721–1723.

Ravasco, J. M., Faustino, H., Trindade, A., & Gois, P. M. (2019). Bioconjugation with maleimides: A useful tool for chemical biology. *Chemistry—A European Journal*, *25*, 43–59.

Schlenoff, J. B. (2014). Zwitteration: Coating surfaces with zwitterionic functionality to reduce nonspecific adsorption. *Langmuir*, *30*, 9625–9636.
Singamaneni, S., Bliznyuk, V. N., Binek, C., & Tsymbal, E. Y. (2011). Magnetic nanoparticles: Recent advances in synthesis, self-assembly and applications. *Journal of Materials Chemistry*, *21*(42), 16819–16845.
Tong, S., Hou, S., Ren, B., & Bao, G. (2011). Self-assembly of phospholipid-PEG coating on nanoparticles through dual solvent exchange. *Nano Letters*, *11*, 3720–3726.
Veitch, N. C. (2004). Horseradish peroxidase: A modern view of a classic enzyme. *Phytochemistry*, *65*(3), 249–259.
Xie, T., Wang, A., Huang, L., Li, H., Chen, Z., Wang, Q., et al. (2009). Recent advances in the support and technology used in enzyme immobilization. *African Journal of Biotechnology*, *8*(19), 4724–4733.
Xie, H. Y., Zhen, R., Wang, B., Feng, Y. J., Chen, P., & Hao, J. (2010). Fe_3O_4/Au core/shell nanoparticles modified with Ni^{2+}–Nitrilotriacetic acid specific to histidine-tagged proteins. *The Journal of Physical Chemistry C*, *114*(11), 4825–4830.
Yang, C., Wu, J., & Hou, Y. (2011). Fe_3O_4 nanostructures: Synthesis, growth mechanism, properties and applications. *Chemical Communications*, *47*(18), 5130–5141.
Zhang, J., Chi, Q., Dong, S., & Wang, E. (1996). In situ electrochemical scanning tunnelling microscopy investigation of structure for horseradish peroxidase and its electrocatalytic property. *Bioelectrochemistry and Bioenergetics*, *39*(2), 267–274.
Zhang, H., Feng, G., Guo, Y., & Zhou, D. (2013). Robust and specific ratiometric biosensing using a copper-free clicked quantum dot–DNA aptamer sensor. *Nanoscale*, *5*, 10307–10315.
Zhang, Y., Guo, Y., Quirke, P., & Zhou, D. (2013). Ultrasensitive single-nucleotide polymorphism detection using target-recycled ligation, strand displacement and enzymatic amplification. *Nanoscale*, *5*(11), 5027–5035.
Zhang, Y., Pilapong, C., Guo, Y., Ling, Z., Cespedes, O., Quirke, P., et al. (2013). Sensitive, simultaneous quantitation of two unlabelled DNA targets using a magnetic nanoparticle–enzyme sandwich assay. *Analytical Chemistry*, *85*(19), 9238–9244.
Zhang, Y., & Zhou, D. (2012). Magnetic particle-based ultrasensitive biosensors for diagnostics. *Expert Review of Molecular Diagnostics*, *12*(6), 565–571.
Zhou, D., Bruckbauer, A., Ying, L., Abell, C., & Klenerman, D. (2003). Building three-dimensional surface biological assemblies on the nanometer scale. *Nano Letters*, *3*(11), 1517–1520.
Zhou, L., Yuan, J., & Wei, Y. (2011). Core–shell structural iron oxide hybrid nanoparticles: From controlled synthesis to biomedical applications. *Journal of Materials Chemistry*, *21*(9), 2823–2840.

Further reading

Gao, L., Wu, J., Lyle, S., Zehr, K., Cao, L., & Gao, D. (2008). Magnetite nanoparticle-linked immunosorbent assay. *Journal of Physical Chemistry C*, *112*, 17357–17361.
Martín, M., Salazar, P., Villalonga, R., Campuzano, S., Pingarrón, J. M., & González-Mora, J. L. (2014). Preparation of core–shell Fe_3O_4@poly (dopamine) magnetic nanoparticles for biosensor construction. *Journal of Materials Chemistry B*, *6*, 739–746.

CHAPTER TWENTY

Enzymes immobilization onto magnetic nanoparticles to improve industrial and environmental applications

Osama M. Darwesh[a,*], Sameh S. Ali[b,c], Ibrahim A. Matter[a], Tamer Elsamahy[c], Yehia A. Mahmoud[b]

[a]Agriculture Microbiology Department, National Research Centre, Dokki, Cairo, Egypt
[b]Botany Department, Faculty of Science, Tanta University, Tanta, Egypt
[c]Biofuels Institute, School of the Environment and Safety Engineering, Jiangsu University, Zhenjiang, China
[*]Corresponding author: e-mail address: darweshosama@yahoo.com

Contents

Abstract

Enzymes as specific natural biocatalysts are present in all living organisms and they play a key role in the biochemical reactions inside, as outside the cell. Despite the wide range of environmental, medical, agricultural, and food applications, the high cost, non-reusability, and limited stability of soluble (non-immobilized) enzymes are considered

Methods in Enzymology, Volume 630
ISSN 0076-6879
https://doi.org/10.1016/bs.mie.2019.11.006

barriers to their commercial application. Immobilization techniques are an effective strategy for solving problems associated with free enzymes in terms of improving the efficiency and stability of catalytic enzymes, as well as enhancing their separation and reusability in continuous industrial applications. Out of different supporting materials, magnetic nanoparticles are considered as the future trend for enzyme immobilization due to their exceptional properties regarding stabilization, easy recovery and reuse. Some recent techniques of enzyme immobilization on magnetic nanoparticles will be detailed hereafter in the chapter.

1. Introduction

Enzymes are the main accelerator for almost all biochemical reactions either inside or outside cells. These biocatalysts are proteins or a glycoproteins macromolecules with active sites and they are produced from all living cells. Enzymes activities have strong relations with the life quality of mankind even before they know the concept of enzymes. For instance, at ancient Egypt, several thousand years ago, people traditionally produced fermentative foods, beverages, textiles, and paper industries through enzymatic fermentations (Dwevedi, 2016; Homaei, Sariri, Vianello, & Stevanato, 2013). With the tremendous progress in science, technology, and industry, the role of fungal, bacterial and plant enzymes became more clearly in different environmental, industrial, and health issues (Emam, Darwesh, & Abdelhameed, 2018). Some of the most industrial applied enzymes are summarized in (Table 1).

Recently, the industrial and agricultural human's activities unfortunately left a large amounts of hazardous materials that cause serious problems to the global ecosystem and all organisms on earth (Darwesh, Moawad, Abd El-Rahim, Barakat, & Sedik, 2014; Hoballah, Saber, Matter, & Zaghloul, 2014). Hence, scientist and researchers are doing their bests to find the most suitable options for solving these problems using ecofriendly and environmental safe technologies (Darwesh et al., 2015; Eida, Darwesh, & Matter, 2018). Many environmental friendly technologies have been applied to remove, detoxify, or degrade xenobiotics, wastes and recalcitrant materials depending on enzymatic activities (Abou-Shanab et al., 2012; Darwesh, Barakat, Mattar, Sabae, & Hassan, 2019; Hasanin, Darwesh, Matter, & El-Saied, 2019; Marrez, Abdelhamid, & Darwesh, 2019; Youssef, Hasanin, Abd El-Aziz, & Darwesh, 2019). Enzymes application for waste treatment is extensive and complicated with degradation of toxic pollutants

Table 1 Some of microbial enzymes and their industrial applications.

Industry	Enzyme	Application	References
Food and Dairy	Acid proteinase	Milk coagulation	Merheb-Dini, Garcia, Penna, Gomes, and Silva (2012)
	Neutral proteinase	Faster cheese ripening, de-bittering	Smit, Smit, and Engels (2005)
	Lipase	Faster cheese ripening, cheese flavor customization	Singh, Kumar, Mittal, and Mehta (2016)
	Lactase (β-galactosidase)	Lactose reduced milk and whey products	Saqib, Akram, Halim, and Tassaduq (2017)
	Aminopeptidase	Faster cheese ripening	Mcsweeney (2004)
	Pectin methyl esterase	Firming of fruit-based products Protein cross linking	Beigi, Ngadi, Holman, and Chénier (2015)
	Transglutaminase	Modifying of visco-elastic properties	Dube, Schäfer, Neidhart, and Carle (2007)
	Catalase	Cheese processing	Afroz, Khan, Pervez, and Uprit (2015)
Detergent	Protease	Protein stain removal	Niyonzima and More (2015)
	Amylase	Carbohydrate stain removal	Lahmar et al. (2017)
	Lipase	Lipid stain removal	Hasan, Shah, Javed, and Hameed (2010)
	Cutinase	Triglyceride removal	Carvalho, Aires-Barros, and Cabral (1998)
	Cellulase	Cleaning, color clarification, anti-redeposition (cotton)	Kuhad, Gupta, and Singh (2011)
	Mannanase	Mannanan stain removal (reappearing stains)	Kumari et al. (2019)

Continued

Table 1 Some of microbial enzymes and their industrial applications.—cont'd

Industry	Enzyme	Application	References
Starch and biofuel	Amylase	Starch liquefaction and saccharification	Saini, Saini, and Dahiya (2017)
	Amyloglucosidase Pullulanase	Saccharification	Hii, Tan, Ling, and Ariff (2012)
	Glucose isomerase	Glucose to fructose conversion	Pawar, Rao, and Deshpande (1996)
	Cyclodextrin-glycosyltransferase	Cyclodextrin production	Feng, Zhuang, and Ran (2011)
	Xylanae	Viscosity reduction (fuel and starch)	Menon and Datta (2017)
	Protease	Yeast nutrition, fuel production	Razzaq et al. (2019)
Baking	Amylase	Flour adjustment, bread softness	Saini et al. (2017)
	Maltogenic α-Amylase	Enhancing of shelf life of breads	Saini et al. (2017)
	Xylanase	Dough conditioning	Ahmad et al. (2014)
	Lipase	Dough stability and conditioning	Sîrbu and Pâslaru (2005)
	Glucose oxidase	Dough strengthening	Kouassi-Koffi et al. (2019)
	Transglutaminase	Laminating of dough strength	Gerrard et al. (2000)
	Phospholipase	Dough stability and conditioning (in situ emulsifier)	Borrelli and Trono (2015)
	Lipoxygenase	Dough strengthening, bread whitening	Borrelli and Trono (2015)
	Protease	Biscuits, cookies	Razzaq et al. (2019)
Animal feed	Phytase	Hydrolyze phytic acid to release phosphorous	Gupta, Gangoliya, and Singh (2013)
	Xylanase	Enhancing of starch digestibility	Cozannet, Kidd, Neto, and Geraert (2017)
	β-glucanase	Digestibility	

Table 1 Some of microbial enzymes and their industrial applications.—cont'd

Industry	Enzyme	Application	References
Pulp and paper	Lipase	Pitch and some contaminants control	Borrelli and Trono (2015)
	Protease	Biofilm removal	Razzaq et al. (2019)
	Amylase	De-inking, drainage improvement, Starch-coating	Saini et al. (2017)
	Xylanase	Bleach boosting	Cozannet et al. (2017)
	Laccase	Non-chlorine bleaching, delignification	Mate and Alcalde (2017)
	Cellulase	De-inking, fiber modification, drainage improvement	Kuhad et al. (2011)
Leather	Protease	Unhearing, bating	Razzaq et al. (2019)
	Lipase	De-pickling	Borrelli and Trono (2015)
	Amylase	Fiber splitting	Saini et al. (2017)
Cosmetics and Personal care	Superoxide dismutase	Free radical scavenging, skin care	Ighodaro and Akinloye (2018)
	Protease	Removal of dead skin	Razzaq et al. (2019)
	laccase	Hair dye	Mate and Alcalde (2017)
	lipase	Skin care	Borrelli and Trono (2015)
	Amyloglucosidase	Antimicrobial (combined with glucose oxidase)	Cawley et al. (2019)
	Glucose oxidase	Bleaching, antimicrobial	Kouassi-Koffi et al. (2019)
	Peroxidase	Antimicrobial	Tonoyan, Fleming, Mc Cay, Friel, and O'Flaherty (2017)

Continued

Table 1 Some of microbial enzymes and their industrial applications.—cont'd

Industry	Enzyme	Application	References
Textile	Cellulase	Denim finishing, cotton softening	Kuhad et al. (2011)
	Amylase	De-sizing	Saini et al. (2017)
	Pectate lyase	Bioscouring	Kalantzi, Mamma, Christakopoulos, and Kekos (2008)
	Catalase	Bleach termination	Kaushal, Mehandia, Singh, Raina, and Arya (2018)
	Protease	Removal of wool fiber scales, degumming of silk	Razzaq et al. (2019)
	Laccase	Bleaching	Mate and Alcalde (2017)
	Lipase	Removal of size lubricants, denim finishing	Borrelli and Trono (2015)
	Peroxidase	Excess dye removal	Darwesh, Moawad, Barakat, and Abd El-Rahim (2015)
	Ligninase and Collagenase	Wool finishing	Singh et al. (2016)
	Cutinase	Cotton scouring, synthetic fiber modification	Carvalho et al. (1998)
Organic synthesis	Lipase	Resolution of chiral alcohols and amides, Synthesis of pharmaceuticals, polymers, biodiesels, biosurfactants	Borrelli and Trono (2015)
	Acylase	Synthesis of semisynthetic penicillin	Singh et al. (2016)
	Glycosyl transferase	Synthesis of oligosaccharides	Weijers, Franssen, and Visser (2008)
	Nitrile hydratase	Synthesis of acrylamide, butyramide, nicotinamide	Bhatia, Mehta, Bhatia, and Bhalla (2014)
	Glucose isomerase	Production of High fructose corn syrup	Pawar et al. (1996)
	Acyl transferase	Synthesis of hydroxamic acids	Singh et al. (2016)
	Laccase	Production of textile dyes, cosmetic pigments, flavor agents, and pesticides	Mate and Alcalde (2017)

(Hussein, Darwesh, & Mekki, 2019; Hussein, Darwesh, Mekki, & El-Hallouty, 2019). The utilization of enzymes in bioremediation can be carried out using whole microbial cells or their excreted enzymes either after isolation and purification or even without. Their action may depend on the presence of other substances like cofactors, co-substrates or mediators (Barakat, Hassan, & Darwesh, 2017; Darwesh, Matter, Eida, Moawad, & Oh, 2019). However, the main limitations for enzymes to be applied at large scale are, the relatively high production and purification costs, difficult be recovered from reaction mixtures and reused (Guo, 2019). Such drawbacks and more could be overcome through enzyme immobilization which allows the easy recovery and reuse of enzymes for further catalytic reaction as well as increasing its stability against storage, pH and operation temperatures (Cui, Ren, Sun, & Jia, 2018; Darwesh, Barakat et al., 2019; Liu, Chen, & Shi, 2018). However, the immobilization of enzymes could couple with the decreases of its activities due to unbalanced carrier-enzyme ration, either in case of cross-linking or encapsulated enzymes (Cao, van Langen, & Sheldon, 2003; Wang, Wan, Liu, Huang, & Xu, 2009).

Nanotechnology deals with the preparation and application of functional materials and structures in nanoscale sizes (Abdelhameed, El-Sayed, El-Shahat, El-Sayed, & Darwesh, 2018; Tseng et al., 2017). By definition, nanostructured materials have at least one dimension (thickness, diameter, length, etc.) between 1 and 100 nm and exhibit a variety of distinctive properties (Fig. 1). Magnetic nanoparticles (MNPs) commonly consist of magnetic elements such as Fe, Co, Cr, Mn, Gd and Ni and their derivatives. The properties of MNPs depends on their synthesis method and chemical structure.

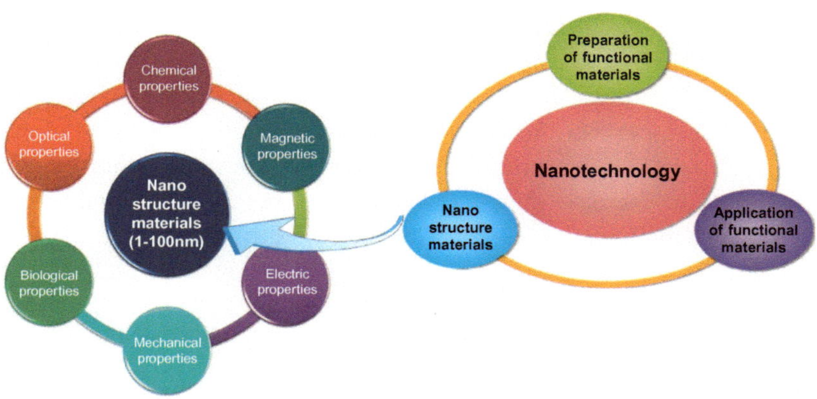

Fig. 1 Diagram of nanotechnology applications and different properties of nano-structure materials.

In the last decade, MNPs have been widely used as supporting materials for immobilization of enzymes because of their unique properties (Wu, He, & Jiang, 2008). In a recent study, in addition to easily and rapid separation (using permanent magnets) and reuse of enzymes immobilized on MNPs, Darwesh, Matter, and Eida (2019) reported high enzymatic stability for more than 3 months. These magnetic supports can be easily installed and stabilized in a fluidized bed reactor to enable continuous operation of enzymatic reactions, which significantly reduces the operation costs. Despite the aforementioned features, there is a problem with the low enzyme loading on the conventional magnetic beads, which led to the using of other alternatives modified MNPs as supporting materials. For instance, a hybrid supporting material from magnetite-mesoporous silica was fabricated by depositing magnetite and MCM-41 nanoparticles onto polystyrene beads using the layer-by-layer (LBL) method. Where, the incorporation of magnetite gives an additional magnetic property to the hollow mesoporous silica shells, resulting in improved enzyme immobilization (Kim, Grate, & Wang, 2006).

2. Magnetic nanoparticles properties

1. MNPs are nano-forms of magnetic elements as iron, gadolinium, cobalt, nickel, manganese, chromium and their chemical compounds. It is superparamagnetic for their nanoscale size, contribution great potentials in deferent applications (Fig. 2). Ferrite nanoparticles are the most discovered magnetic nanoparticles. Magnetic nanoparticles can be selective attached to a functional molecules and allow transportation to a targeted location under an external magnetic field. For prevention of aggregation and minimizing the particles interaction with the system environment, surface coating is the suitable solution. The surface of ferrite nanoparticles is often modified by silicones, surfactants, silica or other materials to increase their stability (Ma et al., 2018; Perera, Zhang, Homer-Vanniasinkam, Coppens, & Edirisinghe, 2018).
2. The MNPs properties depend on the synthesis method and chemical nature. Superparamagnetism is caused by thermal effects that the thermal fluctuations are strong enough to spontaneously demagnetize a previously saturated assembly; therefore, these particles have zero coercivity and have no hysteresis (Fig. 3). An external magnet is able to magnetize

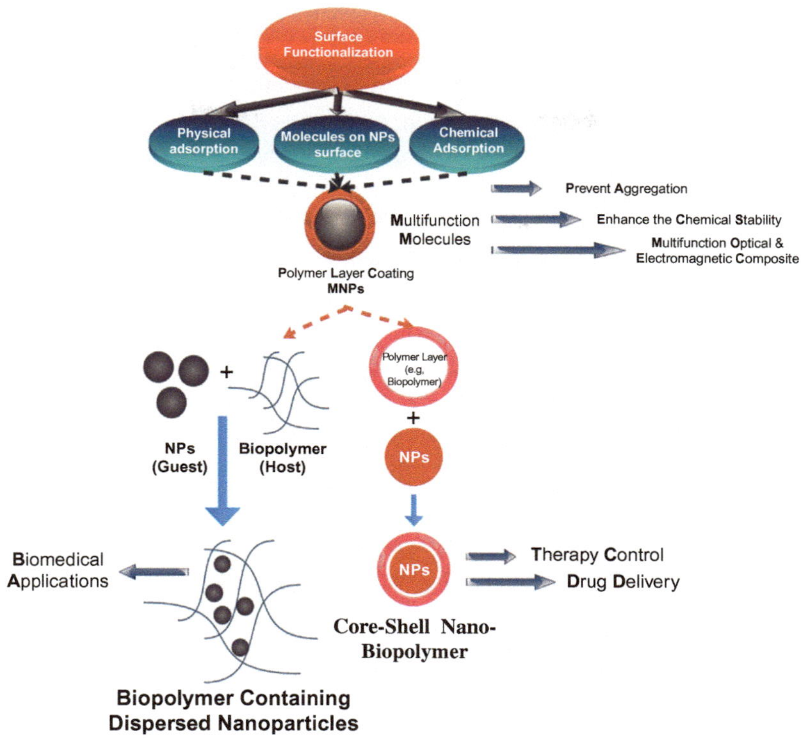

Fig. 2 Advantages of polymer coating magnetic nanoparticles.

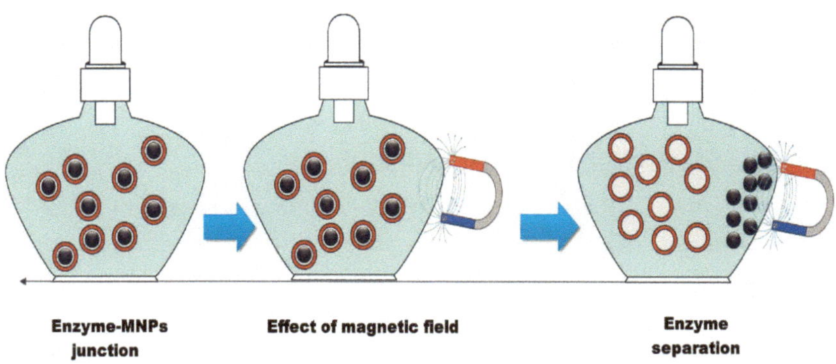

Fig. 3 Effect of magnetic field on the enzyme recovering process.

the nanoparticles with much larger magnetic exposure. When the field is removed, magnetic nanoparticles display no magnetization. This property can be suitable for controlled therapy, targeted drug delivery and environmental applications. Magnetocaloric effect (MCE) is a property for heating up of magnetic materials when they are placed in a magnetic field and cooling down when they are removed from a magnetic field. MNPs deliver a promising alternative to conventional bulk materials because of their particle size-dependent superparamagnetic features. The large surface area of MNPs has the potential to provide better heat exchange with the surrounding environment (Gao, Goriacheva, Tarakina, & Sukhorukov, 2016; Neufeld, Ware, Lutzke, Khetani, & Reynolds, 2016; Zhang et al., 2017).

3. Methodologies for enzyme immobilization on MNPs

3.1 Production and activation of Fe_3O_4 MNPs

The Nano-size Fe_3O_4 particles (at concentration of 300 mM iron ions) could be prepared by the co-precipitation of Fe^{2+} and Fe^{3+} ions (molar ratio 1:2, at 25 °C) in the presence of ammonia solution (29.6%, at pH 10) as precipitating agent. This reaction is carried out under hydrothermal treatment (80 °C for 30 min) to convert the hydroxyl form Fe_3O_4 into oxide with constant stirring until the black precipitate (MNPs) develops. Finally, the MNPs are dried under vacuum at 70 °C after washing several times with water and absolute ethanol (Darwesh et al., 2015). The formation of MNPs can be indicated by changing the color into black (as shown in Fig. 4), and the magnetism can be demonstrated using a magnet, while nanoparticle sizes are measured using an electron microscope or a particle size analyzer.

3.2 Surface modification of MNPs

After preparation of magnetic nanoparticles, some modifications should be done to protect it against aggregation, oxidation and/or changing the surface charge as well as for improving of immobilization capacity. These modifications doesn't only serve to protect it against degradation but also can be used to enable further functionalization with specific components, such as specific active binding sites for enzymes, drugs, or other functional groups. For instant, the application of glutaraldehyde as coupling agent for covalent attachment of lipase and peroxidase to the surface of activated nanoparticle (Fig. 5). Surface modification of MNPs simply could be performed by

Fig. 4 Magnetic nanoparticles (Fe_3O_4) prepared by co-precipitation method.

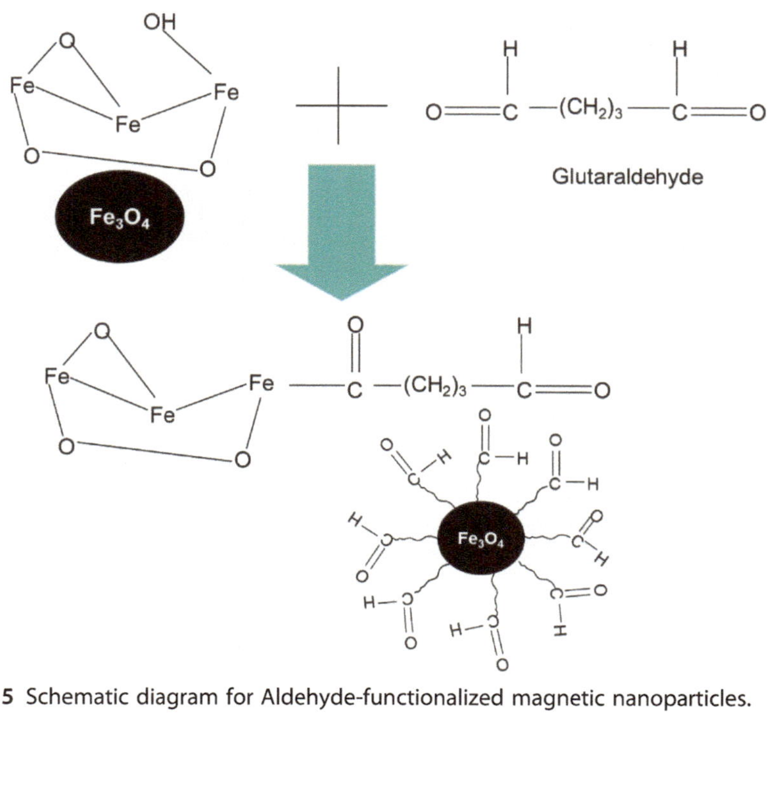

Fig. 5 Schematic diagram for Aldehyde-functionalized magnetic nanoparticles.

dispersing the synthesized MNPs (~5 g) in ethanol with sonication (10 min) for getting the complete dispersion. 10 mL of glutaraldehyde solution (10%) is added to the above nanoparticles and incubated at room temperature (for about 2 h). The particles should be then washed with water to remove the excess of glutaraldehyde (Darwesh, Matter et al., 2019).

4. Mechanisms of enzyme immobilization onto MNPs

Mechanisms of enzyme immobilization onto MNPs includes both nonspecific physical adsorption and covalent binding. Non-covalent binding, depends on the surface interaction of enzymes with the nano-materials through van der Walls and electrostatic forces, hydrophobic or—stacking interactions. Covalent bonding is widely used for immobilization of enzymes on bare or modified MNPs, which can be achieved by creation of covalent bonds between the active groups on the particles and the chemical groups of enzymes (Fig. 6). This method is often associated with some reduced enzyme activity, but it can achieve strong binding of enzyme on MNPs that can prevent the loss of immobilized enzymes.

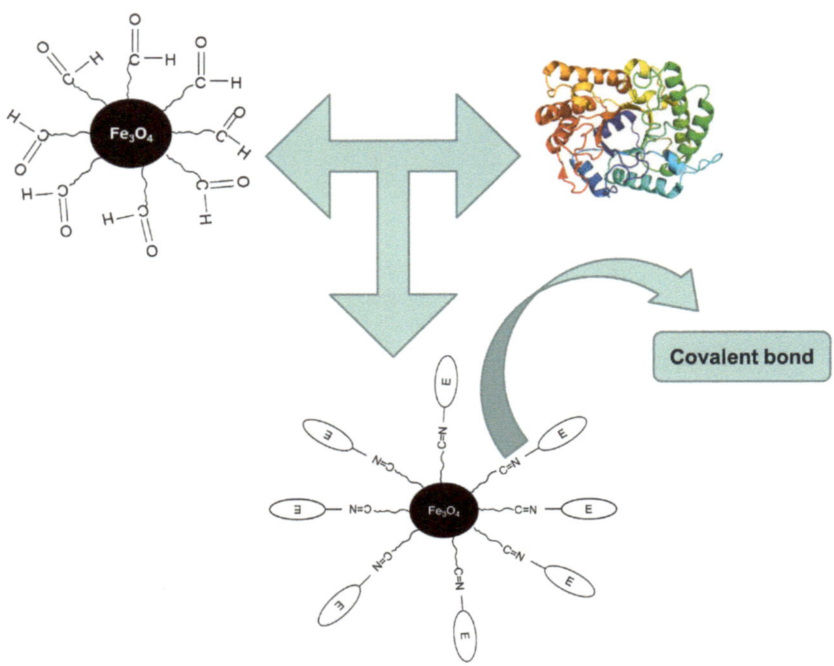

Fig. 6 Covalent binding between enzyme and modified MNPs for immobilization.

5. Immobilization of peroxidase on modified magnetic nanoparticles

Prior enzyme immobilization, MNPs are activated by shaking overnight in phosphate buffer (0.1 M, at 4 °C). After separation of MNPs, 50 mL of phosphate buffer, pH 6.0 containing peroxidase enzyme (5 U/mL) is added to the activated MNPs and incubated under shaking condition at 4 °C for 24 h. The peroxidase-bound MNPs are then decanted using permanent magnet and washed several times with deionized water and stored for use. This technology is good choice to be used in bioreactors for bioremediation of industrial wastewater such as textile wastewater that containing toxic dyes. The MNPs-immobilized peroxidase can withstand extreme conditions, such as high/low temperatures, high salinity, high/low pH and high substrate (pollutant) concentration (Darwesh et al., 2015).

Darwesh et al. (2015) studied the effect of immobilization time on enzyme immobilization process, and the results are illustrated in Fig. 7. The activity of immobilized peroxidase was increased and showed maximum activity after 24 h of immobilization reaction. However, the activity of free enzyme was decreased within the same time. The maximum activity of the immobilized peroxidase was set to 100% as a reference value. However, an immobilization efficiency of 100% means that all of the enzyme added would be immobilized onto the nanoparticles. The immobilization reaction components were checked of their activity to confirm which one response about activity in reaction system. The full activity was for free enzyme before immobilization process, but this activity was

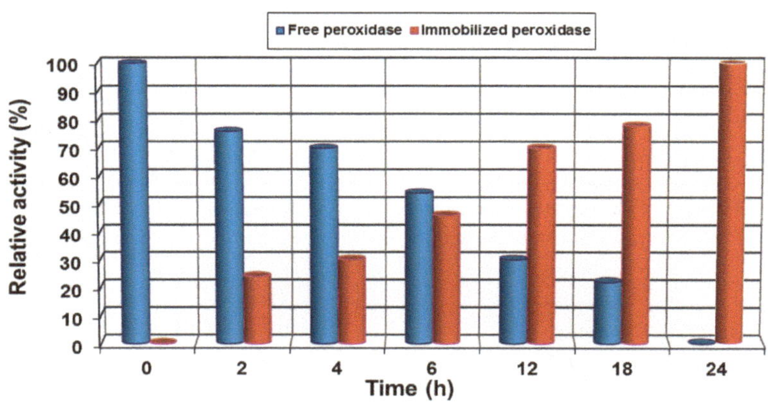

Fig. 7 Immobilization efficiency of peroxidase with respect to the immobilization time.

denied after immobilization technology. After ending of reaction, the activity was come back to immobilized enzyme, while, the Fe_3O_4 didn't have any changes in chemical reaction of enzyme assay.

6. Characterization of immobilized enzymes onto MNPs

Many characterization parameters should be defined for MNPs and MNPs-immobilized enzymes such as size, shape, crystallinity, surface active groups, surface charge and magnetic properties. The most popular characterization techniques and the relevant physicochemical properties are presented in Table 2.

Table 2 Characterization techniques applied to evaluate physicochemical properties of MNPs.

Techniques	Evaluation
Infrared spectroscopy (IR)	Nature of function group
Nuclear magnetic resonance spectroscopy (NMR)	Longitudinal and transverse relaxivity; structure conformation
Fluorescence correlation spectroscopy	Dimension, binding kinetics of hydrodynamic
Superconducting quantum interference device (SQUID); vibrating sample magnetometry (VSM)	Magnetic properties
Electron microscopy (TEM; SEM)	Morphology, size distribution, composition
X-ray diffraction (XRD)	Crystal structure
Surface-enhanced Raman scattering	Size distribution, electronic characteristics
Dynamic light scattering (DLS)	Hydrodynamic diameter
Zeta potential measurement	Surface charge
Circular dichroism	Thermal constancy
Thermal analysis (differential scanning calorimetry, thermogravimetric analysis, etc.)	Thermal stability, surface coverage, nature of surface functionalization
Mass spectroscopy	Molecular weight
Atomic force microscopy	Shape heterogeneity
Scanning tunneling microscopy; small-angle X-ray scattering	Shape heterogeneity, size and size navigation

6.1 Transmission electron microscopy (TEM)

Transmission electron microscope is successfully applied to determine the average particle size, size distribution and morphology of the MNPs and immobilized enzyme as well. JEM-1400 TEM was used to in many studies as following; a drop of well dispersed nanoparticle dispersion was placed onto the amorphous carbon-coated 200 mesh carbon grid, followed by drying the sample at ambient temperature, before it was loaded into the microscope.

6.2 Energy-dispersive X-ray spectroscopy

The structure and their components of magnetic nanoparticles and immobilized enzyme could be measured using Energy-dispersive X-ray spectroscopy (EDX). The dry samples are spread on a conducting adhesive tape, pasted on a metallic stub. The morphologies of the tested samples are investigated and imaged with scanning electron microscope (SEM) (QUANTA FEG 250 ESEM, USA). This is accompanied by energy dispersive X-ray spectroscopy (EDAXAMETEK Inc.; Mahwah, NJ, USA) at an acceleration voltage of 15 kV, the films are fixed on the surface of a sticky tape.

7. Evaluation of immobilized enzymes

It is necessary to know if the enzyme properties enhanced or reduced after immobilization or not? The concept of evaluation technology here differs from characterization, as the characterization can confirm the success of immobilization processes only regardless of whether the immobilized enzymes still activity or not. The immobilized enzymes should be saved their activity against many parameters like low and high temperature, low and high pH values, high concentrations of wastes and storage at deferent temperatures for long time, as well as, allowed of immobilized enzymes for recycling. Thermal stability is one of the significant criteria for industrial using of peroxidase. The thermal stability of peroxidase in free and immobilized states was investigated. The covalent immobilization has a clearly enhancing on the thermal stability of the peroxidase similar to that reported by Darwesh, Matter et al., (2019).

The storage stability of enzyme and their bioactivity are of great importance in industrial sector. Therefore, the storage stability is an important concern in immobilization processes of enzyme particulate when using in industry. The activity of immobilized peroxidase was not affected by storage.

When storage at 25 °C, free enzyme lost activity after only 20 days, while immobilized peroxidase, however, maintained its full activity over 120 days. These observations clearly demonstrate that, MNPs-immobilized peroxidase is remarkably stable on storage than free form enzyme. The enhancement of storage stability for the immobilized peroxidase can be accredited to the multipoint covalent conjugation between activated magnetic nanoparticles by glutaraldehyde and peroxidase enzyme and could create resistance to face environmental changes (Darwesh, Matter et al., 2019).

7.1 Repeated use of the immobilized peroxidase enzyme

The reusability of the immobilized peroxidase is examined by conducting the activity measurement of the immobilized peroxidase at 20 °C. After each activity measurement, the immobilized peroxidase is separated magnetically and washed several times with phosphate buffer solution. Then, the chemicals of enzyme reaction are added to the immobilized peroxidase and the next activity measurement is carried out. The highest value of enzyme activity in the measurements is assigned the value of 100% activity.

7.2 Effect of temperature and pH on the peroxidase enzyme activity

The effect of temperature on the activity of free and immobilized peroxidase is estimated after 3 min incubation of 3 mL of the reaction mixtures at a temperature range of 10–100 °C. The enzyme activity is then determined by measuring the absorbance of the solution at 420 nm. On the other hand, the effect of pH on the activity of the free and immobilized enzyme is investigated in the pH range of 1.5–12 at 20 °C. The enzyme activity is calculated at each pH value. All data used in the experiments should be averages of triplicated experiments. The highest value of enzyme activity in each set will assigned the value of 100% activity.

7.3 Thermal stability of free and immobilized peroxidase

Thermal stability of free and immobilized peroxidase are examined by measuring their residual activities after being incubated for 60 min in the temperature ranging from 10 to 100 °C. and then the reaction mixture is assayed at 20 °C. The highest value of enzyme activity in each set is assigned the value of 100% activity.

7.4 Storage stability of the free and immobilized peroxidase

Storage stability of the free and immobilized peroxidase is examined by measuring their activities after being stored at 4 or 25 °C through 120 days. The highest value of enzyme activity in each set is assigned the value of 100% activity.

8. Enzymatic bioremediation of textile industry wastewater containing direct green or reactive red azo dye

The prototype sequential bioreactor (PSB) was designed to achieve continuous treatment of textile industry wastewater containing direct green or reactive red azo dyes using immobilized peroxidase enzyme. The PSB involves one container as presented in Fig. 8. The jacket of this container was built from glass and may build from stainless steel; the inert part was built from glass. The height of this container was 17 cm and the diameter was 12 cm with working volume of 1.0 L (Fig. 8). The container involves a permanent magnet inters the glass cover.

The enzymatic bioremediation of textile industry wastewater was investigated using prototype sequential bioreactor. This bioreactor was inoculated with peroxidase enzyme immobilized onto Fe_3O_4 magnetic nanoparticles. The system was operated by textile industry wastewater containing direct green or reactive red azo dye; in this case, the magnet was in the top. After complete the wastewater degradation, the magnet was moved down to collect the immobilized peroxidase. The effluents (degraded wastewater) were exit by tap. And then the new treatment cycle was started.

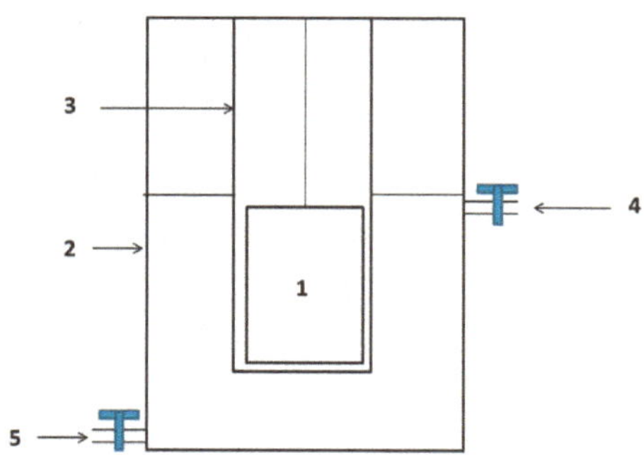

Fig. 8 Schematic diagram of prototype sequential bioreactor. Where 1, permanent magnet; 2, container; 3, glass jacket; 4, influent tap; 5, effluent tap.

The decolorization of wastewater containing direct green (at 385 nm) or reactive red (at 555 nm) was measured by a spectrophotometer (Jenway UV/Visible-2605 spectrophotometer, England) before treatment and every hour after treatment using the following equation:

$$\text{Decolorization activity } \% = \frac{\text{Reading of [C] decolorization} - \text{Reading of [S] decolorization}}{\text{Reading of [C] decolorization}} \times 100$$

Where C = control, S = sample.

9. Conclusion

We can report that the isolation, purification and immobilization technology of peroxidase onto modified iron magnetic nanoparticles is a benefit solution for applying enzymes in environmental remediation. The immobilized peroxidase had obvious advantages over free peroxidase including the ease of enzyme recycling using magnetic field; enhanced tolerance to tested environmental stress conditions; and higher storage stability. In addition, immobilized peroxidase successfully decolorized textile industry wastewater containing different dyes. Importantly, the decolorization process proved scalable to lab scale bioreactor level. We therefore believe that the immobilized peroxidase reported in this study can be employed in industrial bioremediation processes, and perhaps in other potential environmental and industrial applications.

References

Abdelhameed, R. M., El-Sayed, H. A., El-Shahat, M., El-Sayed, A. A., & Darwesh, O. M. (2018). Novel triazolothiadiazole and triazolothiadiazine derivatives containing pyridine moiety: Design, synthesis, bactericidal and fungicidal activities. *Current Bioactive Compounds*, *14*(2), 169–179.

Abou-Shanab, R. A. I., Khalafallah, M. A., Emam, N. F., Aly, M. A., Abou-Sdera, S. A., & Matter, I. A. (2012). Characterisation and identification of carbofuran-utilising bacteria isolated from agricultural soil. *Chemistry and Ecology*, *28*(2), 193–203.

Afroz, Q. M., Khan, K. A., Pervez, A., & Uprit, S. (2015). Enzymes used in dairy industries. *International Journal of Applied Research*, *1*(10), 523–527.

Ahmad, Z., Butt, M. S., Ahmed, A., Riaz, M., Sabir, S. M., Farooq, U., et al. (2014). Effect of *Aspergillus niger* xylanase on dough characteristics and bread quality attributes. *Journal of Food Science and Technology*, *51*(10), 2445–2453.

Barakat, K. M., Hassan, S. W. M., & Darwesh, O. M. (2017). Biosurfactant production by haloalkaliphilic Bacillus strains isolated from Red Sea, Egypt. *Egyptian Journal of Aquatic Research*, *43*(2017), 205–211.

Beigi, T. M., Ngadi, M. O., Holman, D. B., & Chénier, M. R. (2015). Pectin methylesterases: A review. *Journal of Bioprocessing & Biotechniques*, *5*, 228.

Bhatia, S. K., Mehta, P. K., Bhatia, R. K., & Bhalla, T. C. (2014). Simultaneous purification of nitrile hydratase and amidase of *Alcaligenes* sp. MTCC 10674. *3 Biotech*, *4*(4), 375–381.

Borrelli, G. M., & Trono, D. (2015). Recombinant lipases and phospholipases and their use as biocatalysts for industrial applications. *International Journal of Molecular Science*, *16*(9), 20774–20840.

Cao, L., van Langen, L., & Sheldon, R. A. (2003). Immobilised enzymes: Carrier-bound or carrier-free? *Current Opinion in Biotechnology*, *14*(4), 387–394.

Carvalho, C. M. L., Aires-Barros, M. R., & Cabral, J. M. S. (1998). Cutinase structure, function and biocatalytic applications. *EJB Electronic Journal of Biotechnology*, *1*(3), 160–173.

Cawley, A., Golding, S., Goulsbra, A., Hoptroff, M., Kumaran, S., & Marriott, R. (2019). Microbiology insights into boosting salivary defences through the use of enzymes and proteins. *Journal of Dentistry*, *80*(1), S19–S25.

Cozannet, P., Kidd, M. T., Neto, R. M., & Geraert, P. (2017). Next-generation non-starch polysaccharide-degrading, multi-carbohydrase complex rich in xylanase and arabinofuranosidase to enhance broiler feed digestibility. *Poultry Science*, *96*(8), 2743–2750.

Cui, J., Ren, S., Sun, B., & Jia, S. (2018). Optimization protocols and improved strategies for metal-organic frameworks for immobilizing enzymes: Current development and future challenges. *Coordination Chemistry Reviews*, *370*, 22–41.

Darwesh, O. M., Barakat, K. M., Mattar, M. Z., Sabae, S. Z., & Hassan, S. H. (2019). Production of antimicrobial blue green pigment Pyocyanin by marine *Pseudomonas aeruginosa*. *Biointerface Research in Applied Chemistry*, *9*(5), 4334–4339.

Darwesh, O. M., Matter, I. A., & Eida, M. F. (2019). Development of peroxidase enzyme immobilized magnetic nanoparticles for bioremediation of textile wastewater dye. *Journal of Environmental Chemical Engineering*, 7(1), 102805, 1–7.

Darwesh, O. M., Matter, I. A., Eida, M. F., Moawad, H., & Oh, Y. (2019). Influence of nitrogen source and growth phase on extracellular biosynthesis of silver nanoparticles using cultural filtrates of *Scenedesmus obliquus*. *Applied Sciences*, *9*(7), 1465.

Darwesh, O. M., Moawad, H., Abd El-Rahim, W. M., Barakat, O. S., & Sedik, M. Z. (2014). Bioremediation of textile reactive blue (RB) azo dye residues in wastewater using experimental prototype bioreactor. *Research Journal of Pharmaceutical, Biological and Chemical Sciences*, *5*(4), 1203–1219.

Darwesh, O. M., Moawad, H., Barakat, O. S., & Abd El-Rahim, W. M. (2015). Bioremediation of textile reactive blue azo dye residues using nanobiotechnology approaches. *Research Journal of Pharmaceutical, Biological and Chemical Sciences*, *6*(1), 1202–1211.

Dube, M., Schäfer, C., Neidhart, S., & Carle, R. (2007). Texturisation and modification of vegetable proteins for food applications using microbial transglutaminase. *European Food Research and Technology*, *225*(2), 287–299.

Dwevedi, A. (2016). Enzyme immobilization: An important link between agriculture and industries. In *Enzyme immobilization* (pp. 45–64). Cham: Springer.

Eida, M. F., Darwesh, O. M., & Matter, I. A. (2018). Cultivation of oleaginous microalgae *Scenedesmus obliquus* on secondary treated municipal wastewater as growth medium for biodiesel production. *Journal of Ecological Engineering*, *19*(5), 38–51.

Emam, H. E., Darwesh, O. M., & Abdelhameed, R. M. (2018). In-growth metal organic framework/synthetic hybrids as antimicrobial fabrics and its toxicity. *Colloids and Surfaces B: Biointerfaces*, *165*, 219–228.

Feng, T., Zhuang, H., & Ran, Y. (2011). The application of cyclodextrin glycosyltransferase in biological science. *Journal of Bioequivalence & Bioavailability*, *3*, 202–206.

Gao, H., Goriacheva, O. A., Tarakina, N. V., & Sukhorukov, G. B. (2016). Intracellularly biodegradable polyelectrolyte/silica composite microcapsules as carriers for small molecules. *ACS Applied Materials & Interfaces*, *8*(15), 9651–9661.

Gerrard, J. A., Newberry, M. P., Ross, M., Wilson, A. J., Fayle, S. E., & Kavale, S. (2000). Pastry lift and croissant volume as affected by microbial transglutaminase. *Journal of Food Science*, *65*(2), 312–314.

Guo, K. W. (2019). Immobilization methods of enzymes: Part I. In *Approaches to enhance industrial production of fungal cellulases* (pp. 127–136). Cham: Springer.

Gupta, R. K., Gangoliya, S. S., & Singh, N. K. (2013). Reduction of phytic acid and enhancement of bioavailable micronutrients in food grains. *Journal of Food Science and Technology*, *52*(2), 676–684.

Hasan, F., Shah, A. A., Javed, S., & Hameed, A. (2010). Enzymes used in detergents: Lipases. *African Journal of Biotechnology*, *9*(31), 4836–4844.

Hasanin, M. S., Darwesh, O. M., Matter, I. A., & El-Saied, H. (2019). Isolation and characterization of non-cellulolytic *Aspergillus flavus* EGYPTA5 exhibiting selective ligninolytic potential. *Biocatalysis and Agricultural Biotechnology*, *17*, 160–167.

Hii, S. L., Tan, J. S., Ling, T. C., & Ariff, A. B. (2012). Pullulanase: Role in starch hydrolysis and potential industrial applications. *Enzyme Research*, *2012*, 921362.

Hoballah, E., Saber, M., Matter, I., & Zaghloul, A. (2014). Bioremediation of polychlorinated biphenyl (PCBs) in a sewaged soil by certain remediative amendments followed by phytoremediation. *Research Journal of Pharmaceutical, Biological and Chemical Sciences*, *5*(2), 91–103.

Homaei, A. A., Sariri, R., Vianello, F., & Stevanato, R. (2013). Enzyme immobilization: An update. *Journal of Chemical Biology*, *6*(4), 185–205.

Hussein, H. A., Darwesh, O. M., & Mekki, B. B. (2019). Environmentally friendly nano-selenium to improve antioxidant system and growth of groundnut cultivars under sandy soil conditions. *Biocatalysis and Agricultural Biotechnology*, *18*, 101080, 1–7.

Hussein, H. A., Darwesh, O. M., Mekki, B. B., & El-Hallouty, S. M. (2019). Evaluation of cytotoxicity, biochemical profile and yield components of groundnut plants treated with nano-selenium. *Biotechnology Reports*, *24*, e00377.

Ighodaro, O. M., & Akinloye, O. A. (2018). First line defence antioxidants-superoxide dismutase (SOD), catalase (CAT) and glutathione peroxidase (GPX): Their fundamental role in the entire antioxidant defence grid. *Alexandria Journal of Medicine*, *54*(4), 287–293.

Kalantzi, S., Mamma, D., Christakopoulos, P., & Kekos, D. (2008). Effect of pectate lyase bioscouring on physical, chemical and low-stress mechanical properties of cotton fabrics. *Bioresource Technology*, *99*(17), 8185–8192.

Kaushal, J., Mehandia, S., Singh, G., Raina, A., & Arya, S. K. (2018). Catalase enzyme: Application in bioremediation and food industry. *Biocatalysis and Agricultural Biotechnology*, *16*, 192–199.

Kim, J., Grate, J. W., & Wang, P. (2006). Nanostructures for enzyme stabilization. *Chemical Engineering Science*, *61* 1017e1026.

Kouassi-Koffi, J. D., Sturza, A., Păucean, A., Man, S., Mureşan, A. E., Petruţ, G., et al. (2019). Effect of glucose oxidase addition on the textural characteristics of wheat-maize dough and bread. *Food Science and Technology*, *39*(1), 127–133.

Kuhad, R. C., Gupta, R., & Singh, A. (2011). Microbial cellulases and their industrial applications. *Enzyme Research*, *2011*, 280696.

Kumari, U., Singh, R., Ray, T., Rana, S., Saha, P., Malhotra, K., et al. (2019). Validation of leaf enzymes in the detergent and textile industries: Launching of a new platform technology. *Plant Biotechnology Journal*, *17*(6), 1167–1182.

Lahmar, I., El Abed, H., Khemakhem, B., Belghith, H., Ben Abdallah, F., & Belghith, K. (2017). Optimization, purification, and starch stain wash application of two new α-

amylases extracted from leaves and stems of *Pergularia tomentosa*. *BioMed Research International, 2017*, 6712742.
Liu, D. M., Chen, J., & Shi, Y. P. (2018). Advances on methods and easy separated support materials for enzymes immobilization. *TrAC Trends in Analytical Chemistry, 102*, 332–342.
Ma, Y., Wang, X., Chen, H., Miao, Z., He, G., Zhou, J., et al. (2018). Polyacrylic acid functionalized Co0.85Se nanoparticles: An ultrasmall pH-responsive nanocarrier for synergistic photothermal-chemo treatment of cancer. *ACS Biomaterials Science & Engineering, 4*(2), 547–557.
Marrez, D. A., Abdelhamid, A. E., & Darwesh, O. M. (2019). Eco-friendly cellulose acetate green synthesized silver nano-composite as antibacterial packaging system for food safety. *Food Packaging and Shelf Life, 20*(100302), 1–8.
Mate, D. M., & Alcalde, M. (2017). Laccase: A multi-purpose biocatalyst at the forefront of biotechnology. *Microbial Biotechnology, 10*(6), 1457–1467.
Mcsweeney, P. L. H. (2004). Biochemistry of cheese ripening. *International Journal of Dairy Technology, 57*(2/3), 127–145.
Menon, G., & Datta, S. (2017). Xylanases: From paper to fuel. *Microbial Applications, 1*, 153–164.
Merheb-Dini, C., Garcia, G. A. C., Penna, A. L. B., Gomes, E., & Silva, R. (2012). Use of a new milk-clotting protease from *Thermomucor indicae-seudaticae* N31 as coagulant and changes during ripening of Prato cheese. *Food Chemistry, 130*(4), 859–865.
Neufeld, M. J., Ware, B. R., Lutzke, A., Khetani, S. R., & Reynolds, M. M. (2016). Waterstable metal nanoparticles for magnetic resonance imaging guided cancer diagnosis and therapy. *ACS Applied Materials & Interfaces, 8*(30), 19343–19352.
Niyonzima, F. N., & More, S. (2015). Detergent-compatible proteases: Microbial production, properties, and stain removal analysis. *Preparative Biochemistry and Biotechnology, 45*(3), 233–258.
Pawar, S., Rao, M. B., & Deshpande, V. V. (1996). Molecular and industrial aspects of glucose isomerase. *Microbiological Reviews, 60*(2), 280–300.
Perera, A. S., Zhang, S., Homer-Vanniasinkam, S., Coppens, M. O., & Edirisinghe, M. (2018). Polymer–magnetic composite fibers for remote-controlled drug release. *ACS Applied Materials & Interfaces, 10*(18), 15524–15531.
Razzaq, A., Shamsi, S., Ali, A., Ali, Q., Sajjad, M., Malik, A., et al. (2019). Microbial proteases applications. *Front Bioengineering and Biotechnology*, 7, 110.
Saini, R., Saini, H. S., & Dahiya, A. (2017). Amylases: Characteristics and industrial applications. *Journal of Pharmacognosy and Phytochemistry, 6*(4), 1865–1871.
Saqib, S., Akram, A., Halim, S. A., & Tassaduq, R. (2017). Sources of β-galactosidase and its applications in food industry. *3 Biotech*, 7(1), 79.
Singh, R., Kumar, M., Mittal, A., & Mehta, P. K. (2016). Microbial enzymes: Industrial progress in 21st century. *3 Biotech, 6*(2), 174.
Sîrbu, A., & Pâslaru, V. (2005). Influence of lipase products on technological properties of the bread flour. *Agroalimentary Processes and Technologies, XI*(1), 185–192.
Smit, G., Smit, B. A., & Engels, W. J. M. (2005). Flavour formation by lactic acid bacteria and biochemical flavour profiling of cheese products. *FEMS Microbiology Reviews, 29*(3), 591–610.
Tonoyan, L., Fleming, G. T. A., Mc Cay, P. H., Friel, R., & O'Flaherty, V. (2017). Antibacterial potential of an antimicrobial agent inspired by peroxidase-catalyzed systems. *Frontiers in Microbiology, 8*, 680.
Tseng, P., Napier, B., Zhao, S., Mitropoulos, A. N., Applegate, M. B., Marelli, B., et al. (2017). Directed assembly of bio-inspired hierarchical materials with controlled nanofibrillar architectures. *Nature Nanotechnology, 12*, 474–480.

Wang, Z. G., Wan, L. S., Liu, Z. M., Huang, X. J., & Xu, Z. K. (2009). Enzyme immobilization on electrospun polymer nanofibers: An overview. *Journal of Molecular Catalysis B: Enzymatic*, *56*(4), 189–195.

Weijers, C. A., Franssen, M. C., & Visser, G. M. (2008). Glycosyltransferase-catalyzed synthesis of bioactive oligosaccharides. *Biotechnology Advance*, *26*(5), 436–456.

Wu, W., He, Q., & Jiang, C. (2008). Magnetic iron oxide nanoparticles: Synthesis and surface functionalization strategies. *Nanoscale Research Letters*, *3*(11), 397.

Youssef, A. M., Hasanin, M. S., Abd El-Aziz, M. E., & Darwesh, O. M. (2019). Green, economic, and partially biodegradable wood plastic composites via enzymatic surface modification of lignocellulosic fibers. *Heliyon*, *5*, e01332, 1–22.

Zhang, L., Pei, J., Wang, H., Shi, Y., Niu, J., Yuan, F., et al. (2017). Facile preparation of poly(lactic acid)/brushite bilayer coating on biodegradable magnesium alloys with multiple functionalities for orthopedic application. *ACS Applied Materials & Interfaces*, *9*(11), 9437–9448.

FSC
www.fsc.org
MIX
Papier | Fördert gute Waldnutzung
FSC® C083411

Kolibri 360 GmbH
Ferdinand-Jühlke-Straße 7
99095 Erfurt, Deutschland
produktsicherheit@kolibri360.de